REFERENCE
RESERVED
FOR
USE IN LIBRARY
ONLY

European Environmental Information Sourcebook

WA 1118709 3

REFERENCE
RESERVED
FOR
USE IN LIBRARY
ONLY

European Environmental Information Sourcebook

Compiled by
OKSANA NEWMAN and ALLAN FOSTER
Manchester Business School, Library and Information Service

**UNIVERSITY OF GLAMORGAN
LEARNING RESOURCES CENTRE**

Pontypridd, Mid Glamorgan, CF37 1DL
Telephone: Pontypridd (01443) 480480

Books are to be returned on or before the last date below

This publication is a creative work fully protected by all applicable copyright laws, as well as by misappropriation, trade secret, unfair competition and other applicable laws. The authors and editors of this work have added value to the underlying factual material herein through one or more of the following: unique and original selection, coordination, expression, arrangement and classification of the information.

All rights in this publication will be vigorously defended.

Copyright © 1994 by Gale Research International Ltd
P.O. Box 699
Cheriton House
North Way
Andover
Hants SP10 5Y

All rights reserved including the right of reproduction in whole or in part in any form.

While every effort has been made to ensure the reliability of the information presented in this publication, Gale Research International Limited does not guarantee the accuracy of the data contained herein. Gale accepts no payment for listing; and inclusion in the publication of any organisation, agency or institution, publication, service or individual does not imply the endorsement of the editors or publisher. Errors brought to the attention of the publisher and verified to the satisfaction of the publisher will be corrected in future editions.

ISBN 1-873477-20-1

Published in the United Kingdom by Gale Research International Limited and simultaneously in the United States of America by Gale Research Inc.

A CIP catalogue record for this book is available from the British Library.

I(T)P

The trademark ITP is used under license.

Typeset by Hodgson Associates, Tunbridge Wells
Printed in the United Kingdom by Ipswich Book Company Ltd

Contents

Introduction (English)	vii
Introduction (French)	xi
Einleitung	xv
Acronyms	xix
Abbreviations	xxiv

1 Air Quality and Global Atmospheric Change
General	1
Air Quality and Pollution	2
Atmosphere	10
Climate	11
Global Warming	14

2 Agriculture and Farming
General	18
Crop Protection and Agrochemicals	23
Fish and Fisheries	27
Land Use	29
Organic Farming	30

3 Costal and Marine Waters
Costal Environment	33
Marine Environment	34

4 Inland Waters
Acid Rain	39
River Environment	41
Water Management	42
Water Quality	48

5 Waste and Recyling
Chemicals	54
Glass Recycling	56
Landfill	56
Metals Recycling	57
Nuclear Waste	59
Packaging	62
Paper and Pulp	65
Plastics	70
Recycling	72
Textiles	77
Waste Management	78
Wastewater	84

6 Conservation
General	87
Environmental Protection	88
Forests	92
Habitats and Species	97
Heritage	104
Land Use	105
Nature Conservation	108
Sustainability	114

7 Energy (non-nuclear)

General	116
Alternative Energy	124
Energy Conservation and Efficiency	130
Fossil Fuels	133

8 Energy (nuclear)

Nuclear Power	138
Nuclear Safety	149

9 Health and Safety

General	152
Asbestos	156
Biotechnology	156
Environmental Health	159
Food Quality and Safety	161
Radiation Protection	163
Toxicology and Chemicals	166

10 Pollution

General	174
Air Pollution	180
Air Quality and Pollution	187
Land Pollution	188
Marine Pollution	189
Noise Pollution and Abatement	192
Traffic Pollution	194
Water Pollution	195

11 Businesses and Services

Eco Labelling	199
Environment and Industry	201
Environmental Assessment	205
Environmental Auditing	207
Environmental Business	207
Environmental Consultants	211
Environmental Education	213
Environmental Engineering	215
Environmental Groups and Movements	217
Environmental Laws and Legislation	230
Environmental Management	235
Environmental Monitoring	240
Environmental Policy	242
Environmental Science	250
Environmental Standards	256
Environmental Technology	257
Ethical Investment	262

12 Miscellaneous

General	263
Chemicals	269
Population	272
Transport	274

Appendices

Euro Info Centres	283
Correspondence Centres	304
Online Hosts	305
CD-ROM Producers and Distributors	308

Indexes

Geographical Index	313
Alphabetical Index	339

Introduction

In the last ten years we have all been made aware of the importance of the appropriate stewardship and management of the global environment. Increasingly we understand that the ways we multiply, produce energy, use natural resources and produce waste threaten to change fundamentally the balance of the world environment. The recognition of the gravity and responsibility of this issue is not just a recent concern. In 1848 John Stuart Mill wrote:

'Is there not the Earth itself, its forests and waters, above and below the surface? These are the inheritance of the human race... What rights, and under what conditions, a person shall be allowed to exercise over any portion of this common inheritance cannot be left undecided. No function of government is less optional than the regulation of these things, or more completely involved in the idea of a civilised society' (Mill, J.S. Principles of political economy. 1848)

As this book shows, governments *have* to be concerned, as constant pressure is placed on them by individuals and groups of all kinds in order to keep the environment close to the centre of political debate. The environmental movement, with its myriad of pressure and interest groups, is complex and varied. The **European Environmental Information Sourcebook (EEIS)** attempts to chart these groups and the sources of information they generate.

Most of today's decision makers will be dead before the planet suffers the full consequences of acid rain, global warming, ozone depletion, widespread desertification, and species loss. Most of today's young voters will be alive, and will be unforgiving if their ancestors get it wrong. If the worldwide economic recession has slowed some of the economic pressures which accentuate environmental despoliation, this may be a temporary interregnum. As economic growth gains pace once more, environmental issues will once again take centre stage.

EEIS is a wide ranging 'one-stop' guide to over 1900 sources of information. Over half of these are organisations. We have tried to provide an overview of the many and disparate organisations throughout Europe which are concerned with environmental issues. We have also included a selection of the vast array of printed and electronic sources published in the last three to four years which have something important to say about the state of Europe's environment.

EEIS was compiled with the needs of a wide variety of people in mind. It provides an introductory point of contact on general and specific environment issues. It was compiled, in the main, using information provided by these organisations. In addition, information was gathered from contacts in other areas and from secondary sources such as recent directories, journals and bibliographies.

The criteria for inclusion are that the organisation must be concerned with some aspect of the environment or have a marked impact on the environment. All bar three organisations listed have their headquarters in Europe. These include international organisations.

For further information, readers looking for commentary on environmental topics, can consult the numerous journals, reports and books listed in the directory. These have been selected from an almost infinite number of sources. We have attempted to be as objective as possible in deciding which sources should be included. The character of the source was an important criterion in this selection process

EEIS includes electronic databases where these are primarily concerned with some aspect of the environment. Access to this data can be achieved by approaching the database producer or the online host/vendor. The appendix has addresses for all these organisations.

What's in the Directory?

EEIS includes some 1900 national and international sources. This lists covers the breadth of environmental issues from acid rain to wind energy. It includes listings of major organisations, printed and electronic sources for most of the countries in Europe. With the breakup of the former Soviet Union and the Yugoslavia Republic, a large number of new states are represented.

EEIS includes government departments and other statutory bodies, academic, educational and research establishments; trade and professional associations; charities and selected commercial and industrial bodies.

EEIS lists, as far as possible, all significant national organisations for most of the countries in Europe, together with some regional ones. The coverage varies by country and subject area. It is more detailed for topics which are central to the environmental movement such as global warming, while for subjects such as heritage only a small number of organisations have been included. Organisations concerned with acid rain are featured strongly in Northern Europe, whereas in Southern

Europe there are more organisations concerned with nature conservation. Some international organisations have been listed because their activities affect or include Europe.

Sources of further information include reference books, directories, yearbooks, dictionaries, reports, conference papers, periodicals and electronic databases. Again the coverage varies according to country, subject and format. For example there are numerous journals on energy in Germany, whereas in Portugal these are virtually non-existent.

How to find information in the Directory

EEIS is divided into 12 chapters treating different broad topics of the environment. Each chapter begins with a listing of general information sources and is then subdivided into specific areas. All the chapter sections are themselves divided by type of source, in the following order:

- government organisation;
- non-government organisation;
- pressure groups;
- books;
- conference papers;
- directories and yearbooks;
- periodicals;
- reports;
- databases;
- libraries.

Please note that international and pan-European sources are listed by the most common form of their name. Names of organisations and other sources are generally in the language of the country concerned.

EEIS has two indexes, one geographical, one alphabetical. The Geographical Index is divided first by area of coverage: International; European; then country by country. Each of these geographical divisions is in turn subdivided into three categories: organisations; publications; and databases.

The Alphabetical Index is divided by organisations, publications and databases which are then listed alphabetically.

Note: Telephone Numbers

Apart from the UK, telephone numbers are given in standard international format; Country code–area code–local number. For example the telephone number for Ministère de l'Environnement in France is 33 1 47 48 12 12. The international access code varies from country to country. Dialling from the UK – 010 is required for access to the international telephone network.

Note that in Germany during the period 1992–1994, areas formerly in the Democratic Republic with numbers beginning with 37 will be replaced by 49. Also in Sweden, Swedish Telecom are undergoing a major number change. The area codes and customer numbers are subject to change.

Information which may be contained in the entries

Name of organisation
Address
Telephone
Fax
Contact name
Geographical coverage
Aims and objectives
Publications
Research activities

Name of published source
Geographical coverage
Publishers
Address of publisher
Telephone
Fax
Editor/Author
Language
Coverage
Price/Annual subscription

Frequency
ISBN/ISSN

Name of database
Geographical coverage
Database producer
Database hosts
Database timespan
Database update
Coverage

Please note

Every effort has been made to ensure the information in this directory is correct at the time of its compilation. Organisations listed in the directory are not responsible for their inclusion, nor for the way in which they are described. Any suggestions, corrections and requests for information should be made to the Editors at:

Manchester Business School,
Booth Street West,
Manchester M15 6PB,
United Kingdom

Successive editions will chart this dynamic and fast changing European environmental scene.

Acknowledgements

This book could not have been compiled without the co-operation of the environmental organisations and agencies who gave much time and help to us. We give them our profound thanks. We also are indebted to Evelyn Simpson (Systems Librarian at Manchester Business School) for her constant support, expertise and 'techno patience'. Finally, we thank the Gale Research International editorial staff, particularly Lee Ripley Greenfield and Jenni Doig, for their professionalism and encouragement.

Oksana Newman
Allan Foster
August 1993

Introduction

Au cours des dix dernières années nous avons tous été rendus conscients de l'importance d'une administration et d'une gestion responsables de l'environnement mondial. Nous comprenons de plus en plus que la multiplication de notre race, nos façons de produire de l'énergie, d'exploiter les ressources naturelles et de produire des déchets menacent de changer fondamentalement l'équilibre de l'environnement de la planète. L'identification de la gravité et des conséquences de cette situation n'est pas seulement un fait nouveau. En 1848 John Stuart Mill écrivait:

'N'y a-t-il pas la terre elle-même, ses forêts et ses eaux, sa surface aussi bien que ses couches souterraines? Voilà le patrimoine de la race humaine... Quels droits une personne peut-elle exercer sur n'importe quelle partie de cet héritage commun et sous quelles conditions sont des questions qui ne peuvent rester irrésolues. Aucune fonction de gouvernement n'est moins optionnelle que la régulation de ces choses ni plus complètement liée à l'idée d'une société civilisée.' (Mill, J.S. *Principles of Political Economy*, 1848)

Comme le démontre cet ouvrage, les gouvernements sont obligés d'être intéressés puisque des individus et des groupes de toutes sortes exercent une pression constante sur eux afin de garder le sujet de l'environnement près du centre du débat politique. Le mouvement pour l'environnement, avec sa multitude de groupes de pression et d'associations engagées est complexe et varié. The **European Environmental Information Sourcebook (EEIS)** a pour but de dresser une liste de ces groupes et associations et de noter les sources d'information sur l'environnement qu'ils fournissent.

La plupart des décideurs d'aujourd'hui seront morts avant que la planète ne souffre de toutes les suites des pluies acides, du réchauffement mondial, de la diminution de la couche d'ozone, de la désertification étendue et de la disparition de certaines espèces. La plupart des jeunes électeurs d'aujourd'hui seront encore vivants et ne pardonneront pas à leurs ancêtres si ceux-ci se sont trompés. Si la récession économique mondiale a ralenti certaines des pressions économiques qui accentuent la spoliation de l'environnement, cette situation peut s'avérer un interrègne temporaire. Lorsque la croissance économique accèlèrera à nouveau, les questions sur l'environnement reprendront place au centre du débat.

EEIS est un guide qui offre des informations instantanées sur plus de 1.900 sources de renseignements sur l'environnement. Plus de la moitié de ces sources sont des organisations. Nous avons essayé de fournir une vue d'ensemble des nombreuses organisations dispersées à travers l'Europe qui s'occupent des questions de l'environnement. Nous avons également incorporé une sélection de la vaste gamme de sources imprimées et électroniques ayant paru dans les trois ou quatre dernières années qui ont des éléments importants à communiquer sur l'état de l'environnement en Europe.

EEIS a été composé en tenant compte des besoins d'un grand nombre de personnes. Le livre fournit un point de contact préliminaire avec des questions générales et spécifiques sur l'environnement. Il a été composé pour la plupart en utilisant les renseignements fournis par ces organisations. De plus, d'autres renseignements ont été rassemblés grâce à des contacts dans d'autres secteurs et à partir de sources secondaires, telles des annuaires récents, des publications spécialisées et des bibliographies.

Afin d'être sélectionnée dans ce guide, une organisation doit répondre à certains critères tels qu'un engagement dans un ou plusieurs aspects de l'environnement ou doit avoir un impact important sur celui-ci. Toutes les organisations figurant dans l'ouvrage, sauf trois, ont leur siège en Europe. Elles comptent parmi elles des organisations internationales.

Pour trouver des renseignements supplémentaires, le lecteur à la recherche de commentaires sur des questions de l'environnement peut consulter les nombreux publications spécialisées, rapports et livres répertoriés dans cet annuaire. Ceux-ci ont été sélectionnés à partir d'un nombre pratiquemment infini de sources d'informations. Nous avons essayé de rester aussi objectifs que possible en décidant quelles sources inclure; la nature de la source étant un critère important dans ce processus de sélection.

EEIS cite des bases de données dans les cas où celles-ci traitent essentiellement les aspects de l'environnement. L'accès à ces bases de données peut se faire par l'intermédiaire du producteur de la base de données ou de l'hôte ou du vendeur en-ligne. Le lecteur trouvera les adresses de ces organisations dans l'appendice.

Que contient l'annuaire EEIS?

EEIS contient environ 1.900 sources nationales et internationales. Cette liste s'étend à travers la gamme des questions sur l'environnement, des pluies acides à l'énergie éolienne. Elle répertorie les organisations principales ainsi que les sources imprimées et électroniques dans la plupart des pays en Europe. Etant donné le démantèlement de l'ancienne Union Soviétique et de l'ancienne République Yougoslave, un grand nombre d'états nouveaux sont représentés.

EEIS donne des renseignements sur des services gouvernementaux et autres organismes de droit public; des établissements universitaires, d'enseignement et de recherches; des associations commerciales et professionnelles; des fondations de bienfaisance et des ensembles commerciaux et industriels sélectionnés.

EEIS répertorie autant que possible toutes les organisations nationales importantes dans la plupart des pays européens ainsi que des organisations régionales. Le contenu varie de pays en pays et par sujet. Il est plus détaillé dans le cas des thèmes centraux au mouvement de l'environnement, tels que le réchauffement mondial, tandis que pour des sujets tels que le patrimoine un petit nombre seulement d'organisations figure dans l'ouvrage. Les organisations qui se préoccupent des pluies acides sont plus nombreuses dans le nord de l'Europe alors qu'il y a plus d'organisations consacrées à la protection de la nature dans le sud de l'Europe. Certaines organisations internationales figurent dans l'annuaire parce que l'Europe en fait partie ou est concernée par leurs activités.

Les autres sources de renseignements comprennent des livres de référence, des annuaires, des dictionnaires, des rapports, des communications de conférences, des périodiques et des bases électroniques de données. Encore une fois, le contenu varie selon le pays, le sujet et le format. Par exemple, en Allemagne on publie de nombreuses publications spécialisées sur l'énergie, tandis qu'au Portugal il n'en existe presque pas.

Comment chercher des renseignements dans l'annuaire

EEIS est divisé en 12 chapitres qui traitent les grands thèmes de l'environnement. Chaque chapitre commence avec un répertoire de sources de renseignements généraux. Le chapitre est alors subdivisé par aspects particuliers du grand thème. Les sources d'informations citées dans chaque partie de tout chapitre apparaissent dans l'ordre suivant:

- organisations gouvernementales;
- organisations non gouvernementales;
- groupes de pression;
- livres;
- communications de conférences;
- annuaires;
- périodiques;
- rapports;
- bases de données;
- bibliothèques.

EEIS a deux index, l'un géographique, l'autre alphabétique. L'index géographique est divisé en premier lieu par l'étendue géographique de la source: internationale; européenne; et ensuite pays par pays. Chacune de ces divisions est subdivisée en trois catégories: organisations; publications; et bases de données. L'index alphabétique est divisé en premier lieu par les organisations, les publications et les bases de données. Les sources sont ensuite répertoriées par ordre alphabétique sous ces trois rubriques.

Les sources internationales et paneuropéennes paraissent sous la forme la plus commune de leur nom. Les noms des organisations et des sources sont cités de façon générale dans la langue du pays.

A noter pour les numéros de téléphone

Pour les numéros de téléphone et de fax (à part les numéros au Royaume-Uni), nous utilisons le format international courant; l'indicatif pour le pays–suivi de l'indicatif pour la région–suivi du numéro local. Par exemple, le numéro de téléphone du Ministère de l'environnement en France est noté ainsi: 33 1 47 48 12 12. L'indicatif international varie selon le pays.

Veuillez noter qu'au cours de la période 1992–94 l'indicatif 37 pour les anciennes régions d'Allemagne de l'Est sera remplacé par l'indicatif 49. En Suède, les services de télécommunications sont en train de changer leur système d'indicatifs. Les indicatifs régionaux et les numéros de clients sont sujets à des modifications.

Les renseignements que les entrées peuvent fournir

Pour une organisation:
Nom de l'organisation
Adresse
Numéro de téléphone
Numéro de fax
Nom de personne à contacter
Etendue
Activités de l'organisation
Publications de l'organisation
Secteur de recherche

Pour une publication:
Nom de la publication
Etendue
Maison d'édition
Adresse de la maison d'édition
Numéro de téléphone
Numéro de télécopie
Rédacteur/Rédactrice/Auteur
Langue
Etendue
Prix/Abonnement
Fréquence
ISBN/ISSN

Pour une base de données:
Nom de la base de données
Etendue
Producteur de la base de données
Hôte de la base de données
Période de couverture de la base de données
Mise à jour de la base de données

Veuillez noter

Tous nos efforts ont été déployés afin de maintenir l'exactitude des renseignements figurant dans cet annuaire au moment de sa parution. Les organisations qui figurent dans cet annuaire ne sont pas responsables de leur entrée, ni pour la description de leurs activités. Veuillez adresser toute suggestion, correction ou autre correspondance relative à cet ouvrage aux rédacteurs à:

Manchester Business School
Booth Street West
Manchester M15 6PB
United Kingdom

Les éditions prochaines de cet annuaire montreront les changements et le dynamisme dans le domaine de l'environnement en Europe.

Remerciements

Cet ouvrage n'aurait pu être publié sans l'assistance des organisations et des agences pour l'environnement qui nous ont beaucoup aidé et nous ont consacré leur temps. Nous tenons à les remercier vivement. Nous devons beaucoup à Evelyn Simpson (bibliothécaire-programmeuse à la Manchester Business School) pour son soutien constant, son expertise et sa patience face aux problèmes des nouvelles technologies. En dernier lieu nous aimerions remercier l'équipe de rédaction de Gale Research International et Lee Ripley Greenfield et Jenni Doig en particulier pour la haute qualité de leur travail et pour leurs encouragements.

Oksana Newman
Allan Foster
août 1993

Einleitung

In den letzten zehn Jahren wurde uns allen die Bedeutung der Pflege und eines weltweit vernünftigen Umgangs mit der Natur bewußt. Wir sehen immer deutlicher, daß der Bevölkerungszuwachs, die Energiegewinnung, der Verbrauch natürlicher Ressourcen und die Produktion von Abfallstoffen drohen, das Gleichgewicht der Natur grundlegend zu stören. Die Einsicht in die Verantwortung und Schwere dieser Problematik ist kein neumodisches Anliegen. 1848 schrieb John Stuart Mill:

'Gibt es denn nicht die Erde selbst, ihre Wälder und Gewässer, ober- und unterhalb der Scholle? Die ist alles das Erbe der Menschheit. Welche Rechte unter welchen Bedingungen einer Person an irgendeinem Teile dieses gemeinsamen Erbes zugestanden werden sollen, kann nicht ohne Entscheid bleiben. Keine andere Aufgabe einer Regierung ist weniger unabdingbar, oder verwirkter mit der Idee einer zivilisierten Gesellschaft als jene.' (Mill, J.S. *Principles of political economy*. 1848)

Wie dieses Buch zeigt, *müssen* die Regierungen Anteil nehmen, da ständiger Druck von Einzelpersonen und Gruppierungen aller Art auf sie ausgeübt wird, um Umweltfragen auf der politischen Tagesordnung zu halten. Die Umweltbewegung, mit ihren unzähligen Pressure-groups und Interessenvertretungen ist komplex und heterogen. Das **European Environmental Information Sourcebook (EEIS)** versucht einen überblick über diese Gruppen und die Informationen, die sie zur Verfügung stellen, zu bieten.

Die meisten der heutigen Entscheidungsträger werden tot sein, noch bevor sich auf unserem Planeten der saure Regen, der Treibhauseffekt, das Ozonloch, die Versteppung und das Artensterben in ihrer ganzen Konsequenz bemerkbar machen. Die meisten der heutigen Jungwähler werden am Leben sein und Fehler ihrer Vorfahren nicht verzeihen. Die weltweite Wirtschaftsflaute lockert einige der ökonomischen Zwänge, die zum Raubbau an der Natur führen, aber dies dürfte ein Zwischenspiel sein. Sobald das Wirtschaftswachstum wieder einsetzt, werden Umweltfragen erneut in den Mittelpunkt rücken.

Das EEIS ist ein umfassender 'Auf-einen-Blick' Führer zu mehr als 1900 Informationsquellen, von denen mehr als die Hälfte Organisationen sind. Wir haben versucht, einen Überblick über die zahlreichen und unterschiedlichen sich mit Umweltfragen befassenden Organisationen in ganz Europa zu geben. Ebenso ist eine Auswahl aus dem riesigen Feld gedruckter und elektronischer Quellen enthalten, die innerhalb der letzen drei bis vier Jahre etwas wesentliches zum Zustand der europäischen Natur zu sagen hatten.

Das EEIS wurde erstellt, um den Bedürfnissen eines weiten Spektrums unterschiedlicher Menschen gerecht zu werden. Es stellt einen Ausgangspunkt für Kontakte mit Organisationen dar, die sich mit allgemeinen und speziellen Umweltfragen beschäftigen. Es wurde größtenteils unter Verwendung von Informationen dieser Organisationen erstellt, zusätzlich fanden Informationen von Kontakten aus anderen Bereichen und Sekundärquellen wie zum Beispiel neuerer Verzeichnisse, Zeitschriften und Bibliographien Eingang.

Die Hauptauswahlkriterien für die Aufnahme in dieses Werk sind die umweltbezogene Zielsetzung einer Organisation, oder ihr deutlicher Einfluß auf das Umweltgeschehen. Alle bis auf drei der Organisationen (einschließlich der internationalen Organisationen) haben ihren Sitz in Europa.

Weiterführende Informationen findet der an Ausführungen zu Umweltfragestellungen interessierte Leser in den zahlreichen im Verzeichnis aufgeführten Zeitschriften, Berichten und Büchern, die aus einer schier endlosen Anzahl von Quellen ausgewählt wurden. Bei der Auswahl der Quellen versuchten wir so objectiv wie möglich vorzugehen, die Art der Quelle war in diesem Auswahlprozeß ein wichtiges Kriterium.

Das EEIS verzeichnet elektronische Datenbanken, deren Hauptinhalt auf Umweltaspekte ausgerichtet ist. Zugang zu diesen Datenbanken ermöglicht der Hersteller, oder im Falle einer Online-Datenbank die Betreiber. Die jeweiligen Adressen sind im Anhang aufgeführt.

Was bietet dieses Verzeichnis?

Das EEIS führt circa 1900 nationale und internationale Quellen auf, die die ganze Breite der Umweltfragen von saurem Regen bis hin zur Gewinnung von Windenergie abdecken. Es enthält Aufstellungen über die großen Organisationen, Druck- und elektronische Informationsquellen der meisten europäischen Länder. Nach dem Zusammenbruch der früheren Sowjetunion und der Jugoslawischen Republik ist auch eine große Anzahl neuer Staaten vertreten.

Das EEIS bietet eine Aufstellung von Regierungsbehörden und andere offizielle Körperschaften, akademischen, Lehr- und Forschungseinrichtungen, Handels- und Berufsverbänden, gemeinnützigen Organisationen, sowie ausgewählten Handels- und Industriekörperschaften.

Das EEIS führt, soweit möglich, alle wichtigen landesweiten, sowie einige grenzüberschreitend regionale Organisationen für die meisten europäischen Länder auf. Die Exhaustivität ist, je nach Land und Themengebiet unterschiedlich. Themengebiete wie der Treibhauseffekt, die eine zentrale Rolle in der Umweltbewegung spielen, sind stärker vertreten als solche wie zum Beispiel Pflege und Erhalt von Naturdenkmälern, für die nur wenige Organisationen genannt werden. In Nordeuropa finden sich mehr Organisationen, die sich mit saurem Regen, in Südeuropa hingegen mehr Organisationen, die sich mit Landschaftspflege befassen. Einige internationale Organisationen, deren Aktivitäten Europa betreffen, wurden ebenfalls berücksichtigt.

Weiterführende Quellen schließen Nachschlagewerke, Verzeichnisse, Jahrbücher, Wörterbücher, Berichte, Tagungsbeiträge, Zeitschriften und elektronische Datenbanken ein. Auch hier hängt die Ausführlichkeit vom Land, dem Thema und der Art der Quellen ab. Beispielsweise gibt es in Deutschland zahlreiche Zeitschriften zum Thema Energie, in Portugal hingegen praktische keine.

Zur Benutzung des Verzeichnisses

Das EEIS ist in 12 Kapitel eingeteilt, die unterschiedliche, breitgefächerte Umweltfragestellungen behandeln. Eingangs eines jeden Kapitels findet sich eine Aufstellung allgemeiner Informationsquellen, anschließend eine nach Teilgebieten gegliederte. Die einzelnen Abschnitte der Kapitel wiederum sind folgendermaßen nach der Art der Quellen eingeteilt:

- Staatliche Organisationen;
- Nicht-staatliche Organisationen;
- Pressure-Groups;
- Bücher;
- Tagungsbeiträge;
- Verzeichnisse und Jahrbücher;
- Zeitschriften;
- Berichte;
- Datenbanken;
- Bibliotheken.

Das EEIS verfügt über ein geographisches, sowie ein alphabetisches Sachverzeichnis. Das geographische Sachverzeichnis ist nach Themengebieten gegliedert: international; europäisch; Einzelstaaten. Jede dieser drei geographischen Einteilungen ist wiederum in folgende Kategorien untergliedert: Organisationen; Veröffentlichungen; und Datenbanken.

Das alphabetische Sachverzeichnis ist nach Organisationen, Veröffentlichungen und Datenbanken unterteilt, die jeweils alphabetisch aufgeführt werden.

Bitte beachten Sie, daß die internationalen und pan-europäischen Quellen unter der gebräuchlichsten Form ihres Namens aufgeführt sind. Für die Namen von Organisationen und anderer Quellen findet größtenteils die landessprachliche Form Verwendung.

Ein Hinweis zu den Telefonnummern

Mit Ausnahme der britischen Telefonnummern wurde die international gebräuchliche Standardnotation verwendet: Landeskennziffer–Vorwahl–Durchwahl. Ein Beispiel: die Nummer des Ministère de l'Environnement in Frankreich 33 1 47481212. Der Vorwahl für Fernverbindungen ist von Land zu Land verschieden.

Bitte beachten Sie, daß in den Jahren 1992–1994 die Telefonnummern der früheren DDR, die mit 37 beginnen, durch Nummern mit 49 ersetzt werden. Auch in Schweden führt die schwedische Telecom umfangreiche Neustrukturierungen durch. Die Vor- und Durchwahlnummern können sich also ändern.

Aufbau der Einträge

Name der Organisation
Adresse
Rufnummer
Telefaxnummer
Ansprechpartner
Geographischer Umfang
Projekte
Veröffentlichungen
Forschungsvorhaben

Name der Veröffentlichung
Adresse des Verlags
Telefonnummer
Telefaxnummer
Geographischer Umfang
Verlag Herausgeber/Autor
Erscheinungsweise
Sprache
Thematik
Preis/Abonnementspreis
ISBN/ISSN

Name der Datenbank
Geographischer Umfang
Datenbankanbieter
Hostrechner der Datenbank
Zeitliche Abdeckung der Datenbank
Fortschreibungen
Thematik

Zur Beachtung

Es wurde allergrößte Sorgfalt darauf verwendet, den aktuellsten Stand und die Korrektheit der Informationen in diesem Verzeichnis zu gewährleisten. Die einzelnen Organisationen zeichnen weder für ihre Aufnahme, noch für die Art, in der sie im Verzeichnis beschrieben werden verantwortlich. Vorschläge, Verbesserungswünsche und Anfragen jeglicher Art sollten an die Herausgeber gerichtet werden:

Manchester Business School
Booth Street West
Manchester M15 6PB
United Kingdom

Folgeauflagen werden die Dynamik und Veränderungen auf dem europäischen Umweltsektor widerspiegeln.

Danksagung

Dieses Buch wäre ohne die Hilfe und den Zeitaufwand der einzelnen Umweltorganisationen und Behörden nicht zustande gekommen. Wir möchten ihnen dafür aufrichtig danken. Wir schulden Evelyn Simpson (Systems Librarian der Manchester Business School) Dank für ihre ständige Unterstützung, ihre Sachkenntnis und ihre 'Technik-Verständnis'. Schließlich danken wir den Mitarbeitern von Gale Research International, insbesondere Lee Ripley Greenfield und Jenni Doig für ihre Professionalität und die Unterstützung.

Oksana Newman
Allan Foster
August 1993

Acronyms

ACCIS   Advisory Committee for the Coordination of Information Systems
ACEA   European Association of Automobile Manufacturers
ACORD   Automotive Consortium on Recycling and Disposal
ACRA   Aluminium Can Recycling Association
ADECA   Association for the Development of Agricultural Fuel
ADEME   Agence de l'Environnement et de la Maîtrise de l'Energie
AEDENAT   Asociacion Ecologista de Defensa de la Naturaleza
AFRC   Agriculture and Food Research Council
AGALEV   Anders Gaan Leven
AGRIS   Agricultural Sciences and Technology
AGROFARMA   Association Industry Difesa Produzioni Agricole
AHCW   Association of Hungarian Chemical Works
AIA   Asbestos International Association
AIC   Asbestos Information Centre
AICB   Association Internationale Contre le Bruit
AIRP   Associazione Italiana di Protezione Contro le Radiazioni
AIS   International Association of the Soap and Detergent Industry
ANDIN   Associazione Nazionale di Ingegneria Nucleare
ANDRA   Agence Nationale pour la Gestion des Dechets Radioactifs
ANFIMA   Associazione Nazionale Fra Fabbricanti di Imballagi Metallici
ANRED   Agence Nationale de la Recuperation et l'Elimination des Dechets
ANSEAU   Association Nationale des Services d'Eau asbl
APERE   Association Promotion Energies Renouvelables
APESB   Associaçao Portuguesa para Estudos de Saneamento Basico
APME   Association of Plastic Manufacturers in Europe
APPA   Association pour la Prevention de la Pollution Atmospherique
APRH   Associaçao Portuguesa dos Recursos Hidricos
ARGE ALP   Association des Régions des Alpes Centrales
ARIC   Atmospheric Research and Information Centre
ARTP   Association of Reclaimed Textile Processors
ASBL   Forum Nucleaire Belge
AWT   Abwassertechnik
BAA   British Agrochemicals Association Ltd
BANC   British Association of Nature Conservationists
BAUM   Bundesdeutscher Arbeitskreis für Umweltbewusstes Management
BBCG   Belgian Biotechnology Coordinating Group
BBL   Bon Beter Leefmilieu
BBMA   British Battery Manufacturers Association
BCRA   British Coal Research Association
BEMAs   British Environment Media Awards
BEWA   British Effluent and Water Association
BGS   British Geological Survey
BIA   Bioindustry Association
BIAC   Business and Industry Advisory Committee
BIR   Bureau International de la Récupération
BMFT   Bundesministerium für Forschung und Technologie
BNF   British Nuclear Forum
BOF   British Organic Farmers
BPBIF   British Paper and Board Industry Federation
BPF   British Plastics Federation
BPI   Belgian Packaging Institute
BRE   Building Research Establishment
BRISA   Base Relacional de la Industria y Servicios Ambientales
BRS   Bibliographic Retrieval System
BSI   British Standards Institution
BTO   British Trust for Ornithology
BWEA   British Wind Energy Association
BWK   Brennstoff-Waerme-Kraft
BYGGDOK   Swedish Institute of Building Documentation
CA IUAPPA   Ceská Asociace IUAPPA
CA Search   Chemical Abstracts
CABI   Commonwealth Agricultural Bureaux International
CABO   Centruum voor Agrobiologisch Onderzoek
CAN UK   Climate Action Network UK
CAPPA   Croatian air Pollution Prevention Association
CAT   Centre for Alternative Technology
CBI   Confederation of British Industry
CCI   Cotton Council International
CD-ROM   Compact Disc-Read Only Memory
CEA   Commissariat a l'Energie Atomique
CEA   Confederation of European Agriculture
CEC   Commission of the European Communities
CEDE   Conseil Européen du Droit de l'Environnement
CEFIC   Conseil Europeen des Féderations de l'Industries Chimique
CEM   Continuous Emission Monitoring
CEMP   Centre for Environmental Management and Planning
CEN   Comité Européen de Normalisation
CENSOLAR   Centro de Estudios de la Energia Solar
CEN-SCK   Studiecentrum voor Kernenergie-Centre d'Etude de l'Energie Nucleaire
CEPAC   Confédération Européene de l'Industrie des Pâtes Papieres et Cartons
CEPI   Confederation of European Paper Industries
CEPS   Centre for European Policy Studies

Acronym	Expansion
CEST	Centre for Exploitation of Science and Technology
CHPA	Combined Heat and Power Association
CIEL	Centre for International Environmental Law
CIEMAT	Centro de Investigación Energética Medioambiental y Tecnológica
CITEPA	Centre Interprofessionnel Technique d'Etudes de la Pollution Atmospherique
CITES	Convention on International Trade in Endangered Species of Wild Fauna and Flora
CLAN	Society for Clean air in the Netherlands
CNE	Climate Network Europe
COBIOTECH	ICSU Steering Committeee for Biotechnology
COCF	Centre for Our Common Future
CODA	Coordinadora de Organizaciones de Defensa Ambiental
CONCAWE	Oil Companies European Organisation for Environmental Health and Protection
COPA	General Committee of Agricultural Cooperation in Europe/Committee of Agricultural Organisations
COSQUEC	Council for Ocupational Standards and Qualifications in Environmental Conservation
CRA	Chemicals Recovery Association
CRE	Coal Research Establishment
CSERGE	Centre for Social and Economic Research on the Global Environment
CSIA	Comitato di Studio perl'Inquinamento Atmosferico
CTC	Cyclists Touring Club
CUBE	Concentration Unit for Biotechnology in Europe
DAA	Danish Agrochemical Association
DAIF	Deutsches Atomforum eV
DEPANA	Liga Para la Defensa del Patrimonio Natural
DGMA	Ministerio de Obras Públicas y Urbanismo–Dirección General de Medio Ambiente
DIMDI	Deutsches Institut fur Medizinische Dokumentation und Information
DNR	Deutsche Naturschutzring
DNS	Danish Nuclear Society
DoE	Department of the Environment
DSD	Duales System Deutschland GmbH
DTC	Danish Toxicology Centre
DTI	Danish Packaging and Transportation Research Institute
DTI	Department of Trade and Industry
DTp	Department of Transport
DUNDIS	Directory of United Nations Databases and Information Services
DVGW	Deutscherverein des Gas und Wasserfaches eV
DWD	Deutsche Wetterdienst
EAPS	European Association for Population Studies
EASOE	European Arctic Stratospheric Ozone Experiment
EC	European Community
ECDIN	Environmental Chemicals Data and Information Network
ECETOC	European Chemical Industry Ecology and Toxicology Centre
ECHO	European Community Host Organisation
ECMT	European Conference of Ministers of Transport
ECMWF	European Centre for Medium Range Weather Forecasts
ECOROPA	Europäische Ökologische Aktion
ECO	Environmental Education Trust
ECO	Irish Environmental Organisation/Organisation for Youth
ECO–PER	Economy and Performance Service
ECPA	European Crop Protection Association
ECSA	European Chlorinated Solvent Association
EDF	Electricite de France
EDMA	Environmental Detergent Manufacturers Association
EDRC	Environment and Development Resource Centre
EEA	European Environmental Agency
EEA	European Environmental Alliance
EEB	European Environment Bureau
EFMA	European Fertiliser Manufacturers' Association
EFPIA	European Federation of Pharmaceutical Industries
EGARA	European Group of Automotive Recycling Association
EGIS	European Geographical Information Systems Foundation
EHAS	Environmental Health Advisory Service
EIB	European Investment Bank
EIEC	European Institute of Ecology and Cancer
EIRIS	Ethical Investment Research Information Service
ELC	Environment Liason Centre International
EMEP	Cooperative Programme for Monitoring and Evaluation of Long Range Transboundary Air Pollution
ENEA	Comitato Nazionale per la Ricerca e per lo Sviluppo dell' Energia Nucleare e dell' Energie Alternativ
ENDS	Environmental Data Services
ENEL	Ente Nazionale per l'Energia Elettrica
ENERO	European Network of Environmental Research Organisations
ENS	European Nuclear Society
ENVN	Enviroline
EOLAS	Irish Science and Technology Agency
EPAL	Empresa Publica das Aguas Livres
EPOQUE	European Parliament Online Query Service
ERL	Environmental Resources Ltd
ERRA	European Recovery and Recycling Association
ESA–IRS	European Space Agency Information Retrieval System
ESC	Economic and Social Committee
ESI	Ecological Studies Institute
ETA	Environmental Transport Association
ETAD	Ecological and Toxicological Association of the Dyestuff Manufacturing Industry
ETAPS	Environment and Technology Association for Paper Sacks
ETDE	Energy Technology Data Exchange
ETSU	Energy Technology Support Unit
EURASAP	European Association for the Science of Air Pollution
EURATOM	European Atomic Energy Community
EUROPIA	European Petroleum Industries Association
EVA	Electric Vehicle Association of Great Britain
EVA	Studiengesellschaft für die Entsorgung von Altahrrzeugen GmbH
EWPCA	European Water Pollution Control Association
EYDAP	Water Supply and Sewage Committee of Major Athens
EYFA	European Youth Forest Action

Euro ACE European Association for Conservation of Energy
Europile Association of European Dry Battery Manufacturers
FACT Food Additives Campaign Team
FAO Food and Agriculture Organisation of the United Nations
FAPPA Finnish Air Pollution Prevention Society
FBID Federation of Biotechnology Industries of Denmark
FEAD European Federation of Waste Management
FEC Foundation for Environmental Conservation
FEDEREC Industrielle de l'Environnement et du Recyclage
FEDERGASACQUA Federazione Italiana Impresa Pubbliche Gas Acqua e Varie
FEDESA European Federation of Animal Health
FEEE Foundation for Environmental Education in Europe
FEPMA Fundacióon Para la Ecologia y la Proteccion del Medio Ambiente
FESCID European Federation of Chemical and General Workers
FEVE Fédération Européene du Verre d'Emballage
FFC Forests Forever Campaign
FICI Federation of Irish Chemical Industries
FIEN Forum Italiano dell'Energia Nucleare
FNAB Biofranc
FoE Friends of the Earth
FoEI Friends of the Earth International
FORATOM Forum Atomique Europeene
FÖK Független Ökological KÖzpont
FSB Food Safety Briefing
FSBH Fachverband Schweizerischer Becherhersteller
FSSRS Farm Structure Survey Retrieval System
FT Financial Times
GAA Greek Agrochemical Association
GEA Governo Locale ed Economia dell' Ambiente
GEDEA Grupo Ecologista de Agrónomos
GEMS/PAC Global Environmental Monitoring System/Programme Activity Centre
GER Global Environment Research
GESAMP Group of Experts on the Scientific Aspects of Marine Pollution
GET Global Environmnetal Technology Network
GIIN Groupe Intersyndical de l'Industrie Nucleaire
GIFAP International Group of National Associations of Manufacturers of Agro–Chemical Products
GLOBE Global Legislators Organisation for a Balanced Environment
GPCE Pharmaceutical Group of the EC
GPS Grüne Partei der Schweiz/Parti éologiste Suisse
GRAEL Green Alternative European Link
GRAP Groupement d'Activities de Presse
GRE Gruppi Ricerca Ecologica
GRID Global Resource Information Database
GSF Gesellschaft für Strahlen und Umweltforschung GmbH
HDRA Henry Doubleday Research Association
HELCOM Baltic Marine Environment Protection Committee Helsinki Commission
HMI Hahn–Meitner Institut Berlin GmbH
HMIP Her Majesty's Inspectorate of Pollution
HMSO Her Majesty's Stationery Office
HSDB Hazardous Substances Databank
HSE Health and Safety Executive
IAEA Agence Internationale de l'Energie Atomique
IAPRI International Association of Packaging Research Institutes
IARC International Agency for Research on Cancer
IAWPRC International Association on Water Pollution Research and Control
ICAN International Conservation Action Analysis
ICAP International Centre for the Application of Pesticides
ICBP International Council for Bird Preservation
ICC International Chamber of Commerce
ICCET Centre for Environmental Technology
ICEL International Council of Environmental Law
ICER Industry Council for Electronic Recycling
ICES International Council for the Exploration of the Seas
ICOMOS UK International Council on Monuments and Sites
ICPIC International Cleaner Production Information Clearing House
ICRP International Commission on Radiological Protection
ICSU International Council of Scientific Unions
IDW International Dolphin Watch
IE/PAC Industry and Environment Programme Activity Centre
IEA International Energy Agency
IEB International Environment Bureau
IEEP Institute for European Environmental Policy
IESC International Environmental Service Centre
IFA International Fertiliser Industry Association
IFAP International Federation of Agricultural Producers
IFE Institut for Energiiteknikk
IFREMER Institut Francais de Recherche pour l'Exploitation de la Mer
IIASA International Institute for Applied Systems Analysis
IIED International Institute for Environment and Development
IMADOS Institute of Handling Transport Packaging and Storage Systems
IMDG International Maritime Dangerous Goods
IMM Institute of Materials Management
IMO International Maritime Organisation
IMechE Institution of Mechanical Engineers
INCPEN Industry Council for Packaging and the Environment
INERIS Institut National de l'Environnement Industriel et des Risques
INETI Institut Nacional de Engenhara e Tecnologia Industrial
INFOTERRA International Environmental Information System
INIS International Nuclear Information System
INLA International Nuclear Power and Waste
INRA Cellule Environnement Institut National de la Recherche Agronomique
INTECOL International Association for Ecology
IOCU International Organisation of Consumers Unions
IOPC International Oil Pollution Compensation Fund
IPCC Intergovernmental Panel for Climatic Change
IPCS International Programme on Chemical Safety

IPICEA International Petroleum Industry Environmental Conservation Association
IPO Instituut voor Plantenziektenkundig Onderzoek
IPPNW International Physicians for the Prevention of Nuclear War
IPRE International Professional Association for Environmental Affairs
IRC International Water and Sanitation Centre
IRPA International Radiation Protection Association
IRPTC UNEP'S International Register of Potentially Toxic Chemicals
IRRD International Road Research Documentation
IRSA Istituto di Ricerca sulle Acque
ISC Internationale Scharrelvees Controle
ISES International Solar Energy Society
ISH Institut für Strahlenhygiene
ISO International Organisation for Standardisation
ISTER East European Environmental Research
ISWA International Solid Wastes and Public Cleansing Association
ITE Institute of Terrestrial Ecology
ITOPF International Tanker Owners Pollution Federation Ltd
ITTO International Tropical Timber Organisation
IUAPPA International Union of Air Pollution Prevention Associations
IUCN International Union for Conservation of Nature and Natural Resources
IUCT Fraunhofer–Institut fur Umweltchemie und Okotoxikologie
IVA Industrieverband Agrar
IVL Swedish Environmental Research Institute
IVN Institut voor Natuurbeschermingseducatie
IVT Industrin for Vaxt–Och Traskyddsmedel
IWC International Whaling Commision
IWC Irish Wildbird Conservancy
IWEM Institution of Water and Environmental Management
IWF Irish Wildlife Federation
IWM Institute of Wastes Management
IWRB International Waterfowl and Wetlands Research Bureau
IWS International Wool Secretariat
IWT International Water Tribunal Foundation
IYF International Youth Federation for Environmental Studies and Conservation
JRC Joint Research Council
KFA Forschungszentrum Julich GmbH
KRAV Kontrollforeningen for Alternativ Odling
KRdL Kommission Reinhaltung de Luft
KTA Kerntechnische Ausschuss
KYAE Kentron Yginis Keasfalias tis Ergassias
KfK Kernforschungszentrum Karlsruhe GmbH
LCA Life Cycle Analysis
LEAF Linking Environment and Farming
LI Lydteknisk Institut
LIDA Lega Italiana dei Diritti dell'Animale
LNETI Laboratório Nacional de Engenharia e Tecnologia Industrial
LVBW Landelijke Ver tot Behoud van de Waddenzee
MAB Programme on Man and the Biosphere
MAFF Ministry of Agriculture Fisheries and Food
MARC Monitoring and Assessment Research Centre
MCS Marine Conservation Society
MECO Mouvement Ecologique Luxembourg
MHIDAS Major Hazard Incident Data Service
MINATOM Ministry of Atomic Energy of the Russian Federation
MN Media Natura
MP Miljopartiet de GrÖna
MWV Mineralol Wirtschafts Verband eV
NAEE National Association for Environmental Education
NAMAS National Measurement and Accreditation Service
NASCO North Atlantic Salmon Conservation Organisation
NATO North Atlantic Treaty Organisation
NAWDC National Association of Waste Disposal Contractors
NEA Nuclear Energy Agency
NEFYTO Nederlandse Stichting voor Fytofarmacie
NEPP National Environmental Policy Plan Plus
NERC Natural Environment Research Council
NETT Network for Environmental Technology Transfer
NFLA Nuclear Free Local Authorities
NGO Non–Government Organisation
NIABA Nederlandse Industriële en Agarische Biotechnologie Associatie
NILU Norsk Institut for Luftforskning
NIPH Norsk Institut for Public Health
NISK Norsk Institut for Skogforskning
NIVA Norsk Institut for Vannforskning
NKS Nordic Committee for Nuclear Safety Research
NNV Norges Naturvernforbund
NOx Nitrous Oxides
NPHI Kansanterveyslaitos
NRA National Rivers Authority
NRPB National Radiological Protection Board
NRR Natural Resources Research
NSCA National Society for Clean Air and Environmental Protection
NT National Trust
OAL Österreichisches Arbeitsring fur Larmbekampfung
OECD Organisation for Economic Cooperation and Development
OGA Organic Growers Association
ONC Office National de la Chasse
OPOCE Office for Official Publications of the European Communities
ORCA Organic Reclamation and Composting Association
OSCOM Oslo Commission
ÖGNU Österreichisches Ges für Natur und Umweltschutz
PAKOE Panellinio Kentro Oikologikon Erevnon
PIFA Packaging and Industrial Film Association
PIRA Packaging Industry Research Association
PPZ Polska Partia Zielonych
PWMI European Centre for Plastics in the Environment
PZITS Polskie Zresenie Inazynierowi Technikow Sanitarnych
RAPRA Rubber and Plastic Research Association
RCEP Royal Commission on Environmental Pollution
REAG Renewable Energy Advisory Group
REPACAR Asociatión de Recuperadores de Papel y Cartón
RIIA Royal Institute of International Affairs

RIVM Rijksinstituut voor Volksgezondheid en Milieuhygiene
RIVO Rijksinstituut voor Visserijonderzoek
RLF Ren Luft Foreningen
RSC Royal Society of Chemistry
RSNC Royal Society for Nature Conservation
RSPB Royal Society for the Protection of Birds
RTECS Registry of Toxic Effects of Chemicals on Substances
SAFO Swedish Atomic Forum
SAGB Senior Advisory Group Biotechnology
SAGUF Schweizer Arbeitsgemeinschaft für Umweltforschung
SAVE Specific Action for Vigorous Energy Efficiency
SCOPE Scientific Committee on Problems of the Environment
SCPF Swedish Pulp and Paper Association
SCRAM Scottish Campaign to Resist the Atomic Menace
SEO Sociedad Española de Ornitologia
SEPR Sociedad Española de Protección Radiológica
SERIX Swedish Environment Research Index
SETAC Society for Environmental Toxicology and Chemistry
SEVE Stredisko pro Efektivni Vyuzivani Energie
SFT State Pollution Control Authority
SFT Statens Forurensningstilsyn
SGR Scientists for Global Responsibility
SIBH Society for the Interpretation of Britain's Heritage
SIGLE System for Information on Grey Literature in Europe
SIPRI Stockholm International Peace and Research Institute
SKAT Swiss Centre for Appropriate Technology
SMRU Sea Mammal Research Unit
SNE Sociedad Nuclear Espanola
SNIE Syndicat National des Engrais
SNM Stichting Natuur en Milieu
SNV Foundation for Packaging and the Environment
SNV Swedish Environment Protection Agency
SPAID Society for Prevention of Asbestosis and Industrial Diseases
SPFPE Spanish Plastics Foundation for the Protection of the Environment
SPM Svensk Polska Miljoforeningen
SPOLD Society for the Promotion of Life Cycle Development
SSB Statistik Sentralbyra
SSIC Societe Suisse Industries Chimique
SSIGE Societe Suisse de l'Industrie de Gaz et des Eaux
SSLRC Soil Survey and Land Research Centre
STUK Sateilyturvakeskus
SVA Schweizerische Vereingung für Atomenergie
SZOPK Slovak Union of Nature and Landscape Protectors
SZOPK Slovakian Society for Conservation of Nature
T & E European Federation for Transport and Environment
TBG Tidy Britain Group, The
TEC Transport Environment Circulation
TFB Transportforskningsberedningen
TFPL Task Force Pro Libra Ltd
THERMIE New Energy Technologies
TNO Toegepast Natuurwetenschappelijk Onderzoek
TRIS Transport Research Information Services
TRRL Transport and Road Research Laboratory
The Met Office Meteorological Office
TUNCAP Turkish National Committee for Air Pollution Research and Control
UFORDAT Umweltforschungsdatenbank
UI Uranium Institute
UIPP Union of Industries of the Protection of Plants
UKAEA UK Atomic Energy Authority
ULIDAT Umweltliteraturdatenbank
UN United Nations
UNCED UN Conference on Environment and Development
UNEP United Nations Environment Programme
UNICE Union of Industrial and Employers' Confederation of Europe
VAK Latvian Environment Protection Club
VCI Association of Chemical Industries
VDP Verband Deutscher Papierfabriken
VKI Vandkvalitetsinstituttet
VNIIAES All-Union Research Institute for Reactor Operators
VNP Association of Netherlands Paper and Board Manufacturers
VUJE Nuclear Power and Plant Research Institute
WCMC World Conservation Monitoring Centre
WEN Women's Environmental Network
WHO/EHE World Health Organisation Division of Environmental Health
WMIB Waste Management Information Bureau
WMO World Meteorological Organisation
WPO World Packaging Organisation
WRC Water Research Centre
WRI World Resources Institute
WTMU World Trade Monitoring Unit
WWF World Wide Fund for Nature
WZB Wissenschaftszentrum Berlin

List of Abbreviations

Country		Currency		Country		Currency	
Aus	Austria	AS	Austrian Schilling	Lat	Latvia		
Bel	Belgium	Bfr	Belgian Franc	Lit	Lithuania		
Bos	Bosnia Hercegovina			Lux	Luxembourg		
Bul	Bulgaria	Lev	Bulgarian lev	Net	Netherlands	Fl	Dutch Florin
Bye	Byelorussia			Nor	Norway		
Cro	Croatia	D	dinar	Pol	Poland		
Cyp	Cyprus			Por	Portugal		
Cze	Czech Republic			Rom	Romania		
Den	Denmark	Dkr	Danish kroner	Rus	Russia		
Est	Estonia			Slo	Slovakia		
Fin	Finland	FIM	Finnish Marks	Slv	Slovenia	D	dinar
Fra	France	Ffr	French francs	Spa	Spain	Ptas	pesetas
Ger	Germany	DM	Deutschmarks	Swe	Sweden	Skr	Swedish krone
Gre	Greece			Swi	Switzerland	Sfr	Swiss franc
Hun	Hungary	F	forint	Tur	Turkey	TL	Turkish Lira
Ice	Iceland			UK	United Kingdom	£	British pound
Ire	Ireland			Ukr	Ukraine		
Ita	Italy	Lit	Italian Lira	USA	United States of America	$	US dollar

1
Air Quality and Global Atmospheric Change

General

NON-GOVERNMENT ORGANISATIONS

– 1 – Vereniging der Lucht
(Society for Clean Air in the Netherlands) (CLAN)

PO Box 6012
NL-2600 JA Delft
Netherlands
Tel: 31 15 69 68 84
Fax: 31 15 61 31 86
Contact name: Dr J van Ham, Secretary
Geographical coverage: Netherlands
Aims and objectives: CLAN's concerns are air, climate and acid rain

BOOKS

– 2 – Changing Atmosphere, The

Geographical coverage: United Kingdom
Publishers: Yale University Press, 1991
Coverage: The study describes the causes of acid rain, ozone depletion and global warming and the evidence for the recent acceleration of each of these phenomena
Price: £12.95
ISBN: 0300033818
Author: Firor, J.

– 3 – Global Climate Changes and Freshwater Ecosystems

Geographical coverage: International
Publishers: Springer, 1992
Coverage: The contributors assume global warming of 2–5 degrees in the next century and explore the implications of this rise on various freshwater ecosystems

Price: £53.00
ISBN: 354097640X
Author: Firth, P. and Fisher, S. G.

REPORTS

– 4 – Global Environmental Change – UK Research Framework

Polaris House
North Star Avenue
Swindon SN2 1EU
United Kingdom
Tel: 0793 411734
Fax: 0793 411691
Geographical coverage: United Kingdom
Coverage: First report of the IACGEC. It is a compendium of programmes of those government agencies which have responsibilities associated with the global environment
Author: UK Global Environmental Research (GER)

RESEARCH CENTRES

– 5 – Centro di Ecologia e Climatologia/Osservatorio Geofisico Sperimentale Macerata
(Centre of Ecology and Climatology/Geophysical Experimental Observatory)

Via Indipendenza 180
I-62100 Macerata
Italy
Tel: 39 733 30743
Geographical coverage: Italy

Research activities: Dynamic meteorology, atmospheric electricity, acid deposits and climatology

– 6 – Institut de l'Environnement du Centre Commun de Recherche

T P 290
I-21020 Ispra (VA)
Italy
Tel: 39 332 78 93 04
Fax: 39 332 78 92 22
Contact name: Mr C Savatteri
Geographical coverage: Italy
Research activities: Global change and environmental chemicals

– 7 – Institute of Meteorology and Physics of the Atmospheric Environment

National Observatory of Athens
PO Box 20048
GR-118 10 Athens
Greece
Tel: 30 1 3456 257
Fax: 30 1 3463 803
Geographical coverage: Greece
Research activities: Air pollution, solar energy, wind energy, atmospheric electricty and climatology

– 8 – International Institute for Applied Systems Analysis (IIASA)

A-2361 Laxenberg
Austria
Tel: 43 2236 715210
Fax: 43 2236 73149
Contact name: Elisabeth Krippl, Office of Communications
Geographical coverage: International
Aims and objectives: IIASA is part of an international network of scientific institutions working together to study global change. IIASA is supported by scientific organisations in 15 countries. Founded in 1972 on the initiative of the USA and ex-USSR
Publications: Catalogue available on request
Research activities: Conducts environmental studies and projects in acid rain, air pollution and other aspects of the environment

DATABASES

– 9 – Global Environmental Change Report

Geographical coverage: International
Coverage: Deforestation; acid rain; ozone depletion
Database producer: Cutter Information Corporation
Database host: DataStar; Dialog
Coverage period: 1990–
Update frequency: monthly

– 10 – Molars

Geographical coverage: International
Coverage: Climatology, meteorology and general atmospheric sciences
Database producer: National Meteorological Library
Database host: ESA–IRS
Coverage period: 1971–

Air Quality and Pollution

GOVERNMENT ORGANISATIONS

– 11 – Bestuur van de Volksgezondheid – Ministerie van Volksgezondheid

Rijksadministratief Centrum/Vesalius Gebouw
B-1010 Brussels
Belgium
Tel: 32 2 564 11 55
Geographical coverage: Belgium
Aims and objectives: Aims to collect the maximum number of documents concerning air pollution, noise and radioactivity

– 12 – Bundesantstalt für Arbeitsschutz und Unfallforschung

Vogelpothsweg 50–52
Dorstfeld
D-4600 Dortmund
Germany
Tel: 49 231 17631
Geographical coverage: Germany
Aims and objectives: Air quality control

– 13 – Division for Air Monitoring and Industrial Compliance

Telemark County Statens Hus
N-3708 Skien
Norway

Tel: 47 3 58 61 20
Fax: 47 3 53 00 20
Geographical coverage: Norway
Aims and objectives: The Division is responsible for monitoring atmospheric pollution

– 14 – Département de la Pollution Atmosphérique de l'Institut National de Récherche Chimique Appliquée
(Air Pollution Dept of the National Institute of Applied Chemical Research)

BP 1
F-91710 Vert-le-Petit
Essonnée
France
Tel: 33 49 82 475
Geographical coverage: France
Aims and objectives: The Institute's concerns are air pollution and industrial hygiene

– 15 – Secretaria de Estado para las Politicas del Agua y del Medio Ambiente

Plaza San Juan de la Cruz 3
E-28071 Madrid
Spain
Tel: 34 1 534 51 49
Contact name: Vicente Albero Silla, Secretary
Geographical coverage: Spain

NON-GOVERNMENT ORGANISATIONS

– 16 – Association pour la Prevention de la Pollution Atmosphérique (APPA)
(Association for the Prevention of Atmospheric Pollution)

58 rue de Rocher
F-5008 Paris
France
Tel: 33 1 429 369 30
Fax: 33 1 429 341 99
Contact name: M Pierre Gaussens
Geographical coverage: France
Aims and objectives: APPA aims to promote clean air through the reduction of pollution

– 17 – Atmospheric Research and Information Centre (ARIC)

The Manchester Metropolitan University
Chester Street
Manchester M1 5GD
United Kingdom
Tel: 061 247 1590/93
Fax: 061 247 6318
Contact name: Mrs Sue Hare, Information Officer
Geographical coverage: United Kingdom

Aims and objectives: National and international information centre on air pollution, particularly acid rain and global climate change. Also research into the chemistry of acid deposition

– 18 – Ceská Asociace IUAPPA (CA IUAPPA)
(Czech Association of IUAPPA)

Doubravcická 10
Prague 10 100 00
Czech Republic
Tel: 42 2 78 11 674
Fax: 42 2 78 11 674
Geographical coverage: Czech Republic

– 19 – Comitato di Studio perl'Inquinamento Atmosferico (CSIA)
(Air Pollution Study Committee)

Via Assarotti 15/8 SC B
I-16122 Genoa
Italy
Tel: 39 10 893922
Fax: 39 10 887766
Geographical coverage: Italy
Aims and objectives: CSIA deals with scientific, technical, legislative and economic aspects of air pollution control

– 20 – Cooperative Programme for Monitoring and Evaluation of Long Range Transboundary Air Pollution (EMEP)

EMEP West
Norwegian Meteorological Institute
PO Box 43 Blindern
N-0313 Oslo
Norway
Geographical coverage: European
Aims and objectives: EMEP presents information on emissions, transports, concentrations and depositions of sulphur and nitrogen pollutants

– 21 – Coordinadora de Organizaciones de Defensa Ambiental (CODA)

Plaza de Santo Domingo 7
E-2013 Madrid
Spain
Tel: 34 1 559 60 25
Fax: 34 1 559 78 97
Geographical coverage: Spain

– 22 – European Association for the Science of Air Pollution (EURASAP)

Air Pollution Group
Imperial College
London SW7 2AZ
United Kingdom
Tel: 071 589 5111
Fax: 071 584 7596
Geographical coverage: European
Aims and objectives: EURASAP was formed in 1986 to advance education and knowledge in air pollution science and its applications. It brings together in conference scientists throughout Europe working in the relevant disciplines
Publications: Regular newsletter

– 23 – Fundación Para la Ecología y la Protección del Medio Ambiente (FEPMA)

Castellana 8
E-28046 Madrid
Spain
Tel: 34 1 575 41 68
Geographical coverage: Spain

– 24 – Hravtsko Drutvo za Zastitu Zraka

(Croatian Air Pollution Prevention Association) (CAPPA)

University of Zagreb
CAPPA
Ksaverska c 2
41000 Zagreb
Croatia
Tel: 38 41 434 188
Fax: 38 41 274 572
Geographical coverage: Croatia

– 25 – Ilmansuojeluyhdistys ry ISY

(Finnish Air Pollution Prevention Society) (FAPPS)

PO Box 335
SF-00131 Helsinki
Finland
Tel: 358 1 45 66 159
Fax: 358 1 45 52 408
Contact name: Mr Kari Larjava, President
Geographical coverage: Finland

– 26 – Information Centre for Air Protection, Polish Ecological Club

ul Armii Czerwonej
Al Korfantego 2/220
PL-40 960 Katowice
Poland
Tel: 48 32 586071
Geographical coverage: Poland

– 27 – International Union of Air Pollution Prevention Associations (IUAPPA)

136 North Street
Brighton BN1 1RG
United Kingdom
Tel: 0273 326313
Fax: 0273 735802
Contact name: John Langston, Director General
Geographical coverage: United Kingdom
Aims and objectives: The Union was formed in 1964 and now has 28 members. It aims to promote public education worldwide in all matters relating to the value and importance of clean air and methods and consequences of air pollution control
Publications: *Clean Air Around the World*; newsletter, quarterly, members handbook;

– 28 – Kommission Reinhaltung der Luft (KRdL) in VDI und DIN

(Commission on Air Pollution Prevention in VDI and DIN)

Robert-Stolz-Strasse 5
Postfach 10 11 39
D-4000 Düsseldorf 1
Germany
Tel: 49 211 6214 532
Fax: 49 211 6214 575
Contact name: Dr Ing Klaus Grefen, Secretary
Geographical coverage: Germany

– 29 – Liga Ochrony Przyrody

(The Nature Protection League)

ul Reja 3/5
PL-02 053 Warsaw
Poland
Geographical coverage: Poland
Aims and objectives: Public service institution, concerned with the protection of the natural environment

– 30 – National Society for Clean Air and Environmental Protection (NSCA)

136 North Street
Brighton BN1 1RG
United Kingdom
Tel: 0273 326313
Fax: 0273 735802
Contact name: Tim Brown/Mary Stevens, Information Department
Geographical coverage: United Kingdom
Aims and objectives: Founded in 1899 as the Smoke Abatement Society, NCSA's objectives are to secure environmental improvement by promoting clean air through the reduction of air pollution, noise and other contaminants
Publications: *Clean Air*, quarterly; *Pollution Handbook*, annually; Annual report and other leaflets and reports

– 31 – Norwegian Clean Air Campaign

Postboks 94
N-1364 Hvalstad
Norway
Tel: 47 2 78 38 60
Fax: 47 2 90 15 87
Geographical coverage: Norway
Aims and objectives: Works towards promoting clean air through the reduction of air pollution and other contaminants

– 32 – Polish Ecological Club/Information Centre for Air Protection

Grunwaldski 8-10
PL-40 950 Katowice
Poland
Tel: 48 32 594 315
Contact name: Piotr Poborski
Geographical coverage: Poland
Aims and objectives: Concerned with all aspects of air pollution

– 33 – Ren Luft Foreningen (RLF)

(Norwegian Clean Air Association)

PO Box 2312 Solli
N-0201 Oslo
Norway
Tel: 47 2 838330
Contact name: Vigdis Ekeberg, Secretary
Geographical coverage: Norway

BOOKS

– 34 – Air Pollution Control in the European Community

Geographical coverage: European
Publishers: Graham and Trotman Ltd, 1991
Editor: Bennett, G.
Language: English
Coverage: Examines the implication of EC Directives on air pollution and control in the member states
ISBN: 1853335673

– 35 – Air Pollution's Toll on Forests and Crops

Geographical coverage: International
Publishers: Yale University Press, 1990
Coverage: Examines the effects of air pollution on forestry and crops
ISBN: 0300045697
Author: MacKenzie, J. J. and El-Ashry, M. T.

– 36 – Ambient Air Pollutants from Industrial Sources

Geographical coverage: Netherlands
Publishers: Elsevier Science Publishers, 1985, Netherlands
Editor: Suess, M. J., Grefen, K. and Reinisch, D
Language: English
Price: Dfl150
ISBN: 0444806059

– 37 – Clean Air Around the World

Geographical coverage: International
Publishers: IUAPPA, 1991
Editor: Loveday Murley
Language: English
Coverage: A reference book for which specialist organisations from many countries have compiled up to date information concerning their national air pollution problems
Price: £32.00
ISBN: 1871688019

– 38 – Continuous Emission Monitoring

Geographical coverage: United Kingdom
Publishers: Van Nostrand Reinhold International, 1992
Coverage: A guide to the field of continuous emission monitoring (CEM) addressing technologies and practices in the monitoring of pollutants emitted from industrial stacks
Price: £43.50
ISBN: 0442007248
Author: Jahnke, J. A.

– 39 – Occupational Exposure Limits for Airborne Toxic Substances, Third Edition

Geographical coverage: International
Publishers: ILO, 1991 (Occupational Health and Safety Series No 37)
Coverage: The latest available values either prescribed or recommended in 15 countries
ISBN: 9221072932

– 40 – Waste, Wastewater, Air Laws and Technology

Geographical coverage: Germany
Publishers: Bohman Druck und Verlag Gesellschaft mbH & Co KG
Language: German
Author: List, W. and Kuntscher, H.

CONFERENCE PAPERS

– 41 – Desulphurisation

Geographical coverage: United Kingdom
Publishers: Taylor and Francis, 1991
Coverage: Covers current technologies and strategies for reducing sulphur dioxide emissions from both large utility plant and small industrial systems
Price: £25.50
ISBN: 1560322322
Author: Kyte, W. S., Institution of Chemical Engineers

– 42 – North Atlantic Treaty Organisation Expert Panel on Air Pollution Modelling Proceedings

Information and Press Service
B-1110 Brussels
Belgium
Geographical coverage: International
Publishers: North Atlantic Treaty Organisation (NATO)
ISSN: 0377 7669

PERIODICALS

– 43 – Air Pollution and Noise Bulletin

Birmingham Central Libraries
Chamberlain Square
Birmingham B3 3HQ
United Kingdom
Tel: 021 235 4392
Geographical coverage: United Kingdom
Publishers: APNB Science Technology and Management Services
Editor: Pratt, D.
Frequency: 6 issues a year
Coverage: The bulletin is a current awareness service for specialists in the field. Its purpose is to alert people to recent publications in English
Price: free

– 44 – Annals of Occupational Hygiene

Headington Hill Hall
Oxford OX3 0BW
United Kingdom
Tel: 0865 794141
Fax: 0865 743952
Geographical coverage: United Kingdom
Publishers: Pergamon Press plc
Editor: McKellinson, J.
Frequency: 6 issues a year
Coverage: Life sciences including air pollution monitoring
Annual subscription: £260.00
ISSN: 0003 4878

– 45 – Clean Air

136 North Street
Brighton BN1 1RG
United Kingdom
Tel: 0273 326313
Fax: 0273 735802
Geographical coverage: United Kingdom
Publishers: National Society for Clean Air and Environmental Protection
Frequency: quarterly
Annual subscription: £18.00
ISSN: 0300 5143

– 46 – Difesa Ambientale

Via A Capecelatro 5
I-20148 Milan
Italy
Geographical coverage: Italy
Publishers: Centro Informazioni Studi Ambientali
Editor: Serbelloni, L. C.
Frequency: 12 issues a year

– 47 – Etudes de Pollution Atmosphérique à Paris et dans les Départements Periphériques

Préfecture de Police
Laboratoire Ctrl
39 bis rue de Dantzig
F-75015 Paris
France
Geographical coverage: France
Frequency: annual
ISSN: 0071 1942

– 48 – Finishing

Turrett House
171 High Street
Rickmansworth
Hertfordshire WD3 1SN
United Kingdom
Tel: 0923 777000
Fax: 0923 771297
Geographical coverage: United Kingdom
Publishers: Turrett Group plc
Editor: Tomkins, G.
Frequency: monthly
Annual subscription: $118
ISSN: 0264 2506

– 49 – Ilmatieteen Laitos Ilmansvojelun Julkaisvja

(Finnish Meteorological Institute Publications on Air Quality)

PO Box 503
SF-00101 Helsinki
Finland

Fax: 358 1 92 92 18
Geographical coverage: Finland
Publishers: Finnish Meteorological Institute
Frequency: irregular
ISSN: 0782 6095

– 50 – Indoor Environment

Allschwilerstrasse 10
Postfach
CH-4009 Basel
Switzerland
Tel: 41 61 306 11 11
Fax: 41 61 306 12 34
Geographical coverage: Switzerland
Publishers: S Karger AG
Editor: Weetman, D. F.
Frequency: bi monthly
Language: English
Coverage: Focuses on original reports pertaining to the quality of the indoor environment at home and in the workplace

– 51 – Journal of Atmospheric Chemistry

PO Box 17
NL-3300 AA Dordrecht
Netherlands
Tel: 31 78 334911
Fax: 31 78 334254
Geographical coverage: Netherlands
Publishers: D Reidel Publishing Co
Editor: Crutzen, P. J. and Ehhalt, D. H.
Frequency: quarterly
Language: English

– 52 – Lufthygienischer Monatsbericht

Unter den Eichen 7
D-6200 Wiesbaden
Germany
Tel: 49 611 5810
Fax: 49 611 581221
Geographical coverage: Germany
Publishers: Landesanstalt für Umwelt
Frequency: monthly
Annual subscription: DM30

– 53 – MST Luft

Forsoeganlaeg Risoe
DK-4000 Roskilde
Denmark
Geographical coverage: Denmark
Publishers: Miljoestyrelsen, Luftforureningslaboratorium
Editor: Fenger, J.
Frequency: irregular
Language: Danish
Price: free

– 54 – Pollution Atmosphérique

58 rue de Rocher
F-75008 Paris
France
Tel: 33 1 429 369 30
Fax: 33 1 429 341 99
Geographical coverage: France
Publishers: Association pour la Prévention de la Pollution Atmosphérique
Editor: Sommer, M.
Frequency: quarterly
Language: French with English summaries
Annual subscription: Ffr394
ISSN: 0032 3632

– 55 – Refrigeration and Air Conditioning

Maclaren House
19 Scarbrook Road
Croydon
Surrey CR9 1QH
United Kingdom
Tel: 081 760 9690
Fax: 081 681 1672
Geographical coverage: United Kingdom
Publishers: EMAP Vision Ltd
Editor: Bailey, A.
Frequency: monthly
Annual subscription: £34.50
ISSN: 0263 5739

– 56 – Smog

Belvedere Golfo Paradiso 21
I-16036 Recco
Genoa
Italy
Geographical coverage: Italy
Publishers: Lega Italiana Contro Fumi e Rumori
Frequency: 3 issues a year

– 57 – Warren Spring Laboratory UK Smoke and Sulphur Dioxide Monitoring Networks

Gunnels Wood Road
Stevenage
Hertfordshire SG1 2BX
United Kingdom
Tel: 0438 741122
Fax: 0438 360858
Geographical coverage: United Kingdom
Publishers: Warren Spring Laboratory
Frequency: irregular
Price: varies

REPORTS

– 58 – Acid Deposition and Vehicle Emissions: European Environmental Pressures on Britain

Geographical coverage: United Kingdom
Publishers: Royal Institute of International Affairs, Policy Studies Unit
Editor: Brackley, P.
ISBN: 0566051257

– 59 – Air Pollution from Vehicles

Geographical coverage: United Kingdom
Publishers: HMSO
ISBN: 0115510001
Author: Transport and Road Research Laboratory

– 60 – Aircraft Pollution: Environmental Impacts and Future Solutions

Geographical coverage: International
Publishers: WWF, August 1991
Language: English
Coverage: Issues of aircraft emissions pollution
ISBN: 2880850827
Author: Barrett, Dr. M.

– 61 – Bikes Not Fumes

Cotterell House
69 Meadrow
Godalming
Surrey GU7 3HS
United Kingdom
Tel: 0483 417217
Fax: 0483 426994
Geographical coverage: United Kingdom
Publishers: Cycling Tourists Club (CTC)
Price: £8.00
Author: Earth Resources Centre

– 62 – Environment in Czechoslovakia, The

Slezka 9
120 29 Prague 2
Czech Republic
Geographical coverage: Czech Republic
Publishers: Department of the Environment, 1990
Coverage: Report on the state of the country's environment
Author: Federal Committee for the Environment

– 63 – Protecting the Earth: A Status Report with Recommendations for a New Energy Policy: see entry 131

Geographical coverage: Germany

– 64 – Volatile Organic Compound Emissions in Western Europe: Report 6/87

Geographical coverage: European
Publishers: CONCAWE, Brussels
Coverage: Control options and their cost effectiveness for gasoline vehicles, distribution and refining

– 65 – WATT Committee on Energy: Air Pollution Acid Rain and the Environment

Geographical coverage: United Kingdom
Publishers: Elsevier, 1988
ISBN: 1851662227
Author: Watt Committee

RESEARCH CENTRES

– 66 – Centre Interprofessionnel Technique d'Etudes de la Pollution Atmosphérique (CITEPA)

(Atmospheric Pollution Technical Research Centre)

3 rue Henri Heine
F-75016 Paris
France
Tel: 33 1 452 712 88
Geographical coverage: France
Publications: National and international databases; reports; monographs; translations
Research activities: All aspects of industrial and atmospheric pollution, including dispersion and techniques of prevention and monitoring

– 67 – Department of Air Pollution

Faculty of Agricultural Sciences
De Dreijen 12
NL-6703 BC Wageningen
Netherlands
Tel: 31 8370 82106
Geographical coverage: Netherlands
Research activities: Air pollution; mutagenicity of air pollution; aerosol research; transport and chemical transformation of air pollutants

– 68 – Environmental Research Unit

St Martins House
Waterloo Road
Dublin 4
Eire
Tel: 353 1 602511
Fax: 353 1 680009
Geographical coverage: Ireland
Research activities: Air and water quality
Additional information: Library Service

– 69 – Hava Kirlenmesi Arastirmalari ve Denetimi Türk Milli Komitesi

(Turkish National Committee for Air Pollution Research and Control) (TUNCAP)

Dokuz Eylül University
Dept of Environmental Engineering
Bornova 35100 Izmir
Turkey
Tel: 90 51 882108
Fax: 90 51 887864
Contact name: Prof Dr Aysen Müezzinoglu, President
Geographical coverage: Turkey

– 70 – Institut de Recherche des Transports Centre de Documentation d'Evaluation et de Recherches Nuisances

109 avenue Salvador Allendre
F-69500 Bron
France
Tel: 33 78269093
Geographical coverage: France
Research activities: Transport; air pollution; noise pollution

– 71 – Institut für Immission- Arbeits- und Strahlenschutz

(Institute for Emission Labour and Radiation Protection)

Hertzstrasse 173
Baden-Württemberg
D-7500 Karlsruhe 21
Germany
Tel: 49 721 75031
Fax: 49 721 758758
Geographical coverage: Germany
Research activities: Concerned with the emission of substances, radiation and noises

– 72 – Institut National de la Santé et de la Recherche Médicale

Université de Paris
11 rue J B Clement
F-92290 Chatenay-Malabry
France
Tel: 33 1 6604518
Contact name: Mr Boudene
Geographical coverage: France
Research activities: Toxicology and air pollution

– 73 – Istituto Sull'Inquinamento Atmosferico

(Institute of Atmospheric Pollution)

Via Salaria Km 29 300
CP 10
I-00016 Monterotondo Stazione
Rome
Italy
Tel: 39 6 005349
Geographical coverage: Italy
Research activities: Development of instrumentation for measuring and monitoring of air pollution; study of atmospheric acidity

– 74 – Landesanstalt für Immissionsschutz Nordrhein-Westfalen

(North Rhine Westphalia State Centre of Air Quality, Noise and Vibration Control)

Wallneyer Strasse 6
D-4300 Essen-Bredeney 1
Germany
Tel: 49 201 79950
Fax: 49 201 7995446
Geographical coverage: Germany
Publications: Various publications
Research activities: Air quality surveillance; abatement of harmful emissions; protection against noise and vibration

– 75 – Medizinisches Institut für Umwelthygiene

(Medical Institute for Environmental Hygiene)

Auf'm Hennekamp 50
D-4000 Düsseldorf
Germany
Tel: 49 211 33890
Geographical coverage: Germany
Research activities: Injurious effects of air and noise pollution

– 76 – Norsk Institut for Luftforskning (NILU)

(Norwegian Institute for Air Research)

PO Box 64
N-2001 Lillestrom
Norway
Tel: 47 6 814170
Fax: 47 6 819247
Geographical coverage: Norway
Publications: Reports

Research activities: Air quality measurements; air pollution modelling; atmospheric corrosion studies and other activities

DATABASES

– 77 – Air/Water Pollution Report

Geographical coverage: United Kingdom
Coverage: All aspects of air and water pollution in the UK
Database producer: Business Publishers Inc
Database host: DataStar; Dialog
Coverage period: 1988–

– 78 – IBSEDEX

Geographical coverage: International
Coverage: Indoor air pollution; heating; ventilation; energy
Database producer: Building Services Research and Information Association
Database host: ESA–IRS
Coverage period: 1960–
Update frequency: weekly

– 79 – VANYTT

Halsingegatan 49
S-113 31 Stockholm
Sweden
Tel: 46 8 34 01 70
Geographical coverage: Sweden
Coverage: Covers aspects of indoor air pollution and energy conservation in buildings
Database producer: The Swedish Institute of Building Documentation (BYGGDOK)

Atmosphere

NON-GOVERNMENT ORGANISATIONS

– 80 – Bulle Bleue
(Blue Bubble)

12 rue Francis de Pressensé
F-65014 Paris
France
Tel: 33 1 454 548 76
Fax: 33 1 454 572 31
Contact name: Mr Jean Claude Ray, President
Geographical coverage: France
Aims and objectives: Deals with atmospheric issues

BOOKS

– 81 – Biosphere, The

Geographical coverage: United Kingdom
Publishers: Belhaven Press, 1990
Coverage: Introduction to the scientific principles involved in studying the structure and functioning of the biosphere
Price: £27.50
ISBN: 1852930373
Author: Bradbury, I.

PERIODICALS

– 82 – Atmospheric Environment Part A

Headington Hill Hall
Oxford OX3 0BW
United Kingdom
Tel: 0865 794141
Fax: 0865 743952
Geographical coverage: United Kingdom
Publishers: Pergamon Press plc
Editor: Moore, D. J. et al
Frequency: monthly
Coverage: Physical sciences
Annual subscription: £1,190.00
ISSN: 0960 1686

– 83 – Atmospheric Environment Part B

Headington Hill Hall
Oxford OX3 0BW
United Kingdom
Tel: 0865 794141
Fax: 0865 743952
Geographical coverage: United Kingdom
Publishers: Pergamon Press plc
Editor: Moore, D. J. et al
Frequency: quarterly
Coverage: Covers topics on the urban environment such as climate, energy and moisture balances
Annual subscription: £192.00
ISSN: 0957 1272

RESEARCH CENTRES

– 84 – European Arctic Stratospheric Ozone Experiment (EASOE)

Madingley Road
Cambridge CB3 0ET
United Kingdom
Tel: 0223 61188
Fax: 0223 62616
Contact name: British Antarctic Survey Ozone Secretariat
Geographical coverage: European
Research activities: Meteorology and atmospheric research in the Antarctic, South Georgia and the South Sandwich Islands

– 85 – Fraunhofer-Institut für Atmosphärische Umweltforschung

(Fraunhofer Institute for Atmospheric Environmental Research)

Kreuzeckbahnstrasse 19
D-8100 Garmisch-Partenkirchen
Germany
Tel: 49 5 10 56
Geographical coverage: Germany
Research activities: Basic and applied research in atmospheric physics chemistry and meteorology

Climate

GOVERNMENT ORGANISATIONS

– 86 – Intergovernmental Panel for Climatic Change (IPCC)

The Meteorological Office
Hadley Centre
London Road
Bracknell RG12 25Y
United Kingdom
Tel: 0344 856655
Geographical coverage: United Kingdom

– 87 – Meteorological Office, The (The 'Met Office')

The Press Office
London Road
Bracknell
Berkshire RG12 2SZ
United Kingdom
Tel: 0344 856655
Contact name: Derek Hardy, Technical Enquiries
Geographical coverage: United Kingdom
Aims and objectives: The Met Office is the UK Meteorological Service and forms part of the Ministry of Defence. Its functions are to provide meteorological services to the forces, other government departments, public corporations, local authorities and the public
Publications: List available on request

– 88 – World Meteorological Organisation (WMO)

Case Postale 2300
41 avenue Giuseppe Motta
CH-1211 Geneva
Switzerland
Tel: 41 22 7308111
Fax: 41 22 7342326
Geographical coverage: International
Aims and objectives: The WMO started activities as a specialised agency of the UN in 1951. Its aim is to make available meteorological and related geophysical and environmental information
Publications: *WMO Bulletin*; quarterly in English, French, Russian and Spanish. Reports, technical notes and training publications

NON-GOVERNMENT ORGANISATIONS

– 89 – Atmospheric Research and Information Centre (ARIC): *see entry 17*

Geographical coverage: United Kingdom

– 90 – Climate Action Network UK (CAN UK)

21 Tower Street
London WC2 9NS
United Kingdom

Tel: 071 497 2712
Fax: 071 240 2291
Contact name: Ms Sally Cavanagh
Geographical coverage: United Kingdom
Aims and objectives: CAN UK is part of the global Climate Action Network. It provides up to date information to NGO's and to the public concerning developments in climate change
Publications: *Hotnews*

– 91 – Danish Meteorological Institute

Lyngbyvej 100
DK-2100 Copenhagen
Denmark
Tel: 45 31 29 21 00/29 74 59
Fax: 45 31 29 12 12
Contact name: Lars Prahm, Director
Geographical coverage: Denmark

– 92 – European Centre for Medium Range Weather Forecasts (ECMWF)

Shinfield Park
Reading RG2 9AX
United Kingdom
Tel: 0734 499101
Contact name: Dr J A Woods, Scientific Officer
Geographical coverage: European
Aims and objectives: The ECMWF is an international organisation established in 1973 and supported by 17 European countries. The Centre's principal objectives are: the development of numerical methods for medium range forecasting and collection and storage of data

– 93 – Ilmatieteen Laitos

(Finnish Meteorological Institute)

Vuorikatu 24
PO Box 503
SF-00101 Helsinki 10
Finland
Tel: 358 1 90 192 91
Geographical coverage: Finland
Aims and objectives: The Institute is concerned with climatology and air pollution

– 94 – Institut Royal Météorologique de Belgique

(Belgian Royal Meteorological Institute)

3 avenue Circulaire
B-1180 Brussels
Belgium
Tel: 32 2 375 24 78
Geographical coverage: Belgium

– 95 – Meteorological Service

Glasnevin Hill
Dublin 9
Eire
Tel: 353 1 8424411
Fax: 353 1 375557
Geographical coverage: Ireland

– 96 – Organisation Météorologique Mondiale

(World Meteorological Organisation) (WMO)

Case Postale 2300
41 avenue Giuseppe Motta
CH-1211 Geneva
Switzerland
Tel: 41 22 730 81 11
Fax: 41 22 734 23 26
Contact name: Dr John Miller, Chief Environment Division
Geographical coverage: International
Aims and objectives: WMO provides information on atmospheric composition and its physical characteristics

– 97 – Österreichisches Gesellschaft für Natur und Umweltschutz (ÖGNU)

(Austrian Society for Nature and Environment Protection)

Hegelgasse 21
A-1010 Vienna
Austria
Tel: 43 222 513 29 62
Fax: 43 222 512 56 01
Contact name: Mr Wilhelm Linder
Geographical coverage: Austria
Aims and objectives: Coordinates environmental activities of NGOs in Austria
Publications: A newspaper dealing with climatic changes

– 98 – Standing Committee on Urban and Building Climatology

Universität Essen
Fachbereich 9
Postfach 10 37 64
D-4300 Essen 1
Germany
Tel: 49 201 1832734
Geographical coverage: Germany

– 99 – World Ozone and Ultra-Violet Data Centre: Atmospheric Environment Service

4905 Dufferin Street
Downsview
Ontario M3H 5T4
Canada
Geographical coverage: International

BOOKS

– 100 – Climate Change

Geographical coverage: International
Publishers: Cambridge University Press, 1991
Coverage: A statement about the current knowledge of climate change and its effect on the world's environment and the prospects for sustainable socio-economic development
Price: £50.00
ISBN: 0521416310

– 101 – Climate Change and Human Impact on the Landscape

Geographical coverage: United Kingdom
Publishers: Chapman and Hall, 1992
Coverage: Comprises scholarly reviews and case studies in late Quarternary palaeoecology and environmental archaeology
Price: £55.00
ISBN: 0412462001
Author: Chambers, F. M.

– 102 – Climatic Change and the Mediterranean

Geographical coverage: European
Publishers: Edward Arnold, 1992
Coverage: Contains contributions from an international team organised under the auspices of UNEP
Price: £79.50
ISBN: 0340553294
Author: Jeftic, L. and Milliman, J. D.

PERIODICALS

– 103 – Climatic Change

Bioscience Division
PO Box 17
NL-3300 AA Dordrecht
Netherlands
Tel: 31 78 334911
Fax: 31 78 334254
Geographical coverage: International
Publishers: Kluwer Academic Publishers
Editor: Schneider, S. H.
Frequency: quarterly
Language: English
Coverage: An interdisciplinary, international journal devoted to the description, causes and implications of climatic change
Annual subscription: Dfl576
ISSN: 0165 0009

REPORTS

– 104 – Climate Change: Designing a Tradeable Permit System

2 rue André Pascal
F-75775 Paris Cedex
France
Geographical coverage: International
Publishers: OECD, 1992
Language: English
Coverage: This report contains technical papers presented at a workshop on tradeable permits organised by OECD. A second volume discussing the use of taxes will also be published

– 105 – Transport and Climate Change: Cutting Carbon Dioxide Emissions from Cars

Geographical coverage: United Kingdom
Publishers: Friends of the Earth (FoE), 1991
Coverage: Latest fuel efficiency technologies, traffic restraint, fiscal incentives and regulations are examined
Price: £7.00
ISBN: 0905966996

RESEARCH CENTRES

– 106 – Deutscher Wetterdienst (DWD)

(German Meteorological Service)

Frankfurter Strasse 135
Postfach 10 04 65
D-6050 Offenbach am Main
Germany
Tel: 49 69 80620
Fax: 49 69 8062 339
Geographical coverage: Germany
Publications: Several reports issued on a regular basis; annual report
Research activities: Relations between climatology and technical problems e.g. utilisation of wind and solar energy, air conditioning

– 107 – Koninklijk Nederlands Meteorologisch Instituut

(Royal Netherlands Meteorological Institute)

Postbus 201
NL-3730 AE De Bilt
Netherlands
Tel: 31 30 766911
Geographical coverage: Netherlands
Research activities: Meteorology

– 108 – Norwegian Meteorological Institute: Meteorological Synthesising Centre–West of EMEP

PO Box 43
Blindern
N-0313 Oslo 3
Norway
Tel: 47 22 96 30 00
Fax: 47 22 96 30 50
Contact name: Sophia Mylona, Scientific Secretary
Geographical coverage: Norway
Aims and objectives: Studies related to environmental problems

– 109 – Vedurstofa Islands
(Icelandic Meteorological Office)

Bustadavegur 9
150 Reykjavik
Iceland
Tel: 354 1 91 600 600
Fax: 354 1 91 28121
Geographical coverage: Iceland
Publications: *Vedrattan*, monthly; *Sea Ice of the Icelandic Coasts*, annual
Research activities: Weather and climate of Iceland; seismology and ozone

DATABASES

– 110 – Geobase

Geographical coverage: International
Coverage: Geology and related disciplines, including climatology and ecology
Database producer: GeoAbstracts Ltd
Database host: Dialog; Orbit
Coverage period: 1980–
Update frequency: monthly

Global Warming

NON-GOVERNMENT ORGANISATIONS

– 111 – Climate Network Europe (CNE)

44 rue du Taciturne
B-1040 Brussels
Belgium
Tel: 32 2 231 01 80
Fax: 32 2 230 57 13
Contact name: Ms Annie Bonnin-Roncerel
Geographical coverage: European
Aims and objectives: CNE is acting as the focal point for 55 NGOs in Europe which share a common concern for the greenhouse effect and wish to cooperate at international level
Publications: *An annotated bibliography on climate change* and *The EC Budget and Global Warming*

– 112 – Comité Européen de l'Association de l'Ozone
(European Committee of the European Ozone Association)

9 rue de Phalsbourg
F-75854 Paris Cedex 17
France
Tel: 33 1 422 738 91
Geographical coverage: European

– 113 – World Data Centre for Greenhouse Gases

Japan Meteorological Agency
1-3-4 Ote-Machi
Chiyoda-ku
Tokyo 100
Geographical coverage: International

PRESSURE GROUPS

– 114 – Greenpeace International Atmosphere and Energy Campaign

Keizersgracht 176
NL-1016 DW Amsterdam
Netherlands
Tel: 31 20 523 6555
Fax: 31 20 523 6500
Contact name: Ms Kelly Rigg
Geographical coverage: International
Aims and objectives: The atmosphere and energy campaign is focused primarily on global warming and the depletion of the ozone layer

BOOKS

– 115 – Country Studies and Technical Options, Volume 2

Geographical coverage: United Kingdom
Publishers: Royal Institute for International Affairs, 1991
Coverage: Contains in-depth technical, analytical and country analyses
Author: Grubb, M. et al

– 116 – Energy Policies and the Greenhouse Effect Volume 1: Policy Appraisal

Geographical coverage: United Kingdom
Publishers: Royal Institute of International Affairs, 1991
Coverage: Concentrates on policy issues arising from attempts to reduce the emissions of greenhouse gases from the energy sector
Author: Grubb, M.

– 117 – Global Warming

Geographical coverage: United Kingdom
Publishers: The MIT Press, 1991
Coverage: The focus of this book is on the economic effects of global warming. It examines which countries will suffer the most from climatic change and what the prospects are for international cooperation
Price: £26.95
ISBN: 026204126X
Author: Dornbusch, R. and Poterba, J. M.

– 118 – Global Warming: The Debate 1991

Geographical coverage: European
Publishers: On behalf of Strategy Europe Ltd by John Wiley and Sons, 1991
ISBN: 0471931578
Author: Thompson, P.

– 119 – Global Warming Forum, A

Geographical coverage: United Kingdom
Publishers: CRC Press, 1993
Coverage: Considers scientific issues, economics, natural resource management concerns and policy considerations on global warming
Price: £72.50
ISBN: 0849344190
Author: Geyer, R. A.

– 120 – Hothouse Earth

Geographical coverage: United Kingdom
Publishers: Bantam Press, 1990
Coverage: An examination of the greenhouse effect against a broader background of natural climatic processes
Price: £12.95
ISBN: 0593017951
Author: Gribbin, J.

– 121 – Living in the Greenhouse

Geographical coverage: United Kingdom
Publishers: Thorsons, 1990
Coverage: An account of what exactly is meant by the greenhouse effect
Price: £4.99
ISBN: 0722522584
Author: Allaby, M.

– 122 – Nuclear Power and the Greenhouse Effect

Geographical coverage: United Kingdom
Publishers: UK Atomic Energy Authority on behalf of the Nuclear Industry, 1992
Coverage: Account of how nuclear power affects global warming
Author: Donaldson, D., Tolland, H. and Grimston, M.

– 123 – Stratospheric Ozone 1991

Geographical coverage: United Kingdom
Publishers: HMSO
ISBN: 0117524573
Author: UK Stratospheric Ozone Review Group

CONFERENCE PAPERS

– 124 – Energy and the Environment

Geographical coverage: United Kingdom
Publishers: Royal Society of Chemistry, 1990
Coverage: Proceedings of a symposium. The papers address the possible causes of global warming in terms of the relative environmental impact of energy producing processes
Price: £47.50
ISBN: 0851866476
Author: Dunderdale, Dr J.

DIRECTORIES/YEARBOOKS

– 125 – Agricultural and Food Research Council Handbook

Geographical coverage: United Kingdom
Publishers: Agricultural and Food Research Council (AFRC)
Frequency: annual

Coverage: Summary of AFRC research including agriculture, agrochemicals and the agricultural implications of global warming
ISSN: 0961 1010

– 126 – Climate Action Network International NGO Directory 1992

Geographical coverage: International
Publishers: Climate Network Europe, 1992
Frequency: annual
Coverage: Non Government organisations concerned with the greenhouse effect
Price: free

PERIODICALS

– 127 – Greenhouse Gases Bulletin

CRE
Stoke Orchard
Cheltenham
Gloucestershire GL52 4RZ
United Kingdom
Fax: 0242 680758
Contact name: Dr Pierce Riemer
Geographical coverage: United Kingdom
Publishers: IEA Greenhouse Gas R & D Programme
Frequency: every two months
Coverage: Fossil fuels and greenhouse gases
Price: £60 –£180

– 128 – Greenhouse Issues

IEA Coal Research
Gemini House
10–18 Putney Hill
London SW15 6AA
United Kingdom
Tel: 081 780 2111
Fax: 081 780 1746
Contact name: Deborah Norman
Geographical coverage: United Kingdom
Publishers: IEA Coal Research on behalf of IEA Greenhouse Gas R & D Programme
Coverage: Worldwide developments in the field of greenhouse gases and fossil fuels
Price: free

REPORTS

– 129 – Global Status of Peatlands and their Role in Carbon Recycling, The

Geographical coverage: United Kingdom
Publishers: Friends of the Earth (FoE), 1992
Coverage: Assesses the role and significance of peatbogs in relation to the greenhouse effect
Price: £18.00
ISBN: 1857501055
Author: The Wetland Ecosystem Research Group, University of Exeter

– 130 – Greenhouse Earth

Geographical coverage: International
Publishers: John Wiley and Sons, 1992
Coverage: This text is a summary of the findings of two reports produced by the Scientific Committee on Problems of the Environment and the UN IPCC.
Price: £9.95
ISBN: 0471935476
Author: Nilsson, A.

– 131 – Protecting the Earth: A Status Report with Recommendations for a New Energy Policy

Geographical coverage: Germany
Publishers: Deutscher Bundestag, 1991
Coverage: Provides an account of the magnitude of the threat of the greenhouse effect. Recommendations are offered for national actions to reduce energy related emissions of radioactive trace gases
ISBN: 3924521719
Author: Enquête Commission, Germany

– 132 – UK Road Transport's Contribution to Greenhouse Gases

Geographical coverage: United Kingdom
Publishers: Transport and Road Research Laboratory (TRRL), 1990
Coverage: The paper is a response to a request to carry out a literature review of the contribution that UK road transport makes to global warming through the greenhouse effect
ISSN: 0266 7045
Author: Waters, M. H. L.

DATABASES

– 133 – Energy Technology Data Exchange (ETDE)

PO Box 1000
Oak Ridge
Tennessee 37831
United States of America
Tel: 1 615 576 1188
Fax: 1 615 576 2865
Contact name: Dora Moneyhun
Geographical coverage: International

Coverage: Greenhouse effect and other related environment topics
Database producer: International Energy Agency
Database host: Dialog Information Services, STN International
Coverage period: 1973–

– 134 – Greenhouse Effect Report

Geographical coverage: International
Coverage: Global warming; scientific assessments, implications and policy responses relating to business, technology, economics, international action
Database producer: Business Publishers Inc
Database host: via PTS Newsletter on DataStar or Dialog
Coverage period: 1988–
Update frequency: monthly

2
Agriculture and Farming

General

GOVERNMENT ORGANISATIONS

– 135 – Bundesministerium für Land und Forstwirtschaft
(Federal Ministry for Agriculture and Forestry)

Stubenring 1
A-1011 Vienna
Austria
Geographical coverage: Austria

– 136 – Commonwealth Agricultural Bureaux International (CABI)

Wallingford
Oxfordshire OX10 8DE
United Kingdom
Tel: 0491 32111
Fax: 0491 33508
Geographical coverage: International
Aims and objectives: CABI is an intergovernmental body which provides information, scientific and development services on agriculture and related disciplines throughout the world
Publications: Catalogue available on request

– 137 – Confederation of European Agriculture (CEA)

PO Box 87
CH-5200 Brugg
Switzerland
Tel: 41 56 413177
Fax: 41 56 41 71 3174
Contact name: Secretariat General
Geographical coverage: European
Aims and objectives: CEA represents the interests of European agriculture and forestry at international level

– 138 – Department of Agriculture and Food

Kildare Street
Dublin 2
Eire
Tel: 353 1 789011
Geographical coverage: Ireland

– 139 – Landbrugsministeriet
(Ministry of Agriculture)

Slotsholmsgade 10
DK-1216 Copenhagen K
Denmark
Tel: 45 33 92 33 01
Fax: 45 33 14 50 42
Geographical coverage: Denmark

– 140 – Minister of Small and Medium Sized Enterprises and Agriculture

Maria-Theresiastraat 1
B-1040 Brussels
Belgium
Tel: 32 2 238 28 11
Fax: 32 2 230 38 51
Geographical coverage: Belgium

– 141 – Ministerio de Agricultura Pesca y Alimentación
(Ministry of Agriculture, Fisheries and Food)

Paseo Infanta Isabel 1
E-28071 Madrid
Spain

Tel: 34 1 347 5000
Fax: 34 1 468 6888
Contact name: Juan Manuel Garcia Bartolomé
Geographical coverage: Spain

– 142 – Ministry of Agriculture

Hallituskatu 3
SF-00170 Helsinki 17
Finland
Tel: 358 1 1601
Geographical coverage: Finland

– 143 – Ministry of Agriculture

Raudarárstig
150 Reykjavik
Iceland
Tel: 354 1 609750
Fax: 354 1 21160
Geographical coverage: Iceland

– 144 – Ministry of Agriculture

Via XX Settembre
I-00187 Rome
Italy
Tel: 39 6 46651
Fax: 39 6 474 2314
Geographical coverage: Italy

– 145 – Ministry of Agriculture

S-103 33 Stockholm
Sweden
Geographical coverage: Sweden

– 146 – Ministry of Agriculture and Fisheries

Praça do Comérico
P-1100 Lisbon
Portugal
Geographical coverage: Portugal

– 147 – Ministry of Agriculture and Food Industry

ul Wspolna 30
PL-00 930 Warsaw
Poland
Tel: 48 22 21 03 11
Geographical coverage: Poland

– 148 – Ministry of Agriculture and Forestry

78 rue de Varenne
F-75700 Paris
France
Tel: 33 1 495 549 55
Geographical coverage: France

– 149 – Ministry of Agriculture and Viticulture

1 rue de la Congrégation
L-1352 Luxembourg
Luxembourg
Tel: 352 4781
Fax: 352 464027
Geographical coverage: Luxembourg

– 150 – Ministry of Agriculture Fisheries and Food (MAFF)

Whitehall Place
London SW1A 2HH
United Kingdom
Tel: 071 270 8080
Geographical coverage: United Kingdom
Aims and objectives: MAFF administers the Government's agriculture, horticulture and fisheries policies in England and has responsibilites for food, trade and animal health throughout the UK
Publications: *Food Additives: A Balanced Approach*

– 151 – Ministry of Agriculture Fisheries and Food: Conservation Policy and Environmental Protection Division

Nobel/Ergon House
17 Smith's Square
London SW1P 3HX
United Kingdom
Tel: 071 236 6563/5654
Geographical coverage: United Kingdom

– 152 – Ministry of Agriculture Fisheries and Food: Fisheries Radiological Laboratory

Pakefield Road
Lowestoft
Suffolk NR33 0HT
United Kingdom
Tel: 0502 562244
Geographical coverage: United Kingdom

– 153 – Ministry of Agriculture, Nature Conservation and Fisheries

Bezuidenhoutseweg 73
PO Box 20401
NL-2500 EK The Hague
Netherlands
Tel: 31 70 340 7911
Fax: 31 70 340 7834
Geographical coverage: Netherlands

– 154 – Ministry of Food Agriculture and Forestry

Rochusstrasse 1
Postfach 14 02 70
D-5300 Bonn 1
Germany
Tel: 49 228 5291
Fax: 49 228 5294262
Geographical coverage: Germany

NON-GOVERNMENT ORGANISATIONS

– 155 – Association for the Development of Agricultural Fuel (ADECA)

45 rue de Naples
F-75008 Paris
France
Tel: 33 1 429 441 49
Geographical coverage: France

– 156 – Committee of Agricultural Organisations (COPA)/General Committee of Agricultural Cooperation in EC

Rue de la Science 23–25
B-1040 Brussels
Belgium
Tel: 32 2 287 27 30
Fax: 32 2 287 27 00
Contact name: Arnaud Borchard, Head of Environmental Section
Geographical coverage: European
Aims and objectives: COPA/COGECA is the umbrella organisation for national farming organisations within the EC

– 157 – Food and Agriculture Organisation of the United Nations (FAO)

Via delle Terme di Caracalla
I-00100 Rome
Italy
Tel: 39 6 57971
Fax: 39 6 57973152/5782610
Contact name: Edouard Saouma, Director General
Geographical coverage: International
Aims and objectives: FAO was established in 1945. The Organisation aims to increase production of agriculture, forestries and fisheries. It has an environment and energy programme. Environmental concerns are a major component of the organisation's field work
Publications: FAO publishes a number of yearbooks. It also publishes a quarterly *Plant Protection Bulletin*; *Food Outlook Monthly*

– 158 – Grupo Ecologista de Agrónomos (GEDEA)

ETS Ingenieros Agrónomos
E-28040 Madrid
Spain
Tel: 34 1 244 48 07
Contact name: Juan Soares
Geographical coverage: Spain

– 159 – International Federation of Agricultural Producers (IFAP)

21 rue Chaptal
F-75009 Paris
France
Tel: 33 1 452 605 53
Fax: 33 1 487 472 12
Contact name: David King, Secretary General
Geographical coverage: International
Aims and objectives: Founded in 1946, IFAP acts as a forum of exchange and as the recognised spokesman for the world's farmers in international fora and to promote and strengthen independent farmers' organisations throughout the world

– 160 – Internationale Scharrelvees Controle (ISC)

(International Eco Meat Control)
PO Box 649
NL-3500 AP Utrecht
Netherlands
Tel: 31 30 340811
Geographical coverage: European
Aims and objectives: ISC is an independent organisation which controls member farmers and butchers on the observance of the eco regulations

– 161 – Linking Environment and Farming (LEAF)

National Agricultural Centre
Stoneleigh
Warwickshire CV8 2LZ
United Kingdom
Tel: 0203 696969
Fax: 0203 696900
Contact name: Caroline Drummond, Project Coordinator
Geographical coverage: United Kingdom
Aims and objectives: LEAF's aims are to develop and promote farm practices which combine care and concern for the environment, with the responsible use of modern methods to produce safe and wholesome food
Publications: *Leafletter*, three times yearly; brochures and information booklets

BOOKS

- 162 - Council of Europe: Farming and Wildlife

Geographical coverage: European
Publishers: Council of Europe, 1989
Coverage: Includes agricultural pollution
ISBN: 9287116857

DIRECTORIES/YEARBOOKS

- 163 - Agricultural and Food Research Council Handbook

Geographical coverage: United Kingdom
Publishers: Agricultural and Food Research Council (AFRC)
Frequency: annual
Coverage: Summary of AFRC research including agriculture, agrochemicals and the agricultural implications of global warming
ISSN: 0961 1010

- 164 - Agricultural and Veterinary Sciences International Who's Who, 4th ed

Geographical coverage: International
Publishers: Longmans, 1990
Coverage: Lists many international sources relating to agriculture and veterinary sciences
ISBN: 0582041007

- 165 - Agricultural Research Centres: A World Directory of Organisations and Programmes, 10th ed

Geographical coverage: International
Publishers: Longmans, 1990
Language: Lists research centres worldwide
ISBN: 0582061229

- 166 - FAO Yearbook 1990

Geographical coverage: International
Publishers: Food and Agriculture Organisation, 1991
Frequency: annual
Language: English, French and Spanish
Coverage: International trade statistics
ISSN: 0071 7126
ISBN: 9250030851

- 167 - Green Book, The (Authority on Tractors Agricultural and Forestry)

Albany House
Hurst Street
Birmingham B5 4BD
United Kingdom
Tel: 021 622 2828
Fax: 021 622 5304
Geographical coverage: United Kingdom
Publishers: Guardian Communications Ltd
Editor: Catling, H.
Frequency: annual
Coverage: Unique reference yearbook specialist guide to agricultural forestry equipment
Price: £56.00

PERIODICALS

- 168 - Du Sol à la Table: formerly Agriculture et Vie

Le Grand Launay
F-49250 St Rémy La Varenne
France
Fax: 33 40471597
Geographical coverage: France
Publishers: Défense de la Culture Biologique
Editor: Lemaire, A.
Frequency: quarterly
Coverage: Agriculture
Annual subscription: Ffr110
ISSN: 1143 3833

- 169 - Farmers Weekly

Quadrant House
The Quadrant
Sutton
Surrey SM2 5AS
United Kingdom
Tel: 081 652 3500
Geographical coverage: United Kingdom
Publishers: Reed Business Publishing Ltd
Editor: Rowe, S.
Frequency: weekly
Coverage: All areas of the agricultural industry in UK and Europe
Annual subscription: £47.00
ISSN: 0014 8474

- 170 - Gesunde Pflanzen

Seelbuschring 9–17
D-1000 Berlin 42
Germany
Tel: 49 30 70784 0
Fax: 49 30 70784199
Geographical coverage: Germany

22 Agriculture and Farming *General*

Publishers: Verlag Paul Parey (Berlin)
Editor: Pag, H.
Frequency: monthly
Language: Text in German, summaries in English
Annual subscription: DM148
ISSN: 0367 4223

– 171 – Green Europe

25 Frant Road
Tunbridge Wells
Kent TN2 5JT
United Kingdom
Tel: 0892 33813
Fax: 0892 24593
Geographical coverage: European
Publishers: Agra (Europe) Ltd
Editor: Faulkner, G.
Frequency: monthly
Language: English
Coverage: Covers European agriculture
Annual subscription: £150.00

– 172 – SCOOP Dienst Landbouwkundig Onderzoek Staring Centrum: Instituut voor Onderzoek van het Landelijk Gebied

(Agricultural Research Department)

PO Box 125
NL-6700 AC Wageningen
Netherlands
Tel: 31 8370 74200
Fax: 31 8370 24812
Geographical coverage: Netherlands
Publishers: Dienst Landbouwkundig Onderzoek
Frequency: 3 issues a year
Language: Dutch and English
Coverage: Agricultural and hydrological research
Price: free
ISSN: 0924 9370

RESEARCH CENTRES

– 173 – Agricultural and Food Development Authority

19 Sandymount Avenue
Ballsbridge
Dublin 4
Eire
Tel: 353 1 688188
Fax: 353 1 688023
Geographical coverage: Ireland
Research activities: The development of all aspects of agricultural activity in Ireland

– 174 – Agricultural University of Norway, Norwegian Centre for International Agricultural Development

Postboks 2
N-1432 As-NLH
Norway
Norway
Geographical coverage: International
Aims and objectives: The Centre takes an active interest in environmental problems of developing countries

– 175 – Agriculture and Food Research Council (AFRC)

Polaris House
North Star Avenue
Swindon SN2 1UH
United Kingdom
Tel: 0793 413200
Fax: 0793 413201
Geographical coverage: United Kingdom
Aims and objectives: AFRC is an independent body dedicated to basic and strategic research in the biological and related sciences
Publications: *AFRC News*, three times a year; Annual report; list available
Research activities: Basic and strategic research into the biological sciences

– 176 – Cellule Environnement Institut National de la Recherche Agronomique (INRA)

17 rue de l'Université
F-75341 Paris Cedex 07
France
Tel: 33 1 427 590 00
Fax: 33 1 470 599 66
Geographical coverage: France
Publications: *Courier de la Cellule Environnement*

– 177 – Centro Studi per l'Applicazione dell'Informatica in Agricoltura

(Centre for Informatic Application in Agriculture)

Academia dei Georgofili Logge Uffizi Corti
I-50122 Firenze
Italy
Tel: 39 55 213360
Geographical coverage: Italy
Research activities: Environmental and agricultural cartography

– 178 – Centrum voor Landbouwkundig Onderzoek – Gent
(Agricultural Research Centre Ghent)

Burgemeesters van Gansberghelaan 96
B-9220 Merelbeke
Oost-Vlaanderen
Belgium
Tel: 32 91 52 20 81
Fax: 32 91 52 15 83
Geographical coverage: Belgium
Research activities: All aspects of agriculture

DATABASES

– 179 – Agra Europe
Geographical coverage: European
Coverage: EC Common Agricultural Policy, food trade
Database producer: Agra Europe
Database host: Agra Europe via Telecom Gold
Coverage period: 1963–
Update frequency: weekly

– 180 – AGREP
Geographical coverage: European
Coverage: Agricultural research projects with public funding, including information on organisations and staff
Database producer: CEC Agriculture
Database host: DIMDI; Datacentralen
Coverage period: 1975–
Update frequency: annually

– 181 – Agricola
Geographical coverage: International
Coverage: Comprehensive global coverage on all aspects of agriculture and related areas
Database producer: US Department of Agriculture, National Agricultural Library
Database host: BRS; Dialog; DIMDI
Coverage period: 1970–
Update frequency: 9,000 records monthly
Additional information: Available on CD–ROM

– 182 – Agris
Geographical coverage: International
Coverage: Covers all aspects of agriculture, particularly concerning global food supply. Also aquatic sciences and fisheries, soil sciences, forestry, pollution
Database producer: UN Food and Agriculture Association
Database host: Dialog; DIMDI; ESA-IRS
Coverage period: 1975–
Update frequency: 10,000 records monthly

– 183 – Commonwealth and Agriculture Bureau Abstracts
Geographical coverage: International
Coverage: Agricultural and biological information compiled by the Commonwealth Agricultural Bureau
Database producer: Commonwealth Agricultural Bureau International
Database host: BRS; DataStar; Dialog; DIMDI; ESA–IRS; STN
Coverage period: 1972–
Update frequency: 10,000 records monthly

– 184 – ELFIS
Geographical coverage: Germany
Coverage: Food agriculture and forestry
Database producer: German Information Systems on Food, Agriculture and Forestry
Database host: DIMDI
Coverage period: 1984–
Update frequency: quarterly

Crop Protection and Agrochemicals

GOVERNMENT ORGANISATIONS

– 185 – Miljoministeriet Miljostyrelsen Kemikaliekontrollen
(The State Chemical Supervision Service)

12 Skovbrynet
DK-2800 Lyngby
Denmark
Tel: 45 02 87 70 66
Contact name: Hedegaard Poulsen, Director
Geographical coverage: Denmark
Aims and objectives: Controls and observes pesticide and toxic substance legislation

NON-GOVERNMENT ORGANISATIONS

– 186 – Asociación Española de Fabricantes de Agroquímicos para la Protección de las Plantas

Almagro 44, 3 dra
E-28010 Madrid
Spain

24 Agriculture and Farming *General*

Contact name: Mr L Roy Parages
Geographical coverage: Spain

– 187 – Associação Portuguesa das Empresas Indústrias de Produtos Químicos

Avenida D Carlos 1
45-3
P-1200 Lisbon
Portugal
Contact name: Mr J Pais
Geographical coverage: Portugal

– 188 – Association Industry Difesa Produzioni Agricole (AGROFARMA)

33 Via Accademia
I-20131 Milan
Italy
Contact name: Dr P Catelani
Geographical coverage: Italy
Aims and objectives: Agrochemical association equivalent to the BAA in Europe

– 189 – Association of Hungarian Chemical Works (AHCW)

Hungaria Korut 178
H-1146 Budapest
Hungary
Contact name: Mr Z Kolonies
Geographical coverage: Hungary

– 190 – British Agrochemicals Association Ltd (BAA)

4 Lincoln Court
Lincoln Road
Peterborough PE1 2RP
United Kingdom
Tel: 0733 349225
Fax: 0733 62523
Contact name: Isobel V B Maltby, Health Safety and Environmental Manager
Geographical coverage: United Kingdom
Aims and objectives: BAA is the UK Trade Association for the manufacture of pesticides. The Association aims to encourage safe and responsible manufacture, distribution and use of agrochemicals with regard for the interests of the community and environment
Publications: *AGCHEM* Newsletter, four times a year; Annual report; list available on request

– 191 – Danish Agrochemical Association (DAA)

Bella Center
International House
Center Boulevard 5
DK-2300 Copenhagen S
Denmark
Contact name: Mr N Lindemark
Geographical coverage: Denmark
Aims and objectives: Agrochemical association equivalent to the BAA in Europe

– 192 – European Crop Protection Association (ECPA)

Avenue Albert Lancaster 79A
B-1180 Brussels
Belgium
Tel: 32 2 375 68 60
Fax: 32 2 375 27 93
Contact name: Konstantinos (Dino) Vlahodimos, Director General
Geographical coverage: European
Aims and objectives: ECPA deals with all relevant issues to the European crop protection industry

– 193 – European Fertiliser Manufacturers' Association (EFMA)

Bleicherweg 33
CH-8002 Zurich
Switzerland
Tel: 41 1 20 91 517
Fax: 41 1 20 91 506
Geographical coverage: European
Aims and objectives: EFMA is the umbrella organisation for fertiliser manufacturers in Europe
Publications: *Agrien News*, monthly

– 194 – Fachverband des Chemischen Industrien Österreichs

Wiedner Haupstrasse 63
PO Box 325
A-145 Vienna
Austria
Contact name: Mr Turk
Geographical coverage: Austria

– 195 – Federation of Irish Chemical Industries (FICI)

Franklin House
140 Pembroke Road
Dublin 2
Eire
Contact name: Mr D O'Brien
Geographical coverage: Ireland

Aims and objectives: Agrochemical association equivalent to the BAA in Europe

– 196 – Greek Agrochemical Association (GAA)

El-Benizelou Str -Nr 2
Kallithea
Athens
Greece
Contact name: Mr J Hadjiangelidis
Geographical coverage: Greece
Aims and objectives: Agrochemical association equivalent to the BAA in Europe

– 197 – Industrieverband Agrar (IVA)

Karlstrasse 21
D-6000 Frankfurt am Main 1
Germany
Contact name: Dr O Bottcher
Geographical coverage: Germany
Aims and objectives: Agrochemical association equivalent to the BAA in Europe

– 198 – Industrin for Vaxt–Och Traskyddsnedel (IVT)

Storgatan 19
BP 5501
S-11485 Stockholm
Sweden
Contact name: Mrs B Ahlstrom-Resvik
Geographical coverage: Sweden
Aims and objectives: Agrochemical association equivalent to the BAA in Europe

– 199 – International Fertiliser Industry Association (IFA)

28 rue Marbeuf
F-75008 Paris
France
Tel: 33 1 422 527 07
Fax: 33 1 422 524 08
Contact name: Claudine Putz, Information Service
Geographical coverage: International
Aims and objectives: Trade Association

– 200 – International Group of National Associations of Manufacturers of Agro-Chemical Products (GIFAP)

Avenue A Lancaster 79A
B-1180 Brussels
Belgium
Tel: 32 2 375 68 60
Fax: 32 2 375 27 93
Contact name: Hans von Loeper, Director
Geographical coverage: International

Aims and objectives: GIFAP represents the interests of agrochemical product manufacturers worldwide. In 1992, the European Crop Protection Association was set up as a sub group of GIFAP

– 201 – Nederlandse Stichting voor Fytofarmacie (NEFYTO)

Hogeweg 16
Postbus 80523
NL-2508 GM The Hague
Netherlands
Contact name: Mr B L Hoppenbrouwer
Geographical coverage: Netherlands
Aims and objectives: Agrochemical association equivalent to the BAA in Europe

– 202 – PHYTOPHAR

49 square Marie-Louise
B-1040 Brussels
Belgium
Contact name: Ms G Detiege
Geographical coverage: Belgium
Aims and objectives: Agrochemical association equivalent to the BAA in Europe

– 203 – Société Suisse Industries Chimiques (SSIC)

(Swiss Society of Chemical Industries)

15 Nordstrasse PO Box
CH-8035 Zurich
Switzerland
Contact name: Ar A Riggenbach
Geographical coverage: Switzerland
Aims and objectives: Agrochemical association of BAA in Europe

– 204 – Syndicat National des Engrais (SNIE)

Tour Elf
Cedex 45
F-92078 Paris la Défense
France
Tel: 33 1 474 445 46
Fax: 33 1 477 869 10
Geographical coverage: France

– 205 – TISIT

Guneydogu Ishanl Kat 3
Daire 15
80040 Tophane/Istanbul
Turkey
Contact name: Mr T Bakirci
Geographical coverage: Turkey

Agriculture and Farming *General*

– 206 – Union of Industries of the Protection of Plants (UIPP)

2 rue Denfert-Rochereau
F-92100 Boulogne Billancourt
France
Contact name: Mr J P Guillou
Geographical coverage: France
Aims and objectives: Agrochemical association equivalent to the BAA in Europe

BOOKS

– 207 – Agriculture and Fertilisers: Fertilisers in Perspective – Their Role in Feeding the World

Geographical coverage: Norway
Publishers: Norsk Hydro, 1990
Coverage: Survey of environmental issues relating to agriculture and fertiliser use
ISBN: 829086101X
Author: Beckman, O. C. et al

– 208 – Crop Protection Chemicals

Geographical coverage: United Kingdom
Publishers: Ellis Horwood, 1990
Coverage: Overview of chemical crop technology
Price: £69.95
ISBN: 0131942425
Author: Lever, B. G.

DIRECTORIES/YEARBOOKS

– 209 – Crop Protection Directory 1988–89

Geographical coverage: United Kingdom
Publishers: Elaine Warrell
Editor: Warrell, E.
Frequency: irregular
Coverage: All aspects of crop protection
Price: £35.00
ISSN: 0953 2463
ISBN: 0951333003

– 210 – World Directory of Pesticide Control Organisations

Geographical coverage: International
Publishers: Royal Society of Chemistry, 1989
Editor: Kidd, H. et al
Coverage: International organisations and national authorities in 133 countries around the world which are involved in pesticides and their control
ISBN: 0851867235

REPORTS

– 211 – Water Pollution Incidents in England and Wales 1990

Geographical coverage: United Kingdom
Publishers: National Rivers Authority, 1992
Frequency: annual
Coverage: Analysis of water pollution incident statistics in England and Wales
Price: £3.50
ISBN: 1873160143

RESEARCH CENTRES

– 212 – Centre Agronomique de Recherches Appliquées du Hainaut
(Agronomic Centre of Applied Research Hainaut)

11 rue Paul Pastur
B-7800 Ath
Hainaut
Belgium
Tel: 32 68 28 12 81
Geographical coverage: Belgium
Research activities: Plant production and protection; fertilisation; weed control; methane production

– 213 – Centrum voor Agrobiologisch Onderzoek (CABO)
(Centre for Agrobiological Research)

PO Box 14
NL-6700 AA Wageningen
Netherlands
Tel: 31 8370 19012
Geographical coverage: Netherlands
Research activities: Basic and applied physiological research on problems of crop production and quality

– 214 – Espoo Research Centre

Box 44
SF-02271 Espoo
Finland
Tel: 358 90 804 71
Geographical coverage: Finland
Research activities: Fertilisers; agrochemicals; speciality chemicals for pulp and paper; chemical and biological plant protection

– 215 – Fraunhofer-Institut für Umweltchemie und Ökotoxikologie (IUCT)

(Fraunhofer Institute for Environmental Chemistry and Toxicology)

Postfach 1260
Grafschaft
D-5948 Schmallenberg
Germany
Tel: 49 29 72 30 2 0
Fax: 49 29 72 30 23 19
Geographical coverage: Germany
Research activities: Exposure to and ecotoxicological effects of environmental chemicals including pesticides and other chemical products

– 216 – Instituut voor Plantenziektenkundig Onderzoek (IPO)

(Plant Protection Research Institute)

Binnenhaven 12
Postbus 9060
NL-6700 GW Wageningen
Netherlands
Tel: 31 8370 19151
Geographical coverage: Netherlands
Publications: Annual report
Research activities: Research into control methods for plant diseases and pests; crop protection; diagnostic methods in virus diseases

– 217 – International Centre for the Application of Pesticides (ICAP)

Cranfield Institute of Technology
Cranfield
Bedfordshire MK43 0AL
United Kingdom
Tel: 0234 750111
Geographical coverage: International
Research activities: Pesticide application technology; pest bionomics monitoring and behaviour

– 218 – Station Fédérale de Recherches Agronomiques de Zurich Reckenholz

(Swiss Federal Research Station for Agronomy Zurich)

Reckenholzstrasse 191–211
CH-8046 Zurich
Switzerland
Tel: 41 1 3777 111
Geographical coverage: Switzerland
Publications: Biannual report
Research activities: Soil science, recultivation and improvement

DATABASES

– 219 – Toxline

Geographical coverage: International
Coverage: Adverse effects of chemicals and related products on humans and animals. Includes data from Pesticide Abstracts
Database producer: National Library of Medicine and Food Science
Database host: Blaise; DataStar; Dialog; DIMDI; National Library of Medicine
Coverage period: 1940–
Update frequency: 12,000 records monthly
Additional information: Available on CD–ROM

Fish and Fisheries

BOOKS

– 220 – Introduction to Aquaculture

Geographical coverage: United Kingdom
Publishers: John Wiley and Sons, 1991
Coverage: Topics discussed include water quality, pumps and the measurement of flow, biological concepts, seaweed cultivation, the raising of fresh and salt water fish and engineering
Price: £42.10
ISBN: 0471611468
Author: Landau, M.

REPORTS

– 221 – European Inland Fisheries Advisory Commission: Water Quality Criteria for European Freshwater Fish

Geographical coverage: European
Publishers: Food and Agriculture Organisation (FAO), 1984
Coverage: Report on nitrite and freshwater fish
ISBN: 9251021775

28 Agriculture and Farming *Fish and Fisheries*

– 222 – Net Losses Gross Destruction – Fisheries in the European Community

Geographical coverage: European
Publishers: Greenpeace International, 1991
Coverage: Highlights the problems of over-fishing in Europe
Price: £2.50

– 223 – Poisonous Fish

Geographical coverage: United Kingdom
Publishers: Friends of the Earth (FoE), 1991
Coverage: A survey of available data on the contamination of freshwater fish flesh with persistent toxic chemicals
Price: £13.00
ISBN: 1857500350

– 224 – Prevalence of Fish Diseases with Reference to Dutch Coastal Waters

Geographical coverage: Netherlands
Publishers: Netherlands Institute for Fishery Investigations, 1986

RESEARCH CENTRES

– 225 – Aquário Vasco da Gama

(Vasco da Gama Aquarium)

Dafundo
P-1495 Lisbon
Portugal
Tel: 351 1 4196337
Geographical coverage: Portugal
Publications: *Relatórios de Actividades do Aquário Vasco da Gama*
Research activities: Fish pathology, aquaculture and marine biology/ecology

– 226 – Bundesforschungs Anstalt für Fischerei

(Federal Research Centre for Fisheries)

Palmaille 9
D-2000 Hamburg 50
Germany
Tel: 49 40 389 05113
Geographical coverage: Germany
Research activities: All aspects of fisheries including marine pollution and environmental protection analysis

– 227 – Danmarks Fiskeri–og Havundersogelser

(Danish Institute for Fisheries and Marine Research)

Charlottenlund Slot
DK-2920 Charlottenlund
Denmark
Tel: 45 31 62 85 50
Fax: 45 31 62 85 36
Geographical coverage: Denmark
Research activities: Freshwater ecology; marine ecology

– 228 – Directorate of Fisheries Research

Fisheries Laboratory
Pakefield Road
Lowestoft
Suffolk NR33 62244
United Kingdom
Tel: 0502 62244
Geographical coverage: United Kingdom
Publications: Fisheries research technical reports; laboratory leaflets; aquatic environment monitoring reports
Research activities: Distribution and effects of radionuclides, metals, pesticides and other pollutants on the aquatic environment

– 229 – Fiskeridirektoratets Ernaeringsinstitutt

(Fisheries Directorate Nutrition Institute)

Postboks 1900
Nordnes
N-5024 Bergen
Norway
Tel: 47 5 238000
Fax: 47 5 238095
Geographical coverage: Norway
Research activities: Nutrition; fish breeding; new technical processes for nutrients

– 230 – Fiskeridirektoratets Havforskningsinstitutt

(Marine Research Institute)

Postboks 1870
Nordnes
N-5024 Bergen
Norway
Tel: 47 5 238500
Geographical coverage: Norway
Publications: *Report on Norwegian Fishery and Marine Investigations*
Research activities: Aquaculture and fish farming; fisheries science

– 231 – Greek National Centre for Marine Research

Aghios Kosmas
GR-166 04 Hellinikon
Athens
Greece
Tel: 30 1 9820 214
Geographical coverage: Greece
Publications: *Thalassographica*, annual
Research activities: Carries out research in the fields of oceanography, fisheries, inland waters, aquaculture and water pollution

– 232 – Irish Sea Fisheries Board

PO Box 12
Crofton Road
Dun Laoghaire
Dublin 4
Eire
Tel: 353 1 841544
Geographical coverage: Ireland
Publications: Market research series; resource and record series; annual report
Research activities: Fisheries and mariculture development; marine resources; food technology; engineering

– 233 – Rannsóknastofnun Fiskidnadarins

(Icelandic Fisheries Laboratories)

PO Box 1390
Skulagata 4
121 Reykjavik
Iceland
Tel: 354 1 20240
Fax: 354 1 623790
Geographical coverage: Iceland
Aims and objectives: Independent and sponsored research and services for the fishing industry
Publications: Biannual report

– 234 – Rijksinstituut voor Visserijonderzoek (RIVO)

(Netherlands Institute for Fishery Investigations)

PO Box 68
NL-1970 AB Ijmuiden
Netherlands
Tel: 31 2550 64646
Geographical coverage: Netherlands

DATABASES

– 235 – Aquatic Sciences and Fisheries Abstracts

Geographical coverage: International
Coverage: Aquatic pollution and environmental quality; marine and freshwater science technology and management; ecology and ecosytems
Database producer: ASFIS/Cambridge Scientific Abstracts
Database host: BRS; Dialog; DIMDI; ESA–IRS
Coverage period: 1975–
Update frequency: 1,000 records monthly

Land Use

BOOKS

– 236 – Farming and the Countryside

Geographical coverage: United Kingdom
Publishers: C A B International, 1991
Coverage: This text brings together economic analyses of the side effects of agriculture and land use
Price: £35.00
ISBN: 0851987133
Author: Hanley, N.

PERIODICALS

– 237 – Ecological Engineering

PO Box 211
NL-1000 AE Amsterdam
Netherlands
Tel: 31 20 5803911
Fax: 31 20 5803705
Geographical coverage: Netherlands
Publishers: Elsevier Science Publishers BV
Editor: Costanza, R.
Frequency: quarterly
Language: English

Coverage: Publishes contributions in ecotechnology including bioengineering, pollution control and sustainable agriculture
Annual subscription: Dfl290
ISSN: 0925 8574

RESEARCH CENTRES

– 238 – Institut für Betriebswirtschaft Bundesforschungsanstalt für Landwirtschaft

(Institute for Farm Economics Federal Agricultural Research Centre)

Bundesallee 50
D-3300 Braunschweig
Germany

Tel: 49 531 596 545
Fax: 49 531 596 367
Geographical coverage: Germany
Research activities: Impact of land use in agriculture and of livestock production on the landscape and on the groundwater

DATABASES

– 239 – Farm Structure Survey Retrieval System (FSSRS)

Geographical coverage: European
Coverage: Contains results of five surveys in the EC
Database producer: EUROSTAT
Database host: EUROSTAT
Coverage period: 1975–

Organic Farming

NON-GOVERNMENT ORGANISATIONS

– 240 – AGROBIO

Rue D Diniz 2-R/C
P-1200 Lisbon
Portugal
Geographical coverage: Portugal
Aims and objectives: Promotes organic farming

– 241 – Asociación Vida Sana

Clot 39
E-08026 Barcelona
Spain
Geographical coverage: Spain
Aims and objectives: Promotes organic farming

– 242 – Associazione per l'Agricultura Biodinamica

Via Privata Vasto 4
I-20121 Milan
Italy
Geographical coverage: Italy
Aims and objectives: Promotes organic farming

– 243 – Biofranc (FNAB)

9 rue Cels
F-75014 Paris
France
Geographical coverage: France
Aims and objectives: FNAB promotes organic farming

– 244 – Biokultura Association

Török u 7
H-1023 Budapest
Hungary
Geographical coverage: Hungary
Aims and objectives: Promotes organic agriculture, and supports the needs of its members

– 245 – Biokultura Egyelulet

Arany JU 25
H-1051 Budapest
Hungary
Geographical coverage: Hungary
Aims and objectives: Promotes organic farming

– 246 – Centre for Alternative Technology (CAT)

Machynlleth
Powys
Wales SY20 9AZ
United Kingdom
Tel: 0654 702400
Contact name: Roger Kelly, Director
Geographical coverage: United Kingdom
Aims and objectives: The Centre founded in 1973, demonstrates and promotes sustainable technologies and ways of living, including renewable energy sources, energy conservation and organic growing. It provides working displays of a wide range of alternative technologies
Publications: *Clean Slate* newsletter, many pamphlets, information sheets, resource booklets and technical papers

– 247 – Dansk Naturkost

Vendersgade 29
DK-1363 Copenhagen
Denmark
Geographical coverage: Denmark

– 248 – DEBIO

Taietveien 1
N-1940 Bjorkelangen
Norway
Geographical coverage: Norway

– 249 – Forschungsring für Biologisch–Dynamische Landwirtschaft

Wirtschaftweise
Baumschulenweg 11
D-6100 Darmstadt
Germany
Geographical coverage: Germany
Aims and objectives: Promotes organic farming

– 250 – Irish Organic Farmers and Growers Association

Killegland Farm
Ashbourne
Co Meath
Eire
Geographical coverage: Ireland
Aims and objectives: Promotes organic farming

– 251 – Kontrollforeningen for Alternativ Odling (KRAV)

Brogarden
Jalla
S-75594 Uppsala
Sweden
Geographical coverage: Sweden

– 252 – Nature et Progrès

Rue Basse Marcelle 26
B-5000 Namur
Belgium
Geographical coverage: Belgium
Aims and objectives: Promotes organic farming

– 253 – Organic Food and Farming Centre

86 Colston Street
Bristol BS1 588
United Kingdom
Tel: 0272 290661
Fax: 0272 252504
Geographical coverage: United Kingdom
Aims and objectives: The Centre incorporates three of the key organisations in the organic movement: Soil Association, British Organic Farmers and Organic Growers Association
Publications: *New Farmer and Grower*, quarterly; Catalogue available on request

– 254 – Organic Reclamation and Composting Association (ORCA)

Temselaan 100
B-1853 Strombeek-Bever
Belgium
Tel: 32 2 456 24 58
Fax: 32 2 456 28 84
Contact name: Nick de Oude, Chairperson
Geographical coverage: European
Aims and objectives: Professional association concerned with the reclamation of organic material

– 255 – Schweizerische Gesellschaft für Biologischen Landbau

CH-6611 Mosogno
Switzerland
Geographical coverage: Switzerland

– 256 – SLVY-liitto

Partala
SF-51900 Juva
Finland
Geographical coverage: Finland
Aims and objectives: Promotes organic farming

– 257 – Verband Organisch–Biologisch Wirtschaftender Bauern Österreichs

Rosensteingasse 43
A-1170 Vienna
Austria
Geographical coverage: Austria
Aims and objectives: Promotes organic farming

– 258 – Vereniging Voor Biologische–Dynamische Landbouw

Traay 128A
NL-3971 GS Drierbergen
Netherlands
Geographical coverage: Netherlands
Aims and objectives: Promotes organic farming

BOOKS

– 259 – Environmental Effects of Conventional and Organic Biological Farming Systems, The

Geographical coverage: United Kingdom
Publishers: Soil Association
Coverage: A detailed review in four parts on the environmental impact of conventional farming
Author: Arden-Clarke, C.

– 260 – Kind Food Guide, The

Geographical coverage: United Kingdom
Publishers: Penguin Books, 1991
Coverage: An A–Z Guide to the ways in which animals are reared for food
ISBN: 0140139842
Author: Eyton, A.

PERIODICALS

– 261 – New Farmer and Grower

BOF/OGA
86 Colston Street
Bristol BS1 5BB
United Kingdom
Geographical coverage: United Kingdom
Publishers: BOF/OGA
Frequency: quarterly
Coverage: Technical journal which includes articles on organic food production
Annual subscription: £35.00

REPORTS

– 262 – Organic Produce in Europe Special Report no 2128

Geographical coverage: European
Publishers: Economist Intelligence Unit, March 1991
Language: English
Coverage: The development of organic farming in Europe
ISBN: 0850585317
Author: Tate, W.

RESEARCH CENTRES

– 263 – Henry Doubleday Research Association (HDRA)

Ryton Gardens
Coventry CV8 3LG
United Kingdom
Tel: 0203 303517
Fax: 0203 639229
Contact name: Jackie Gear, Executive Director
Geographical coverage: United Kingdom
Aims and objectives: HDRA is Britain's membership organisation for organic gardeners
Publications: *HDRA Newsletter*, four times a year; Annual report; List available on request
Research activities: Organic soil cultivation methods

– 264 – Organic Advisory Service

Elm Farm Research Centre
Hamstead Marshall
nr Newbury
Berkshire RG15 OHR
United Kingdom
Tel: 0488 58298
Fax: 0488 58503
Geographical coverage: United Kingdom
Aims and objectives: Research and information centre

3
Coastal and Marine Waters

Coastal Environment

NON-GOVERNMENT ORGANISATIONS

– 265 – Irish Coastal Environment Group

8 Belgrave Square
Monkstown
Co Dublin
Eire
Geographical coverage: Ireland

BOOKS

– 266 – Good Beach Guide to over 170 of Britain's Best Beaches 1990

Geographical coverage: United Kingdom
Publishers: Elbury Press
Coverage: A guide to the pollution free beaches of Britain
Price: £59.50
ISBN: 0852238525

– 267 – Impact of Sea Level Changes on European Coastal Lowlands, The

Geographical coverage: European
Publishers: Blackwell Publishing, 1992
Coverage: Specialists examine the likely impacts of sea level rises on European coastal lowlands
Price: £60.00
ISBN: 0631181830
Author: Tooley. M. J. and Jelgersma, S.

– 268 – Recreational Water Quality Management: Vol 1 Coastal Waters

Geographical coverage: European
Publishers: Ellis Horwood, 1992
Coverage: This study describes management strategies for the coastal water environment and its waste disposal and pollution problems
Price: £40.00
ISBN: 0137700253
Author: Kay, D.

– 269 – Submerging Coasts

Geographical coverage: United Kingdom
Publishers: John Wiley and Sons, 1993
Coverage: This text describes the changes that occur in coastal landforms during a rising sea level
ISBN: 0471938076
Author: Bird, E. C. F.

PERIODICALS

– 270 – Blue Flag Campaign, The

Friluftsradet
Olof Palmes Gade 10
DK-2100 Copenhagen Ø
Denmark
Geographical coverage: European
Publishers: Foundation for Environmental Education in Europe (FEEE)
Frequency: annual
Language: English
Coverage: Raises consciousness and encourages growth and understanding of the coastal environment

34 Coastal and Marine Waters *Marine Environment*

REPORTS

– 271 – Seaside Award

Lion House
26 Muspole Street
Norwich NR3 1 DJ
United Kingdom
Tel: 0603 762888
Fax: 0603 760580
Geographical coverage: United Kingdom
Publishers: The Seaside Award
Frequency: annual
Coverage: Developed by the Tidy Britain Group, it promises a high standard of cleanliness, water quality and safety

RESEARCH CENTRES

– 272 – Deutsches Hydrographisches Institut

(German Hydrographic Institute)

Transport
Bernhard-Nocht-Strasse 78
D-2000 Hamburg 36
Germany
Tel: 49 40 31901
Fax: 49 40 31905150
Geographical coverage: Germany
Research activities: Monitoring seawater for radioactivity and other noxious substances

Marine Environment

GOVERNMENT ORGANISATIONS

– 273 – Ministry of Agriculture Fisheries and Food (MAFF): Marine Environmental Protection Division

Nobel/Ergon House
17 Smith's Square
London SW1P 3HX
United Kingdom
Tel: 071 238 5876
Geographical coverage: United Kingdom

– 274 – Water Directorate Marine Branch Department of the Environment

Romney House
43 Marsham Street
London SW1P 3PY
United Kingdom
Tel: 071 276 8504
Geographical coverage: United Kingdom

NON-GOVERNMENT ORGANISATIONS

– 275 – Arbeitsgemeinschaft Information Meeresforschung Meerestechnik

Stilleweg 2
D-3000 Hannover 51
Germany
Tel: 49 511 6468655
Geographical coverage: Germany
Aims and objectives: Marine pollution and sea water quality

– 276 – Baltic Marine Environment Protection Committee Helsinki Commission (HELCOM)

Mannerheimintie 12 A
SF-00100
Helsinki 10
Finland
Tel: 358 1 90 602
Fax: 358 1 90 644
Geographical coverage: Finland
Aims and objectives: HELCOM aims to protect the marine environment of the Baltic Sea from all types of pollution

– 277 – Bonn Commission

New Court
48 Carey Street
London WC2A 2JE
United Kingdom
Geographical coverage: European
Aims and objectives: The Bonn Commission aims to protect the North Sea from pollution

– 278 – Centre National pour l'Exploitation des Océans

Boîte Postale 337
F-29273 Brest
Finistère
France

Tel: 33 98804650
Geographical coverage: France
Aims and objectives: Concerns marine pollution

– 279 – Federation Seas at Risk

Vossiusstraat 20
NL-1071 AD Amsterdam
Netherlands
Tel: 31 20 676 1477
Geographical coverage: Netherlands

– 280 – Institut Atlantique

120 rue de Longchamp
F-75116 Paris
France
Tel: 33 1 7272436
Geographical coverage: France
Aims and objectives: Pollution control and environmental information

– 281 – International Council for the Exploration of the Seas (ICES): Conseil International pour l'Exploration de la Mer

Palaegade 2–4
DK-1261 Copenhagen K
Denmark
Tel: 45 33 15 42 25
Fax: 45 33 93 42 15
Geographical coverage: International
Publications: Journals; bulletins; newsletters; reports; data lists
Research activities: Coordinates marine scientific activities of member countries

– 282 – International Maritime Organisation (IMO)

4 Albert Embankment
London SE1 7SR
United Kingdom
Tel: 071 735 7611
Fax: 071 587 3210
Contact name: Roger Kohn, Information Officer
Geographical coverage: International
Aims and objectives: The aims of the IMO are to increase the safety of international maritime shipping and prevent pollution of the seas from ships, by the adoption and implementation of conventions, codes and other instruments

– 283 – Landelijke Vereniging tot Behoud van de Waddenzee (LVBW)

(Dutch Society for the Wadden Sea Conservation)

PO Box 90
NL-8860 AB Harlingen
Netherlands
Tel: 31 51 78 15541
Fax: 31 51 78 17977
Contact name: Mr Hans Revier, Coordinator LVBW
Geographical coverage: Netherlands
Aims and objectives: LVBW aims to protect and restore the natural environment of the International Wadden Sea area of the Netherlands, Germany and Denmark

– 284 – Marine Conservation Society (MCS)

9 Gloucester Road
Ross on Wye
Herefordshire HR9 5BU
United Kingdom
Tel: 0989 66017
Contact name: John Parsler, Campaigns Manager
Geographical coverage: United Kingdom
Aims and objectives: MCS aims to protect the marine environment for both wildlife and future generations by promoting its sustainable and environmentally sensitive management
Publications: *Marine Conservation*; *The Good Beach Guide*; *Coastal Directory*; other publications

– 285 – North Atlantic Salmon Conservation Organisation (NASCO)

11 Rutland Square
Edinburgh EH1 2AS
United Kingdom
Tel: 031 228 2551
Geographical coverage: United Kingdom
Aims and objectives: NASCO aims to promote the conservation of salmon

– 286 – Oslo Commission (OSCOM)

New Court
48 Carey Street
London WC2 2JE
United Kingdom
Tel: 071 242 9927
Fax: 071 831 7427
Geographical coverage: European
Aims and objectives: The Commission was set up under the Convention for the Prevention of Marine Pollution by Dumping from Ships and Aircraft in Oslo 1974, and now works jointly with the Paris Commission set up under the Convention for the Prevention of Marine Pollution

– 287 – Schutzgemeinschaft Deutscher Nordseeküste

Weserstrasse 78
D-2940 Wilhelmshaven
Germany
Geographical coverage: Germany

– 288 – Secretariat for the Protection of the Mediterranean Sea

Placa Lesseps 1
E-08023 Barcelona
Spain
Tel: 34 3 217 16 95
Geographical coverage: European
Aims and objectives: The Secretariat, established in 1983, seeks to protect the sea from pollution by information exchange and action amongst its members

– 289 – Wereld Natuur Fonds

PO Box 7
NL-3700 AA Zeist
Netherlands
Geographical coverage: Netherlands

PRESSURE GROUPS

– 290 – Friends of the Earth (FoE)

Heemraadsingel 193
NL-3023 CB Rotterdam
Netherlands
Tel: 31 10 4779360
Fax: 31 10 4779356
Contact name: Gerard Peet, Lobbyist
Geographical coverage: Netherlands

– 291 – Werkgroep Noordzee
(North Sea Foundation)

Vossiusstraat 20-111
NL-1071 AD Amsterdam
Netherlands
Tel: 31 20 676 1477
Geographical coverage: Netherlands
Aims and objectives: FoE campaigns on the North Sea

BOOKS

– 292 – Climatic Change and the Mediterranean

Geographical coverage: European
Publishers: Edward Arnold, 1992
Coverage: Contains contributions from an international team organised under the auspices of UNEP
Price: £79.50
ISBN: 0340553294
Author: Jeftic, L. and Milliman, J. D.

CONFERENCE PAPERS

– 293 – Pollution of the Mediterranean Sea

Geographical coverage: European
Publishers: Pergamon Books, 1987
Editor: Miloradov, M.
ISBN: 0080355781

PERIODICALS

– 294 – Marine Environmental Research

Crown House
Linton Road
Barking
Essex G11 8JU
United Kingdom
Tel: 081 594 7272
Fax: 081 594 5942
Geographical coverage: United Kingdom
Publishers: Elsevier Science Publishers Ltd
Editor: Goesijadi, G. and Spies, R. B.
Frequency: 8 issues a year
Coverage: Provides research papers on chemical, physical and biological interactions in the oceans and coastal waters
Annual subscription: £269.00
ISSN: 0141 1136

– 295 – Techniques in Marine Environmental Sciences

Palaegade 2–4
DK-1261 Copenhagen K
Denmark
Fax: 45 33 93 42 15
Geographical coverage: Denmark
Publishers: International Council for the Exploration of the Sea
Editor: Pawlak, J.
Frequency: irregular
Language: Danish
Price: Dkr40
ISSN: 0903 2606

– 296 – WWF Baltic Bulletin

WWF Baltic Bulletin
c/o Box 26044
S-750 26 Uppsala
Sweden
Fax: 46 18 46 95 59
Geographical coverage: European
Publishers: WWF
Frequency: quarterly
Language: English
Coverage: Marine life in the Baltic Sea
Price: free

REPORTS

– 297 – Baltic Environment Proceedings no 39: Baltic Marine Environment Protection Commission

Mannerheimintie 12 A
SF-00100
Helsinki 10
Finland
Tel: 358 1 90 602
Fax: 358 1 90 644
Geographical coverage: Finland

– 298 – Common Future of the Wadden Sea, The

Geographical coverage: European
Publishers: World Wide Fund for Nature, 1991
Language: English
Coverage: Wadden Sea Conservation

– 299 – International Maritime Dangerous Goods (IMDG)

IMO
4 Albert Embankment
London SE1 7SR
United Kingdom
Geographical coverage: International
Aims and objectives: IMDG Code includes regulations for the prevention of pollution by harmful substances carried by sea in packaged form

– 300 – Prevalence of Fish Diseases with Reference to Dutch Coastal Waters

Geographical coverage: Netherlands
Publishers: Netherlands Institute for Fishery Investigations, 1986

RESEARCH CENTRES

– 301 – Aquário Vasco da Gama
(Vasco da Gama Aquarium)

Dafundo
P-1495 Lisbon
Portugal
Tel: 351 1 4196337
Geographical coverage: Portugal
Publications: *Relatórios de Actividades do Aquário Vasco da Gama*
Research activities: Fish pathology, aquaculture and marine biology/ecology

– 302 – Danmarks Fiskeri–og Havundersogelser
(Danish Institute for Fisheries and Marine Research)

Charlottenlund Slot
DK-2920 Charlottenlund
Denmark
Tel: 45 31 62 85 50
Fax: 45 31 62 85 36
Geographical coverage: Denmark
Research activities: Freshwater ecology; marine ecology

– 303 – Greek National Centre for Marine Research

Aghios Kosmas
GR-166 04 Hellinikon
Athens
Greece
Tel: 30 1 9820 214
Geographical coverage: Greece
Publications: *Thalassographica*, annual
Research activities: Carries out research in the fields of oceanography, fisheries, inland waters, aquaculture and water pollution

– 304 – Group of Experts on the Scientific Aspects of Marine Pollution (GESAMP)

4 Albert Embankment
London SE1 7SR
United Kingdom
Tel: 071 735 7611
Fax: 071 587 3210
Geographical coverage: United Kingdom
Aims and objectives: GESAMP was formed in 1969 to serve as a mechanism for encouraging coordination, collaboration and harmonisation of activities related to marine pollution of common interest to the sponsoring bodies
Publications: Reports and studies
Research activities: Oceanographic research aspects of marine pollution including monitoring

– 305 – Institut Français de Recherche pour l'Exploitation de la Mer (IFREMER)
(French Research Institute for Ocean Utilisation)

66 avenue d'Iena
F-75116 Paris
France
Tel: 33 1 472 355 28
Fax: 33 1 472 302 79
Geographical coverage: France
Research activities: Living resources, engineering and technology, and environmental and ocean research

– 306 – Institute for Marine Environmental Research

Prospect Place
The Hoe
Plymouth PL1 3DH
United Kingdom
Tel: 0752 221371
Geographical coverage: United Kingdom
Research activities: NERC research institute

– 307 – Institutl Roman de Cercetari Marine
(Marine Research Institute of Romania)

vul Lenin 300
8700 Constanta
Romania
Tel: 40 16 43288
Geographical coverage: Romania
Publications: Annual report
Research activities: Studies in the Black Sea and Atlantic Ocean on water pollution; marine technology and other aspects

– 308 – Instytut Meteorologii i Gospodarki Wodnej
(Meteorology and Water Management Institute)

Podlesna Street 61
PL-01 673 Warsaw
Poland
Tel: 48 22 352813
Geographical coverage: Poland
Publications: Publications catalogue available on request
Research activities: Work is conducted in the fields of hydrology, meteorology, oceanology, water management and engineering

– 309 – International Laboratory of Marine Radioactivity

2 avenue Prince Hereditaire Albert
MC-9800 Monaco
Monaco
Tel: 33 93 50 44 88
Fax: 33 93 25 73 46
Geographical coverage: International
Research activities: Occurrence and behaviour of radioactive substances and other forms of pollution in the marine environment

– 310 – Irish Sea Fisheries Board

PO Box 12
Crofton Road
Dun Laoghaire
Dublin 4
Eire
Tel: 353 1 841544
Geographical coverage: Ireland
Publications: Market research series; resource and record series; annual report
Research activities: Fisheries and mariculture development; marine resources; food technology; engineering

– 311 – Merentutkimuslaitos
(Finnish Institute of Marine Research)

PO Box 33
Lyypekinkuja 3A
SF-00931 Helsinki
Finland
Tel: 358 1 90 33 10 44
Fax: 358 1 90 33 13 76
Geographical coverage: Finland
Publications: *Finnish Marine Research*
Research activities: Physical, chemical and biological oceanographic and marine environmental studies; monitoring research in the Black Sea

– 312 – Scottish Marine Biological Association

Dunstaffnage
Marine Research Laboratory
PO Box 3
Oban
Argyll PA34 4AD
United Kingdom
Tel: 0631 62244
Fax: 0631 65518
Geographical coverage: United Kingdom
Aims and objectives: Promotes marine science in Scotland through research and education
Research activities: Oceanography and ecology of the marine environment, particularly in Scottish coastal waters

DATABASES

– 313 – Baltic

Geographical coverage: Sweden
Coverage: Aspects of pollution in the Baltic
Database producer: Swedish National Environment Protection Agency
Database host: DIMDI
Coverage period: 1980–

– 314 – Oceanic Abstracts

Geographical coverage: International
Coverage: Marine biology; Pollution of oceans and estuaries
Database producer: Cambridge Scientific Abstracts
Database host: BRS; Dialog; ESA–IRS
Coverage period: 1964–
Update frequency: 1,500 records monthly

4
Inland Waters

Acid Rain

GOVERNMENT ORGANISATIONS

– 315 – Nordic Council

Box 19506
S-104 32 Stockholm
Sweden
Tel: 46 8 1434 20
Fax: 46 8 117536
Geographical coverage: European
Aims and objectives: The Nordic Council is the official body for cooperation between parliaments and governments in Denmark, Finland, Iceland, Norway, Sweden, Greenland and the Faroe and Aland Islands

NON-GOVERNMENT ORGANISATIONS

– 316 – Atmospheric Research and Information Centre (ARIC)

The Manchester Metropolitan University
Chester Street
Manchester M1 5GD
United Kingdom
Tel: 061 247 1590/93
Fax: 061 247 6318
Contact name: Mrs Sue Hare, Information Officer
Geographical coverage: United Kingdom
Aims and objectives: National and international information centre on air pollution, particularly acid rain and global climate change. Also research into the chemistry of acid deposition

– 317 – Swedish NGO Secretariat on Acid Rain

Box 245
S-401 24 Gothenburg
Sweden
Tel: 46 31 15 39 55
Fax: 46 31 15 09 33
Geographical coverage: Sweden

BOOKS

– 318 – Acid Deposition: Volume 1 Sources Effects and Controls

Geographical coverage: United Kingdom
Publishers: The British Library/Technical Communications, 1989
Editor: Longhurst, J. W. S.
Coverage: A collection of papers, with a European perspective on acid deposition. Monitoring freshwater acidification, soils and forest systems, structural materials and control technologies
ISBN: 0946655332

– 319 – Acid Deposition: Volume 2 Origins Impacts and Abatement Strategies

Geographical coverage: United Kingdom
Publishers: Springer Verlag/The British Library, 1991
Editor: Longhurst, J. W. S.
Coverage: A collection of papers, with a European perspective on acid deposition
ISBN: 3540537414

– 320 – Acid Politics

Geographical coverage: United Kingdom
Publishers: Pinter Publishers, 1990
Coverage: A comparison of different policy approaches to acid rain recognition and control in the UK and West Germany
Price: £30.00
ISBN: 1852931167

– 321 – Acid Rain and the Environment, Volume 3: 1988–91

Geographical coverage: United Kingdom
Publishers: The British Library/Technical Communications, 1992
Editor: Grayson, L.
Coverage: A select bibliography of literature on acid rain, focusing on the effects, the research and the attempts to control the problem
Price: £32.00 in UK
ISBN: 0946655421

– 322 – Bowker A & I Acid Rain Abstracts Annual

Geographical coverage: International
Publishers: R R Bowker
Coverage: Focuses on the causes, and aquatic and terrestrial effects of acid deposition as well as precursor gases and related oxidants.
Price: £140.00
ISBN: 0835226778

– 323 – Mapping Critical Loads for Europe

Coordination Centre for Effects
PO Box 1
NL-3720 BA Bilthoven
Netherlands
Geographical coverage: Netherlands
Language: English
Coverage: RIVM has presented a series of maps showing the critical loads for acidity, sulphur and nitrogen in Europe
Author: Rijksinstituut voor Volksgezondheid en Milieuhygiene (RIVM)

PERIODICALS

– 324 – Acid News

Box 245
S-401 24 Gothenburg
Sweden
Tel: 46 31 15 09 33
Fax: 46 31 15 09 33
Geographical coverage: Sweden
Publishers: Swedish Society for Nature Conservation
Editor: Agren, C.
Frequency: monthly
Language: English
Coverage: Acid rain and the acidification of the environment
Price: free
ISSN: 0281 5087

– 325 – Acqua Aria

Via Casella 16
I-20156 Milan
Italy
Tel: 39 2 330221
Fax: 39 2 39214341
Geographical coverage: Italy
Publishers: Editrice Arti Poligrafiche Europee
Frequency: 10 issues a year
Coverage: Aspects concerning acid rain
Annual subscription: L90000
ISSN: 0391 5557

REPORTS

– 326 – WATT Committee on Energy: Air Pollution Acid Rain and the Environment

Geographical coverage: United Kingdom
Publishers: Elsevier, 1988
ISBN: 1851662227
Author: Watt Committee

DATABASES

– 327 – Acid Rain

Geographical coverage: International
Coverage: Causes, effects and counter measures to acid rain
Database producer: Bowker Electronic Publishing
Database host: DataStar; DIMDI; ESA–IRS
Coverage period: 1984–
Update frequency: 100 records monthly
Additional information: Available on CD–ROM

River Environment

GOVERNMENT ORGANISATIONS

– 328 – National Rivers Authority (NRA)

30–34 Albert Embankment
London SE1 7TL
United Kingdom
Tel: 071 820 0101
Fax: 071 820 1603
Geographical coverage: United Kingdom
Aims and objectives: The NRA was established by the Water Act 1989. It represents something quite new in protecting the water environment; an independent watchdog with clearly defined duties and the resources to put them into effect
Publications: Various publications

BOOKS

– 329 – Global Climate Changes and Freshwater Ecosystems

Geographical coverage: International
Publishers: Springer, 1992
Coverage: The contributors assume global warming of 2–5 degrees in the next century and explore the implications of this rise on various freshwater ecosystems
Price: £53.00
ISBN: 354097640X
Author: Firth, P. and Fisher, S. G.

– 330 – River Conservation and Management

Geographical coverage: International
Publishers: John Wiley and Sons, 1991
Coverage: This study discusses ways of assessing the conservation potential of rivers and offers legal means of river protection
Price: £65.00
ISBN: 0471929468
Author: Boon, P. J. and Calow, P.

– 331 – Zahlentafeln der Physikalisch Chemischen Untersuchungen des Rheinwassers

(Tableaux Numeriques des Analyses Physico Chimiques des Eaux du Rhin)

Hohenzollernstrasse 18
Postfach 309
D-5400 Koblenz
Germany
Fax: 49 261 36572
Geographical coverage: Germany
Publishers: International Commission for the Protection of the Rhine against Pollution
Language: German and French
Price: DM10
ISSN: 0173 6507

PERIODICALS

– 332 – A E S Ambiente e Sicurezza: Rivista Dell'Antiquinamento

Piazza della Repubblica 26
I-20124 Milan
Italy
Geographical coverage: Italy
Publishers: Eris SpA
Editor: Meinardi, S.
Frequency: 12 issues a year
Language: Italian with summaries in English
Coverage: Environmental studies
Annual subscription: Lit43000
ISSN: 0391 7339

– 333 – Journal of Aquatic Ecosystem Health

PO Box 17
NL-3300 AA Dordrecht
Netherlands
Tel: 31 78 334911
Fax: 31 78 334254
Geographical coverage: Netherlands
Publishers: Kluwer Academic Publishers
Editor: Munawar, M.
Frequency: quarterly
Language: English
Coverage: Covers topics relating to the health of biological components of aquatic ecosystems
Annual subscription: Dfl354
ISSN: 0925 1014

– 334 – Regulated Rivers Research and Management

Baffins Lane
Chichester
West Sussex
United Kingdom
Tel: 0243 779777
Fax: 0243 775878
Geographical coverage: United Kingdom
Publishers: Ian Wiley and Sons Ltd

Editor: Petts, G.
Frequency: quarterly
Coverage: Devoted to interdisciplinary research
Annual subscription: $250
ISSN: 0886 9375

REPORTS

– 335 – Poisonous Fish

Geographical coverage: United Kingdom
Publishers: Friends of the Earth (FoE), 1991
Coverage: A survey of available data on the contamination of freshwater fish flesh with persistent toxic chemicals
Price: £13.00
ISBN: 1857500350

RESEARCH CENTRES

– 336 – Greek National Centre for Marine Research

Aghios Kosmas
GR-166 04 Hellinikon
Athens
Greece
Tel: 30 1 9820 214
Geographical coverage: Greece
Publications: *Thalassographica*, annual
Research activities: Carries out research in the fields of oceanography, fisheries, inland waters, aquaculture and water pollution

Water Management

GOVERNMENT ORGANISATIONS

– 337 – Ministère de l'Environnement: Direction de la Prévention des Pollutions de l'Eau

14 boulevard du Général Leclerc
F-92512 Neuilly-sur-Seine
France
Geographical coverage: France

– 338 – Ministère de l'Environnement et des Eaux et Forêts

51 rue de Prague
L-2918 Luxembourg-Ville
Luxembourg
Tel: 352 478 870
Geographical coverage: Luxembourg

– 339 – Ministry for Environment and Water Management

Fo utca 46–50
Pf 351
H-1011 Budapest
Hungary
Tel: 36 1 201 3843
Fax: 36 1 201 2846
Geographical coverage: Hungary

– 340 – National Water Authority Department for International Relations

Fo utca 44–50
H-1011 Budapest
Hungary
Geographical coverage: Hungary

NON-GOVERNMENT ORGANISATIONS

– 341 – Association Nationale des Services d'Eau asbl (ANSEAU)

Chausssée Waterloo 255
B-1060 Brussels
Belgium
Geographical coverage: Belgium

– 342 – British Effluent and Water Association (BEWA)

5 Castle Street
High Wycombe
Buckinghamshire HP13 6RZ
United Kingdom
Tel: 0494 444544
Geographical coverage: United Kingdom
Aims and objectives: BEWA is the national association for British process contractors, manufacturers and suppliers of water and effluent treatment plant equipment

and associated chemicals. Provides a focal point for government and other bodies in the water sector
Publications: List available on request

– 343 – Compagnie Générale des Eaux

52 rue d'Anjou
F-75384 Paris
France
Geographical coverage: France

– 344 – Deutscherverein des Gas- und Wasserfaches eV (DVGW)

Haupstrasse 71–79
D-8236 Eschborn 1
Germany
Geographical coverage: Germany

– 345 – Empresa Publica das Aguas Livres (EPAL)

Avenida da Liberadade 24
P-1200 Lisbon
Portugal
Tel: 351 1 346 13 56
Fax: 351 1 346 31 26
Geographical coverage: Portugal

– 346 – FABRIMETAL

Rue des Drapiers 21
B-1050 Brussels
Belgium
Tel: 32 2 510 23 11
Fax: 32 2 510 23 01
Geographical coverage: Belgium
Aims and objectives: Waste and water trade association

– 347 – Federazione Italiana Impresa Pubbliche Gas Acqua e Varie (FEDERGASACQUA)

Piazza Cola di Rienzo 80
I-00192 Rome
Italy
Geographical coverage: Italy

– 348 – Hydroprojekt

Taborska 31
4-Nusle
140 43 Prague
Czech Republic
Geographical coverage: Czech Republic

– 349 – Institution of Water and Environmental Management (IWEM)

15 John Street
London WC1N 2EB
United Kingdom
Tel: 071 831 3110
Fax: 071 405 4967
Contact name: Howard Evans, Executive Director
Geographical coverage: International
Aims and objectives: Founded in 1987, the IWEM is composed of professional individuals involved in water and environmental management issues. IWEM aims to advance the science and practice of effective environmental management
Publications: *Water and Environment Management Journal*, six times a year; *Journal of IWEM*, bi monthly; Newsletter bi monthly; yearbook

– 350 – International Institute for Water: Centre International de l'Eau

149 rue Gabriel Péri
Box 290
F-54515 Vandœuvre Cedex
France
Tel: 33 831 587 87
Fax: 33 831 587 99
Contact name: Jean-Claude Block, Director
Geographical coverage: International
Aims and objectives: The Institute's principal aim is to foster exchanges, information dissemination, and international cooperation in the field of water and sanitation

– 351 – International Water Supply Association

1 Queen Anne's Gate
London SW1H 9BT
United Kingdom
Tel: 071 222 8111
Fax: 071 222 7243
Contact name: Leonard Bays, Secretary General
Geographical coverage: International
Aims and objectives: Aims to improve knowledge of public water supplies and promote the exchange of information

– 352 – International Water Tribunal Foundation (IWT)

Damrak 83
NL-1012 LN Amsterdam
Netherlands
Tel: 31 20 6240 610
Fax: 31 20 6228 384
Contact name: Arthur van Norden, Director
Geographical coverage: International
Aims and objectives: The IWT was created in 1981. This tribunal works to ensure that consumer groups, farmers

44 Inland Waters *Water Management*

and environmental organisations which aim at integrated rural development, are able to find a platform to express their concerns

– 353 – Norwegian Institute of Technology

Division of Hydraulic and Sanitary Engineering
N-7034 Trondheim
Norway
Tel: 47 7 59 47 59
Contact name: Prof Hallvard Odegaard
Geographical coverage: Norway

– 354 – Oslo Yann Og Avlop

Trondheimsveien 5
N-Oslo 5
Norway
Tel: 47 2 662020
Fax: 47 2 664083
Geographical coverage: Norway
Aims and objectives: Trade association for waste and water

– 355 – Österreichisches Vereinigung für das Gas und Wasserfach

Schubertring 14
A-1010 Vienna
Austria
Geographical coverage: Austria

– 356 – Polskie Zresenie Inazynierowi Technikow Sanitarnych (PZITS)

ul Czackiego 3/5
PL-00 043 Warsaw
Poland
Geographical coverage: Poland

– 357 – Secretaria Estado Ambiente Hidrologicos

A V Aimirante Gago Coutinha 30
P-1000 Lisbon
Portugal
Tel: 351 1 847 0080
Geographical coverage: Portugal
Aims and objectives: Trade association for waste and water

– 358 – Société Suisse de l'Industrie de Gaz et des Eaux (SSIGE)

Grutlistrasse 44
CH-8027 Zurich
Switzerland
Geographical coverage: Switzerland

– 359 – Water Supply and Sewage Committee of Major Athens (EYDAP)

156 Oropou Street
GR-111 46 Galatsi
Greece
Geographical coverage: Greece

BOOKS

– 360 – Digest of Environmental Protection and Water Statistics

Geographical coverage: United Kingdom
Publishers: HMSO, 1992
Frequency: annual
Coverage: Provides an extensive set of statistics and describes broad environmental trends. This is a key statistical source on the UK environment
ISBN: 0117526347

– 361 – Environmental Protection through Sound Water Management in the Pulp and Paper Industry

Geographical coverage: United Kingdom
Publishers: Pira, 1992
Coverage: Overview of water management and the environment
Price: £65.00
ISBN: 0902799940
Author: Webb, L.

– 362 – Rational Use of Water and its Treatment in the Chemical Industry

Geographical coverage: International
Publishers: UN/ECE, 1991
Coverage: Examines the methods of rational use of water and its treatment in specific industrial areas
ISBN: 9211165008

– 363 – Water and Drainage Law

Geographical coverage: United Kingdom
Publishers: Sweet and Maxwell, 1990
ISBN: 0421387009
Author: Bates, J. H.

– 364 – Water and the Environment

Geographical coverage: United Kingdom
Publishers: Ellis Horwood, 1992
Coverage: A volume of essays addressing key concerns in the area of water management, including coastal sewage discharges, groundwater abstraction, river and tidal engineering and others
Price: £55.00

ISBN: 0139506926
Author: Currie, J. C. and Pepper, A. T.

DIRECTORIES/YEARBOOKS

– 365 – European Water and Wastewater Industry

Geographical coverage: European
Publishers: Longman, 1990
Coverage: Contains information on the structure of the water supply and wastewater treatment practised throughout Western Europe
Price: £85.00
ISBN: 0582050804
Author: Sobey, T. and Hay, A. R.

– 366 – Water Supply and Wastewater Disposal International Almanac

Gooiland 11
NL-2716 BP Zoetermeer
Netherlands
Geographical coverage: International
Publishers: International Institute for Water Supply and Wastewater Disposal
Editor: Kepinski, A. and Kepinski, W. A. S.
Frequency: annual
Language: English, French and German
Price: $70
ISSN: 0169 2577

– 367 – Who's Who in European Water 1992

Geographical coverage: European
Publishers: Sterling Publications, 1992
Editor: Harris, R.
Coverage: Organisations in Europe which are concerned with all aspects of water
ISSN: 0966 7083

– 368 – Who's Who in the Water Industry

Geographical coverage: United Kingdom
Publishers: Turret Group plc, 1992
Editor: Garrett, W. (WSA)
Frequency: annual
Coverage: Organisations related to the water industry in Europe

PERIODICALS

– 369 – Aqua Fennica

PO Box 436
SF-00101 Helsinki 10
Finland
Fax: 358 1 73 14 41 88
Geographical coverage: Finland
Publishers: Aqua Fennica Publishing Board
Editor: Seuna, P.
Frequency: semi annual
Language: Text in English, summary in Finnish
Coverage: Water management
Price: FIM 120
ISSN: 0356 7133

– 370 – Journal of the Institution of Water and Environmental Management

15 John Street
London WC1N 2EB
United Kingdom
Tel: 071 831 3110
Fax: 071 405 4967
Geographical coverage: International
Editor: Evans, H. R.
Frequency: bi monthly
Coverage: Aspects of water and environmental management worldwide
ISSN: 0951 7359

– 371 – Korrespondez Abwasser

Markt 71
Postfach 1160
D-5205 St Augustin 1
Germany
Tel: 49 2241 232 0
Fax: 49 2241 232 51
Geographical coverage: Germany
Publishers: Gesellschaft zur Foerderung der Abwassertechnik
Frequency: monthly
Language: German with summaries in English, French and German
Annual subscription: DM169
ISSN: 0341 1478

– 372 – Vandteknik

Vilh Becks Vej 60
DK-8260 Viby J
Denmark
Tel: 45 86 11 23 33 609
Geographical coverage: Denmark
Publishers: Dansk Vandteknisk Forening
Editor: Muncle, E.
Frequency: 10 issues a year

Coverage: Focuses on water resources, groundwater protection and water quality
Annual subscription: Dkr240
ISSN: 0106 3677

– 373 – Wasserwirtschaft

Pfizerstrasse 5–7
Postfach 10 60 11
D-7000 Stuttgart 1
Germany
Tel: 49 711 2191 332
Fax: 49 711 2191 350
Geographical coverage: Germany
Publishers: Franckh-Kosmos Verlags GmbH und Co
Editor: Marotz, G.
Frequency: monthly
Coverage: Features technology of water purification, groundwater protection and sewage treatment
Annual subscription: DM164.40
ISSN: 0043 0978

– 374 – Wasserwirtschaftliche Mitteilungen

Marc-aurel Strasse 5
A-1010 Vienna
Austria
Geographical coverage: Austria
Publishers: Oesterreichischer Wasserwirtschaftsverband
Frequency: monthly
ISSN: 0043 0994

– 375 – Water and Environment International

Queensway House
2 Queensway
Redhill
Surrey RH1 1QS
United Kingdom
Tel: 0737 768611
Fax: 0737 761989
Geographical coverage: International
Publishers: International Trade Publications Ltd
Editor: Manson, J.
Frequency: 6 issues a year
Coverage: Covers all aspects of water supply and sewage/effluent treatment
Annual subscription: £46.00

– 376 – Water and Waste Treatment

111 St James Road
Croydon
Surrey CR9 2TH
United Kingdom
Tel: 081 684 4082
Fax: 081 684 9729
Geographical coverage: United Kingdom
Publishers: Faversham House Group Ltd
Editor: Mitchell, V.
Frequency: monthly
Annual subscription: £53.00
ISSN: 0043 1133

– 377 – Water Resources Management

PO Box 17
NL-3300 AA Dordrecht
Netherlands
Tel: 31 78 334911
Fax: 31 78 334254
Geographical coverage: Netherlands
Publishers: Kluwer Academic Publishers
Editor: Tsakiris, G.
Frequency: quarterly
Language: English
Annual subscription: Dfl256
ISSN: 0920 4741

– 378 – Water Supply

Osney Mead
Oxford OX2 0EL
United Kingdom
Tel: 0865 240201
Fax: 0865 721205
Geographical coverage: United Kingdom
Publishers: Blackwell Scientific Publications Ltd
Editor: Bays, L. R.
Frequency: quarterly
Language: English and French
Annual subscription: £278.00
ISSN: 0735 1917

– 379 – Wetlands Ecology and Management

PO Box 97747
NL-2509 GC The Hague
Netherlands
Geographical coverage: Netherlands
Publishers: S P B Academic Publishing bv
Editor: Sharitz, R. R.
Frequency: quarterly
Language: English
Coverage: Publishes research and review papers on fundamental and applied aspects of wetlands of freshwater, brackish or marine origin
Annual subscription: $120
ISSN: 0923 4861

– 380 – World Water

Thomas Telford House
1 Heron Quay
London E14 4JD
United Kingdom

Tel: 071 987 6999
Fax: 071 537 2443
Geographical coverage: International
Publishers: Thomas Telford Services Ltd
Editor: Opie, R.
Frequency: monthly
Language: English
Coverage: Covers international water and wastewater industry with emphasis on European issues
Annual subscription: £69.00
ISSN: 0140 9050

– 381 – Zeitschrift für Wasser und Abwasserforschung
(Journal for Water and Wastewater Research)

Postfach 10 11 61
D-6940 Weinheim
Germany
Tel: 49 6201 602 0
Fax: 49 6201 602328
Geographical coverage: Germany
Publishers: V C H Verlagsgesellschaft mbH
Frequency: 6 issues a year
Language: English and German
ISSN: 0044 3727

RESEARCH CENTRES

– 382 – Centre d'Enseignement et de Recherche pour la Gestion des Ressources Naturelles et l' Environnement
(Management of Natural Resources and Environment Study Centre)

Le Palatino
17 avenue de Choisy
F-75013 Paris
France
Tel: 33 1 458 412 55
Geographical coverage: France
Research activities: Urban hydrology; surface hydrology; environmental management; urban ecology

– 383 – Eidgenössische Anstalt für Wasserversorgung Abwasserreiningung und Gewässerschutz
(Swiss Federal Institute for Water Resources and Water Pollution Control)

Überlandstrasse 133
CH-8600 Dübendorf
Switzerland
Tel: 41 1 823 55 11
Geographical coverage: Switzerland
Publications: Various publications
Research activities: Quantitative evaluation of chemical biological and physical processes of natural waters

– 384 – Institutul de Cercetäri si Projectäri pentru Gospodärirea Apelor
(Research and Design Institute for Water Resources Engineering)

Splaiul Independentei 294
Bucharest Sector 6
Romania
Tel: 40 0 492037
Geographical coverage: Romania
Research activities: Water quality protection; water supply and treatment; wastewater treatment; sludge treatment and use

– 385 – Institutul de Cercetäri pentru Pedologie si Agrochimie
(Soil Science and Agrochemistry Research Institute)

Belevardul Märästi 61
71331 Bucharest
Romania
Tel: 40 0 172180
Geographical coverage: Romania
Publications: Anale ICPA, annual
Research activities: Soil physics; soil pollution control

– 386 – Instytut Meteorologii i Gospodarki Wodnej
(Meteorology and Water Management Institute)

Podlesna Street 61
PL-01 673 Warsaw
Poland
Tel: 48 22 352813
Geographical coverage: Poland
Publications: Publications catalogue available on request
Research activities: Work is conducted in the fields of hydrology, meteorology, oceanology, water management and engineering

– 387 – Istituto di Ricerca sulle Acque (IRSA)
(Water Research Institute)

Via Reno 1
I-00198 Rome
Italy
Tel: 39 6 8841451
Geographical coverage: Italy
Publications: Quaderni dell'Istituto di Ricerca sulle Acque, serial; Notizario Metodi Analittici per le Acque, quarterly
Research activities: Water supply and water pollution

Inland Waters — Water Quality

– 388 – Soil Survey and Land Research Centre (SSLRC)

Cranfield Institute of Technology
Silsoe Campus
Silsoe
Bedfordshire MK45 4DT
United Kingdom
Tel: 0525 60428
Fax: 0525 61147
Geographical coverage: United Kingdom
Aims and objectives: The Centre's work relates to sustainable soil and water management, pollution prevention and the monitoring of soil quality
Publications: Large number of soil maps and books

– 389 – Vesien ja ympäristöntutkimuslaitos

(Water and Environment Research Institute)

PO Box 436
SF-00101 Helsinki
Finland
Tel: 358 1 19 291
Geographical coverage: Finland
Publications: Scientific papers, instructions, recommendations and annual reports
Research activities: Water pollution control; water supply; wastewater treatment

– 390 – Water Research Centre (WRC)

Henley Road
Medmenham
PO Box 16
Marlow
Buckinghamshire SL7 2HD
United Kingdom
Tel: 0491 571531
Geographical coverage: United Kingdom

DATABASES

– 391 – Aqualine

WRC
Henley Road
Medmenham
PO Box 16
Marlow
Buckinghamshire SL7 2HD
United Kingdom
Tel: 0491 571531
Geographical coverage: International
Coverage: Water and waste technology, including the effects of pollution, environmental protection, industrial effluents and appropriate technology
Database producer: Water Research Centre
Database host: Orbit
Coverage period: 1960–
Update frequency: fortnightly

Water Quality

GOVERNMENT ORGANISATIONS

– 392 – Camara Municipal de Lisboa Direcção Municipal de Infraestructuras e Saneamento

Department de Saneamento
Rua da Cruz Vermelha 12
P-1600 Lisbon
Portugal
Geographical coverage: Portugal

– 393 – Department of the Environment, Water Directorate

Romney House
43 Marsham Street
London SW1P 3PY
United Kingdom
Tel: 071 276 8277
Geographical coverage: United Kingdom

– 394 – French Association for the Protection of Water

4 rue Ménard 4
F-78000 Versailles
France
Tel: 33 39 51 88 94
Fax: 33 39 02 37 11
Geographical coverage: France

– 395 – Ministerie van de Vlaamse Gemeenschap Administrie voor Ruimtelijke Dienst Groen Waters en Bossen

Markiesskomplx Markiesstraat 1
B-1000 Brussels
Belgium
Geographical coverage: Belgium

– 396 – Ministero dei Lavori Pubblici

Seizone 11
Piazzale Porta Pia 1
I-00198 Rome
Italy
Tel: 39 6 84821
Geographical coverage: Italy

– 397 – Ministry of Environment and Territorial Administration

5a rue de Prague
Luxembourg
Luxembourg
Tel: 352 4781
Fax: 352 400410
Geographical coverage: Luxembourg

– 398 – Ministry of Housing Physical Planning and Environment

Van Alkemadelaan 85
PO Box 20951
NL-2500 EZ The Hague
Netherlands
Tel: 31 70 335 3535
Fax: 31 70 335 3502
Geographical coverage: Netherlands

– 399 – Ministry of the Environment National Agency of Environmental Protection: Water Resources Division

Strandgade 29
DK-1401 Copenhagen K
Denmark
Geographical coverage: Denmark

– 400 – Ministry of the Environment Town Planning and Public Works

Amaliados 17
Athens
Greece
Tel: 30 1 643 1461 9
Geographical coverage: Greece

– 401 – Principal Water Authority Corporation of Dublin

City Hall
Dublin 2
Eire
Geographical coverage: Ireland
Aims and objectives: The Corporation is responsible for the provision of services related to water quality

– 402 – Secrétaire d'Etat auprès du Premier Ministre de la Prevention des Risques Technologiques

345 avenue Georges-Mandel
F-75016 Paris
France
Tel: 33 1 408 121 22
Fax: 33 1 464 738 95
Geographical coverage: France

– 403 – Secretaria de Estado para las Politicas del Agua y del Medio Ambiente

Plaza San Juan de la Cruz 3
E-28071 Madrid
Spain
Tel: 34 1 534 51 49
Contact name: Vicente Albero Silla, Secretary
Geographical coverage: Spain

– 404 – Work Group for Water

Drapstraat 142
B-2640 Mortsel
Belgium
Tel: 32 3 449 79 45
Fax: 32 3 448 08 80
Geographical coverage: Belgium

NON-GOVERNMENT ORGANISATIONS

– 405 – Associação Portuguesa dos Recursos Hidricos (APRH)

c/o LNEC
Avenida do Brasil 101
P-1799 Lisbon Codex
Portugal
Tel: 351 1 848 21 31
Fax: 351 1 89 76 60
Geographical coverage: Portugal

– 406 – Deutsche Verkehrswissenschaftliche Gesellschaft eV

Apostelnstrasse 9
D-5000 Cologne 1
Germany
Tel: 49 221 241193
Geographical coverage: Germany
Aims and objectives: Concerned with traffic, traffic noise, water quality and noise reduction

– 407 – EUREAU

Chaussée de Waterloo 255
Bte 6
B-1060 Brussels
Belgium
Tel: 32 2 537 4302
Fax: 32 2 539 2142
Contact name: F Rillaerts, Secretary General
Geographical coverage: European

– 408 – European Institute for Water

181 rue de la Pompe
F-75116 Paris
France
Tel: 33 1 475 562 20
Fax: 33 1 475 562 21
Contact name: General Secretariat
Geographical coverage: European

– 409 – Federation of the Water Protection Association in Finland

c/o Lounais-Suomen Vesiensuojeluyhdistys ry
Telekatu 16
SF-20360 Turku
Finland
Tel: 358 21 53 82 22
Fax: 358 21 38 18 38
Geographical coverage: Finland
Aims and objectives: Trade association

– 410 – International Association on Water Pollution Research and Control (IAWPRC)

1 Queen Anne's Gate
London SW1H 9BT
United Kingdom
Tel: 071 222 38 48
Fax: 071 233 1197
Contact name: Anthony Milburn, Executive Director
Geographical coverage: International
Aims and objectives: Founded in 1965, IAWPRC works to encourage international communication, cooperative efforts and exchange of information on water pollution control research and control and water quality management
Publications: *Water Research*, monthly; *Water Science and Technology*, monthly; *Water Quality International*, quarterly

– 411 – International Water and Sanitation Centre (IRC)

PO Box 93190
NL-2509 AD The Hague
Netherlands
Tel: 31 70 331 4133
Fax: 31 70 381 4034
Geographical coverage: International
Aims and objectives: IRC is concerned with knowledge generation and transfer and technical information exchange for water supply and sanitation improvement in developing countries

– 412 – Water Resources Division National Technical University of Athens

5 Iroon Polytechnion
GR-157 73 Athens
Greece
Geographical coverage: Greece

BOOKS

– 413 – Eutrophication of Freshwaters

Geographical coverage: International
Publishers: Chapman and Hall, 1991
Coverage: Explains causes and effects of eutrophication worldwide
Price: £35.00
ISBN: 0412329700
Author: Harper, D.

– 414 – Introduction to Aquaculture

Geographical coverage: United Kingdom
Publishers: John Wiley and Sons, 1991
Coverage: Topics discussed include water quality, pumps and the measurement of flow, biological concepts, seaweed cultivation, the raising of fresh and salt water fish and engineering
Price: £42.10
ISBN: 0471611468
Author: Landau, M.

– 415 – Technologie Appropriée de Déminéralisation de l'Eau Potable, La

Geographical coverage: France
Publishers: WHO, 1985
Language: French
Price: Swfr16
ISBN: 9289020121

PERIODICALS

– 416 – Environmental Toxicology and Water Quality

Baffins Lane
Chichester
West Sussex
United Kingdom
Tel: 0243 770351
Fax: 0243 775878
Geographical coverage: United Kingdom
Publishers: John Wiley and Sons Ltd
Editor: Dutka, B.
Frequency: quarterly
Coverage: Covers all aspects of environmental toxicology and water quality
Annual subscription: $179

– 417 – Feld Wald Wasser

Vordergasse 58
CH-8201 Schaffhausen
Switzerland
Geographical coverage: Switzerland
Publishers: Meier und Cie AG
Editor: Keller, Dr W.
Frequency: monthly
Price: Sfr45

– 418 – SCOOP Dienst Landbouwkundig Onderzoek Staring Centrum: Instituut voor Onderzoek van het Landelijk Gebied

(Agricultural Research Department)

PO Box 125
NL-6700 AC Wageningen
Netherlands
Tel: 31 8370 74200
Fax: 31 8370 24812
Geographical coverage: Netherlands
Publishers: Dienst Landbouwkundig Onderzoek
Frequency: 3 issues a year
Language: Dutch and English
Coverage: Agricultural and hydrological research
Price: free
ISSN: 0924 9370

– 419 – Vatten

Nyckelkroken 22
S-226 47 Lund
Sweden
Geographical coverage: Sweden
Publishers: Swedish Association for Water Hygiene
Frequency: quarterly
Language: Text in Swedish, summaries in English and German
Price: Skr580
ISSN: 0042 2886

– 420 – Wasser und Abwasser in Forschung und Praxis

Viktoriastrasse 44A
Postfach 7330
D-4800 Bielefeld
Germany
Tel: 49 521 583080
Geographical coverage: Germany
Publishers: Erich Schmidt Verlag GmbH & Co
Frequency: irregular
Language: German with English and French summaries
Price: varies
ISSN: 0512 5030

– 421 – Water Bulletin

Turrett House
171 High Street
Rickmansworth
Hertfordshire WD3 1SN
United Kingdom
Tel: 0923 777000
Fax: 0923 771297
Geographical coverage: United Kingdom
Publishers: Turret Group plc
Editor: Garrett, P.
Frequency: weekly
Price: £1.00
Annual subscription: £45.00
ISSN: 0262 9909

– 422 – Water Quality International

Headington Hill Hall
Oxford OX3 0BW
United Kingdom
Tel: 0865 794141
Fax: 0865 743911
Geographical coverage: International
Publishers: Pergamon Press plc
Editor: Horobin, W. A.
Frequency: 5 issues a year
Language: English
Coverage: Worldwide developments in the scientific and technical aspects of water pollution control
Annual subscription: £75.00
ISSN: 0892 211X

– 423 – Waterschapsbelangen

Johan van Oldenbarneveltlaan 5
PO Box 80 200
NL-2508 GE The Hague
Netherlands
Tel: 31 70 3519751
Fax: 31 70 3544642
Geographical coverage: Netherlands

Publishers: Unie van Waterschappen
Language: Dutch
Coverage: Covers current news and information concerning the waterboards. Features include environment protection
Annual subscription: Dfl102
ISSN: 0043 1486

REPORTS

– 424 – European Inland Fisheries Advisory Commission: Water Quality Criteria for European Freshwater Fish

Geographical coverage: European
Publishers: Food and Agriculture Organisation (FAO), 1984
Coverage: Report on nitrite and freshwater fish
ISBN: 9251021775

– 425 – Research and Technological Development for the Supply and Use of Freshwater Resources

Geographical coverage: European
Publishers: EC Office of Publications, Luxembourg, 1992
Language: English
Coverage: Study of water resources in the European Community

– 426 – Trace Element Occurrence in British Groundwaters

Geographical coverage: United Kingdom
Publishers: British Geological Survey, 1989
Editor: Edmunds, W. M., Cook, J. M., Miles, D. G
Coverage: The BGS reveals zones of acidified groundwater with abnormal concentrations, mostly of aluminium, but also of metals such as zinc, copper and nickel
ISBN: 0852721897

RESEARCH CENTRES

– 427 – Environmental Research Unit Water Resources

St Martins House
Waterloo Road
Dublin 4
Eire
Tel: 353 1 602511
Fax: 353 1 680009
Geographical coverage: Ireland

– 428 – Institut für Betriebswirtschaft Bundesforschungsanstalt für Landwirtschaft

(Institute for Farm Economics Federal Agricultural Research Centre)

Bundesallee 50
D-3300 Braunschweig
Germany
Tel: 49 531 596 545
Fax: 49 531 596 367
Geographical coverage: Germany
Research activities: Impact of land use in agriculture and of livestock production on the landscape and on the groundwater

– 429 – Institut für Wasser- Boden- und Lufthygiene

(Institute for Water Soil and Air Hygiene)

Corrensplatz 1
Bundesgesundheitsamt Postfach
D-1000 Berlin 33
Germany
Tel: 49 30 83082313
Geographical coverage: Germany
Research activities: Drinking water quality and treatment, wastewater and environmental hygiene, water pollution control

– 430 – NIPH Department of Toxicology

Geitmyrsveien 75
N-0462 Oslo 4
Norway
Tel: 47 2 35 6020
Geographical coverage: Norway
Research activities: Toxicological evaluation and consultation related to contaminants in food and drinking water

– 431 – Norsk Institut for Vannforskning (NIVA)

(Norwegian Institute for Water Research)

PO Box 69
Korsvoll
N-0808 Oslo
Norway
Tel: 47 2 235280
Fax: 47 2 394189
Geographical coverage: Norway
Publications: Annual report
Research activities: Water quality management

– 432 – Vandkvalitetsinstituttet (VKI)
(Water Quality Institute)

Agern Allee 11
DK-2970 Horsholm
Denmark
Tel: 45 42 86 52 11
Fax: 45 42 86 72 73
Geographical coverage: Denmark
Research activities: Soil and groundwater impact assessment; wastewater treatment technology; water quality modelling data processing

DATABASES

– 433 – Water Resources Abstracts
Geographical coverage: International
Coverage: Water quality, pollution, waste treatment
Database producer: US Department of the Interior Geological Survey
Database host: Dialog
Coverage period: 1971–
Update frequency: every two months
Additional information: Available on CD–ROM

5
Waste and Recycling

Chemicals

NON-GOVERNMENT ORGANISATIONS

– 434 – Bundesarbeitgeberverband Chemie

Abraham Lincoln St 24
D-6200 Wiesbaden
Germany
Tel: 49 611 778810
Geographical coverage: Germany
Aims and objectives: Promotes chemical recycling

– 435 – Chemical and Oil Recycling Association

O'r Diwedd
Parc Lund
Kinmel Bay
Rhyl
Clwyd
Wales LL18 5JG
United Kingdom
Tel: 0745 332427
Contact name: Mr J Looker
Geographical coverage: United Kingdom
Aims and objectives: The Association aims to promote and represent the interest of members in the recovery and re-use of oils and solvents

– 436 – Chemicals Recovery Association (CRA)

5 Coopers Close
The Laurels
Borrowash
Derby DE7 3XW
United Kingdom
Tel: 0332 677236
Geographical coverage: United Kingdom
Aims and objectives: CRA aims to promote and protect the interests of members in the recovery of recycling and resuse of solvents oils and chemicals

– 437 – Economy and Performance Service (ECO-PER)

SARP Industries
Zone Portuaire
F-78520 Limay
France
Contact name: G Felten or S Pequignot
Geographical coverage: France
Aims and objectives: ECO-PER is a collaboration between Atochem and SARP industries which offers the metals and dry cleaning industries more rational management of solvents through recycling and reduced waste

– 438 – Environmental Detergent Manufacturers Association (EDMA)

Mouse Lane
Steyning BN44 3DG
United Kingdom
Tel: 0903 879077
Fax: 0903 879052
Contact name: Robin Bines, Chairperson
Geographical coverage: European
Aims and objectives: The Association debates environmental consequences of detergents and lobbies for high standards on EC legislation. Also provides information to consumers, manufacturers, the media and national regulatory bodies

– 439 – Lega per l'Ambiente

(League for the Environment)

Via Salaria 280
I-00199 Rome
Italy
Tel: 39 6 884 1552
Fax: 39 6 844 3504
Contact name: Ms Giovanna Melandri
Geographical coverage: Italy
Aims and objectives: The League works on both political and technical levels, mainly issues linked with energy, the chemical industry and waste
Publications: *Ambienta Italia 1991*; *Il Futuro del Sole – Atmosphera Clima e Uomo*; *La carbon diet*; *La Città Amica*

– 440 – Society for Environmental Toxicology and Chemistry (SETAC)

Avenue Prekelinden 149
B-1200 Brussels
Belgium
Fax: 32 2 462 28 84
Geographical coverage: European
Aims and objectives: SETAC has formed a Life Cycle Analysis group

BOOKS

– 441 – Effluent Treatment and Waste Disposal

Geographical coverage: European
Publishers: Taylor and Francis, 1990
Coverage: This text deals with the way chemical engineers can ensure that the effluent and waste from their plants meet both present and future standards
Price: £34.50
ISBN: 1560320753
Author: European Federation of Chemical Engineer

– 442 – Multimedia Transport and Fate of Pollutants

Geographical coverage: International
Publishers: Prentice Hall, 1993
Coverage: Coverage includes the effects of transport of meteorological conditions, and modes of transport as determined by the physiochemical properties of the chemical in question
Price: £68.50
ISBN: 0136057349
Author: Cohan, Y.

– 443 – Solvent Recovery Handbook

Geographical coverage: United Kingdom
Publishers: Edward Arnold, 1993
Coverage: A guide to the effective recovery of solvents
Price: £75.00
ISBN: 0340574674
Author: Smallwood, I.

PERIODICALS

– 444 – Solvents and the Environment: Industry and Environment Vol 14 No 4 1991

Geographical coverage: International
Publishers: UNEP
Editor: Aloisi de Larderel, J.
Language: English
Coverage: Journal concentrates on all aspects of solvents
ISSN: 0378 9993

– 445 – Solvents Digest

Avenue E Van Nieuwenhuyse 4
B-1160 Brussels
Belgium
Tel: 32 2 676 72 11
Fax: 32 2 676 73 00
Geographical coverage: European
Publishers: Euro Chlor
Editor: Ginn, N.
Frequency: quarterly
Language: French, English, Dutch and Italian
Coverage: All topics related to chlorinated solvents in Europe
Price: free

– 446 – Waste Management: *see entry 486*

Geographical coverage: United Kingdom

Glass Recycling

NON-GOVERNMENT ORGANISATIONS

– 447 – Asociación Nacional de Recuperadores de Vidrio

69 Calle Muntaner
entlo 2
E-08011 Barcelona
Spain
Tel: 34 3 253 42 07
Geographical coverage: Spain
Aims and objectives: Association for glass recycling

– 448 – British Glass Recycling Department

Northumberland Road
Sheffield S10 2UA
United Kingdom
Tel: 0742 686201
Fax: 0742 681073
Contact name: Sally Whyte, Information Officer
Geographical coverage: United Kingdom
Aims and objectives: Promote glass recycling. It also coordinates the National Bottle Bank Scheme

– 449 – Bundesverband Glasindustrie

Stresemannstrasse 26
Postfach 10 17 53
D-4000 Düsseldorf 1
Germany
Geographical coverage: Germany
Aims and objectives: Promotes glass recycling

– 450 – Fédération Européenne du Verre d'Emballage (FEVE)

(The European Glass Container Federation)
Avenue Louise 89
B-1050 Brussels
Belgium
Tel: 32 2 539 34 34
Fax: 32 2 539 37 52
Contact name: Guy Robyns, Chairman
Geographical coverage: European
Aims and objectives: FEVE aims to promote glass recycling
Publications: *Glass Gazette*

– 451 – NV Vereenigde Glasfabrieken

Postbus 46
NL-3100 AA Schiedam
Netherlands
Geographical coverage: Netherlands
Aims and objectives: Promotes glass recycling

PERIODICALS

– 452 – Glass Gazette

Avenue Louise 89
B-1050 Brussels
Belgium
Geographical coverage: European
Publishers: Fédération Européene du Verre
Editor: Somogyi, A.
Language: English and French
Coverage: Glass productivity, energy conservation, recycling and the environment

Landfill

GOVERNMENT ORGANISATIONS

– 453 – Wastes Technical Policy Unit

Local Environment Quality Division
Romany House
London SWIP 3YP
United Kingdom
Tel: 071 276 8063
Geographical coverage: United Kingdom

NON-GOVERNMENT ORGANISATIONS

– 454 – WATT Committee's Working Group on Methane

Savoy Hill House
Savoy Hill
London WC2R OBU
United Kingdom
Tel: 071 379 6875
Fax: 071 497 9315
Contact name: Max Wallis
Geographical coverage: United Kingdom

Aims and objectives: The Working Group are compiling a report for government on UK emissions of methane and ways to limit them

BOOKS

– 455 – Landfill Gas

Publishers: Commission of the European Communities
Coverage: A study which reviews the state of the art technologies for landfill sites and methane recovery
Price: Ecu67

– 456 – Sanitary Landfill

Geographical coverage: United Kingdom
Publishers: John Wiley and Sons, 1990
Coverage: Describes examining site selection criteria, generating leachate, characterising waste and designing gas venting systems
Price: £39.30
ISBN: 047161386X
Author: Bagchi, A.

DIRECTORIES/YEARBOOKS

– 457 – Sitefile Digest: A Digest of Authorised Waste Treatment and Disposal Sites in Great Britain

Geographical coverage: United Kingdom
Publishers: Aspinwall and Company, 1991
ISBN: 0951287818

REPORTS

– 458 – Survey of Gassing Landfill Sites in England and Wales, A

Geographical coverage: United Kingdom
Publishers: Friends of the Earth (FoE), 1992
Coverage: Identifies over 450 gassing landfill sites in England and Wales
Price: £8.00
ISBN: 1857501101

Metals Recycling

NON-GOVERNMENT ORGANISATIONS

– 459 – Aluminium Can Recycling Association (ACRA)

1 Mex House
52 Blucher Street
Birmingham B1 1QU
United Kingdom
Tel: 021 633 4656
Fax: 021 633 4698
Geographical coverage: United Kingdom
Aims and objectives: ACRA aims to maximise the UK drinks can recycling rate by raising national awareness of aluminium can recycling
Publications: *Alu Can Recycling Times*, newsletter twice a year

– 460 – Aluminium Federation Ltd

Broadway House
Calthorpe Road
Five Ways
Birmingham B15 1TN
United Kingdom
Tel: 021 456 1103
Fax: 021 456 2274
Contact name: Mrs G C Robinson, Executive
Geographical coverage: United Kingdom
Aims and objectives: The Federation represents most sectors of the UK aluminium industry. Provides technical information. Also has the largest library of documents about aluminium in the world

– 461 – Aluminium Foil Recycling Campaign

38–42 High Street
Bidford on Avon
Warwickshire B50 4AA
United Kingdom
Tel: 0800 626287
Fax: 0789 490391
Geographical coverage: United Kingdom
Aims and objectives: AFRC aims to establish an integrated system which links charity collectors to metal merchants, via support systems supplied by local authorities and other nationally structured organisations, such as supermarkets
Publications: *Update Progress Report*

– 462 – Association of European Dry Battery Manufacturers (Europile)

PO Box 5032
CH-3001 Berne
Switzerland
Tel: 41 31–922 13 33
Fax: 41 31– 21 6961
Contact name: Secretariat, ATAG Ernst & Young Ltd
Geographical coverage: European
Aims and objectives: Europile represents twelve countries with manufacturing locations in Belgium, Denmark, France, Germany, Greece, Italy, Portugal, Spain, Switzerland and UK

– 463 – Associazione Nazionale Fra Fabbricanti di Imballaggi Metallici (ANFIMA)

Via Pirelli 27
I-20124 Milan
Italy
Tel: 39 2 66981877
Fax: 39 2 6706285
Geographical coverage: Italy
Aims and objectives: ANFIMA promotes the development and image of metal packaging and follows standardised processes

– 464 – British Battery Manufacturers Association (BBMA)

Cowley House
9 Little College Street
London SW1P 3XS
United Kingdom
Tel: 071 222 0666
Fax: 071 233 0335
Geographical coverage: United Kingdom
Aims and objectives: BBMA promotes recycling of batteries. Encourages used battery schemes which will comply with the EC Directive Proposals

– 465 – Can Makers, The

c/o GCI London Ltd
1 Chelsea Manor Gardens
London SW3 5PN
United Kingdom
Tel: 071 351 2400
Fax: 071 352 6244
Contact name: Patricia Braun or Nicky Harris
Geographical coverage: United Kingdom
Aims and objectives: The Can Makers was formed in 1981 as the body to represent the UK manufacturers of beer and soft drink cans, together with their suppliers of aluminium and tinplate and the country's specialist detinning company
Publications: Can Makers Update

– 466 – Eurométaux

6th Floor
Avenue de Broqueville 12
B-1150 Brussels
Belgium
Tel: 32 2 775 63 11
Fax: 32 2 779 05 23
Geographical coverage: European
Aims and objectives: European Association representing the metal industries

– 467 – European Aluminium Association

6th Floor
Avenue de Broqueville 12
B-1150 Brussels
Belgium
Tel: 32 2 775 63 11
Fax: 32 2 779 05 23
Geographical coverage: European
Aims and objectives: European Association representing the aluminium industry

– 468 – European Association of Automobile Manufacturers (ACEA)

20011 rue du Noyer
B-1040 Brussels
Belgium
Tel: 32 2 732 55 50
Geographical coverage: European
Aims and objectives: ACEA works in promoting community wide car recycling scheme

– 469 – European Group of Automotive Recycling Association (EGARA)

PO Box 3281
NL-5203 DG Den Bosch
Netherlands
Tel: 31 73 424086
Fax: 31 73 426405
Contact name: Peter Goedbloed
Geographical coverage: European
Aims and objectives: EGARA seeks to promote the interests of the scrap car industry at European Commission level

– 470 – Save-a-Can

Kingsgate House
536 Kings Road
London SW10 0TE
United Kingdom
Tel: 071 351 5208
Fax: 071 351 7676
Contact name: Nick Hadjinikos/Georgina Rodgers
Geographical coverage: United Kingdom

Aims and objectives: Save-a-Can's aims are to increase awareness of and expand the available facilities for the recycling of all types of used cans.

– 471 – Studiengesellschaft für die Entsorgung von Altahrrzeugen GmbH (EVA)

Umweltschutzbeauftragten
Voest Alpine Krems GmbH
Postfach 2
A-4031 Linz
Austria
Geographical coverage: Austria
Aims and objectives: EVA examines waste management policies for cars at the end of their life

PERIODICALS

– 472 – Automotive Engineering

400 Commonwealth Drive
Warrendale PA 15096
United States of America
Geographical coverage: International
Publishers: Society of Automotive Engineers
Frequency: monthly
Coverage: Car recycling
Annual subscription: $72
ISSN: 0098 2571

REPORTS

– 473 – Design for Recyclability

Geographical coverage: United Kingdom
Publishers: Institute of Metals on behalf of the Materials Forum, 1988
Coverage: This report probes changes in engineering design and changes in materials to improve recyclability
ISBN: 0901462462
Author: Henstick, M. E.

– 474 – Disposal of Vehicles: Issues and Actions

5 Berners Road
London N1 0PW
United Kingdom
Geographical coverage: United Kingdom
Publishers: UK Centre for the Exploitation of Science and Technology (CEST)
Price: £15.00

– 475 – Recycling Lead and Zinc: The Challenge of the 1990's

Geographical coverage: International
Publishers: International Lead and Zinc Study Group, 1991
Language: English
Coverage: This report covers the developments in the main aspects of recycling lead and zinc, together with assessments of expected trends in world supply and demand for lead and zinc during the 1990s
Price: £65.00

Nuclear Waste

GOVERNMENT ORGANISATIONS

– 476 – Service Central de Protection Contre les Rayonnements Ionisants

BP 35
F-78110 Le Vesinet
France
Tel: 33 1 397 604 32
Fax: 33 1 397 608 96
Contact name: Pierre Pellerin, Director
Geographical coverage: France
Aims and objectives: Government body of surveyors of pollution due to radioactive substances

NON-GOVERNMENT ORGANISATIONS

– 477 – Agence Nationale pour la Gestion des Déchets Radioactifs (ANDRA)

Route de Panorama R Schuman
BP 38
F-92266 Fontenaty-aux-Roses Cedex
France
Tel: 33 1 465 470 80
Fax: 33 1 465 499 26
Geographical coverage: France
Aims and objectives: ANDRA is the French national agency for radioactive waste management and disposal

– 478 – International Nuclear Power and Waste (INLA)

22 square de Meeus
B-1040 Brussels
Belgium
Tel: 32 2 513 68 45
Fax: 32 2 518 38 33
Contact name: Frédéric Vandenabeele, General Secretary
Geographical coverage: International
Aims and objectives: INLA carries out international research into the legal problems relating to the civil use of nuclear power

PRESSURE GROUPS

– 479 – Hazardous Export/Import Prevention Project Greenpeace International

Keizersgracht 176
NL-1016 DW Amsterdam
Netherlands
Tel: 31 20 523 6555
Fax: 31 20 523 6500
Geographical coverage: International

BOOKS

– 480 – Britain's Nuclear Waste: Safety/Sting

Geographical coverage: United Kingdom
Publishers: Belhaven Press, 1989
ISBN: 1852930055
Author: Openshaw, S., Carver, S. and Fernie, J.

– 481 – International Politics of Nuclear Waste, The

Geographical coverage: International
Publishers: Macmillan Press, 1990
Coverage: Looking at the politics of nuclear waste, this book examines the subject from an international standpoint
Price: £35.00
ISBN: 033349363X
Author: Blowers, A. and Lowry, D.

– 482 – Radioactive Aerosols

Geographical coverage: United Kingdom
Publishers: Cambridge University Press, 1991
Coverage: A study of radioactive gases and particles which are dispersed in the environment, either from natural causes or following nuclear tests and accidental emissions
Price: £50.00

ISBN: 0521401216
Author: Chamberlain, A. C.

– 483 – Radioactive Waste

Geographical coverage: European
Publishers: Routledge, 1991
Coverage: This analysis of nuclear strategy uses as case studies the present practices employed by Germany, Sweden and the UK
Price: £35.00
ISBN: 0415054923
Author: Berkhout, F.

– 484 – Radioactive Waste Management

Geographical coverage: United Kingdom
Publishers: Hemisphere Publishing Corporation, 1990
Coverage: Examines the entire spectrum of radioactive waste
Price: £61.00
ISBN: 0891166661
Author: Tang, Y. S. and Saling, J. H.

– 485 – Radioactive Waste Management

Geographical coverage: International
Publishers: British Nuclear Energy Society, 1989
Coverage: Disposal of radioactive waste
Price: £95.00
ISBN: 0727715259

PERIODICALS

– 486 – Waste Management

Headington Hill Hall
Oxford OX3 0BW
United Kingdom
Tel: 0865 794141
Fax: 0865 743952
Geographical coverage: United Kingdom
Publishers: Pergamon Press plc
Editor: Moghissi, A. A.
Frequency: monthly
Coverage: Nuclear and chemical waste
Annual subscription: £380.00

REPORTS

– 487 – Safety Assessment of Radioactive Waste Repositories: Systematic Approaches to Scenario Development

Geographical coverage: European
Publishers: OECD Publications, 1992

Language: English
Coverage: Analysis of the long term safety of radioactive waste
Price: £21; Ffr150; $38
ISBN: 9264136053

– 488 – Single European Dump, The: Free Trade in Hazardous and Nuclear Wastes in the New Europe

Greenpeace EC Unit
Avenue de Tervuren 36
B-1040 Brussels
Belgium
Tel: 32 2 736 99 27
Fax: 32 2 736 44 60
Geographical coverage: European
Publishers: Greenpeace, December 1991
Language: English
Author: Carroll, S.

– 489 – Working for Public Safety and Environmental Protection

Nuclear Policy Information Unit
Town Hall
Manchester M60 2LA
United Kingdom
Tel: 061 234 3244
Geographical coverage: United Kingdom
Publishers: Nuclear Free Local Authorities (NFLA)
Coverage: Review of safe nuclear waste management options for the future

RESEARCH CENTRES

– 490 – Deutsches Hydrographisches Institut

(German Hydrographic Institute)

Transport
Bernhard-Nocht-Strasse 78
D-2000 Hamburg 36
Germany
Tel: 49 40 31901
Fax: 49 40 31905150
Geographical coverage: Germany
Research activities: Monitoring seawater for radioactivity and other noxious substances

– 491 – Eidgenössische Kommission für die Sicherheit von Kernanlagen

(Federal Commission for the Safety of Nuclear Installations)

c/o KSA Sekretariat
CH-5303 Würenlingen
Switzerland
Tel: 41 56 99 39 43
Geographical coverage: Switzerland
Research activities: Licensing of nuclear power plants, nuclear installations, research reactors and radioactive waste disposal

– 492 – European Institute for Transuranium Elements

Postfach 2340
D-7500 Karlsruhe
Germany
Tel: 49 7247 841
Fax: 49 7247 4046
Geographical coverage: European
Publications: *Nuclear Fuels and Actinide Research*, biannual progress report; annual report
Research activities: Nuclear fuels and actinide research; reactor safety; radioactive waste management

– 493 – International Laboratory of Marine Radioactivity

2 avenue Prince Hereditaire Albert
MC-9800 Monaco
Monaco
Tel: 33 93 50 44 88
Fax: 33 93 25 73 46
Geographical coverage: International
Research activities: Occurrence and behaviour of radioactive substances and other forms of pollution in the marine environment

– 494 – Joint Research Centre Ispra Establishment

I-21020 Ispra (Varese)
Italy
Tel: 39 332 789111
Geographical coverage: Italy
Research activities: Reactor and Nuclear Plant Safety; treatment and storage of radioactive wastes

Packaging

NON-GOVERNMENT ORGANISATIONS

– 495 – Belgian Packaging Institute (BPI)

Rue Picard 15
B-1210 Brussels
Belgium
Tel: 32 2 427 2583/427 2593
Fax: 32 2 425 9975
Contact name: J V Maciels, General Services Manager
Geographical coverage: Belgium
Aims and objectives: The BPI was established in 1954 by the Government, Packaging and Transport Industries. It represents and promotes the interests of the Belgian packaging industry worldwide

– 496 – Comité Européen de Normalisation (CEN)

Central Secretariat
Rue de Stassart 36
B-1050 Brussels
Belgium
Geographical coverage: European
Aims and objectives: CEN has a working group looking at terminology, symbols and criteria for life cycle assessment of packaging

– 497 – European Packaging Federation

c/o CPC/Europe Consumer R & D Centre
CH-8240 Thayngen
Switzerland
Tel: 41 53 396666
Contact name: J Abspoel, Chairperson
Geographical coverage: European

– 498 – European Union for Packaging and the Environment

26 avenue Livingstone
Bte 3
B-1040 Brussels
Belgium
Tel: 32 2 231 12 99
Fax: 32 2 230 76 58
Geographical coverage: European

– 499 – Finnish Packaging Association

Ritarikartu 3 b A
SF-00170 Helsinki
Finland
Tel: 358 0 65 13 44
Fax: 358 0 66 68 99
Contact name: Jorma Hamaiainen, Managing Director
Geographical coverage: Finland

– 500 – Foundation for Packaging and the Environment (SNV)

Prinses Beatrixlaan 5
The Hague
Netherlands
Tel: 31 70 381 19091
Contact name: R Mulder, Director
Geographical coverage: Netherlands

– 501 – Fraunhofer-Institut für Lebensmitteltechnologie und Verpackung

(Food Technology and Packaging)

Schragenhofstrasse 35
D-8000 Munich 50
Germany
Tel: 49 89 149 0090
Fax: 49 89 149 00980
Geographical coverage: Germany

– 502 – Hungarian Institute for Materials Handling and Packaging

Rigo utca 3
PO Box 189
H-1431 Budapest
Hungary
Tel: 36 1 1137460
Fax: 36 1 1338170
Contact name: Gyozo Polhammer
Geographical coverage: Hungary
Aims and objectives: The Institute provides information and reference service, postgraduate courses, laboratory and quality tests

– 503 – Industry Council for Packaging and the Environment (INCPEN)

Premier House
10 Greycoat Place
London SW1P 1SB
United Kingdom
Tel: 071 222 8866
Fax: 071 976 7178
Contact name: K Banks, Director General
Geographical coverage: United Kingdom
Aims and objectives: INCPEN was formed to bring together all sectors of industry involved in the manufacture of packaging. Its aims are to further the protection of the environment in so far as it is affected by packaging and to promote the environmental benefits

Publications: *INCPEN Journal*, four times a year; *INCPEN Newsletter*, monthly; Annual report

– 504 – Institute of Handling Transport Packaging and Storage Systems (IMADOS)

Konevova 131
130 83 Prague 3
Czech Republic
Tel: 42 2 26 41 41
Fax: 42 2 235 29 93
Contact name: V Hejna, Director
Geographical coverage: Czech Republic
Aims and objectives: IMADOS is a research, development, consulting, testing and information centre

– 505 – Nederlands Verpakkingscentrum

(Netherlands Packaging Industry)

PO Box 164
NL-2800 AD Gouda
Netherlands
Tel: 31 1820 12411
Fax: 31 1820 12769
Contact name: G Schaap
Geographical coverage: Netherlands
Aims and objectives: Provides information and documentation, education and training and trade promotion on packaging

– 506 – Norske Emballasjeforening, Den

(Norwegian Packaging Association)

Sorkedalsveien 6
N-0369 Oslo
Norway
Tel: 47 2 296 5076
Fax: 47 2 260 0067
Contact name: Jan Bjerk, Managing Director
Geographical coverage: Norway

– 507 – Österreichisches Verpackungszentrum

Wiedner Haupstrasse 63
A-1045 Vienna
Austria
Tel: 43 1 50105 3044 3045
Fax: 43 1 50206 253
Contact name: Ing A Horlezeder
Geographical coverage: Austria
Aims and objectives: Trade association

– 508 – Packaging and Industrial Film Association (PIFA)

Premier House
15 Wheeler Gate
Nottingham NG1 2NN
United Kingdom
Tel: 0602 484525
Geographical coverage: United Kingdom
Aims and objectives: Trade Association

– 509 – World Packaging Organisation (WPO)

42 avenue de Versailles
F-75016 Paris
France
Tel: 33 1 428 829 74
Fax: 33 1 452 502 73
Contact name: Pierre J Louis, General Secretary
Geographical coverage: International
Aims and objectives: WPO is a unique centre for packaging improvement and facilitating the permanent exchange of technical and marketing information

BOOKS

– 510 – Packaging in the Environment

Geographical coverage: United Kingdom
Publishers: Blackie Academic and Professional, 1992
Coverage: Appraises the key environmental issues raised by the packaging industry
ISBN: 0751400912
Author: Levy, G. M.

– 511 – Verpackungsverordnung

(Packaging Ordinance A Practical Handbook)

Geographical coverage: Germany
Publishers: Behr's Verlag GmbH
Coverage: Objectives of the ordinance as well as solution strategies for industry
Author: Rummler, T. and Schutt, W.

DIRECTORIES/YEARBOOKS

– 512 – Packaging Industry Directory 1993

Geographical coverage: United Kingdom
Publishers: Benn Business Information Services Ltd
Frequency: annual
Coverage: Wide coverage of the packaging industry in the UK. Also lists associations worldwide
Price: £68.00
ISSN: 0269 9834
ISBN: 0863821626

PERIODICALS

– 513 – Packaging

Turrett House
171 High Street
Rickmansworth
Hertfordshire WD3 1SN
United Kingdom
Tel: 0923 777000
Fax: 0923 771297
Geographical coverage: United Kingdom
Publishers: Turrett Group plc
Editor: Shepherd, N.
Frequency: bi monthly
Coverage: All aspects of packaging
Price: $94
ISSN: 0030 9060

REPORTS

– 514 – Bates and Wacker Packaging Report 93

Bates and Wacker SC
9 rue du Moniteur
B-1000 Brussels
Belgium
Tel: 32 2 219 03 05
Geographical coverage: Belgium
Coverage: Packaging and the Environment

– 515 – German Chamber of Industry and Commerce: German Packaging Laws

Mecklenburg House
16 Buckingham Gate
London SW1E 6LB
United Kingdom
Tel: 071 233 5656
Geographical coverage: Germany

– 516 – Packaging in Europe: France Special Report no 2046

Geographical coverage: France
Publishers: The Economist Intelligence Unit, Business International
Language: English
Coverage: The market and the suppliers in France, in the 1990s
Author: Briant, Y.

– 517 – Packaging in Europe: Italy Special Report no 2049

Geographical coverage: Italy
Publishers: The Economist Intelligence Unit, Business International
Language: English
Coverage: The market and the suppliers in Italy, in the 1990s
Author: Reeves, T.

– 518 – Packaging in Europe: Spain and Portugal Special Report no R162

Geographical coverage: Spain
Publishers: The Economist Intelligence Unit, Business International
Language: English
Coverage: The market and the suppliers in Spain and Portugal in the 1990s
Author: Lewis, G.

– 519 – Packaging in Europe: West Germany Special Report no 2047

Geographical coverage: Germany
Publishers: The Economist Intelligence Unit, Business International
Language: English
Coverage: The market and the suppliers in West Germany in the 1990s
Author: Lushington, R.

– 520 – Packaging Waste and the Polluter Pays Principle: A Taxation Solution

CSERGE
University College
London
United Kingdom
Tel: 071 380 7037
Geographical coverage: United Kingdom
Publishers: Centre for Social and Economic Research on the Global Environment

– 521 – Returnable and Non-Returnable Packaging

Johannes Kepler University
A-4040 Linz
Austria
Tel: 43 732 2468 235
Fax: 43 732 2468 209
Geographical coverage: Austria
Publishers: Institute for Economics
Editor: Poll, Prof G. and Schneider, Prof F.
Coverage: Case Study of the economic and waste disposal implications of fruit juice packs

RESEARCH CENTRES

– 522 – Danish Packaging and Transportation Research Institute

Gregersensvej
PO Box 141
DK-2630 Taastrup
Denmark
Tel: 45 43 99 66 11
Fax: 45 43 71 37 98
Contact name: Kirsten Nielsen, Manager
Geographical coverage: Denmark
Research activities: To initiate and carry out research and development in the fields of packaging and transportation

– 523 – International Association of Packaging Research Institutes (IAPRI)

Flamingo Straat
B-V900 Ghent
Belgium
Contact name: Dr Franz Lox, Secretary General
Geographical coverage: International
Aims and objectives: IAPRI aims to promote the exchange and collaboration of R & D and testing between packaging scientists and technologists in all parts of the world

– 524 – Swedish Packaging Research Institute – PACKFORSK

PO Box 9
S-164 93 Kista
Sweden
Tel: 46 8 7525700
Fax: 46 8 751 3889
Contact name: Stig Bergstedt
Geographical coverage: Sweden
Research activities: Research that promotes development within the packaging and distribution areas

Paper and Pulp

GOVERNMENT ORGANISATIONS

– 525 – Association for the Promotion of Recycled Paper

63 avenue du Maréchal Douglas Haig
F-78000 Versailles
France
Tel: 33 1 396 333 69
Fax: 33 1 462 318 40
Geographical coverage: France

– 526 – Minelsbumprom

(Ministry of Timber, Pulp and Paper and Wood Processing)

GSP Telegrafnyy pereulok 1
Moscow 101934
Russia
Geographical coverage: Russia

– 527 – Tidy Britain Group, The (TBG)

The Pier
Wigan
Lancashire WN3 4EX
United Kingdom
Tel: 0942 824620
Fax: 0942 824778
Contact name: J Wheeler, Information Officer
Geographical coverage: United Kingdom

Aims and objectives: TBG is the government's official agency campaigning for litter reduction and environmental improvement.
Publications: *Clean Nineties Newsletter*; educational and promotional material; annual report

NON-GOVERNMENT ORGANISATIONS

– 528 – Asociación de Recuperadores de Papel y Cartón (REPACAR)

Avenida de Manzanares 212
8 A
E-28026 Madrid
Spain
Tel: 34 1 475 64 85
Fax: 34 1 475 64 85
Contact name: Ángel Merino, Secretary General
Geographical coverage: Spain
Aims and objectives: REPACAR is a paper and board recycling association

– 529 – Associação Portuguesa de Fabricantes de Papel e Cartão

Rua de S Nicolau
26-3
P-1100 Lisbon
Portugal

Tel: 351 1 87 94 84/5
Geographical coverage: Portugal

– 530 – Association of Netherlands Paper and Board Manufacturers (VNP)

Postbus 3009
NL-2002 DA Haarlem
Netherlands
Tel: 31 23 31 91 25
Geographical coverage: Netherlands
Aims and objectives: Trade association

– 531 – Associazione Italiana fra gli Industriali della Carta Cartoni e Paste per Carta

Via Giovanni da Procida 3
I-20149 Milan
Italy
Tel: 39 2 342008/39 2 342010
Geographical coverage: Italy
Aims and objectives: Trade association

– 532 – British Paper and Board Industry Federation (BPBIF)

Papermakers House
Rivenhall Road
Westlea
Swindon SN5 7BD
United Kingdom
Tel: 0793 886086
Fax: 0793 886182
Geographical coverage: United Kingdom
Aims and objectives: Trade association

– 533 – Confédération Européenne de l'Industrie des Pâtes Papiers et Cartons (CEPAC)

Rue Washington 40
Bte 7
B-1050 Brussels
Belgium
Tel: 32 2 649 67 09/649 67 25
Geographical coverage: European

– 534 – Confederation of European Paper Industries (CEPI)

306 avenue Louise
B-1050 Brussels
Belgium
Tel: 32 2 627 49 11
Fax: 32 2 646 81 37
Geographical coverage: European
Aims and objectives: CEPI provides information on the paper industry in Europe

– 535 – Conservation Papers Ltd

228 London Road
Reading RG6 1AH
United Kingdom
Tel: 0734 668611
Fax: 0734 351605
Contact name: Ian Green
Geographical coverage: United Kingdom
Aims and objectives: Marketing and Information service connected with the environment in the paper and computer industry
Publications: *Paper Round*; *Green Paper 92*; *Green Labelling Update*; *Talking Conservation Newsletter*

– 536 – Copacel

(French Federation of the Paper Board and Cellulose Industry)
154 boulevard Haussmann
F-75008 Paris
France
Tel: 33 1 456 287 07
Geographical coverage: France

– 537 – Environment and Technology Association for Paper Sacks (ETAPS)

Papermakers House
Rivenhall Road
Westlea
Swindon SN5 7BD
United Kingdom
Tel: 0793 886086
Fax: 0793 886182
Geographical coverage: United Kingdom
Aims and objectives: ETAPS is a trade association which aims to assist all sectors of the British paper and board industry

– 538 – Fachverband Schweizerischer Becherhersteller (FSBH)

Waldeggstrasse 22 B
CH-3800 Interlaken
Switzerland
Tel: 41 36 22 22 26
Fax: 41 36 22 40 41
Contact name: Dr Herman M Reber, Advokat
Geographical coverage: Switzerland
Aims and objectives: FSBH examines problems and opportunities associated with promotion of paper and plastic cups and studies ecological and environmental problems which may arise through their manufacture and sale

– 539 – Papir–es Nyomdaipari Muszaki Egyesulet

(Technical Association of the Paper and Printing Industry)

Kossuth Lajos ter 6–8
H-1055 Budapest V
Hungary
Tel: 36 1 321 748
Geographical coverage: Hungary
Aims and objectives: Trade association

– 540 – Papirindustriens Sentralforbund

(Norwegian Pulp and Paper Association)

Drammensveien 30
PO Box 2446
Solli
N-0202 Oslo 2
Norway
Tel: 47 2 554210
Geographical coverage: Norway
Aims and objectives: Trade Association

– 541 – Promysi Papiru a Celulozy Generalni Reditelstvi

Stepanska 3
112 77 Prague 1
Czech Republic
Tel: 42 2 2133
Geographical coverage: Czech Republic

– 542 – Pulp and Paper Information Centre

Papermakers House
Rivenhall Road
Westlea
Swindon SN5 7BE
United Kingdom
Tel: 0793 886086
Fax: 0793 886182
Geographical coverage: United Kingdom
Aims and objectives: Provides information on the paper industry worldwide

– 543 – Stowarzyszenie Inzynierow i Technikow Rezemyslu Papierniczego w Polsce

(Technical Association of the Polish Paper Industry)

Pac Komuny Paryskiej 5a
PL-90 950 Lodz
Poland
Tel: 48 42 243 65
Geographical coverage: Poland

– 544 – Swedish Pulp and Paper Association (SCPF)

Box 26210
S-100 71 Stockholm
Sweden
Tel: 46 8 789 2883
Geographical coverage: Sweden

– 545 – Verband der Schweizerischen Sellstoff-, Papier- und Kartonindustrie

Bergstrasse 110
Postfach 134
CH-8030 Zurich
Switzerland
Tel: 41 1 47 97 47
Geographical coverage: Switzerland
Aims and objectives: Trade association

– 546 – Verband Deutscher Papierfabriken (VDP)

Adenauerallee 55
D-5300 Bonn 1
Germany
Tel: 49 228 267 05–0
Fax: 49 228 267 05 62
Geographical coverage: Germany
Aims and objectives: Trade Association

– 547 – Verband Deutscher Windelservice Lücke GmbH

(German Nappy Service Association)

Westfalenstrasse 98
D-4300 Essen 1
Germany
Tel: 49 201 51 11 11
Geographical coverage: Germany

– 548 – Vereinigung Österreichischer Papierindustrierhertsteller

Gumpendorfer Strasse 6
A-1061 Vienna
Austria
Tel: 43 1 58 886 0
Geographical coverage: Austria
Aims and objectives: Trade association

BOOKS

– 549 – Environmental Issues in the Pulp and Paper Industries

Geographical coverage: United Kingdom

Publishers: Pira, 1991
Coverage: Review environmental aspects of forestry, pulping, paper making, bleaching and waste treatment practices
Price: £60.00
ISBN: 0902799606
Author: Kirkpatrick, N.

– 550 – Recycled Papers: The Essential Guide

Geographical coverage: United Kingdom
Publishers: The MIT Press, 1992
Coverage: A guide to recycled printing and writing papers
Price: £35.95
ISBN: 0262200899
Author: Thompson, C.

PERIODICALS

– 551 – Paper and Packaging Analyst

Geographical coverage: United Kingdom
Publishers: The Economist Intelligence Unit, Business International, 1992
Frequency: quarterly
Coverage: Paper and packaging worldwide
ISSN: 0959 9266

– 552 – Pulp and Paper International

Ch de Charleroi 123a
Bte 5
B-1060 Brussels
Belgium
Tel: 32 2 538 60 40
Fax: 32 2 537 56 26
Geographical coverage: European
Frequency: monthly
Language: English
Coverage: All aspects of the pulp and paper industry

– 553 – Pulp and Paper Week

Miller Freeman Publications Inc
500 Howard Street
San Francisco
United States of America
Tel: 1 415 397 1881
Geographical coverage: International
Publishers: Miller Freeman Publications Inc
Frequency: weekly
Language: English
Coverage: Pulp and paper markets

– 554 – Talking Conservation

Geographical coverage: United Kingdom
Publishers: Conservation Papers Ltd
Editor: Bird, H
Coverage: All aspects of paper conservation

REPORTS

– 555 – Eco-Labelling of Paper Products

Miljoministeriet Miljostyrelsen
Strandgade 29
DK-1401 Copenhagen
Denmark
Tel: 45 31 57 83 10
Fax: 45 31 57 24 49
Geographical coverage: Denmark
Publishers: Danish Environmental Protection Agency
Coverage: Guidelines for the awarding of eco-labels for paper products
Price: Dkr95.00

– 556 – Environmental Impact of Paper Recycling

University of Oxford
Dept of Nuclear Physics
Keble Road
Oxford OX1 3RH
United Kingdom
Geographical coverage: United Kingdom
Editor: Koay, J
Price: £5.00

RESEARCH CENTRES

– 557 – Asociación de Investigación Técnica de la Industria Papelera Española

(Spanish Paper Industry Research Association)
PO Box 33045
Carretera de la Coruna km 7
E-28080 Madrid
Spain
Tel: 34 1 207 09 77
Fax: 34 1 357 28 28
Geographical coverage: Spain
Publications: *Investigación y Técnica del Papel*
Research activities: Materials for pulp and paper manufacture

– 558 – Instytut Celulozowo-Papierniczy

(Pulp and Paper Research Institute)
Ulica Gdanska 121
PL-90 950 Lodz
Poland
Tel: 48 42 365300
Geographical coverage: Poland
Aims and objectives: Offical research centre

– 559 – Oy Keskuslaboratorio – Centrallaboratorium Ab

(Finnish Pulp and Paper Research Institute)

PO Box 136
SF-00101 Helsinki 10
Finland
Tel: 358 1 90 43 711
Fax: 358 1 90 464 305
Geographical coverage: Finland
Publications: Annual report
Research activities: Technical research in the pulp, paper and board industry; packaging and printing and the protection of the environment

– 560 – Packaging Industry Research Association (Pira)

Randalls Road
Leatherhead
Surrey KT22 7RU
United Kingdom
Tel: 0372 376161
Fax: 0372 360104
Contact name: Brian Blunden, Managing Director
Geographical coverage: International
Aims and objectives: Pira International is the leading independent centre for research, consultancy, training and information services for the paper, packaging, printing and publishing industries
Publications: Various publications
Research activities: Include improving the recyclability of papers, solid waste management, life cycle analysis and environmental legislation

– 561 – Papiertechnische Stiftung für Papierezeugung und Papierverarbeitung

(Paper Technology Foundation for Paper Production and Conversion)

Hess-Strasse 130a
D-8000 Munich 40
Germany
Tel: 49 89 126001
Fax: 49 89 1236592
Geographical coverage: Germany
Publications: Research reports
Research activities: Environmental protection for the chemical pulp, mechanical pulp, waste paper pulp, manufacturing industry

– 562 – Papirindustriens Forskningsinstitutt

(Norwegian Pulp and Paper Research Institute)

Forskningsveien 3
Postboks 115
Vinderen
N-0319 Oslo 3
Norway
Tel: 47 2 140090
Fax: 47 2 468014
Geographical coverage: Norway
Research activities: Paper and pulp research

– 563 – Tampere University of Technology

PO Box 527
SF-33101 Tampere
Finland
Tel: 358 31 163 852
Geographical coverage: Finland

DATABASES

– 564 – Paperchem

Geographical coverage: International
Coverage: All aspects of paper and pulp industries
Database producer: Institute of Paper Science and Technology
Database host: Dialog
Coverage period: 1967–
Update frequency: monthly

– 565 – Pira Abstracts

Geographical coverage: International
Coverage: Recycling of packaging wastes
Database producer: Research Association for the Paper and Board industry
Database host: Orbit; Pergamon
Coverage period: 1975–
Update frequency: 500 records fortnightly

Plastics

NON-GOVERNMENT ORGANISATIONS

– 566 – Association des Transformateurs de Matières Plastiques (Fechiplast): Fédération des Industries Chimiques

Square Marie-Louise 49
B-1040 Brussels
Belgium
Tel: 32 2 238 97 11
Fax: 32 2 231 13 01
Contact name: G Scheys, Secretariat
Geographical coverage: Belgium
Aims and objectives: Fechiplast is an association of plastic processors which promote recycling

– 567 – Association of Plastic Manufacturers in Europe (APME)

Avenue E Van Nieuwenhuyse 4
Box 3
B-1160 Brussels
Belgium
Tel: 32 2 675 32 97
Fax: 32 2 675 39 35
Contact name: Nancy Russotto, Director
Geographical coverage: European
Aims and objectives: In 1990, APME set up the European Centre for Plastics in the Environment (PWMI) to act as the European centre for all matters relating to plastic waste

– 568 – British Plastics Federation (BPF)

5 Belgrave Square
London SW1X 8PD
United Kingdom
Tel: 071 235 9483
Fax: 071 235 8045
Contact name: T W Moffitt, Environmental Project Manager
Geographical coverage: United Kingdom
Aims and objectives: BPl promotes the use of plastics and emphasises their energy savings over other materials
Publications: Annual report; series of fact sheets

– 569 – British Plastics Reclamation Centre

c/o Sheffield Central Supplies Organisation
Staniforth Road
Sheffield S9 3HD
United Kingdom
Tel: 0742 444187
Geographical coverage: United Kingdom
Aims and objectives: In 1989, BPF joined in the first Recycling City project in Sheffield, which is part of a national initiative coordinated by FoE. The BPF's objective was to innovate and initiate cost effective methods of collecting and sorting plastics for recycling
Publications: Fact Sheets available

– 570 – Fachverband Schweizerischer Becherhersteller (FSBH): *see entry 538*

Geographical coverage: Switzerland

– 571 – RECOUP

9 Metro Centre
Welbeck Way
Shrewsbury Avenue
Peterborough PE2 7WH
United Kingdom
Tel: 0733 390021
Fax: 0733 390031
Geographical coverage: United Kingdom
Aims and objectives: Promotes the recycling of plastics

– 572 – Spanish Plastics Foundation for the Protection of the Environment (SPFPE)

c/o BASF Española
Spain
Tel: 34 1 562 8641
Geographical coverage: Spain

BOOKS

– 573 – Biodegradable Polymers and Plastics

Geographical coverage: United Kingdom
Publishers: Royal Society of Chemistry, 1992
Coverage: An exploration of the environmental problems caused by durable polymer materials
Price: £45.00
ISBN: 0851862071
Author: Vert, M. and Feijen, J.

CONFERENCE PAPERS

– 574 – Plastics Recycling 1991

Scandinavia Section
Mikhel Bryggers Gade 10
DK-1460 Denmark
Denmark
Geographical coverage: European
Publishers: Society of Plastic Engineers, 1991
Editor: Skov, H.
Coverage: Twenty five papers from a conference involving twelve European countries covering different aspects of recycling of plastics

DIRECTORIES/YEARBOOKS

– 575 – European Index of Key Plastics Recycling Schemes

Avenue E Van Nieuwenhuyse 4
B-1160 Brussels
Belgium
Tel: 32 2 675 32 58
Fax: 32 2 675 40 02
Geographical coverage: European
Publishers: APME
Language: English
Coverage: Detailed information of about 100 plastics reclamation schemes in 14 European countries

– 576 – Guide de l'Industrie Belge du Recyclage des Matières Plastiques 1991

(Guide to the Belgian Plastic Recycling Industry)

Geographical coverage: Belgium
Publishers: Fechiplast
Language: English, Dutch and French
Coverage: Lists associations and organisations in the Belgian packaging industry

PERIODICALS

– 577 – Plastics Technology

633 Third Avenue
New York
NY 10017
United States of America
Tel: 1 212 986 4800
Geographical coverage: International
Publishers: Bill Communications Inc

– 578 – Reinforced Plastics

Crown House
Linton Road
Barking
Essex IG11 8JU
United Kingdom
Tel: 081 594 7272
Fax: 081 594 5942
Geographical coverage: United Kingdom
Publishers: Elsevier Science Publishers Ltd
Frequency: 11 issues a year
Coverage: All aspects concerning plastics
Annual subscription: £59.00
ISSN: 0034 3617

REPORTS

– 579 – Management of Waste Plastic Packaging Films

65 rue de Prony
F-75854 Paris Cedex 17
France
Geographical coverage: European
Publishers: Plasteurofilm
Coverage: A study on the valorisation of waste in the light of proposals in EC Draft Packaging Directive
Price: £10.00

– 580 – Plastics Recovery in Perspective

Geographical coverage: European
Publishers: European Centre for Plastics in the Environment (PWMI)
Language: English
Coverage: Information and statistical data on recycling and recovery of plastics waste in Western Europe

– 581 – Progress in the Recycling of Plastics in the UK

Geographical coverage: United Kingdom
Publishers: British Plastics Federation, December 1991
Coverage: The development of plastic wastes recycling in the UK
Author: Moffitt, T. W.

– 582 – Recyclage des Matières Plastiques, Le

24 rue du Quatre Septembre
F-75002 Paris
France
Geographical coverage: France
Publishers: Centre de Prospective de d'Etudes Ministère de la Recherche et de la Technologie

– 583 – Technical Innovation in the Plastics Industry and its Influence on the Environmental Problem of Plastic Waste

Publications Sales
CEC
2 rue Mercier
L-2985 Luxembourg
Luxembourg
Geographical coverage: European
Publishers: Commission of the European Communities
Coverage: Case Study; evaluation of eco balances
Price: Ecu22.50

– 584 – Worldwide Trends and Opportunities in Automotive Passenger Vehicles 1995–2005

Dick Mann Associates
Henegar House
St Andrews Road
Coventry CV5 6FP
United Kingdom
Tel: 0203 712424
Fax: 0203 714108
Geographical coverage: International
Publishers: MIC
Coverage: Polymer trends in autos
Price: $12,500 for 3

RESEARCH CENTRES

– 585 – Instituto de Plásticos y Caucho
(Plastics and Rubber Institute)

Juan de la Cierva 3
E-28006 Madrid
Spain
Tel: 34 1 262 29 00
Geographical coverage: Spain
Research activities: Plastics recycling

DATABASES

– 586 – RAPRA Technology

Geographical coverage: International
Coverage: Rubber and plastics industry and related aspects such as land use and pollution
Database producer: RAPRA Technology Ltd
Database host: Orbit; Pergamon
Coverage period: 1972–
Update frequency: fortnightly

Recycling

NON-GOVERNMENT ORGANISATIONS

– 587 – Agence Nationale de la Récupération et l'Elimination des Déchets (ANRED)

2 square Lafayette
BP 406
F-49004 Angers Cedex
France
Tel: 33 4188 9825
Contact name: Press and Information Officer
Geographical coverage: France
Aims and objectives: ANRED is the offical reclamation body in France
Additional information: Scarabee Database

– 588 – British Scrap Federation

16 High Street
Brampton
Huntingdon
Cambridgeshire PE18 8TU
United Kingdom
Tel: 0480 455249
Fax: 0480 453680
Contact name: J A Clubb, Executive Director
Geographical coverage: United Kingdom
Aims and objectives: The Federation advises on health and safety and new Government legislation
Publications: *News Review Newsletter*

– 589 – Bureau International de la Récupération (BIR)
(International Federation of Recovery and Recycling)

Rue du Lombard 24
Bte 14
B-1000 Brussels
Belgium

Tel: 32 2 514 21 80
Fax: 32 2 514 12 26
Contact name: Francis Veys, Secretary General
Geographical coverage: International
Aims and objectives: Created in 1948, the BIR is the world federation of traders and industrialists which are involved in and which promote the interests of the recovery and recycling industry on an international level
Publications: Annual Report

– 590 – Civil Forsvaret

Aarhus
Denmark
Tel: 45 86 78 53 44
Geographical coverage: Denmark

– 591 – Confédération Belge de la Récupération

Place du Samedi 13
PO Box 5-6
B-1000 Brussels
Belgium
Tel: 32 2 217 99 93
Geographical coverage: Belgium
Aims and objectives: Recycling Association for non-ferrous metals, paper and textiles

– 592 – Duales System Deutschland GmbH (DSD)

Rochusstrasse 2–6
D-5300 Bonn 1
Germany
Tel: 49 2 28 979 2255
Fax: 49 2 28 979 2195
Contact name: Petra Rob, Director of Communication Department
Geographical coverage: Germany
Aims and objectives: DSD aims to solve the problem of packaging waste by collecting packaging materials for recycling using economically viable methods

– 593 – European Recovery and Recycling Association (ERRA)

83 avenue E Mountier
Box 14
B-1200 Brussels
Belgium
Tel: 32 2 772 52 52
Fax: 32 2 772 54 19
Contact name: Jacques Fonteyne, Managing Director
Geographical coverage: European
Aims and objectives: The ERRA aims to demonstrate whether comprehensive multi material household waste recovery and recycling schemes can be technically, financially and environmentally viable

– 594 – Industrielle de l'Environnement et du Recyclage (FEDEREC)

101 rue de Prony
F-75017 Paris
France
Tel: 33 1 405 401 94
Fax: 33 1 405 477 88
Contact name: Information Officer
Geographical coverage: France

– 595 – Industry Council for Electronic Recycling (ICER)

6 Bath Place
Rivington Street
London EC2A 3JE
United Kingdom
Tel: 071 729 4766
Contact name: Clare Snow
Geographical coverage: United Kingdom
Aims and objectives: ICER aims to promote recycling in the electronics industry

– 596 – Waste Watch

68 Grafton Way
London W1P 5LE
United Kingdom
Tel: 071 383 3320
Fax: 071 383 3364
Contact name: Jo Gordon
Geographical coverage: United Kingdom
Aims and objectives: Waste Watch is the national agency for the promotion and development of action on waste reduction and recycling
Publications: Produces many publications including a national directory of recycling information

– 597 – World Action for Recycling Materials and Energy from Waste: WARMER

83 Mount Ephraim
Tunbridge Wells
Kent TN4 8BS
United Kingdom
Tel: 0892 524626
Fax: 0892 525287
Contact name: Sue Dixon
Geographical coverage: United Kingdom
Aims and objectives: WARMER is a worldwide information service. It aims to encourage the recycling of materials and energy from post consumer waste
Publications: *Warmer Bulletin*, quarterly; information sheets and pamphlets

BOOKS

– 598 – Recycling: Energy from Community Waste

Geographical coverage: International
Publishers: The British Library, 1991
Editor: Grayson, L.
Coverage: A bibliography examining the literature of the last decade on recycling. There is a coverage of UK, USA and European sources
Price: £25.00 in UK
ISBN: 0712307818

– 599 – Recycling: New Materials for Community Waste

Geographical coverage: International
Publishers: The British Library, 1991
Editor: Grayson, L.
Coverage: A bibliography examining the literature of the last decade on recycling. There is a coverage of UK, USA and European sources
Price: £25.00 in UK
ISBN: 0712307788

– 600 – Reprocessing of Tyres and Rubber Wastes

Geographical coverage: United Kingdom
Publishers: Ellis Horwood, 1991
Coverage: Looks at the methods used for processing scrap tyres and production wastes from the rubber industry
Price: £45.00
ISBN: 0139329488X
Author: Makarov, V. Mi. and Drozdovskiy, V.

– 601 – Waste Not Want Not: The Production and Dumping of Toxic Waste

Geographical coverage: United Kingdom
Publishers: Earthscan Publications Ltd, 1992
Coverage: Campaigning book on the production and dumping of toxic waste in Britain and Ireland
Price: £9.95
ISBN: 185383095X
Author: Allen, R.

DIRECTORIES/YEARBOOKS

– 602 – Furniture Recycling Network

SOFA
Unit 3
Pilot House
41 King Street
Leicester LE1 6RN
United Kingdom
Geographical coverage: United Kingdom
Publishers: SOFA
Coverage: Directory of furniture recycling schemes in the UK
Price: £5.00

– 603 – National Directory of Recycling Information

Geographical coverage: United Kingdom
Publishers: Waste Watch, 1991
Frequency: annual
Coverage: Provides a county by county list of the recycling facilities run by local authorities and by voluntary groups
ISBN: 0719913268

– 604 – Salvo Demolition Directory 93/94

PO Box 1295
Bath BA1 3TJ
United Kingdom
Tel: 0225 445387
Geographical coverage: European
Publishers: Salvo
Coverage: Covers UK, Belgium and France extensively with a few entries for Netherlands, Italy, Australia and Canada

– 605 – UK Recycling Directory, The

Geographical coverage: United Kingdom
Publishers: Recycling and Resource Management Magazine
Coverage: Contains information on local authorities and personnel by country and by county
Price: £19.00

– 606 – Who is Who in Recycling Worldwide 1991

Trend Associates
PO Box 80
CH-1255 Veyrier Geneva
Switzerland
Geographical coverage: International
Publishers: Trend Associates, 1991
Frequency: annual
Language: English
Coverage: Directory of recycling companies, associations, distributors, processors, experts and consultants. Listed by country

PERIODICALS

– 607 – Deutsche Dependance

Postfach 1553
D-5407 Boppard am Rhein 1
Germany
Tel: 49 6742 3017
Fax: 49 6742 81358
Geographical coverage: Germany
Publishers: The Warmer Campaign
Editor: Polster, R.
Coverage: Recycling and waste management

– 608 – Materials Reclamation Weekly

PO Box 109
MacLaren House
Scarbrook Road
Croydon CR9 1QH
United Kingdom
Tel: 081 688 7788
Fax: 081 760 0473
Geographical coverage: United Kingdom
Publishers: EMAP MacLaren Publishers Ltd
Editor: Martin, I.
Frequency: weekly
Coverage: Reclamation of metals, paper and waste
Price: £1.00
Annual subscription: £49.50
ISSN: 0025 5386

– 609 – Recycling

Kasernenstrasse 67
Postfach 10 27 17
D-4000 Düsseldorf 1
Germany
Tel: 49 211 8870
Geographical coverage: Germany
Publishers: (Bundersverband der Deutschen Schrottwirtschaft) Handelsblatt GmbH
Editor: Willeke, R.
Frequency: 3 issues a year
Annual subscription: DM24
ISSN: 0174 1446

– 610 – Resource Recycling

Box 10540
Portland
Oregon 97210
United States of America
Fax: 1 503 227 36135
Geographical coverage: International
Publishers: Resource Recycling Inc
Frequency: monthly
ISSN: 0744 4710

– 611 – Resources Conservation and Recycling

Elsevier Science Publishers BV
PO Box 211
NL-1000 AE Amsterdam
Netherlands
Tel: 31 20 5803911
Fax: 31 20 5803598
Geographical coverage: Netherlands
Publishers: Elsevier Science Publishers BV
Frequency: quarterly
Language: English
Annual subscription: Dfl812
ISSN: 0921 3449

– 612 – Warmer Bulletin

83 Mount Ephraim
Tunbridge Wells
Kent TN4 8BS
United Kingdom
Tel: 0892 524626
Fax: 0892 525287
Geographical coverage: United Kingdom
Publishers: The Warmer Campaign
Editor: Thurgood, M
Frequency: quarterly
Coverage: Recycling of materials
Price: free

– 613 – Waste Disposal and Recycling Bulletin

Birmingham Central Library
Chamberlain Square
Birmingham B3 3HQ
United Kingdom
Tel: 021 235 4392
Geographical coverage: United Kingdom
Publishers: WDRB Science Technology and Management Section
Editor: Pratt, D.
Frequency: 6 issues a year
Coverage: The bulletin is a current awareness service for specialists in the field. Its purpose is to alert people to recent publications in English
Price: free

REPORTS

– 614 – Automotive Consortium on Recycling and Disposal Proposals

Forbes House
Halkin Street
London SW1X 7DS
United Kingdom
Geographical coverage: United Kingdom

Publishers: ACORD
Coverage: End of Life Vehicle Disposal concept
Price: free

– 615 – CFCs and Halons: Alternatives and the Scope for Recovery for Recycling and Destruction

Geographical coverage: United Kingdom
Publishers: HMSO
Price: £21.00

– 616 – Environmental Impact of Recycling, The

Geographical coverage: United Kingdom
Publishers: Warren Spring
Editor: Ogilvie.S. M.
Coverage: Attempts to quantify the environmental burdens created by the recycling process of various materials taken from the household waste stream
Price: £25.00

– 617 – Kirkgate Market: A Feasibility Study

PO Box 19
6–8 Great George Street
Leeds LS1 6TP
United Kingdom
Tel: 0532 438777
Geographical coverage: United Kingdom
Publishers: SWAP Recycling
Price: £6.95

– 618 – Realized and Projected Recycling Processes for Used Batteries

PO Box 5032
CH-3001 Berne
Switzerland
Tel: 41 31 922 13 33
Fax: 41 31 21 69 61
Geographical coverage: European
Publishers: EUROPILE/EUROBAT, June 1991
Language: English

– 619 – Recycling in Member Countries

Bremerholm 1
DK-1069 Copenhagen K
Denmark
Geographical coverage: European
Publishers: International Solid Wastes Association
Coverage: Detailed studies of recycling in Denmark, Italy, Norway, UK, Spain and Sweden
Price: Dkr 45.00

– 620 – Recycling: Public Attitudes and Market Reality

Mintel International Group
18-19 Long Lane
London EC1A 9HE
United Kingdom
Geographical coverage: United Kingdom
Publishers: Mintel, 1992
Price: £895.00

DATABASES

– 621 – Environmental Bibliography

Geographical coverage: International
Coverage: Pollution; waste management; health hazards
Database producer: Environmental Studies Institute, International Academy, Santa Barbara
Database host: Dialog; Mead Data Central
Coverage period: 1973–
Update frequency: 4,000 records every two months
Additional information: Available on CD–ROM

– 622 – Scarabee

Geographical coverage: International
Coverage: Waste; including management, treatment and recycling of domestic, agricultural and industrial waste
Database producer: Agence National de la Récuperation et l'Elimination des Déchets
Database host: ANRED (direct from producer)
Coverage period: 1975–
Update frequency: weekly

– 623 – Wasteinfo

Geographical coverage: International
Coverage: Non-radioactive waste management, including disposal, treatment, recycling and environmental impact
Database producer: Waste Management Information Bureau, UKAEA
Database host: Orbit
Coverage period: 1973–
Update frequency: monthly

Textiles

NON-GOVERNMENT ORGANISATIONS

– 624 – Association of Reclaimed Textile Processors (ARTP)

60 Toller Lane
Bradford BD8 9BZ
United Kingdom
Tel: 0274 491241
Contact name: R Charnley, Secretary
Geographical coverage: United Kingdom

– 625 – Biotecno SA

Via Collina 9
CH-6962 Lugano
Switzerland
Geographical coverage: Switzerland

– 626 – Cotton Council International (CCI)

239 Old Marylebone Road
London NW1 5QT
United Kingdom
Tel: 071 402 0029
Fax: 071 724 8979
Geographical coverage: International

– 627 – Ecological and Toxicological Association of the Dyestuff Manufacturing Industry (ETAD)

PO Box CH-4005 Basle 5
Switzerland
Tel: 41 61 681 22 30
Fax: 41 61 691 42 78
Geographical coverage: International
Aims and objectives: ETAD helps to coordinate the industry's efforts to minimise impact of organic colourants on health and environment
Publications: Annual report; various booklets

– 628 – Evergreen Recycled Fashions

Albert Mills
Bradford Road
Batley Carr
Dewsbury
West Yorkshire WF13 2HE
United Kingdom
Tel: 0924 453419
Geographical coverage: United Kingdom

– 629 – International Cleaner Production Information Clearing House (ICPIC)

Tour Mirabeau
39–43 quai André Citroen
F-75739 Paris Cedex 15
France
Geographical coverage: International

– 630 – International Wool Secretariat (IWS)

6–7 Carlton Gardens
London SW1Y 5AE
United Kingdom
Tel: 071 930 7300
Fax: 071 930 8884
Geographical coverage: International

– 631 – Ozono Elettronica Internazionale

Via G Pelizza da Volpedo 55
I-20092 Cinisello Balsamo
Milan
Italy
Geographical coverage: Italy

– 632 – Society of Dyers and Colourists

PO Box 244
Perkin House
Gratton Road
Bradford BD1 2JB
United Kingdom
Tel: 0274 7215138
Fax: 0274 392888
Geographical coverage: United Kingdom

– 633 – Textile Institute, The

International Headquarters
10 Blackfriars Street
Manchester M3 5DR
United Kingdom
Tel: 061 834 8457
Fax: 061 835 3087
Geographical coverage: International
Aims and objectives: The Institute represents the interests of its members. Offers an information service and organises conferences and exhibitions
Publications: Various

CONFERENCE PAPERS

– 634 – Environmental Issues in Wool Processing: Are Textiles Finishing the Environment?

Textiles Institute
10 Blackfriars Street
Manchester M3 5DR
United Kingdom
Tel: 061 834 8457
Geographical coverage: International
Publishers: Textiles Institute, 1990
Author: Shaw, T.

PERIODICALS

– 635 – Society of Dyers and Colourists

PO Box 244
Perkin House
Gratton Road
Bradford BD1 2JB
United Kingdom
Tel: 0274 7215138
Fax: 0274 392888
Geographical coverage: United Kingdom
Publishers: Society of Dyers and Colourists
Editor: Dinsdale, P.
Frequency: 10 issues a year
Coverage: Presents research and practical papers on coloration
Annual subscription: £120.00
ISSN: 0037 9859

REPORTS

– 636 – Textiles and the Environment Special Report no 2150

Geographical coverage: United Kingdom
Publishers: Economist Intelligence Unit
ISBN: 0850585473
Author: Watson, J.

RESEARCH CENTRES

– 637 – British Textile Technology Group

Wira House
West Park Ring Road
Leeds LS16 6QL
United Kingdom
Tel: 0532 781381
Geographical coverage: United Kingdom
Research activities: Energy audits; water economy and recovery; effluent treatment; environmental monitoring in textile industries

– 638 – Istituto di Ricerche e Sperimentazione Laniera 'O. Rivetti

(O. Rivetti Wool Industry Research and Experimental Institute)

Piazza Lamarmora 5
I-13051 Biella
Italy
Tel: 39 15 20490; 21655
Geographical coverage: Italy
Research activities: Energy saving; textile waste and water treatment

Waste Management

GOVERNMENT ORGANISATIONS

– 639 – Department of the Environment, Waste Management Division

Romney House
43 Marsham Street
London SW1P 3PY
United Kingdom
Tel: 071 276 8461
Geographical coverage: United Kingdom

NON-GOVERNMENT ORGANISATIONS

– 640 – Association of Waste Management and Entrepreneurs in Finland

Nuijamiestentie 7
SF-00400 Helsinki
Finland
Tel: 358 1 57 56 00
Fax: 358 1 57 14 05
Geographical coverage: Finland
Aims and objectives: Trade association

– 641 – Association Vincotte ASBL

125 rue de Rhode
B-1630 Linkebeek
Brabant
Belgium
Tel: 32 2 358 3580
Contact name: Mr M Warzee
Geographical coverage: Belgium
Aims and objectives: The Association carries out control measurements and studies on air and water pollutants, noise and solid wastes

– 642 – Commission for Municipal Solid Waste: Department of Environmental Studies

University of the Aegean
Lesvos
Greece
Contact name: C P Halvadakis, Associate Prof of Environmental Engineering
Geographical coverage: Greece

– 643 – European Federation of Waste Management (FEAD)

Avenue des Gaulois 19
B-1040 Brussels
Belgium
Tel: 32 2 732 32 13
Fax: 32 2 734 95 92
Contact name: Dieter Vogt, Secretary General
Geographical coverage: European
Aims and objectives: FEAD is the umbrella organisation of the national associations of private waste management companies. Its aims are to represent the members' interests and exchange information worldwide

– 644 – FABRIMETAL

Rue des Drapiers 21
B-1050 Brussels
Belgium
Tel: 32 2 510 23 11
Fax: 32 2 510 23 01
Geographical coverage: Belgium
Aims and objectives: Waste and water trade association

– 645 – Institut National de l'Environnement Industriel et des Risques (INERIS)

(National Institute for Industrial Environment and Hazards)

Parc Technologique ALATA BP N 2
F-60550 Verneuil-en-Halette
France
Tel: 33 1 445 566 77
Fax: 33 1 445 566 99
Contact name: Ms Christine Heuraux
Geographical coverage: France
Aims and objectives: INERIS focuses on all chemical pollutions and technical hazards except nuclear hazards

– 646 – Institute of Wastes Management (IWM)

3 Albion Place
Northampton NN1 1UD
United Kingdom
Tel: 0604 20426
Contact name: C J Murphy, Technical Secretary
Geographical coverage: United Kingdom
Aims and objectives: IWM aims to promote all scientific, technical and practical aspects of wastes management, including collection, storage, treatment and disposal with special emphasis on safeguarding the environment
Publications: List available on request

– 647 – International Organisation of Consumer Unions (IOCU)

Emmastraat 9
NL-2595 EG The Hague
Netherlands
Tel: 31 70 347 63 31
Fax: 31 70 383 49 76
Contact name: Marjolijn Peters, Coordinator Environmental Programmes
Geographical coverage: International
Aims and objectives: Established in 1960, IOCU aims to link the activities of consumer organisations and to promote the expansion of the consumer movement worldwide. Concerns include hazardous products and technologies, wastes and the environmental aspects of consumption

– 648 – International Solid Wastes and Public Cleansing Association (ISWA)

Bremerholm 1
DK-1069 Copenhagen K
Denmark
Tel: 45 33 91 44 91
Fax: 45 33 91 91 88
Contact name: Jeanne Moller, Secretary General
Geographical coverage: International
Aims and objectives: ISWA's objective is to promote the adoption of acceptable systems of solid wastes management, through technological development and improvement of practices, for the protection of the environment and conservation of materials and energy sources
Publications: *Waste Management and Research*, bi monthly; *ISWA Times*, quarterly; *Yearbook*, annual; Reports and Book series

– 649 – Istituto del Terziario per l'Ambiente

Corso Cavour 2
I-70123 Bari
Italy
Tel: 39 80 274111
Fax: 39 80 274228
Geographical coverage: Italy
Aims and objectives: An information exchange for industrial wastes, with the objective of increasing the volume of waste materials being diverted for re-use and recycling

– 650 – National Association of Waste Disposal Contractors (NAWDC)

Mountbarrow House
6–20 Elizabeth Street
London SW1 9RB
United Kingdom
Tel: 071 824 8882
Fax: 071 824 8753
Contact name: David Boyd or Steve Webb
Geographical coverage: United Kingdom
Aims and objectives: The Association's prime objectives are to represent the UK waste management industry and promote and encourage high standards of waste management
Publications: *NAWDC News*, six times a year; *NAWDC Newsletter*, six times a year; *NAWDC Directory*, annually; Annual Report

– 651 – Oslo Yann Og Avlop

Trondheimsveien 5
N-Oslo 5
Norway
Tel: 47 2 662020
Fax: 47 2 664083
Geographical coverage: Norway
Aims and objectives: Trade association for waste and water

– 652 – Photographic and Waste Management Association

Carolyn House
22–26 Dingwall Road
Croydon CR0 9XF
United Kingdom
Tel: 081 681 1680
Geographical coverage: United Kingdom
Aims and objectives: The Association aims to promote photographic waste recycling

– 653 – Secretaria Estado Ambiente Hidrologicos

A V Aimirante Gago Coutinha 30
P-1000 Lisbon
Portugal
Tel: 351 1 847 0080
Geographical coverage: Portugal
Aims and objectives: Trade association for waste and water

– 654 – UNEP/IEO Cleaner Production Working Group on Halogenated Solvents: Department of Environmental Technology

Gregersensvej
PO Box 141
DK-25630 Taastrup
Denmark
Contact name: Aage S Hillersborg
Geographical coverage: International
Aims and objectives: Strives to change current thinking about how and what products are produced to minimise emissions and wastes

– 655 – Waste Management Information Bureau (WMIB)

AEA Environment and Energy
B7.12 Harwell
Didcot
Oxfordshire OX11 0RA
United Kingdom
Tel: 0235 433442/432334
Fax: 0235 432854
Geographical coverage: United Kingdom
Aims and objectives: WMIB is the national referral centre on the management of non-radioactive wastes. Provides information and advice on all aspects of waste management
Additional information: WASTEINFO Database service

– 656 – Waste Watchers Deutschland eV

Grosse Elbstrasse 160
D-2000 Hamburg 50
Germany
Tel: 49 40 3891 400
Fax: 49 40 385089
Geographical coverage: Germany
Aims and objectives: Promotion of environmentally sound policies, the most environmentally benign products and the most environmentally responsible lifestyles

BOOKS

– 657 – 1,000 Terms in Solid Waste Management

Geographical coverage: International
Publishers: International Solid Wastes Association
Editor: Skitt, J.
Language: English, Italian, German, French and Spanish
Coverage: Comprehensive international glossary for waste management professionals
Price: Dkr150, £15.00

– 658 – Hazardous Materials: Sources of Information on their Transportation

Geographical coverage: United Kingdom
Publishers: The British Library, 1990
Coverage: Covering the years 1979–1990, this literature guide looks at the technical, social and legal aspects of transporting hazardous materials within Europe and the UK
Price: £25.00
ISBN: 0712307737
Author: Lees, N.

– 659 – Incineration and the Environment – A Source Book

Tel: 071 222 7899
Fax: 071 222 4557
Geographical coverage: United Kingdom
Publishers: Institution of Mechanical Engineers
Coverage: Has information on organisation, conferences, consultancies, training courses, reference books and technical reports
Price: £25.00

– 660 – UK Waste Law

Geographical coverage: European
Publishers: Sweet and Maxwell, 1992
Coverage: Provides an overview of waste management law in the EC and the UK
Price: £35.00
ISBN: 0421430109
Author: Bates, J. H.

– 661 – Urban Solid Waste Management

Geographical coverage: European
Publishers: IRIS, 1991, Italy
Editor: Pescod, M. B.
Price: $170

DIRECTORIES/YEARBOOKS

– 662 – Entsorgung 92

(Waste Management '92)

Friedhelm Merz-Verlag
Alberichstrasse 15
D-5300 Bonn
Germany
Geographical coverage: Germany
Publishers: Bundesverband der Deutschen Entsorgungswirtschaft
Frequency: annual
Coverage: Provides information for waste management professionals
Price: DM24.50

– 663 – POLMARK–The European Pollution Control and Waste Management Industry Directory

Geographical coverage: European
Publishers: ECOTEC Research and Consulting Ltd, 1992
Frequency: annual
Language: English
Coverage: Comprehensive directory of European pollution control – includes some market analysis information
Price: $425
ISBN: 0863545890

– 664 – Waste Management International: Vol 1 Directory of Manufacturers and Services

Geographical coverage: International
Publishers: John Wiley and Sons, 1993
Coverage: Describes available products and services in all areas of waste management worldwide
Price: £95.00
ISBN: 0471933287
Author: Gambrill, J.

– 665 – Waste Management Yearbook

Geographical coverage: United Kingdom
Publishers: Longman, 1992
Coverage: A guide to waste management
Price: £57.50
ISBN: 0582057310

PERIODICALS

– 666 – Biocycle

419 State Avenue
Emmaus
PA 18049
United States of America
Tel: 1 215 967 4135
Geographical coverage: International
Publishers: J G Press Inc
Editor: Goldstein, J.
Frequency: monthly
Coverage: Focuses on composting and recycling techniques, from collection and processing to materials and marketing
Price: $58
ISSN: 0276 5055

– 667 – Croners Waste Management

Croner House
London Road
Kingston upon Thames
Surrey KT2 6SR
United Kingdom
Tel: 081 547 3333
Fax: 081 547 2637
Geographical coverage: United Kingdom
Publishers: Croner Publications Ltd
Editor: Hand, C.
Frequency: quarterly
Coverage: Specific and practical direction to those dealing with particular kinds of waste on a daily basis
Price: £123.90
ISBN: 185524117X

– 668 – ENDS

Unit 24
Finsbury Business Centre
40 Bowling Green Lane
London EC1R 0NE
United Kingdom
Tel: 071 278 4745
Fax: 071 837 7612
Geographical coverage: United Kingdom
Publishers: Environmental Data Services Ltd
Editor: Mayer, M.
Frequency: monthly
Coverage: Environmental policy in the UK and EEC; pollution control, waste management and the profiles of key environmental organisations. This is a key environmental information resource
ISSN: 0260 1249

– 669 – Environmental Policy and Practice

52 Kings Road
Richmond
Surrey TW10 6EP
United Kingdom
Tel: 081 995 0877
Fax: 081 747 9663
Geographical coverage: United Kingdom
Publishers: EDD Publications
Editor: Aston, G.
Frequency: quarterly
Coverage: Pollution control and waste management environmental law
Price: £20.00
Annual subscription: £66.00

– 670 – Haznews

140 Battersea Park Road
London SW11 4NB
United Kingdom
Tel: 071 498 2511
Fax: 071 498 2343
Geographical coverage: International
Publishers: Profitastral Ltd
Editor: Coleman, D.
Frequency: monthly
Language: English
Coverage: Provides news and features on technology, research, legislation, company developments on hazardous waste management
Annual subscription: £176.00
ISSN: 0953 5357

– 671 – Info Déchets Environnement et Technique

21 rue d'Hauteville
F-75010 Paris
France
Tel: 33 1 424 632 32
Fax: 31 1 452 341 44
Geographical coverage: France
Publishers: Groupement d'Activités de Presse (GRAP)
Editor: Biery, G.
Frequency: monthly
Annual subscription: Ffr700
ISSN: 0241 7375

– 672 – Journal of Hazardous Materials

PO Box 211
NL-1000 AE Amsterdam
Netherlands
Tel: 31 20 5803911
Geographical coverage: International
Publishers: Elsevier Scientific Publishing Co
Frequency: quarterly
ISSN: 0304 3894

– 673 – Müllmagazin

Kurfürstenstrasse 14
D-1000 Berlin 30
Germany
Tel: 49 30 2616854
Fax: 49 30 2650366
Geographical coverage: Germany
Publishers: Institut für Ökologisches Recycling eV
Frequency: quarterly
Annual subscription: DM60
ISSN: 0934 3482

– 674 – Wasser Luft und Boden

Lise-Meitner Strasse 2
Postfach 2760
D-6500 Mainz 1
Germany
Tel: 49 6131 992 01
Fax: 49 6131 992 100
Geographical coverage: Germany
Publishers: Vereinigte Fachverlage GmbH
Frequency: 9 issues a year
Coverage: Technical journal concerned with industrial water economics, purification of air and waste disposal
Annual subscription: DM220
ISSN: 0938 8303

– 675 – Waste Management

9 Saxon Court
St Peters Gardens
Northampton NN1 I5X
United Kingdom
Tel: 0604 20426
Fax: 0604 21339
Geographical coverage: United Kingdom
Publishers: Institute of Waste Management
Editor: Porter, C.
Frequency: monthly
Coverage: Concerned with the collection and disposal of waste including refuse collection, transportation, disposal and street sweeping
Annual subscription: £46.00

– 676 – Waste Management and Research

24–28 Oval Road
London NW1 7DX
United Kingdom
Tel: 071 267 4466
Fax: 071 482 2293
Geographical coverage: International
Publishers: Academic Press Ltd
Editor: Rushbrook, P.
Frequency: bi monthly
Coverage: Presents papers on all aspects of solid waste management
Annual subscription: $228
ISSN: 0734 242X

REPORTS

– 677 – EC Waste Management Policy: Transforming from Waste Treatment to Waste Prevention

Aloys-Schulte Strasse 6
D-5300 Bonn 1
Germany
Geographical coverage: European
Publishers: Institut für Europäische Umweltpolitik eV
Price: free
Author: Webber, A.

– 678 – Energy from Waste State-of-the-Art Report

Bremerholm 1
DK-1069 Copenhagen K
Denmark
Geographical coverage: International
Publishers: International Solid Wastes Association
Language: English
Coverage: ISWA's working group on waste incineration provides detailed national surveys
Price: £40.00

– 679 – UK Waste Management Industry

Geographical coverage: United Kingdom
Publishers: Institute of Waste Management, January 1992
Editor: Holmes, J. R.
Coverage: Waste management industry in the UK
Price: £25.00
ISBN: 0902944258

RESEARCH CENTRES

– 680 – Bundesamt für Umweltschutz
(Federal Office of Environmental Protection)

Hallwylstrasse 4
CH-3003 Berne
Switzerland
Tel: 41 31 61 93 11
Geographical coverage: Switzerland
Research activities: Water protection; air pollution; noise control; soil pollution; control of chemicals and waste

– 681 – Centrum voor de Studie van Water Bodem en Lucht (Becewa Vzw)

271 Krijgslaan
B-9000 Ghent
Belgium
Tel: 32 91 22 77 59
Contact name: Mr Vercruysse
Geographical coverage: Belgium
Research activities: Soil, water and air pollution and waste management

DATABASES

– 682 – GeoRef

Geographical coverage: International
Coverage: Geology and earth sciences, including geophysics; energy; mineral resources; waste disposal; sewage and radioactive waste
Database producer: American Geological Institute
Database host: Dialog; Orbit; STN
Coverage period: 1933–
Update frequency: 7,000 records monthly
Additional information: Available on CD-ROM

– 683 – Major Hazard Incident Data Service (MHIDAS)

Geographical coverage: International
Coverage: Major Hazard Incident Data Service; including health and safety, all incidents involving hazardous materials in the UK and internationally
Database producer: Health & Safety Executive, IAEA
Database host: ESA–IRS
Coverage period: 1964–
Update frequency: quarterly

– 684 – Waste

Environmental Safety Centre
B7.12 Harwell
Didcot
Oxfordshire OX11 0RA
United Kingdom
Tel: 0235 433442
Fax: 0235 432854
Contact name: Diana Silver, Manager
Geographical coverage: International
Coverage: International database on waste management
Database producer: Waste Management Information Bureau
Database host: Orbit
Coverage period: 1968–

Wastewater

NON-GOVERNMENT ORGANISATIONS

– 685 – British Effluent and Water Association (BEWA)

5 Castle Street
High Wycombe
Buckinghamshire HP13 6RZ
United Kingdom
Tel: 0494 444544
Geographical coverage: United Kingdom
Aims and objectives: BEWA is the national association for British process contractors, manufacturers and suppliers of water and effluent treatment plant equipment and associated chemicals. Provides a focal point for government and other bodies in the water sector
Publications: List available on request

– 686 – Water Supply and Sewage Committee of Major Athens (EYDAP)

156 Oropou Street
GR-111 46 Galatsi
Greece
Geographical coverage: Greece

BOOKS

– 687 – Waste, Wastewater, Air Laws and Technology

Geographical coverage: Germany
Publishers: Bohman Druck und Verlag Gesellschaft mbH & Co KG
Language: German
Author: List, W. and Kuntscher, H.

– 688 – Water and the Environment

Geographical coverage: United Kingdom
Publishers: Ellis Horwood, 1992
Coverage: A volume of essays addressing key concerns in the area of water management, including coastal sewage discharges, groundwater abstraction, river and tidal engineering and others
Price: £55.00
ISBN: 0139506926
Author: Currie, J. C. and Pepper, A. T.

DIRECTORIES/YEARBOOKS

– 689 – European Water and Wastewater Industry

Geographical coverage: European
Publishers: Longman, 1990
Coverage: Contains information on the structure of the water supply and wastewater treatment practised throughout Western Europe
Price: £85.00
ISBN: 0582050804
Author: Sobey, T. and Hay, A. R.

– 690 – Water Supply and Wastewater Disposal International Almanac

Gooiland 11
NL-2716 BP Zoetermeer
Netherlands
Geographical coverage: International
Publishers: International Institute for Water Supply and Wastewater Disposal
Editor: Kepinski, A. and Kepinski, W. A. S.
Frequency: annual
Language: English, French and German
Price: $70
ISSN: 0169 2577

PERIODICALS

– 691 – Abwassertechnik (AWT)

Postfach 1460
D-6200 Wiesbaden
Germany
Tel: 49 611 791 0
Fax: 49 611 791 285
Geographical coverage: Germany
Publishers: Bauverlag GmbH
Frequency: 6 issues a year
Coverage: Devoted to treatment of sewerage, recycling and water pollution
Annual subscription: DM114
ISSN: 0932 3708

– 692 – Wasserwirtschaft

Pfizerstrasse 5–7
Postfach 10 60 11
D-7000 Stuttgart 1
Germany
Tel: 49 711 2191 332
Fax: 49 711 2191 350
Geographical coverage: Germany
Publishers: Franckh-Kosmos Verlags GmbH und Co
Editor: Marotz, G.
Frequency: monthly
Coverage: Features technology of water purification, groundwater protection and sewage treatment
Annual subscription: DM164.40
ISSN: 0043 0978

– 693 – Water and Waste Treatment

111 St James Road
Croydon
Surrey CR9 2TH
United Kingdom
Tel: 081 684 4082
Fax: 081 684 9729
Geographical coverage: United Kingdom
Publishers: Faversham House Group Ltd
Editor: Mitchell, V.
Frequency: monthly
Annual subscription: £53.00
ISSN: 0043 1133

– 694 – World Water

Thomas Telford House
1 Heron Quay
London E14 4JD
United Kingdom
Tel: 071 987 6999
Fax: 071 537 2443
Geographical coverage: International
Publishers: Thomas Telford Services Ltd
Editor: Opie, R.
Frequency: monthly
Language: English
Coverage: Covers international water and wastewater industry with emphasis on European issues
Annual subscription: £69.00
ISSN: 0140 9050

– 695 – Zeitschrift für Wasser und Abwasserforschung

(Journal for Water and Wastewater Research)

Postfach 10 11 61
D-6940 Weinheim
Germany
Tel: 49 6201 602 0
Fax: 49 6201 602328
Geographical coverage: Germany
Publishers: V C H Verlagsgesellschaft mbH
Frequency: 6 issues a year
Language: English and German
ISSN: 0044 3727

RESEARCH CENTRES

– 696 – British Textile Technology Group: see entry 637

Geographical coverage: United Kingdom

– 697 – Institut für Wasser- Boden- und Lufthygiene
(Institute for Water Soil and Air Hygiene)

Corrensplatz 1
Bundesgesundheitsamt Postfach
D-1000 Berlin 33
Germany
Tel: 49 30 83082313
Geographical coverage: Germany
Research activities: Drinking water quality and treatment, wastewater and environmental hygiene, water pollution control

– 698 – Institutul de Cercetäri pentru Pedologie si Agrochimie
(Soil Science and Agrochemistry Research Institute)

Belevardul Märästi 61
71331 Bucharest
Romania
Tel: 40 0 172180
Geographical coverage: Romania
Publications: *Anale ICPA*, annual
Research activities: Soil physics; soil pollution control

– 699 – Vandkvalitetsinstituttet (VKI)
(Water Quality Institute)

Agern Allee 11
DK-2970 Horsholm
Denmark
Tel: 45 42 86 52 11
Fax: 45 42 86 72 73
Geographical coverage: Denmark
Research activities: Soil and groundwater impact assessment; wastewater treatment technology; water quality modelling data processing

– 700 – Vesien ja ympäristöntutkimuslaitos
(Water and Environment Research Institute)

PO Box 436
SF-00101 Helsinki
Finland
Tel: 358 1 19 291
Geographical coverage: Finland
Publications: Scientific papers, instructions, recommendations and annual reports
Research activities: Water pollution control; water supply; wastewater treatment

---

DATABASES

– 701 – Water Resources Abstracts

Geographical coverage: International
Coverage: Water quality, pollution, waste treatment
Database producer: US Department of the Interior Geological Survey
Database host: Dialog
Coverage period: 1971–
Update frequency: every two months
Additional information: Available on CD–ROM

6
Conservation

General

GOVERNMENT ORGANISATIONS

– 702 – National Federation for the Defence of the Environment

PO Box 3046
F-24003 Périgueux
France
Tel: 33 53 08 29 01
Fax: 33 53 09 52 52
Geographical coverage: France
Aims and objectives: Protection of landscape and decrease of waste production

NON-GOVERNMENT ORGANISATIONS

– 703 – International Conservation Action Analysis (ICAN)

36 St Mary's Street
Wallingford
Oxfordshire OX10 0EU
United Kingdom
Geographical coverage: International

– 704 – Irish Environmental Conservation: Organisation for Youth (ECO)

Cope Street 10
Dublin 2
Eire
Tel: 353 1 679673
Geographical coverage: Ireland

PERIODICALS

– 705 – Environmental Conservation

PO Box 564
CH-1001 Lausanne
Switzerland
Tel: 41 21 207381
Fax: 41 21 235444
Geographical coverage: International
Publishers: Elsevier Sequoia SA
Editor: Polunin, N.
Frequency: quarterly
Language: English
Annual subscription: Sfr260
ISSN: 0376 8929

– 706 – Instituto Nacional de Investigaciónes Agrarias Comunicaciónes Serie: Recursos Naturales

Paseo de la Infanta Isabel 1
E-28071 Madrid
Spain
Tel: 34 1 441 31 93
Geographical coverage: Spain
Publishers: Centro de Publicaciones
Frequency: irregular
Language: Text in Spanish, summaries in English
Price: ptas175
ISSN: 0210 3338

– 707 – Instituto Nacional Para la Conservación de la Naturaleza Monografías

Gran via San Francisco 35
Madrid 5
Spain
Geographical coverage: Spain
Frequency: irregular

– 708 – Ochrona Przyrody

Państowe Wydawnictwo Naukowe
ul Miodowa 10
PL-00 251 Warsaw
Poland
Geographical coverage: Poland
Publishers: Polska Akademia Nauk Zaklad Ochrony Przyrody
Editor: Zarzycki, K.
Frequency: annual
Language: Text in Polish, summaries in English and French
Coverage: Conservation
Price: varies
ISSN: 0078 3250

– 709 – Präparator, Der

Postfach 250260
D-4630 Bochum 1
Germany
Tel: 49 2041 29716
Geographical coverage: Germany
Publishers: Verband Deutscher Praeparatoren eV
Editor: Eckardt, S.
Frequency: quarterly
Language: Text in English and German
Annual subscription: DM45
ISSN: 0032 6542

– 710 – Vida Silvestre

Paseo de la Infanta Isabel 1
E-28071 Madrid
Spain
Tel: 34 1 227 39 39
Geographical coverage: Spain
Publishers: Ministerio de Agricultura Pesca y Alimentación
Editor: Hernandez, R.
Price: ptas1500
ISSN: 0210 3605

Environmental Protection

GOVERNMENT ORGANISATIONS

– 711 – Bundesministerium für Umwelt Naturschutz und Reaktorsicherheit

Kennedyallee 10
D-5300 Bonn 1
Germany
Tel: 49 228 3052010
Contact name: Marlene Mühe, Public Relations
Geographical coverage: Germany

– 712 – Ministry of Agriculture Fisheries and Food: Conservation Policy and Environmental Protection Division

Nobel/Ergon House
17 Smith's Square
London SW1P 3HX
United Kingdom
Tel: 071 236 6563/5654
Geographical coverage: United Kingdom

– 713 – National Agency for Environmental Protection

Strandgade 29
DK-1401 Copenhagen K
Denmark
Tel: 45 31 57 83 10
Fax: 45 31 57 24 49
Geographical coverage: Denmark
Aims and objectives: The Agency administers many environmental acts, particularly the Environmental Protection Act, the Marine Environment Act, the Watercourse and Water Supplies Acts and others

– 714 – National Council for Environmental Protection

Piaţa Victoriei 1
Bucharest
Romania
Geographical coverage: Romania

– 715 – Office for the Protection of the Environment

Customs House
Dublin 1
Eire
Tel: 353 1 679 3377 ext 2477
Contact name: Minister for Environmental Protection
Geographical coverage: Ireland

– 716 – State Committee for Environmental Protection

11 Nezholanovoy ulitsa
Moscow
Russia
Geographical coverage: Russia

NON-GOVERNMENT ORGANISATIONS

– 717 – Austrian Environmental Protection Association

Abteilung für Sozialmedizin
Sonnenburgstrassse 16/1
A-6020 Innsbruck
Austria
Tel: 43 512 507 2500
Fax: 43 512 507 2491
Geographical coverage: Austria

– 718 – Cyprus Association for the Protection of the Environment

Chanteclare Building
PO Box 3810
Nicosia
Cyprus
Geographical coverage: Cyprus

– 719 – European Environmental Agency (EEA)

200 rue de la Loi
B-1049 Brussels
Belgium
Tel: 32 2 235 11 28
Fax: 32 2 235 01 44
Contact name: Philippe Bourdeau, Director
Geographical coverage: European
Aims and objectives: The Regulation creating the EEA was adopted by the Council in 1990. Its main objective is to provide technical and scientific support to the Community and its Member States in the area of environmental protection

– 720 – International Petroleum Industry Environmental Conservation Association (IPIECA)

1 College Hill
1st Floor
London EC4R 2RA
United Kingdom
Tel: 071 248 3447
Fax: 071 489 9067
Geographical coverage: International
Aims and objectives: IPIECA was formed in 1974, by BP, Exxon, Mobil and Shell, to be the point of contact between the international petroleum industry and UNEP.
Publications: *IPIECA Newsletter*, 3 times a year

– 721 – Lega per l'Ambiente
(League for the Environment)

Via Salaria 280
I-00199 Rome
Italy
Tel: 39 6 884 1552
Fax: 39 6 844 3504
Contact name: Ms Giovanna Melandri
Geographical coverage: Italy
Aims and objectives: The League works on both political and technical levels, mainly issues linked with energy, the chemical industry and waste
Publications: *Ambienta Italia 1991*; *Il Futuro del Sole – Atmosphera Clima e Uomo*; *La carbon diet*; *La Città Amica*

– 722 – Panellinio Kentro Oikologikon Erevnon (PAKOE)
(Panhellenic Centre of Environmental Studies)

GR-115 27 Athens
Greece
Tel: 30 1 7770198
Fax: 30 1 7752060
Contact name: Ms Maria Psaltakis
Geographical coverage: Greece
Aims and objectives: PAKOE focuses its activities on environmental protection

– 723 – Umweltbundesamt
(German Environmental Agency)

Bismarckplatz 1
D-1000 Berlin 33
Germany
Tel: 49 30 8030
Fax: 49 30 8903 2285
Geographical coverage: Germany
Aims and objectives: The Umweltbundesamt was established in 1974 as a scientific federal authority responsible for questions relating to environmental protection. It provides data and services in the field of environmental research

Publications: Publications of the Umweltbundesamt are also available in English

– 724 – Union of the Societies for the Protection of the Environment in Slovenia

Celovska 43
Ljubljana 61000
Slovenia
Geographical coverage: Slovenia

PRESSURE GROUPS

– 725 – Latvian Environment Protection Club (VAK)

Smilsu Str 12
226050 Riga
Latvia
Tel: 7 0132 229081
Fax: 7 0132 212917
Contact name: Eriks Leitis
Geographical coverage: Latvia

BOOKS

– 726 – Digest of Environmental Protection and Water Statistics

Geographical coverage: United Kingdom
Publishers: HMSO, 1992
Frequency: annual
Coverage: Provides an extensive set of statistics and describes broad environmental trends. This is a key statistical source on the UK environment
ISBN: 0117526347

– 727 – International Protection of the Environment

Geographical coverage: International
Publishers: Oceana Publications. 1990
Editor: Ruster, B. and Seama, B.
ISBN: 037910086X

– 728 – Taxation for Environmental Protection

Geographical coverage: European
Publishers: Quorum books, 1992
Coverage: This text examines the ways in which industrialised nations have used and are developing tax laws to help alleviate environmental problems
Price: £39.85
ISBN: 089930575X
Author: Gaines, S. E. and Westin, R. A.

– 729 – Umweltschutz in Deutschland

(Environmental Protection in Germany)
Geographical coverage: Germany
Publishers: Economica Verlag, Bonn
Language: German
Coverage: Gives an overview of the newly united Germany. It includes reports from a National Committee for the UN conference in Brazil, and provides a survey of all sectors of German environmental issues
Price: free
Author: Environment Ministry

PERIODICALS

– 730 – Environmental Protection Bulletin

Davis Building
165–171 Railway Terrace
Rugby
Warwickshire CV21 3HQ
United Kingdom
Tel: 0788 578214
Fax: 0788 560833
Geographical coverage: United Kingdom
Publishers: Institution of Chemical Engineers
Editor: Gardner, D.
Frequency: bi monthly
Coverage: An exchange of information and experience on environmental matters related to the chemical and process industries
Annual subscription: £105.00
ISSN: 0957 9052

– 731 – Milieu en Bedrijf

Louisalaan 485
B-1050 Brussels
Belgium
Tel: 32 2 723 11 11
Fax: 32 2 649 84 80
Geographical coverage: Belgium
Publishers: C E D Samson (Subsidiary of Woltens Belge nv)
Language: Flemish
Coverage: Publishes the latest laws and activities in environmental protection

– 732 – Problemy Bolshikh Gorodov

(Problems of Large Metropolitan Areas)

PR Serova 5
101958 Moscow
Russia
Tel: 7 095 921 67 05
Geographical coverage: Russia
Publishers: Moskovskii Gorodskoi Territorial'ny Tsentr Nauchno Tekhnicheskoi Informatsii
Editor: Mikhailova, I. P.

Language: Russian
Coverage: Covers urban environmental protection
Price: Rb10
ISSN: 0233 5816

– 733 – Protec

Via Tagliamento 29
I-00198 Rome
Italy
Tel: 39 6 8543603
Fax: 39 6 8440697
Geographical coverage: Italy
Publishers: Publi & Consult SpA
Frequency: 9 issues a year
Language: Italian text with summaries in English and Italian
Annual subscription: Lit65000
ISSN: 1120 1681

– 734 – Quercus

La Pedriza 1
E-28002 Madrid
Spain
Tel: 34 1 413 40 75
Fax: 34 1 519 21 94
Geographical coverage: Spain
Editor: Varillas, B.

– 735 – Umwelt

Heinrichstrasse 24
Postfach 10 10 54
D-4000 Dusseldorf 1
Germany
Tel: 49 211 61880 0
Fax: 49 211 6188 112
Geographical coverage: Germany
Publishers: Verein Deutscher Ingenieure
Editor: Firnhaber, H.
Frequency: 10 issues a year
Coverage: Specialists and engineers in charge of environmental protection in industry, administrative bodies, communal authorities, service companies, engineering consultancies and technical schools
Annual subscription: DM216
ISSN: 0041 6355

– 736 – Waterschapsbelangen

Johan van Oldenbarneveltlaan 5
PO Box 80 200
NL-2508 GE The Hague
Netherlands
Tel: 31 70 3519751
Fax: 31 70 3544642
Geographical coverage: Netherlands
Publishers: Unie van Waterschappen
Language: Dutch
Coverage: Covers current news and information concerning the waterboards. Features include environment protection
Annual subscription: Dfl102
ISSN: 0043 1486

REPORTS

– 737 – Environmental Protection in Germany

Geographical coverage: Germany
Publishers: Economica Verlag, 1992
Frequency: annual
Language: English
Coverage: National report of the Federal Republic of Germany for the United Nations conference
Price: free
Author: Federal Ministry for the Environment, Public Relations Division

– 738 – Swedish Environmental Protection Agency Report

PO Box 1302
S-17185 Solna
Sweden
Fax: 46 8 984513
Geographical coverage: Sweden
Publishers: Swedish Environmental Protection Agency
Frequency: irregular
Language: Text in Swedish, occasionally in English
Price: varies
ISSN: 0282 7298

– 739 – Working for Public Safety and Environmental Protection

Nuclear Policy Information Unit
Town Hall
Manchester M60 2LA
United Kingdom
Tel: 061 234 3244
Geographical coverage: United Kingdom
Publishers: Nuclear Free Local Authorities (NFLA)
Coverage: Review of safe nuclear waste management options for the future

RESEARCH CENTRES

– 740 – Bundesamt für Umweltschutz
(Federal Office of Environmental Protection)

Hallwylstrasse 4
CH-3003 Berne
Switzerland

Tel: 41 31 61 93 11
Geographical coverage: Switzerland
Research activities: Water protection; air pollution; noise control; soil pollution; control of chemicals and waste

– 741 – Bundesforschungs Anstalt für Fischerei
(Federal Research Centre for Fisheries)

Palmaille 9
D-2000 Hamburg 50
Germany
Tel: 49 40 389 05113
Geographical coverage: Germany
Research activities: All aspects of fisheries including marine pollution and environmental protection analysis

– 742 – Instytut Ksztaltowania Srodowiska
(Environmental Protection Institute, Laboratory of Environmental Monitoring)

Krucza 5/11
PL-00 548 Warsaw
Poland
Tel: 48 22 295263
Geographical coverage: Poland
Research activities: Nature conservation; ecology; atmospheric monitoring; soil conservation; monitoring equipment;

DATABASES

– 743 – Swedish Environment Research Index (SERIX)

Library and Documentation Section
Box 1302
S-17125 Solna
Sweden
Tel: 46 8 799 1000
Geographical coverage: Sweden
Coverage: Environmental research projects in Sweden
Database producer: Swedish National Environment Protection Board

Forests

GOVERNMENT ORGANISATIONS

– 744 – Bundesministerium für Land und Forstwirtschaft
(Federal Ministry for Agriculture and Forestry)

Stubenring 1
A-1011 Vienna
Austria
Geographical coverage: Austria

– 745 – Confederation of European Agriculture (CEA)

PO Box 87
CH-5200 Brugg
Switzerland
Tel: 41 56 413177
Fax: 41 56 41 71 3174
Contact name: Secretariat General
Geographical coverage: European
Aims and objectives: CEA represents the interests of European agriculture and forestry at international level

– 746 – Forestry Commission

231 Corstorphine Road
Edinburgh EH12 7AT
United Kingdom
Tel: 031 334 0303
Geographical coverage: United Kingdom
Aims and objectives: The Forestry Commission is the Government Department responsible for promoting the interests of forestry, the development of afforestation and the production and supply of timber and other forest products in Britain
Publications: *Forest Life*, four times a year; Annual report; List available on request

– 747 – Inspection Générale de l'Environnement et des Forêts

Avenue Albert 1 er 187
B-5000 Namur
Belgium
Tel: 32 81 24 66 11
Geographical coverage: Belgium

– 748 – Ministère de l'Environnement et des Eaux et Forêts

51 rue de Prague
L-2918 Luxembourg-Ville
Luxembourg
Tel: 352 478 870
Geographical coverage: Luxembourg

– 749 – Ministry of Agriculture and Forestry

78 rue de Varenne
F-75700 Paris
France
Tel: 33 1 495 549 55
Geographical coverage: France

– 750 – Ministry of Food Agriculture and Forestry

Rochusstrasse 1
Postfach 14 02 70
D-5300 Bonn 1
Germany
Tel: 49 228 5291
Fax: 49 228 5294262
Geographical coverage: Germany

NON-GOVERNMENT ORGANISATIONS

– 751 – European Youth Forest Action (EYFA)

PO Box 566
NL-6130 Sittard
Netherlands
Tel: 31 46 513045
Fax: 31 46 516460
Geographical coverage: European
Publications: *Green Tree News*

– 752 – Food and Agriculture Organisation of the United Nations (FAO)

Via delle Terme di Caracalla
I-00100 Rome
Italy
Tel: 39 6 57971
Fax: 39 6 57973152/5782610
Contact name: Edouard Saouma, Director General
Geographical coverage: International
Aims and objectives: FAO was established in 1945. The Organisation aims to increase production of agriculture, forestries and fisheries. It has an environment and energy programme. Environmental concerns are a major component of the organisation's field work
Publications: FAO publishes a number of yearbooks. It also publishes a quarterly *Plant Protection Bulletin*; *Food Outlook Monthly*

– 753 – Forests Forever Campaign (FFC)

4th Floor
Clareville House
26/27 Oxendon Street
London SW1Y 4EL
United Kingdom
Tel: 071 839 1891
Contact name: Michael James, Director
Geographical coverage: United Kingdom
Aims and objectives: FFC is an initiative of the Timber Trades Federation, with the aim of helping to safeguard the world's forests, future timber supplies and to promote the cause of wood
Publications: *FFC Campaign News*, three times a year; *Real Wood Guide*

– 754 – International Tropical Timber Organisation (ITTO)

Sangyo Boeki Centre Building
2 Yamashita-cho
Naka-ku
Yokohama 231
Tel: 81 45 671 7045
Fax: 81 45 671 7007
Geographical coverage: International
Aims and objectives: The Organisation was founded in 1985. It promotes research and development in forest management and wood use and encourages development of national policies aimed at sustainable utilisation of tropical forests and their genetic resources

– 755 – National Forest and Nature Agency

Slotsmarken 13
DK-2970 Horsholm
Denmark
Tel: 45 45 76 53 76
Fax: 45 45 76 54 77
Geographical coverage: Denmark

BOOKS

– 756 – Air Pollution's Toll on Forests and Crops

Geographical coverage: International
Publishers: Yale University Press, 1990
Coverage: Examines the effects of air pollution on forestry and crops
ISBN: 0300045697
Author: MacKenzie, J. J. and El-Ashry, M. T.

– 757 – Future Forest Resources of the Former European USSR

Geographical coverage: European
Publishers: Parthenon Publishing, 1992
Coverage: Presents the timber assessment of the IIASA Forest Study for the former European USSR
Price: £48.00
ISBN: 1850704252
Author: Nilsson, S. and Salinas, O.

– 758 – Future Forest Resources of Western and Eastern Europe

Geographical coverage: European
Publishers: Parthenon Publishing, 1991
Coverage: Presents the timber assessment of the IIASA Forest Study for Western and Eastern Europe, not including the Soviet Union
Price: £48.00
ISBN: 1850704244
Author: Nilsson, S. and Sallinas, O.

– 759 – Global Forests: Issues for Six Billion People

Geographical coverage: International
Publishers: McGraw Hill Book Company, 1992
Coverage: Evaluates issues linking forests to the biosphere and the management of forests and wildlands in the global context
Price: £31.95
ISBN: 0070357021
Author: Laarman, J. and Sedjo, R.

– 760 – Silvicultural Systems

Geographical coverage: European
Publishers: Clarendon Press, 1991
Coverage: This guide aims to provide foresters and land managers with the theoretical basis and practical applications of 20 different silvicultural systems
Price: £16.50
ISBN: 019854670X
Author: Matthews, J. D.

– 761 – Skogsskjötsel i en Utslippstid

(Forest Management in a time of emissions)

Geographical coverage: Norway
Publishers: Det Norske Skogselskap, 1992
Coverage: Describes the effects of air pollution on forests and what can be done to mitigate the effects
Price: Nkr80
Author: Braadland, B. and Rognerud, P. A.

– 762 – Skogsutsikter

Göterborgs Universitet
Medicinaregatan 20B
S-413 90 Gothenburg
Sweden
Geographical coverage: Sweden
Publishers: Institutionen för Miljövard, 1992
Language: Swedish
Coverage: Reviews state of European forests and the problems of air pollution
Author: Elvingson, P.

– 763 – Tropical Rain Forest

Geographical coverage: United Kingdom
Publishers: Facts on File, 1990
Coverage: Survey of the world's rain forests. The nature of the forests and the ecosystems within them are defined, as are the threats and consequence of extinction
Price: £19.95
ISBN: 0816019444
Author: Newman, A.

– 764 – Tropical Rain Forest Ecology

Geographical coverage: United Kingdom
Publishers: Blackie Academic, 1991
Coverage: Illustrates the importance of change and diversity in the rain forest
Price: £33.00
ISBN: 0216931479
Author: Mabberley, D. J.

– 765 – Who's Hand on the Chainsaw? UK Government Policy and the Tropical Rain Forests

Geographical coverage: United Kingdom
Publishers: Friends of the Earth (FoE), 1992
Price: £8.95
ISBN: 1857501705

DIRECTORIES/YEARBOOKS

– 766 – Green Book, The (Authority on Tractors Agricultural and Forestry)

Albany House
Hurst Street
Birmingham B5 4BD
United Kingdom
Tel: 021 622 2828
Fax: 021 622 5304
Geographical coverage: United Kingdom
Publishers: Guardian Communications Ltd
Editor: Catling, H.
Frequency: annual

Coverage: Unique reference yearbook specialist guide to agricultural forestry equipment
Price: £56.00

PERIODICALS

– 767 – Arboriculture Journal

PO Box 42
Bicester
Oxfordshire OX6 7NW
United Kingdom
Tel: 0869 320949
Geographical coverage: International
Publishers: AB Academic Publishers
Editor: Hall, T. H. R.
Frequency: quarterly
Coverage: Science of tree care and management worldwide
Annual subscription: £89.00
ISSN: 0307 1375

– 768 – Forest Damage in Europe – Environmental Fact Sheet No 1

Swedish NGO Secretariat on Acid Rain
Box 245
S-401 24 Gothenburg
Sweden
Tel: 46 31 153955
Fax: 46 31 150933
Geographical coverage: European

– 769 – Journal of Tropical Ecology

Edinburgh Building
Shaftesbury Road
Cambridge CB2 2RU
United Kingdom
Tel: 0223 312393
Fax: 0223 315052
Geographical coverage: United Kingdom
Publishers: Cambridge University Press
Frequency: quarterly
Coverage: Papers in the field of ecology of tropical regions original research and reviews
Annual subscription: $61
ISSN: 0266 4674

REPORTS

– 770 – Forests Foregone: The EC's Trade in Tropical Timbers and the Destruction of the Rainforests

Geographical coverage: European
Publishers: Friends of the Earth (FoE), 1993

Price: £13.95
ISBN: 1857502302

– 771 – Risk Assessment of Ecological Effects and Economic Impacts of Acidification on Forestry in Sweden, A

Lund Institute of Technology
Box 124
S-221 00 Lund
Sweden
Tel: 46 46 108274
Geographical coverage: Sweden
Publishers: Department of Chemical Engineering
Editor: Sverdrup, H. et al
Coverage: Acidification of Sweden's forest area

RESEARCH CENTRES

– 772 – Centre de Recerca Ecológica i Aplicacions Forestals

(Centre of Ecological Research and Forestry Application)

Universitat Autònoma de Barcelona (Ciencias) E-08193
Bellaterra
Spain
Tel: 34 3 692 02 00
Geographical coverage: Spain
Research activities: Land ecology and forest application, particularly concerning the Mediterranean forest

– 773 – Centro de Investigacões Florestais

(Forestry Research Centre)

Tapada das Necessidades
P-1300 Lisbon
Portugal
Tel: 351 1 664222
Geographical coverage: Portugal
Aims and objectives: Official research centre
Publications: *Publicacões Florestais, Estudos e Informação*

– 774 – Department of Environmental Forestry

Sveriges Lantbruksuniversitet
Box 7078
S-750 07 Uppsala
Sweden
Tel: 46 18 67 10 00
Geographical coverage: Sweden
Research activities: Ecological effects of silviculture on ground vegetation

– 775 – Erdészeti Tudományos Intézet
(Forest Research Institute)

Frankel Leo Utca 42–44
H-1023 Budapest
Hungary
Tel: 36 1 150 624
Geographical coverage: Hungary
Publications: Various publications
Research activities: Tree breeding; seedling production; afforestation; environmental protection

– 776 – Forstliche Bundesversuchsanstalt
(Federal Forest Research Institute)

Seckendorff-Gudent-Weg 8
A-1131 Vienna
Austria
Tel: 43 1 222 823638
Geographical coverage: Austria
Publications: Various publications
Research activities: Fertilisation; forest ecology; genetic variability

– 777 – Irish Forestry Board

1–3 Sidmonton Place
Bray
Co Wicklow
Eire
Tel: 353 1 867751
Fax: 353 1 868126
Geographical coverage: Ireland
Aims and objectives: The Board's concerns are forestry management, engineering and research services
Research activities: Environmental effects of forest operations; nutriton; cultivation; biomass

– 778 – Istituto per le Piante da Legno e l'Ambiente SpA
(Timber and Environment Institute)

Corso Casale 476
I-10132 Turin
Italy
Tel: 39 11 890044; 890515
Geographical coverage: Italy
Research activities: Improvement of timber production in accordance with the principles of land conservation and environment protection

– 779 – Metsäntutkimuslaitos
(Finnish Forest Research Institute)

Unioninkatu 40A
SF-00170 Helsinki
Finland
Tel: 358 1 66 14 01
Fax: 358 1 62 53 08
Geographical coverage: Finland
Publications: Various publications
Research activities: Monitoring of state of the forests and development of methods of preventing forest damage

– 780 – Norsk Institut for Skogforskning (NISK)
(Norwegain Forest Research Institute)

Hogskolevein 12
N-1432 As
Norway
Tel: 47 9 949660
Geographical coverage: Norway
Publications: Various
Research activities: Regeneration, yield, operations, economics, nature conservancy and public utilisation of forest areas

– 781 – Rannsóknarstöd Skógraektar Rikisins
(Iceland Forestry Research Station)

Mógilsá Mosfellsbaer
Iceland
Tel: 354 1 666014
Geographical coverage: Iceland
Research activities: Silvicultural problems

DATABASES

– 782 – Agris

Geographical coverage: International
Coverage: Covers all aspects of agriculture, particularly concerning global food supply. Also aquatic sciences and fisheries, soil sciences, forestry, pollution
Database producer: UN Food and Agriculture Association
Database host: Dialog; DIMDI; ESA-IRS
Coverage period: 1975–
Update frequency: 10,000 records monthly

– 783 – ELFIS

Geographical coverage: Germany
Coverage: Food agriculture and forestry
Database producer: German Information Systems on Food, Agriculture and Forestry
Database host: DIMDI
Coverage period: 1984–
Update frequency: quarterly

Habitats and Species

GOVERNMENT ORGANISATIONS

– 784 – Department of the Environment, Endangered Species Branch

Tollgate House
Houlton Street
Bristol BS2 9DJ
United Kingdom
Tel: 0272 218202
Geographical coverage: United Kingdom

NON-GOVERNMENT ORGANISATIONS

– 785 – Bureau of the Convention on Wetlands of International Importance Especially as Waterfowl Habitat: Ramsar Convention Bureau

Avenue du Mont Blanc
CH-1196 Gland
Switzerland
Tel: 41 22 64 91 14
Fax: 41 22 64 46 15
Geographical coverage: International
Aims and objectives: The convention signed at Ramsar provides the framework for international cooperation for the conservation of wetland habitats. Its objectives are to stem the loss of wetlands and ensure their conservation
Publications: Annual report; newsletter, proceedings and a Convention brochure

– 786 – Convention on International Trade in Endangered Species of Wild Fauna and Flora (CITES)

CITES Secretariat
Rue de Maupas
CP 78
CH-1000 Lausanne 9
Switzerland
Tel: 41 21 200081
Geographical coverage: International
Aims and objectives: CITES came into force in 1975 and now has over 100 member countries. These countries act by banning commercial trade in an agreed list of currently endangered species and by regulating and monitoring trade in others that might become endangered

– 787 – Countryside Commission: *see entry 900*

Geographical coverage: United Kingdom

– 788 – Earthwatch: *see entry 949*

Geographical coverage: International

– 789 – Earthwatch Europe: *see entry 950*

Geographical coverage: European

– 790 – European Federation of Animal Health (FEDESA)

Rue Defacqz 1
Box 8
B-1050 Brussels
Belgium
Tel: 32 2 537 21 25
Fax: 32 2 537 00 49
Contact name: Dr Johan Vanhemelrijk, Secretary General
Geographical coverage: European

– 791 – Faunistisch–Ökologische Arbeitsgemeinschaft

Zoologisches Institut Universität
Olshausenstrasse 40
D-2300 Kiel 1
Germany
Geographical coverage: Germany

– 792 – Hellenic Ornithological Society

PO Box 64057
GR-157 01 Athens
Greece
Geographical coverage: Greece

– 793 – Hessische Gesellschaft für Ornithologie und Naturschutz eV

Lindenstrasse 5
D-6363 Echzell
Germany
Geographical coverage: Germany

– 794 – International Council for Bird Preservation (ICBP)

32 Cambridge Road
Girton
Cambridge CB3 OPJ
United Kingdom
Tel: 0223 277318
Contact name: Melvyn Risebrow, Development Director
Geographical coverage: International

Aims and objectives: ICBP is dedicated to saving the world's birds and their habitats
Publications: *World Birdwatch Newletter*, four times a year; Annual Report and various technical publications

– 795 – International Dolphin Watch (IDW)

Parklands
North Ferriby
Humberside HU14 3ET
United Kingdom
Tel: 0482 643403
Fax: 0482 634914
Contact name: Dr H E Hobbs, Honorary Director
Geographical coverage: International
Aims and objectives: IDW is a non-profit organisation for the study and conservation of dolphins
Publications: *Dolphin Newsletter*, three to four times a year; annual report

– 796 – International Whaling Commision (IWC)

The Red House
Station Road
Histon
Cambridge CB4 4NP
United Kingdom
Tel: 0223 233971
Fax: 0223 232876
Geographical coverage: International
Aims and objectives: The objectives of the IWC were defined by the International Convention of the Regulation of Whaling 1946 as conservation of whale stocks and the orderly development of the whaling industry. The Commission also conducts research on whales
Publications: Publications list available on request

– 797 – Irish Wildbird Conservancy (IWC)

Ruttledge House
8 Longford Place
Monkstown
Co Dublin
Eire
Tel: 353 1 804322
Fax: 353 1 844407
Geographical coverage: Ireland
Aims and objectives: IWC aims to promote conservation, education and research in relation to wildbirds and their environment

– 798 – Irish Wildlife Federation (IWF)

The Conservation Centre
132a East Wall Road
Dublin 3
Eire
Tel: 353 1 366821
Geographical coverage: Ireland
Aims and objectives: IWF campaigns on national and international issues. Youth training and employment in conservation work
Publications: *Badger*, quarterly

– 799 – Koninklijke Maatschappij voor Dierkunde van Antwerpen

26 Koningin Astridplein
B-2000 Antwerp
Belgium
Tel: 32 3 31 16 40
Geographical coverage: Belgium
Aims and objectives: Contributes to the promotion of zoological and botanical sciences, the conservation of endangered animal species and nature conservation

– 800 – Lega Italiana dei Diritti dell'Animale (LIDA)

(Italian Animal Rights Association)

Vle del Vignola 75
Rome
Italy
Tel: 39 6 761 612075
Geographical coverage: Italy
Aims and objectives: LIDA promotes laws for the environment and non-violent alternative life style proposals

– 801 – Lega Italiana Protezione Uccelli

(Italian Bird Protection Society)

Vicolo San Tiburzio
5/A Parma
Italy
Tel: 39 521 33414/27116
Geographical coverage: Italy

– 802 – Ligue Luxembourgeoise pour la Protection de la Nature et des Oiseaux

BP 709
L-2017 Luxembourg
Luxembourg
Tel: 352 47 23 64
Geographical coverage: Luxembourg
Publications: *Regulus*, quarterly

– 803 – Marine Conservation Society (MCS): Let Coral Reefs Live

9 Gloucester Road
Ross on Wye
Herefordshire HR9 5BU
United Kingdom
Tel: 0989 66017
Geographical coverage: United Kingdom

Aims and objectives: The main aim of the MCS is to protect the marine environment for both wildlife and future generations by promoting its sustainable and environmentally sensitive management
Publications: *Marine Conservation*, magazine regularly; *The Good Beach Guide*; *Coastal Directory*; annual report and others

– 804 – Norges Jeger og Fiskerforbund
(Norwegian Association of Anglers and Hunters)

Postboks 94
N-1364 Hvalstad
Norway
Tel: 47 2 783860
Geographical coverage: Norway

– 805 – Norsk Ornitologisk Forening
(Norwegian Ornithological Society)

PO Box 2207
N-7001 Trondheim
Norway
Tel: 47 7 525242
Geographical coverage: Norway
Aims and objectives: Works towards the advancement of ornithology

– 806 – North Atlantic Salmon Conservation Organisation (NASCO)

11 Rutland Square
Edinburgh EH1 2AS
United Kingdom
Tel: 031 228 2551
Geographical coverage: United Kingdom
Aims and objectives: NASCO aims to promote the conservation of salmon

– 807 – Nucleo Portugues de Estudio e Protecção da Vida Selvagem
(Portuguese Core Group for Wildlife Studies and Protection)

Bairro Fundo de Fomento de Habitação
Bloco DR/C
P-5300 Braganca
Portugal
Geographical coverage: Portugal

– 808 – Office National de la Chasse (ONC)

Boite Postale 236
F-75822 Paris Cedex 17
France
Tel: 33 1 422 781 75
Fax: 33 1 476 379 13
Geographical coverage: France

Aims and objectives: ONC is dedicated to the conservation of wildlife

– 809 – Ornithologische Arbeitsgemeinschaft für Schleswig-Holstein und Hamburg

Institut für Haustierkunde
Olsausenstrasse 40
D-2300 Kiel 1
Germany
Geographical coverage: Germany
Aims and objectives: Concerned with the welfare of birds

– 810 – Otter Trust

Earsham
nr Bungay
Suffolk
United Kingdom
Tel: 0986 893470
Contact name: Barbara Franklin, Personal assistant to director
Geographical coverage: United Kingdom
Aims and objectives: The Otter Trust's aims are to promote conservation of otters, whenever and wherever necessary for their survival
Publications: *Otters Magazine*, annually; *Operation Otter*; and others

– 811 – Royal Society for the Protection of Birds (RSPB)

The Lodge
Sandy
Bedfordshire SG12 2DL
United Kingdom
Tel: 0767 680551
Fax: 0767 692365
Contact name: Anne Scott, Enquiry Officer
Geographical coverage: United Kingdom
Aims and objectives: The RSPB was founded in 1889. It received its Royal Charter in 1904. The RSPB exists to conserve wild birds and the environment in which they live
Publications: *Birds* Magazine, quarterly; *Bird Life*, six times a year; annual report; newsletter; numerous publications

– 812 – RSNC Wildlife Trusts Partnership: Formerly Royal Society for Nature Conservation

The Green
Whitham Park
Waterside South
Lincoln LN5 7JR
United Kingdom
Tel: 0522 544400
Fax: 0522 511616
Contact name: Douglas Brown, Publicity Manager

Geographical coverage: United Kingdom
Aims and objectives: The RSNC Wildlife Trusts Partnership is the largest voluntary organisation in the UK concerned with all aspects of wildlife protection
Publications: *Natural World*, magazine three times a year; annual review and numerous reports

– 813 – Sea Mammal Research Unit (SMRU)

High Cross
Madingley Road
Cambridge CB3 0ET
United Kingdom
Tel: 0223 311354
Fax: 0223 328927
Geographical coverage: United Kingdom
Aims and objectives: SMRU provides scientific advice to government departments on the conservation and management of seals and whales
Publications: Publications list is available on request; Annual report

– 814 – Sea Turtle Protection Society

PO Box 51154
GR-145 10 Kisissia
Greece
Geographical coverage: Greece

– 815 – Sociedad Española de Ornitologia (SEO)

Ciudad Universitaria s/n Facultad de Biologia
E-28040 Madrid
Spain
Tel: 34 1 549 35 54
Geographical coverage: Spain

– 816 – SOS Sea Turtles

PO Box 350
CH-8810 Horgen
Switzerland
Contact name: Kurt Ansler
Geographical coverage: International

– 817 – World Wide Fund for Nature – Austria: *see entry 913*

Geographical coverage: Austria

– 818 – World Wide Fund for Nature – Belgium: *see entry 914*

Geographical coverage: Belgium

– 819 – World Wide Fund for Nature – Central and East Europe: *see entry 915*

Geographical coverage: European

– 820 – World Wide Fund for Nature – Denmark: *ee entry 916*

Geographical coverage: Denmark

– 821 – World Wide Fund for Nature – Finland: *see entry 917*

Geographical coverage: Finland

– 822 – World Wide Fund for Nature – France: *see entry 918*

Geographical coverage: France

– 823 – World Wide Fund for Nature – Germany: *see entry 919*

Geographical coverage: Germany

– 824 – World Wide Fund for Nature – Greece: *see entry 920*

Geographical coverage: Greece

– 825 – World Wide Fund for Nature – Italy: *see entry 921*

Geographical coverage: Italy

– 826 – World Wide Fund for Nature – Netherlands: *see entry 922*

Geographical coverage: Netherlands

– 827 – World Wide Fund for Nature – Norway: *see entry 923*

Geographical coverage: Norway

– 828 – World Wide Fund for Nature – Spain: *see entry 924*

Geographical coverage: Spain

– 829 – World Wide Fund for Nature – Sweden: *see entry 925*

Geographical coverage: Sweden

– 830 – World Wide Fund for Nature – Switzerland: *see entry 926*

Geographical coverage: Switzerland

– 831 – World Wide Fund for Nature – UK: *see entry 927*

Geographical coverage: United Kingdom

– 832 – World Wide Fund for Nature (WWF): *see entry 928*

Geographical coverage: International

– 833 – Zoologische Gesellschaft Frankfurt von 1858 eV

Alfred-Brehm-Platz 16
D-6000 Frankfurt
Germany
Geographical coverage: Germany

BOOKS

– 834 – Atlas of Endangered Species, The

Geographical coverage: International
Publishers: David & Charles, 1992
Coverage: Shows many thousands of animals, insects and plants that are at risk
Price: £16.99
ISBN: 0715300741
Author: Burton, J. A.

– 835 – Effects of Acidification on Bird and Mammal Populations

Geographical coverage: Sweden
Publishers: Swedish Environmental Protection Agency
Coverage: A review of the literature from North America and Europe, principally Scandinavia
Author: Eriksson, M. O. G.

– 836 – Elephants, Ivory and Economics

Geographical coverage: United Kingdom
Publishers: Earthscan Publications, 1990
Coverage: Covering the economy of the ivory trade
Price: £6.95
ISBN: 1853830739
Author: Barbier, E. B. and Burgess, J. C.

– 837 – Global Biodiversity

Geographical coverage: International
Publishers: Chapman and Hall, 1992
Coverage: A compendium of conservation information, it provides standardised and comparable data for 205 countries of the world
Price: £39.95
ISBN: 0412472406
Author: Groombridge, B.

– 838 – Last Chance to See...

Geographical coverage: International
Publishers: W Heinemann, 1990
ISBN: 0434009245
Author: Adams, D. and Carwardine, W.

– 839 – Red Data Birds in Britain

Geographical coverage: United Kingdom
Publishers: T & A D Poyser, 1990
Coverage: Focusing on vulnerable species of birds living in or visiting Britain, this book examines populations found and the need to recognise responsibilities for bird conservation
ISBN: 0856610569
Author: Batten, L. A. and Bibby, C. J.

– 840 – Tropical Rain Forest: *see entry 763*

Geographical coverage: United Kingdom

– 841 – Tropical Rain Forest Ecology: *see entry 764*

Geographical coverage: United Kingdom

DIRECTORIES/YEARBOOKS

– 842 – Agricultural and Veterinary Sciences International Who's Who, 4th ed

Geographical coverage: International
Publishers: Longmans, 1990
Coverage: Lists many international sources relating to agriculture and veterinary sciences
ISBN: 0582041007

PERIODICALS

– 843 – Ecosystems of the World

PO Box 211
NL-1000 AE Amsterdam
Netherlands
Tel: 31 20 5803911
Geographical coverage: International
Publishers: Elsevier Science Publishers BV
Frequency: irregular
Language: English

102 Conservation *Habitats and Species*

– 844 – Greenbits
Old Town Hall
Mandale Road
Thornaby
Cleveland TS17 6AW
United Kingdom
Tel: 0642 608405
Fax: 0642 671519
Geographical coverage: United Kingdom
Publishers: Cleveland Wildlife Trust
Editor: Gordon, T. and Straughan, N.
Frequency: 3 issues a year
Coverage: Environmental wildlife
Annual subscription: £10.00

– 845 – Horticulture Française
BP 30
F-34471 Perols Cedex
France
Geographical coverage: France
Publishers: Edition du Lien
Frequency: monthly
Annual subscription: Ffr325
ISSN: 0395 8531

– 846 – Horticulture Week
38–42 Hampton Road
Teddington
Middlesex TW11 0JE
United Kingdom
Tel: 081 943 5000
Geographical coverage: United Kingdom
Publishers: Haymarket Magazines Ltd
Frequency: weekly
Annual subscription: $250

– 847 – Journal of Tropical Ecology: *see entry 769*
Geographical coverage: United Kingdom

– 848 – Oryx
Fauna and Flora Preservation Society
Osney Mead
Oxford OX2 0EL
United Kingdom
Tel: 0865 240201
Geographical coverage: United Kingdom
Publishers: Blackwell Scientific Publishers Ltd
Editor: Morris, J.
Frequency: quarterly
Coverage: International conservation
Annual subscription: £44.00
ISSN: 0030 6053

– 849 – Österreichisches Tierschutzzeitung: *see entry 943*
Geographical coverage: Austria

– 850 – Revue d'Ecologie: La Terre et la Vie
57 rue Cuvier
F-75005 Paris
France
Tel: 33 1 470 731 95
Geographical coverage: France
Publishers: Société Nationale de Protection de la Nature
Editor: Bourliere, F.
Frequency: quarterly
Coverage: Discusses relationships between plant and animal life with the intention of conserving and maintaining the delicate balance of nature
Annual subscription: Ffr400
ISSN: 0249 7395

– 851 – Species
World Conservation Centre
Avenue du Mont Blanc
CH-1196 Gland
Switzerland
Tel: 41 22 649114
Fax: 41 22 644615
Geographical coverage: International
Publishers: International Union for Conservation of Nature and Natural Resources
Language: English
Price: Sfr25
ISSN: 1016 927X

– 852 – Sveriges Djurskyddsfoereningars Riksfoerbund
Markvardsgatan 10
4 Tr
S-113 53 Stockholm
Sweden
Tel: 46 8 323 666
Geographical coverage: Sweden
Publishers: Djurskyddet
Language: Swedish

REPORTS

– 853 – Net Losses Gross Destruction – Fisheries in the European Community
Geographical coverage: European
Publishers: Greenpeace International, 1991
Coverage: Highlights the problems of over-fishing in Europe
Price: £2.50

RESEARCH CENTRES

– 854 – British Trust for Ornithology (BTO)

The Nunnery
Nunnery Place
Thetford
Norfolk IP24 2PU
United Kingdom
Tel: 0842 750050
Fax: 0842 750030
Contact name: Dr Rowena Langston, Research Contracts Officer
Geographical coverage: United Kingdom
Aims and objectives: The BTO exists to promote research on British birds and their habitats
Publications: *BTO News*, six times a year; *Ringing and Migration*, three times a year; *Bird Study*, three times a year
Additional information: Information avaialable on several databases

– 855 – Bundesforschungsanstalt für Naturschutz und Landschaftsökologie: see entry 946

Geographical coverage: Germany

– 856 – Department of Plant Ecology

Lund University
POB 117
S-221 00 Lund
Sweden
Tel: 46 46 107000
Contact name: Ursula Falkengren-Grerup
Geographical coverage: Sweden
Research activities: The relationship between soils and plants, especially with regard to the acidified forest soils of southern Sweden

– 857 – Department of Wildlife Ecology

Sveriges Lantbruksuniversitet
Box 7002
S-750 07 Uppsala
Sweden
Tel: 46 18 67 10 00
Geographical coverage: Sweden
Research activities: Analyses of habitat selection; knowledge of the environment needs of birds and mammals

– 858 – Direktoratet for Vilt og Ferskvannsfisk

(Directorate for Wildlife and Freshwater Fish)
Tungasletta 2
N-7000 Trondheim
Norway
Tel: 47 7 913020
Geographical coverage: Norway
Research activities: Wildlife Management

– 859 – Institute of Terrestrial Ecology (ITE)

Monks Wood Experimental Station
Abbots Ripton
Huntingdon PE17 2LS
United Kingdom
Tel: 04873 381
Fax: 04873 467
Geographical coverage: United Kingdom
Aims and objectives: The ITE is part of the Natural Environment Research Council (NERC). It undertakes specialist ecological research in all aspects of terrestrial environment and seeks to understand the ecology of species and of natural and man-made communities
Publications: Annual report; list of publications available on request
Research activities: Studies the factors determining the structure, composition and processes of land and freshwater systems

– 860 – International Waterfowl and Wetlands Research Bureau (IWRB)

Slimbridge
Gloucestershire GL2 7BX
United Kingdom
Tel: 0453 890624/634
Fax: 0453 890697
Contact name: Simon Nash, Administrator
Geographical coverage: International
Aims and objectives: The IWRB was established to stimulate and coordinate research and conservation concerning waterfowl and their wetland habitats
Publications: *IWRB News*, newsletter twice a year; occasional publications

– 861 – World Conservation Monitoring Centre (WCMC)

219 Huntingdon Road
Cambridge CB3 0DL
United Kingdom
Tel: 0223 277314
Fax: 0223 277136
Contact name: Jo Taylor, Information Officer
Geographical coverage: International

Aims and objectives: WCMC is a joint venture between the three partners who developed the World Conservation Strategy: IUCN, UNEP and WWF. Its mission is to support conservation and sustainable development through the provision of information
Publications: *Red Data Books*
Research activities: On behalf of CITES, data on world trade in wildlife are collected and analysed by the WTMU, which is part of this centre

DATABASES

– 862 – Aquatic Sciences and Fisheries Abstracts

Geographical coverage: International
Coverage: Aquatic pollution and environmental quality; marine and freshwater science technology and management; ecology and ecosytems
Database producer: ASFIS/Cambridge Scientific Abstracts
Database host: BRS; Dialog; DIMDI; ESA–IRS;
Coverage period: 1975–
Update frequency: 1,000 records monthly

Heritage

NON-GOVERNMENT ORGANISATIONS

– 863 – Architectural Heritage Fund

17 Carlton House Street
London SW1Y 5AW
United Kingdom
Tel: 071 925 0199
Fax: 071 321 0180
Contact name: Barbara Wright
Geographical coverage: United Kingdom
Aims and objectives: The Heritage Fund promote the preservation and rehabilitation of buildings of historical, architectural or constructional interest
Publications: *Preservation in Action*, twice a year

– 864 – Central Office of Historic Monuments

Dronningensgt 13
PO Box 8196
N-0034 Oslo
Norway
Tel: 47 2 94 04 00
Geographical coverage: Norway

– 865 – International Council on Monuments and Sites (ICOMOS UK)

10 Barley Mow Passage
Chiswick
London W4 4PH
United Kingdom
Tel: 081 994 6477
Fax: 081 747 8464
Contact name: Secretary
Geographical coverage: United Kingdom

Aims and objectives: ICOMOS was founded to promote an understanding of the ethics and practice of conservation internationally. It is based in Paris. In the UK, committees of experts have been set up on the conservation of historic parks and gardens and historic buildings
Publications: *ICOMOS Information*; magazine quarterly; *ICOMOS UK Newsletter*, twice a year; conference and seminar reports

– 866 – International Federation of Associations for the Protection of Europe's Cultural and National Heritage: Europa Nostra

9 Buckingham Gate
London SW1E 6JP
United Kingdom
Tel: 071 821 11 71
Fax: 071 828 69 48
Geographical coverage: International
Aims and objectives: Founded in 1963, Europa Nostra is a confederation of independent conservation associations which are working throughout Europe, towards an improvement in the quality of life in the environment

– 867 – Liga Para la Defensa del Patrimonio Natural (DEPANA)

Aragón 281
E-08009 Barcelona
Spain
Tel: 34 3 215 14 84
Contact name: Miquel Rafa, Secretary General
Geographical coverage: Spain

– 868 – National Trust (NT)

36 Queen Anne's Gate
London SW1H 9AS
United Kingdom
Tel: 071 222 9251
Fax: 071 222 5097
Contact name: Warren Davis, Press and Public Relations Manager
Geographical coverage: United Kingdom
Aims and objectives: The National Trust is a charity. It works for the preservation of places of historic interest and natural beauty. It is Britain's biggest private land owner and the world's largest conservation organisation with over 2 million members
Publications: *The National Trust Magazine*, three times a year; annual report; list available on request

– 869 – Natuurmonumenten: *see entry 906*

Geographical coverage: Netherlands

– 870 – Swiss League for the Protection of National Heritage

Merkurstrasse 45
CH-8032 Zurich
Switzerland
Tel: 41 1 252 26 60
Fax: 41 1 252 28 70
Geographical coverage: Switzerland
Aims and objectives: The League was formed in 1905 and has over 23,000 members. It is concerned with the preservation of the environment, historical sites and monuments against alteration and destruction
Publications: *Heimatschutz/Sauvegarde*, four times a year

BOOKS

– 871 – SAVE Action Guide, The

Geographical coverage: United Kingdom

Publishers: Collins and Brown, 1991
Coverage: This book explains how the organisation goes about its campaigns and looks at schemes through which architectural conservation can be actively promoted
Price: £7.99
ISBN: 1855850567
Author: Binney, M., OBE, and Watson-Smyth, M.

PERIODICALS

– 872 – Interpretation Heritage

201 Buryfield Road
Solihull
West Midlands B91 2BB
United Kingdom
Tel: 021 704 3961
Geographical coverage: United Kingdom
Publishers: Society for the Interpretation of Britain's Heritage (SIBH)
Editor: Jackson, K.
Frequency: 3 issues a year
Annual subscription: £20.00

– 873 – Steine Sprechen

Karlsplatz 5
A-1010 Vienna
Austria
Tel: 43 222 587 96630
Geographical coverage: Austria
Publishers: Österreichische Gesellschaft für Denkmal und Ortsbidpflege
Editor: Schwartz, M.
Frequency: quarterly
Coverage: Covers the history the care and upkeep of historical monuments, buildings and gardens
Price: AS240
ISSN: 0039 1026

Land Use

GOVERNMENT ORGANISATIONS

– 874 – Norwegian Mapping Authority

Kartverksvein
N-3500 Honefoss
Norway
Tel: 47 32 11 81 00
Fax: 47 32 11 81 01
Geographical coverage: Norway
Aims and objectives: Ministry Agency provides geographic information and services designed to help users to promote safety and efficiency in conservation and the use of land and sea

BOOKS

– 875 – Fens and Bogs in the Netherlands: Vegetation, History, Nutrient Dynamics and Conservation

Geographical coverage: Netherlands
Publishers: Kluwer Academic Publishers, 1992
Coverage: Focuses on the geology, land use history, palaeontology, ecology and conservation of peatlands in the Netherlands
Price: £135.00
ISBN: 0792313879
Author: Verhoeven, J. T. A.

– 876 – Land Use Change: The Causes and Consequences

Geographical coverage: United Kingdom
Publishers: Commission of the European Communities
Editor: Whitby, M. C.
ISBN: 0117015539

– 877 – Landscape Changes in Britain

Geographical coverage: United Kingdom
Publishers: Institute of Terrestrial Ecology
Coverage: Landscape change and the long term environmental effects of differing uses of land
ISBN: 0904282953
Author: Barr, C. et al

– 878 – Vanishing Tuscan Landscapes

Geographical coverage: Italy
Publishers: Pudoc, 1991
Coverage: This integrated ecological survey of the Solano Basin shows the impact of recent changes in the character and intensity of land use
ISBN: 9022009645

DIRECTORIES/YEARBOOKS

– 879 – Green Belt, Green Fields and the Urban Fringe: The Pressures of Land in the 1980's

Geographical coverage: United Kingdom
Publishers: The British Library, 1990
Coverage: A guide to sources on the literature of issues such as location pattern of new industries, pressure of development on land and changes in rural economy
Price: £25.00 in UK
ISBN: 0712307702
Author: Grayson, L.

PERIODICALS

– 880 – Land Use Policy

PO Box 63
Westbury House
Bury Street
Guildford GU2 FBH
United Kingdom
Geographical coverage: United Kingdom
Publishers: Butterworth
Frequency: quarterly
ISSN: 0264 8377

– 881 – Landscape Ecology

PO Box 97747
NL-2509 GC The Hague
Netherlands
Geographical coverage: Netherlands
Publishers: S P B Academic Publishing bv
Editor: Golley, F. B.
Frequency: quarterly
Language: Text in English
Coverage: Publishes papers on fundamental and applied research, as well as papers in land use, nature management and environmental conservation
Annual subscription: Dfl245
ISSN: 0921 2973

RESEARCH CENTRES

– 882 – Centre de Recerca Ecológica i Aplicacions Forestals: *see entry 772*

Geographical coverage: Spain

– 883 – Department of Plant Ecology: *see entry 856*

Geographical coverage: Sweden

– 884 – Institute of Terrestrial Ecology (ITE)

Merlewood Research Station
Grange over Sands
Cumbria LA11 6JU
United Kingdom
Tel: 05395 32264
Fax: 05395 34705
Contact name: Dr R Runce
Geographical coverage: United Kingdom
Research activities: Surveys of land cover, landscape features and vegetation using 1km^2 sample unit

– 885 – Instytut Podstaw Inzynierii Srodowiska PAN

(Environmental Engineering Institute)

Ulica M Curie Sklodowskiej 34
PL-41 800 Zabrze
Poland
Tel: 48 32 716481
Geographical coverage: Poland
Publications: *Archives of Environmental Protection*; *Prace i Studia*
Research activities: Atmospheric pollution protection; water pollution protection; reclamation of soil; biological reclamation of dumps

– 886 – Istituto per le Piante da Legno e l'Ambiente SpA: *see entry 778*

Geographical coverage: Italy

– 887 – Norsk Polar Institut

PO Box 158
Rolfstangveien 12
N-1330 Oslo Lufthavn
Norway
Tel: 47 2 123650
Fax: 47 2 123854
Contact name: Nils Are Oeritsland, Director
Geographical coverage: Norway
Aims and objectives: Central government institution responsible for mapping, surveys and practical scientific investigations of the polar region

– 888 – Soil Survey and Land Research Centre (SSLRC)

Cranfield Institute of Technology
Silsoe Campus
Silsoe
Bedfordshire MK45 4DT
United Kingdom
Tel: 0525 60428
Fax: 0525 61147
Geographical coverage: United Kingdom
Aims and objectives: The Soil Survey and Land Research Centre provides research, development and consultancy service in all aspects of land management, land use planning and land policy development
Publications: List available on request

DATABASES

– 889 – Farm Structure Survey Retrieval System (FSSRS)

Geographical coverage: European
Coverage: Contains results of five surveys in the EC
Database producer: EUROSTAT
Database host: EUROSTAT
Coverage period: 1975–

– 890 – International Road Research Documentation (IRRD)

Geographical coverage: International
Coverage: All aspects of road transport and land use
Database producer: OECD Road Transport Research Programme
Database host: ESA–IRS
Coverage period: 1972–
Update frequency: 1,200 records monthly

– 891 – Transdoc

Geographical coverage: European
Coverage: Transport and land use
Database producer: European Conference of Ministers of Transport
Database host: ESA–IRS
Coverage period: 1970–
Update frequency: 250 records monthly

– 892 – Transport Research Information Services (TRIS)

Geographical coverage: International
Coverage: Transport, land use, energy, economics
Database producer: US Transportation Research Board
Database host: Dialog
Coverage period: 1968–
Update frequency: 750 records monthly

Nature Conservation

GOVERNMENT ORGANISATIONS

– 893 – Center for Nature Protection and Education

Ommeganckstraat 20
B-2018 Antwerp
Belgium
Tel: 32 3 226 02 91
Geographical coverage: Belgium

– 894 – Directorate for Nature Management, The

Tungasletta 2
N-7005 Trondheim
Norway
Tel: 47 73 58 05 00
Fax: 47 73 91 54 33
Geographical coverage: Norway
Aims and objectives: Ministry agency responsible for implementing nature management policy

– 895 – English Nature

Northminster House
Peterborough PE1 1UA
United Kingdom
Tel: 0733 340345
Fax: 0733 68834
Contact name: Librarian
Geographical coverage: United Kingdom
Aims and objectives: English Nature was established by an Act of Parliament and is responsible for advising government on nature conservation in England
Publications: Wide range of publications

– 896 – Ministry for the Environment Nature Conservation and Nuclear Reactors Safety

Kennedyallee 5
Postfach 12 06 29
D-5300 Bonn 1
Germany
Tel: 49 228 3050
Fax: 49 228 305 3225
Geographical coverage: Germany

– 897 – Ministry of Agriculture, Nature Conservation and Fisheries

Bezuidenhoutsweg 73
PO Box 20401
NL-2500 EK The Hague
Netherlands
Tel: 31 70 340 7911
Fax: 31 70 340 7834
Geographical coverage: Netherlands

– 898 – Nature Conservation Council

Hverfisgate 26
101 Reykjavik
Iceland
Tel: 354 1 27855
Contact name: Eythor Einarsson, Chairperson
Geographical coverage: Iceland
Aims and objectives: The Council is a governmental organisation working under an act which was set up in 1971. The Act was to ensure, as far as possible, the course of natural processes according to their own laws

NON-GOVERNMENT ORGANISATIONS

– 899 – Council of Europe

BP 431
F-67006 Strasbourg Cedex
France
Tel: 33 88 41 20 00
Fax: 33 88 41 27 81
Contact name: Anders Björck, President
Geographical coverage: European
Aims and objectives: The Council was established in 1949. One of its steering committees is the European Committee for the Conservation of Nature and Natural Resources. The Council administers the Berne Convention on the Conservation of European Wildlife and Habitats
Publications: *Naturopa*, three times a year; *Forum*, quarterly; *A Future for our Past*; Catalogue available

– 900 – Countryside Commission

John Dower House
Crescent Place
Cheltenham
Gloucestershire GL50 3RA
United Kingdom
Tel: 0242 521381
Fax: 0242 584270
Contact name: Liz Prynne, Communications Branch
Geographical coverage: United Kingdom

Aims and objectives: The Commission is an advisory and promotional body, working in partnership with others – and providing grants and advice for projects which conserve the natural beauty of the countryside
Publications: *Countryside Newspaper*, six times a year; leaflets, books, reports, and many others
Additional information: Library Service

– 901 – Danmarks Naturfredningsforening
(Danish Society for the Conservation of Nature)

Norregade 2
DK-1165 Copenhagen
Denmark
Tel: 45 33 32 20 21
Fax: 45 33 32 22 02
Geographical coverage: Denmark
Aims and objectives: The Society appeals against decisions made by local or regional authorities, if those decisions do not take into account environmental considerations
Publications: *Natur of Miljo*, quarterly

– 902 – Federazione Italiana Pro Natura
(Italian Federation for the Protection of Nature)

c/o Isea
Via Marchensana 12
Bologna
Italy
Tel: 39 51 231999
Geographical coverage: Italy
Aims and objectives: The Federation aims to promote the conservation of nature

– 903 – Hungarian Society of Nature Conservationists

Kolto ul 21
H-1121 Budapest
Hungary
Tel: 36 1 1750684
Geographical coverage: Hungary

– 904 – International Union for Conservation of Nature and Natural Resources (IUCN): World Conservation Union

World Conservation Centre
Avenue du Mont Blanc
CH-1196 Gland
Switzerland
Tel: 41 22 64 71 81
Fax: 41 22 64 29 26
Geographical coverage: International
Aims and objectives: IUCN is a union of sovereign states, government agencies and non-governmental organisations concerned with the initiation and promotion of scientifically based actions that will ensure the perpetuation of man's natural environment

– 905 – Liga Ochrony Przyrody
(The Nature Protection League)

ul Reja 3/5
PL-02 053 Warsaw
Poland
Geographical coverage: Poland
Aims and objectives: Public service institution, concerned with the protection of the natural environment

– 906 – Natuurmonumenten
(The Society for the Preservation of Nature in the Netherlands)

Schaep en Burgh
Noordereinde 60
NL-1243 JJ 'S-Graveland
Netherlands
Tel: 31 35 62004
Fax: 31 35 63174
Contact name: Dr P Winsemius, President
Geographical coverage: Netherlands
Aims and objectives: The Society was formed in 1905 and is now the largest independent organisation for nature conservation in the Netherlands. Its main objective is the conservation and management of nature reserves. It also preserves many buildings as national monument
Publications: *Natuurbehoud*, magazine quarterly; books and other information

– 907 – Norges Naturvernforbund (NNV)
(The Norwegian Society for the Conservation of Nature)

PO Box 2113
Grünerlokka
N-0505 Oslo 5
Norway
Tel: 47 2 715520
Fax: 47 2 715640
Geographical coverage: Norway
Aims and objectives: The Society was formed in 1914. It has a youth organisation "Nature and Youth". It works for the prevention of over exploitation of natural resources and harmful influence on nature
Publications: *Natur and Miljo*; *Natur and Miljo Bulletin*

– 908 – Österreichisches Gesellschaft für Natur und Umweltschutz (ÖGNU)
(Austrian Society for Nature and Environment Protection)

Hegelgasse 21
A-1010 Vienna
Austria
Tel: 43 222 513 29 62
Fax: 43 222 512 56 01
Contact name: Mr Wilhelm Linder
Geographical coverage: Austria

Aims and objectives: Coordinates environmental activities of NGOs in Austria
Publications: A newspaper dealing with climatic changes

– 909 – Ramsar Convention Bureau

Avenue du Mont Blanc
CH-1196 Gland
Switzerland
Tel: 41 22 64 71 81
Geographical coverage: International
Aims and objectives: The Bureau endeavours to protect, manage and enhance the wetlands. This includes all areas outside the oceans which are permanently or periodically covered in water

– 910 – Slovakian Society for Conservation of Nature (SZOPK)

Gorkeho 6
811 01 Bratislava
Slovakia
Tel: 42 7 506 65
Geographical coverage: Slovakia

– 911 – Stichting Natuur en Milieu (SNM)
(Netherlands Society for Nature and the Environment)

Donkerstraat 17
NL-3511 KB Utrecht
Netherlands
Tel: 31 30 33 13 28
Fax: 31 30 33 13 11
Contact name: Mr Ralph Hallo, Coordinator International Affairs
Geographical coverage: Netherlands
Aims and objectives: SNM is a coordinating body for many environmental and nature conservation groups
Publications: Books, articles and reports

– 912 – Suomen Luonnonsuojeluitto ry
(Finnish Association for Nature Conservation)

Peramiehenkatv 11 A 8
SF-00150 Helsinki
Finland
Tel: 358 1 0642 881
Geographical coverage: Finland
Publications: *Suomen Luonto*, 8 times a year; *Luonnonsvojelija*, 10 times a year

– 913 – World Wide Fund for Nature – Austria

Ottakringerstrasse
114–116/9 Postfach 1
A-11622 Vienna
Austria
Tel: 43 1 46 14 630
Geographical coverage: Austria

– 914 – World Wide Fund for Nature – Belgium

608 chaussée de Waterloo
B-1060 Brussels
Belgium
Tel: 32 2 347 30 30
Geographical coverage: Belgium

– 915 – World Wide Fund for Nature – Central and East Europe

WWF International
CH-1196 Gland
Switzerland
Tel: 41 22 64 92 87
Contact name: Jan Habrovsky, WWF coordinator
Geographical coverage: European

– 916 – World Wide Fund for Nature – Denmark

Ryesgade 3F
DK-2200 Copenhagen N
Denmark
Tel: 45 35 36 36 35
Geographical coverage: Denmark

– 917 – World Wide Fund for Nature – Finland

Uudenmaankatu 40
SF-00120 Helsinki 12
Finland
Tel: 358 1 644 511
Geographical coverage: Finland

– 918 – World Wide Fund for Nature – France

151 boulevard de la Reine
F-78000 Versailles
France
Tel: 33 1 395 075 14
Geographical coverage: France

– 919 – World Wide Fund for Nature – Germany

Hedderichstrasse 110
Postfach 70 11 27
D-6000 Frankfurt am Main 70
Germany
Tel: 49 69 60 50 030
Geographical coverage: Germany

– 920 – World Wide Fund for Nature – Greece

Askelpiou Street 14
GR-106 80 Athens
Greece
Tel: 30 1 362 33 42
Geographical coverage: Greece

– 921 – World Wide Fund for Nature – Italy

Via Salaria 290
I-00199 Rome
Italy
Tel: 39 6 854 24 92
Geographical coverage: Italy

– 922 – World Wide Fund for Nature – Netherlands

PO Box 7
NL-3700 AA Zeist
Netherlands
Tel: 31 3404 22 164
Geographical coverage: Netherlands

– 923 – World Wide Fund for Nature – Norway

PO Box 7065 Homansbyen
N-0306 Oslo 3
Norway
Tel: 47 2 69 61 97
Geographical coverage: Norway

– 924 – World Wide Fund for Nature – Spain

ADENA
Santa Engracia 6
E-28010 Madrid
Spain
Tel: 34 1 308 2309
Geographical coverage: Spain

– 925 – World Wide Fund for Nature – Sweden

Ulriksdals Slott
S-171 71 Solna
Sweden
Tel: 46 8 85 01 20
Geographical coverage: Sweden

– 926 – World Wide Fund for Nature – Switzerland

Förrlibuckstrasse 66
Postfach
CH-8037
Switzerland
Tel: 41 1 272 20 44
Geographical coverage: Switzerland

– 927 – World Wide Fund for Nature – UK

Panda House
Weyside Park
Godalming
Surrey GU7 1XR
United Kingdom
Tel: 0483 426444
Fax: 0483 426409
Geographical coverage: United Kingdom

– 928 – World Wide Fund for Nature (WWF)

Avenue du Mont Blanc
CH-1196 Gland
Switzerland
Tel: 41 22 364 91 11
Fax: 41 22 364 53 58
Contact name: Charles de Haes, Director
Geographical coverage: International
Aims and objectives: WWF is the largest private international nature conservation organisation. It promotes awareness of conservation problems and raises funds for the protection of threatened species and environments
Publications: *WWF News*; and an annual review. Book list available on request

PRESSURE GROUPS

– 929 – Slovak Union of Nature and Landscape Protectors (SZOPK)

Gorkeho 6
811 01 Bratislava
Slovakia
Tel: 42 7 506 65
Fax: 42 7 506 65
Contact name: Elena Vartikaya, International Affairs of SZOPK
Geographical coverage: Slovakia
Aims and objectives: Environmental pressure group

BOOKS

– 930 – International Handbook of National Parks and Nature Reserves

Geographical coverage: International
Publishers: Greenwood Press, 1990
Coverage: A comparative exploration of global parks preservation, it describes and evaluates national parks and nature reserves in over 25 nations
Price: £65.00
ISBN: 0313249024
Author: Allin, C. W.

– 931 – Nature Conservation in Europe – Agenda 2000

Geographical coverage: European
Publishers: World Wide Fund For Nature, 1991
Editor: McGillivray, Z.
Language: English
Coverage: This publication was produced on the occasion of the "Nature Conservation – Europe 2000" conference, organised by the European Parliament in collaboration with WWF
ISBN: 2880850800

PERIODICALS

– 932 – Amenagement et Nature

21 rue du Conseiller Collignon
F-75116 Paris
France
Fax: 33 1 452 045 36
Geographical coverage: France
Publishers: Association pour les Espaces Naturels
Editor: Bechmann, R.
Frequency: quarterly
Annual subscription: Ffr150
ISSN: 0044 7463

– 933 – Ecos

c/o Nature Conservation Bureau
36 Kingfisher Court
Hambridge Road
Newbury
Berkshire RG14 5SJ
United Kingdom
Tel: 0635 550380
Geographical coverage: United Kingdom
Publishers: British Association of Nature Conservationists (BANC)
Frequency: quarterly
Coverage: Review of conservation
ISSN: 0143 9073

– 934 – Ecos

Zubovsky Bulvar 4
119021 Moscow
Russia
Tel: 7 201 85 10
Fax: 7 230 21 70
Geographical coverage: Russia
Editor: Rudenko, V. B.
Language: Text in English, German and Russian
Coverage: Concerned with issues of nature conservation and ecology around the world. Includes articles by foreign authors

– 935 – Ludwig Boltzmann Institut für Umweltwissenschaften und Naturschutz Mitteilungen

Heinrichstrasse 5
A-8010 Graz
Austria
Geographical coverage: Austria
Publishers: Österreichischen Akademie der Wissenschaften
Frequency: irregular

– 936 – Milieudefensie

POB 20050
NL-1000 HB Amsterdam
Netherlands
Geographical coverage: Netherlands
Publishers: Vereniging Milieudefensie
Frequency: monthly
Price: Dfl45

– 937 – Natur og Miljoe

Norregade 2
DK-1165 Copenhagen K
Denmark
Geographical coverage: Denmark
Publishers: Danmarks Naturfredingsforening
Frequency: quarterly
Coverage: Covers many aspects of nature conservation
Annual subscription: Dkr135

– 938 – Natur og Miljoe

Postboks 2113
Grunnerhoka N-0505 Oslo 5
Norway
Geographical coverage: Norway
Publishers: Norges Naturvernforbund
Frequency: 6 issues a year
Coverage: Covers many aspects of nature conservation
Price: $27

– 939 – **Natura e Societa**

Borgo Felino 54
Parma
Italy
Tel: 39 521 481871
Geographical coverage: Italy
Publishers: Federazione Nazionale pro Natura
Editor: de Marchi, A.
Frequency: quarterly
Coverage: Presents articles on the protection of nature, naturalistic education, current events in nature, conservative and legislative laws for nature
Annual subscription: Lit5000
ISSN: 0393 8875

– 940 – **Nature et Mieux-Vivre**

BP 3
F-32350 Gavarret
France
Geographical coverage: France
Publishers: Sodeco
Editor: Rousseau, J. P.
Frequency: monthly
Coverage: Includes articles on many aspects of nature conservation

– 941 – **Naturopa**

BP 431 R6
F-67006 Strasbourg Cedex
France
Tel: 33 88 41 20 20
Fax: 33 88 41 27 84
Geographical coverage: European
Publishers: European Information Centre for Nature Conservation Council of Europe
Editor: Hoekstra, H. H.
Frequency: quarterly
Language: English, French, German, Italian, Spanish
Coverage: Wildlife conservation in the member countries of the Council of Europe
ISSN: 0250 7102

– 942 – **Natuur en Milieu**

Donkerstraat 17
NL-3511 KB Utrecht
Netherlands
Tel: 31 30 331328
Fax: 31 30 331311
Geographical coverage: Netherlands
Publishers: Association for Nature and the Environment
Frequency: monthly
Language: Dutch
Coverage: All aspects of conservation
Price: Dfl50

– 943 – **Österreichisches Tierschutzzeitung**

Neuer Markt 9
A-1010 Vienna
Austria
Geographical coverage: Austria
Publishers: Franz Abele
Editor: Abele, F.
Frequency: monthly
Coverage: Conservation
Price: AS92

– 944 – **Sveriges Natur**

Box 4625
S-116 91 Stockholm
Sweden
Tel: 46 8 7026500
Fax: 46 8 7022702
Geographical coverage: Sweden
Publishers: Svenska Naturskyddsfoereningen
Editor: Vaste, L.
Frequency: 6 issues a year
Language: Text in Swedish, summaries in English
Price: Skr210
ISSN: 0349 5264

– 945 – **Varstvo Narave**

Plecnikov trg 2
Box 176
61001 Ljubljana
Slovenia
Fax: 38 61 213 120
Geographical coverage: Slovenia
Publishers: Zavod Slovenuje za Varstvo Naravne in Kulturne Dediscine
Editor: Habjan, J.
Frequency: annual
Language: Text in Slovenian, summaries in English
Price: DM10
ISSN: 0506 4252

RESEARCH CENTRES

– 946 – **Bundesforschungsanstalt für Naturschutz und Landschaftsökologie**

(Federal Research Centre for Nature Conservation and Landscape Ecology)

Konstantinstrasse 110
D-5300 Bonn 2
Germany
Tel: 49 228 84910
Fax: 49 228 8491200
Geographical coverage: Germany
Publications: Annual report
Research activities: Nature conservation; animal ecology; landscape management

– 947 – Institut für Umweltwissenschaften und Naturschutz
(Institute of Environmental Sciences and Conservation)

Heinrichstrasse 5
A-8010 Graz
Austria
Tel: 43 316 36068
Geographical coverage: Austria
Research activities: Ecology; nature conservation; plant geography; entomology and landscaping

Sustainability

NON-GOVERNMENT ORGANISATIONS

– 948 – Centre for Alternative Technology (CAT): see entry 246
Geographical coverage: United Kingdom

– 949 – Earthwatch
PO Box 403
680 Mt Auburn Street
Watertown
MA 02272
United States of America
Tel: 1 617 926 8200
Fax: 1 617 926 8532
Geographical coverage: International
Aims and objectives: Founded in 1971, Earthwatch is an international association of citizens and scientists working to sustain the world's environment, to monitor global change, to conserve endangered species and to foster world health and international cooperation
Publications: *Earthwatch*, magazine six times a year; annual report

– 950 – Earthwatch Europe
Belsyre Court
57 Woodstock Road
Oxford OX2 6HU
United Kingdom
Tel: 0865 311600
Fax: 0865 311383
Contact name: Brian Walker, Executive Director
Geographical coverage: European
Aims and objectives: Earthwatch Europe, set up in conjunction with Earthwatch USA, aims to fund scientific field work by finding members of the public prepared to share both the costs and the labour of field research
Publications: *Earthwatch Europe Bulletin*; annual report

– 951 – Environment and Development Resource Centre (EDRC)
Boulevard Louis Schmidt 26
B-1040 Brussels
Belgium
Tel: 32 2 736 80 50
Fax: 32 2 733 97 67
Contact name: Ronald Kingham
Geographical coverage: International
Aims and objectives: The EDRC is an international organisation established in 1988. The purpose of the Centre is to promote global sustainable development which is ecologically sound

– 952 – Institute for Sustainable Development
ul Krzywickiego 9
Warsaw
Poland
Tel: 48 22 02078
Contact name: Andrzej Kassenberg
Geographical coverage: Poland

– 953 – Panos Institute
31 rue de Reuilly
F-75012 Paris
France
Tel: 33 1 437 929 35
Fax: 33 1 497 991 35
Geographical coverage: France
Aims and objectives: The Institute works to promote sustainable development

– 954 – Panos Institute
Alfred Place 8
London WC1E 7EB
United Kingdom
Tel: 071 278 11 11
Fax: 071 278 03 45
Contact name: Jon Tinker, President
Geographical coverage: United Kingdom

Aims and objectives: This independent information and policy studies institute, exists as three separate organisations in France, the UK and the US. Combining research with dissemination, the organisation works to promote sustainable development
Publications: *Panoscope*, bi monthly; and various studies

– 955 – Regional Environment Centre for Central and Eastern Europe

Miklós tér 1
H-1035 Budapest
Hungary
Tel: 36 1 168 6284
Fax: 36 1 168 7851
Contact name: Janusz Kindler, Chairperson
Geographical coverage: European
Aims and objectives: Promote ecologically sustainable development and increasing environmental awareness in the central/east European region
Publications: *Information Bulletin*, quarterly

BOOKS

– 956 – Sustainability and Environmental Policy

Geographical coverage: International
Publishers: Sigma
Editor: Dietz, F. J. et al
Language: English
Coverage: Discusses the concept of sustainability in relation to economic theory and environmental and economic policy
Price: DM 36.00
ISBN: 3894043431

– 957 – Third Revolution, The: Environment, Population and a Sustainable World: *see entry 2298*

Geographical coverage: United Kingdom

REPORTS

– 958 – Green Paper on the Environment – A Community Strategy for Sustainable Mobility

Geographical coverage: European
Publishers: Commission of the EC, 1992
Coverage: The Green Paper proposes introducing a framework for a Community strategy for sustainable mobility which would set down broad guidelines to contain the impact of transport on the environment

– 959 – State of the World 1992: A Worldwatch Institute Report on Progress Towards a Sustainable Society

1776 Massachusetts Avenue NW
Washington DC 20077-6628
United States of America
Geographical coverage: International
Publishers: Worldwatch Institute, 1992
Price: $10 95
ISBN: 0393308340
Author: Brown, L. R. et al

– 960 – Sustainable Development: An Imperative for Environment Protection

Economic Affairs Division
Marlborough House
Pall Mall
London SW1Y 5HX
United Kingdom
Geographical coverage: International
Publishers: Commonwealth Secretariat Publications, 1991
Coverage: The report identifies ways of facilitating the participation of developing countries in global agreements and in action to protect the environment
ISBN: 085092376X

RESEARCH CENTRES

– 961 – East European Environmental Research (ISTER)

Vaci u 62–64
H-1056 Budapest
Hungary
Tel: 36 1 136 70
Fax: 36 1 136 70
Contact name: Mr Janos Vargha
Geographical coverage: Hungary
Research activities: Research and consultancy on activities in the interests of sustainable development and new environmental policies

7
Energy (Non-Nuclear)

General

GOVERNMENT ORGANISATIONS

– 962 – Agence de l'Environnement et de la Maîtrise de l'Energie (ADEME)
(French Environment and Energy Management Agency)

27 rue Louis Vicat
F-75015 Paris
France
Tel: 33 1 476 520 00
Fax: 33 1 464 552 36
Contact name: Olivier Hertz, Director
Geographical coverage: France

– 963 – Bundesamt für Energiewirtschaft
(Federal Office of Energy)

Kapellenstrasse 14
CH-3003 Berne
Switzerland
Tel: 41 31 61 56 11
Fax: 41 31 26 43 07
Contact name: E Kiener, Director
Geographical coverage: Switzerland
Aims and objectives: Swiss energy administration
Publications: Technical reports in German and French

– 964 – Centro de Investigación Energética Medioambientale y Tecnológica (CIEMAT)

Avenida Compultense 22
E-28040 Madrid
Spain
Tel: 34 1 346 60 00
Fax: 34 1 346 60 05
Contact name: José Ángel Azuara Solis, Director General
Geographical coverage: Spain

– 965 – Danish Energy Agency: Ministry of Energy

Landemaerket 11
DK-1119 Copenhagen K
Denmark
Tel: 45 33 92 67 00
Fax: 45 33 11 47 43
Geographical coverage: Denmark

– 966 – Department of Energy

Clare Street
Dublin 2
Eire
Tel: 353 1 715233
Geographical coverage: Ireland

– 967 – Department of Energy

1 Palace Street
London SW1E 5HQ
United Kingdom
Tel: 071 215 5000
Geographical coverage: United Kingdom
Aims and objectives: Development of national policies in relation to all forms of energy, including energy efficiency and the development of new sources of energy, and international aspects of energy policy

General Energy (Non-Nuclear)

– 968 – Direção Geral de Energia
(General Directorate for Energy)

241 Rua da Beneficencia
P-1093 Lisbon
Portugal
Tel: 351 1 771091
Geographical coverage: Portugal
Aims and objectives: Energy planning; legislation; alternative sources of energy; energy policy; energy conservation; energy statistics

– 969 – Energiministeriet
(Ministry of Energy)

Slotsholmsgade 1
DK-1216 Copenhagen K
Denmark
Tel: 45 33 92 75 00
Geographical coverage: Denmark

– 970 – Ministerium für Soziale Gesundheit und Energie des Landes Schleswig-Holstein

Kronshagener Weg 130a
D-2300 Kiel 1
Germany
Tel: 49 431 1695 0
Fax: 49 431 15169
Contact name: Dr Klaus Rave
Geographical coverage: Germany
Aims and objectives: The Ministry is responsible for energy policy, energy technology and price regulations

– 971 – Ministry of Economic Affairs: Directorate General for Energy

Bezuidenhoutseweg 6
PO Box 20201
NL-2500 EC The Hague
Netherlands
Tel: 31 70 3796320
Fax: 31 70 3796358
Geographical coverage: Netherlands

– 972 – Ministry of Energy and Natural Resources

Ataturk Blvd
143 Daekanluklar
Turkey
Tel: 90 4 1174455
Fax: 90 4 1177971
Geographical coverage: Turkey

– 973 – Ministry of Industry, Energy and Technology and Commerce

80 Odos Mihalakopoulou
GR-115 28 Athens
Greece
Tel: 30 1 775 7657
Geographical coverage: Greece

– 974 – Ministry of Transport and Energy

19–21 boulevard Royal
Luxembourg
Luxembourg
Tel: 352 47 94 1
Fax: 352 46 43 15
Geographical coverage: Luxembourg

– 975 – Turkish Ministry of Energy and Natural Resources

Kanya Yolu
Bestpe
Ankara
Turkey
Tel: 90 4 2126420
Fax: 90 4 2229405
Geographical coverage: Turkey

NON-GOVERNMENT ORGANISATIONS

– 976 – Energik

Brouwersyliet 15
bus 7
B-2000 Antwerp
Belgium
Tel: 32 2 231 16 60
Fax: 32 2 233 76 60
Geographical coverage: Belgium

– 977 – Institut Français de l'Energie

3 rue Henri-Heine
F-75016 Paris
France
Tel: 33 1 452 446 14
Fax: 33 1 405 007 54
Contact name: Yves Chainet, General Manager
Geographical coverage: France
Aims and objectives: Environmental effects of thermal processes

– 978 – International Energy Agency (IEA)

2 rue André Pascal
F-75775 Paris Cedex 16
France

118 Energy (Non-Nuclear) *General*

Tel: 33 1 452 482 00
Fax: 33 1 452 499 88
Contact name: Helga Steeg, Executive Director
Geographical coverage: International
Aims and objectives: The IEA, founded in 1974 as an autonomous body within the OECD, is the energy Forum for twenty one countries. Its objective is to maintain and improve energy security of its member countries

– 979 – NOVEM

PO Box 17
NL-6130 AA Sittard
Netherlands
Tel: 31 4490 28260
Fax: 31 4490 28260
Contact name: Jack de Leeuw
Geographical coverage: Netherlands
Aims and objectives: Dutch organisation for energy and environment

– 980 – Stredisko pro Efektivni Vyuzivani Energie (SEVE)
(Energy Efficiency Centre)

Slezka 9
120 29 Prague 2
Czech Republic
Tel: 42 2 25 61 04
Fax: 42 2 25 85 56
Contact name: Ms Susan Legro
Geographical coverage: Czech Republic
Aims and objectives: SEVE aims to achieve economic development and environmental protection through improved energy efficiency

– 981 – WATT Committee on Energy

Savoy Hill House
Savoy Hill
London WC2R OBU
United Kingdom
Tel: 071 379 6875
Fax: 071 497 9315
Contact name: Geraldine Oliver, Information Manager
Geographical coverage: United Kingdom
Aims and objectives: The Watt Committee provides a forum for discussion of matters relating to energy. Many of the UK's top energy experts work for the Committee. Aims to promote research and development concerning all aspects of energy and to encourage informed opinion
Publications: Various reports. List available on request

BOOKS

– 982 – Digest of UK Energy Statistics

1 Palace Street
London SW1E 5HE
United Kingdom
Tel: 071 873 0011
Geographical coverage: United Kingdom
Publishers: Department of Energy
Frequency: annual
Price: £16.50
ISSN: 0307 0603

– 983 – Finland Kauppa–Jateollisuusministerioe Energiatilastot

Pohjoinene Makasiinikatu 6
SF-00130 Helsinki
Finland
Fax: 358 0 160 26 95
Geographical coverage: Finland
Publishers: Ministry of Trade and Industry Energy Department
Editor: Tervo, P.
Frequency: annual
Language: English Finnish and Swedish
Price: FIM118
ISSN: 0785 3165

– 984 – Introduction to Energy

Geographical coverage: United Kingdom
Publishers: Cambridge University Press, 1990
Coverage: Explores energy issues and the benefits and problems technology has brought
Price: £30.00
ISBN: 0521350913
Author: Cassedy, E. S. and Grossman, P. Z.

– 985 – Netherlands Centraal Bureau voor de Statistiek: Nederlandse Energiehuishouding

Prinses Beatrixlaan 428
Voorburg
Netherlands
Geographical coverage: Netherlands
Publishers: Centraal Bureau voor de Statistiek
Frequency: quarterly
Language: Dutch
ISSN: 0168 5236

– 986 – OECD World Energy Statistics

2 rue André Pascal
F-75775 Paris Cedex 16
France

Tel: 33 1 452 482 00
Fax: 33 1 452 485 00
Geographical coverage: International
Publishers: Organisation for Economic Cooperation and Development (OECD)
Frequency: annual
Language: English
Price: varies

DIRECTORIES/YEARBOOKS

– 987 – Annuaire de l'Administration des DRIR

22 rue Monge
F-75005 Paris
France
Tel: 33 1 458 362 02
Geographical coverage: France
Publishers: Annales des Mines
Editor: Matheu, M.
Frequency: annual
Coverage: Directory of the public agencies dealing with energy, industry, research and environment
Annual subscription: Ffr295
ISSN: 1140 7123

PERIODICALS

– 988 – Brennstoff-Wärme-Kraft (BWK)

Heinrichstrasse 24
Postfach 10 10 54
D-4000 Dusseldorf 1
Germany
Tel: 49 211 6188 0
Fax: 49 211 6188 112
Geographical coverage: Germany
Publishers: V D I Verlag GmbH
Editor: Pohl, W.
Frequency: monthly
Language: German text, summaries in English
Annual subscription: DM356
ISSN: 0006 9612

– 989 – Dansk Energi Tidsskrift

Rosenborggade 1130
Copenhagen K
Denmark
Tel: 45 33 15 22 77
Geographical coverage: Denmark
Publishers: Forlaget Beilin og Johansen ApS
Editor: Hesselea, S. J.
Frequency: 10 issues a year
Coverage: Focuses on energy in Danish trade and industry
Annual subscription: Dkr490
ISSN: 0108 8068

– 990 – Dimensione Energia

Via G B Martini 3
I-00198 Rome
Italy
Tel: 39 6 850 92465
Geographical coverage: Italy
Publishers: Ente Nazionale per l'Energia Elettrica
Frequency: bi monthly
Language: Italian text, summaries in English
Annual subscription: Lit40000

– 991 – EC Energy Monthly

Tower House
Southampton St
London WC2E 7HA
United Kingdom
Tel: 071 240 9391
Fax: 071 240 7946
Geographical coverage: European
Publishers: Financial Times Business Information Ltd
Frequency: monthly
Language: English
Coverage: Reports on energy related developments within the European Community
Annual subscription: £273.00

– 992 – Ekspress–Informatsiya Pryamoe Preobrazovanie Teplovoi i Khimicheskoi Energii V Elekricheskuyu

Baltiskaya ulitsa 14
Moscow A-219
Russia
Geographical coverage: Russia
Publishers: VINITI
Frequency: weekly
Language: Russian
Price: Rb38
ISSN: 0207 5032

– 993 – ENEA Notiziario Energia e Innovazione

Via Regina Margherita 125
I-00198 Rome
Italy
Tel: 39 6 85282778
Geographical coverage: Italy
Publishers: Redazione Nitiziario
Editor: Pierantoni, F. and Ceroni, A.
Frequency: monthly
Language: Italian text, summaries in English and French
Annual subscription: Lit50000
ISSN: 0393 716X

120 Energy (Non-Nuclear) *General*

– 994 – Energetica
Piaţa Presei Libere 1
71341 Bucharest
Romania
Tel: 40 0 18 06 30
Fax: 40 0 18 48 03
Geographical coverage: Romania
Publishers: Editura Technica
Editor: Pavel, E.
Frequency: monthly
Language: Romanian, occasionally in English, French or German
Coverage: Covers scientific and related economic and ecological problems stemming from the utilisation of all forms of energy
Price: lei800
ISSN: 0421 1715

– 995 – Energetyka
ul Biala 4
PO Box 1004
PL-00 950 Warsaw
Poland
Geographical coverage: Poland
Publishers: Wydawnictwo Czasopism i Ksiazek Technicznych SIGMA–NOT
Frequency: monthly
Language: Polish text, summaries in English
Price: $61
ISSN: 0013 7294

– 996 – Energia
Via Regina Margherita 290
I-00198 Rome
Italy
Tel: 39 6 4402061
Fax: 39 6 8840926
Geographical coverage: Italy
Publishers: Editrice dell'Automobile s r l
Editor: Clo, A.
Frequency: quarterly
Annual subscription: Lit200000
ISSN: 0392 7911

– 997 – Energia
Triana 51–53
E-28016 Madrid
Spain
Tel: 34 1 4576400
Fax: 34 1 4573945
Geographical coverage: Spain
Publishers: Ingenieria Quimica SA
Editor: de la Pezuela, P.
Frequency: bi monthly
Annual subscription: ptas1000
ISSN: 0210 2056

– 998 – Energiagazdalkodas
Lenin-korut 9–11
H-1072 Budapest 7
Hungary
Tel: 36 1 222 408
Geographical coverage: Hungary
Publishers: Lapkiado Vallalat
Editor: Rapp, T.
Frequency: monthly
Language: Hungarian text, contents page in English, German and Russian
Price: $50

– 999 – Energie & Milieutechnologie
(Energy and Environmental Technology)

Postbus 235
NL-2280 AE Rijswijk
Netherlands
Tel: 31 70 3988100
Fax: 31 70 3988276
Geographical coverage: Netherlands
Publishers: Stam Tijdschriften BV
Frequency: 8 issues a year
Language: Dutch
Price: Dfl94
ISSN: 0925 2924

– 1000 – Energieanwendung
Karl-Heinestrasse 27
D-7031 Leipzig
Germany
Tel: 37 41 4081011
Fax: 37 41 4012571
Geographical coverage: Germany
Publishers: Deutscher Verlag für Grundstoffindustrie
Frequency: monthly
Coverage: Covers the technology of energy and energy economics in industry agriculture and transportation
Annual subscription: DM198
ISSN: 0013 7405

– 1001 – Energiministeriet Energiforskningsprogram
Landemoerket 11
DK-1119 Copenhagen K
Denmark
Geographical coverage: Denmark
Publishers: Energistyrelsen
Frequency: quarterly
Price: free
ISSN: 0108 4011

General Energy (Non-Nuclear)

– 1002 – Energy Digest
177 Hagden Lane
Watford
Hertfordshire WD1 8LW
United Kingdom
Geographical coverage: United Kingdom
Publishers: Comprint Ltd
Editor: Perkins, J.
Frequency: bi monthly
Annual subscription: £35.00
ISSN: 0367 1119

– 1003 – Energy Economist
Tower House
Southampton Street
London WC2E 7HA
United Kingdom
Tel: 071 240 9391
Fax: 071 240 7946
Geographical coverage: International
Publishers: Financial Times Business Information Ltd
Frequency: monthly
Language: English
Coverage: International energy news, statistics, prices and market commentary on all major energy sources
Annual subscription: £330.00
ISSN: 0262 7108

– 1004 – Energy in Europe
L-2985 Luxembourg
Luxembourg
Geographical coverage: European
Publishers: Office for Official Publications of the European Communities
Editor: Lanham, Dr
Frequency: 3 issues a year
Language: English, French German and Spanish
Coverage: Concerned with the management of the energy sector in Europe
Annual subscription: $47
ISSN: 0256 6141

– 1005 – Enerpresse
142 rue Montmarte
F-75002 Paris
France
Fax: 33 1 403 997 52
Geographical coverage: France
Publishers: Société d'Information et de Documentation Bureau d'Information Professionnelles
Editor: Lavilleon, P.
Coverage: News and comments in the energy field
Annual subscription: Ffr11600
ISSN: 0153 9442

– 1006 – Engineering and Automation
Postfach 3240
D-8520 Erlangen 2
Germany
Geographical coverage: Germany
Publishers: Siemens Verlag AG
Frequency: bi monthly
Language: English
Coverage: Covers automation in the energy sector
Annual subscription: DM102
ISSN: 0939 205X

– 1007 – Forum
Postbus 1
NL-1755 ZG Petten
Netherlands
Tel: 31 2246 4413
Fax: 31 2246 3053
Geographical coverage: Netherlands
Publishers: Netherlands Energy Research Foundation
Editor: Slijkerman, A.
Frequency: monthly
Language: Dutch

– 1008 – Fusion
Dotzheimerstrasse 166
D-6200 Wiesbaden
Germany
Tel: 49 6121 806955
Fax: 49 6121 884101
Geographical coverage: Germany
Publishers: Dr Boettiger Verlags GmbH
Frequency: quarterly
Annual subscription: DM35
ISSN: 0173 9387

– 1009 – Great Britain Department of Energy Publications
1 Palace Street
London SW1E 5HE
United Kingdom
Fax: 071 834 3771
Geographical coverage: United Kingdom
Publishers: Department of Energy
Frequency: annual
Coverage: Lists annual department publications
Price: free
ISSN: 0951 855X

– 1010 – Green Magazine
PO Box 381
Millharbour
London E14 9TW
United Kingdom

Tel: 071 987 5090
Fax: 071 987 2160
Geographical coverage: United Kingdom
Publishers: Northern and Shell plc
Editor: Deeson, M.
Frequency: monthly

– 1011 – Informazione Innovativa

Via dell Palme 13
I-35137 Padua
Italy
Tel: 39 49 36435
Geographical coverage: Italy
Publishers: Centro Studi "l'Uomo e l'Ambiente"
Editor: Merlin, T.
Language: Text in Italian, summaries in English
Coverage: Clean technologies, energy sources, industrial health and safety
Annual subscription: Lit400000

REPORTS

– 1012 – Energi och Miljö i Norden 1991

Store Strandstraede 18
DK-1255 Copenhagen K
Denmark
Geographical coverage: Denmark
Publishers: Nordic Council of Ministers, 1991
Language: Danish
Coverage: Report on the way taxes and charges can be used as levers to control developments in the energy sector

– 1013 – Energy and Environmental Conflicts in East/Central Europe: The Case of Power Generation

Royal Institute of International Affairs
10 St James Square
London SW1Y 4LE
United Kingdom
Tel: 071 957 5711
Geographical coverage: European
Publishers: EEP/RIIA
Coverage: Suggest a range of policy options available in the power generation sectors of the different countries
ISBN: 0905031393
Author: Russell, J.

– 1014 – Energy in Denmark

Strandgade 29
DK-1901 Copenhagen
Denmark
Geographical coverage: Denmark
Publishers: Ministry of Energy
Frequency: annual

Price: free
ISSN: 0901 3768

– 1015 – Energy in the European Community

L-2985 Luxembourg
Luxembourg
Geographical coverage: European
Publishers: Office for Official Publications of the European Communities
Language: English
ISBN: 9282617378

– 1016 – Energy Taxation and Environmental Policy in EFTA Countries 1991

Rue de Varembé 9-11
CH-1211 Geneva 20
Switzerland
Geographical coverage: European
Publishers: EFTA, 1991
Language: English
Coverage: Summarises the experience of energy taxation in EFTA countries

– 1017 – European Energy Report

Tower House
Southampton Street
London WC2E 7HA
United Kingdom
Tel: 071 240 9391
Fax: 071 240 7946
Geographical coverage: European
Publishers: Financial Times Business Information Ltd
Frequency: fortnightly
Language: English
Coverage: Provides news and analyses from the European energy industries
Annual subscription: £449.00
Additional information: Also available online

RESEARCH CENTRES

– 1018 – Institut for Energiiteknikk (IFE)

PO Box 40
N-2007 Kjeller
Norway
Tel: 47 6 8060 00
Fax: 47 6 8163 56
Contact name: N G Aamodt, Managing Director
Geographical coverage: Norway
Research activities: Nuclear and non-nuclear fields including petroleum technology

– 1019 – Orkustonfnun
(National Energy Authority)

Grensäsvegur 9
108 Reykjavik
Iceland
Tel: 354 1 83600
Geographical coverage: Iceland
Publications: *Orkumál*; energy statistics; annual report
Research activities: Geological, geodetic, hydrological and hydraulic research for hydro-power developments

– 1020 – Toegepast Natuurwetenschappelijk Onderzoek (TNO)

PO Box 6013
NL-2600 JA Delft
Netherlands
Tel: 31 15 69 68 86
Fax: 31 15 61 31 86
Contact name: Prof Foppe De Walle
Geographical coverage: Netherlands
Research activities: TNO's expertise and research capacity is focused on the issue of energy and the environment and on process technology

DATABASES

– 1021 – DOE Energy

Geographical coverage: International
Coverage: All aspects of energy including nuclear, fossil, solar, tidal, wind, geothermal energy; policy; environmental impact and conservation
Database producer: US DOE in association with European agencies
Database host: Dialog
Coverage period: 1974–
Update frequency: fortnightly

– 1022 – EDF-DOC

Geographical coverage: International
Coverage: Electricity sources, production, transmission, distribution and applications of electricity
Database producer: Electricité de France Information Systems Research
Database host: ESA–IRS
Coverage period: 1970–
Update frequency: 2,000 records monthly

– 1023 – ENEL

Geographical coverage: Italy
Coverage: All aspects of the energy industry in Italy
Database producer: Ente Nazionale per l'Energia Elettrica (ENEL)
Database host: ESA–IRS
Coverage period: 1980–

– 1024 – Energie

Geographical coverage: Germany
Coverage: Covers German language literature on energy from Austria, Germany and Switzerland
Database producer: FIZ Karlsruhe
Database host: STN
Coverage period: 1976–

– 1025 – EUREKA

Geographical coverage: European
Coverage: Research projects including biotechnology, transport and energy technology applications
Database producer: EUREKA
Database host: ECHO
Coverage period: 1987–

Energy Conservation and Efficiency

REPORTS

– 1026 – Economic Restructuring in Eastern Europe and Acid Rain Abatement Strategies 1991

IIASA
A-2361 Laxenberg
Austria

Geographical coverage: European
Publishers: IIASA, 1991
Coverage: Report presents an energy scenario for Eastern European countries in which a gradually improved energy efficiency is assumed
Author: Amann, M., Hordijk, L. et al

Alternative Energy

GOVERNMENT ORGANISATIONS

– 1027 – Center for Energy Saving and Clean Technology: *see entry 1083*

Geographical coverage: Netherlands

– 1028 – Renewable Energy Advisory Group (REAG)

Electricity Division
1 Palace Street
London SW1E 5HE
United Kingdom
Tel: 071 215 3484
Geographical coverage: United Kingdom

NON-GOVERNMENT ORGANISATIONS

– 1029 – Asociación Ecologista de Defensa de la Naturaleza (AEDENAT)

c/o Campomanes 13
2 izq
E-28013 Madrid
Spain
Tel: 34 1 541 1071
Fax: 34 1 571 7108
Contact name: Dr J Carlos Rodrigues Murillo
Geographical coverage: Spain
Aims and objectives: AEDENAT presented an alternative energy plan for Spain in 1991

– 1030 – Association Promotion Energies Renouvelables (APERE)

(Association for the Promotion of Renewable Energies)

18 rue de la Sablonnière
B-1000 Brussels
Belgium
Tel: 32 2 218 18 96
Fax: 32 2 223 14 96
Contact name: Mr Jean Marc Van Nypelseer
Geographical coverage: Belgium
Aims and objectives: APERE promotes increased renewable energies

– 1031 – British Wind Energy Association (BWEA)

4 Hamilton Place
London W1V 0BQ
United Kingdom
Tel: 071 499 6230
Fax: 071 499 6230
Geographical coverage: United Kingdom
Aims and objectives: BWEA promotes discussion and disseminates information on wind energy studies, projects and products through publications and meetings
Publications: *Wind Directions*, quarterly; *Wind Engineering*, quarterly
Research activities: To promote, discuss and to disseminate information on wind energy

– 1032 – Brontosaurus Movement/EcoWatt

Bubenska 6
170 00 Prague 7
Czech Republic
Tel: 42 2 80 29 10
Fax: 42 2 80 29 06
Contact name: Mr Jan Hrebec
Geographical coverage: Czech Republic
Aims and objectives: EcoWatt is the Brontosaurus working group dealing with alternative energy strategies

– 1033 – Centre for Alternative Technology (CAT)

Machynlleth
Powys
Wales SY20 9AZ
United Kingdom
Tel: 0654 702400
Contact name: Roger Kelly, Director
Geographical coverage: United Kingdom
Aims and objectives: The Centre founded in 1973, demonstrates and promotes sustainable technologies and ways of living, including renewable energy sources, energy conservation and organic growing. It provides working displays of a wide range of alternative technologies
Publications: *Clean Slate* newsletter, many pamphlets, information sheets, resource booklets and technical papers

– 1034 – Danish Organisation for Renewable Energy

Willemoesgade 14
DK-21000 Copenhagen Ø
Denmark
Tel: 45 31 42 90 91
Fax: 45 31 42 90 95
Contact name: Mr Gunnar Boye Olesen
Geographical coverage: Denmark
Aims and objectives: The Organisation supports the development and use of renewable energy

Alternative Energy **Energy (Non-Nuclear)** 125

– 1035 – European Wind Energy Association

Via Bormida 2
I-00198 Rome RM
Italy
Fax: 39 6 841 1933
Geographical coverage: European

– 1036 – Göncöl Alliance

Ilona u 3 Pf 184
H-2600 Vac
Hungary
Tel: 36 27 11179
Fax: 36 11 568957
Contact name: Mr Agoston Nagy
Geographical coverage: Hungary
Aims and objectives: Local and national campaign for renewable energy and energy efficiency

– 1037 – International Solar Energy Society (ISES)

Twyford High School
Twyford Crescent
London W3 9PP
United Kingdom
Tel: 081 993 6999
Geographical coverage: International
Aims and objectives: ISES is a forum for all those interested in utilising renewable energy sources. Promotes the study of the applications of the sun's power
Publications: List available on request

– 1038 – New Energy Technologies (THERMIE)

DGXVII THERMIE Program
Rue de la Loi 200
B-1049 Brussels
Belgium
Tel: 32 2 236 04 36
Fax: 32 2 235 01 50
Contact name: Michael Thomas Gowen
Geographical coverage: European
Aims and objectives: THERMIE was set up to support projects for the application of new energy technologies, whose realisation is associated with a considerable degree of risk

– 1039 – Renewable Energy Enquiries Bureau

Building 149
Harwell
Oxfordshire OX11 0RA
United Kingdom
Tel: 0235 432350
Geographical coverage: United Kingdom
Aims and objectives: Information service

– 1040 – Solar Energy Society of Ireland

University College of Dublin
School of Engineering
Bellfield
Dublin 4
Eire
Tel: 353 1 693244
Fax: 353 1 830921
Geographical coverage: Ireland

– 1041 – SOLEREC Département Solaire

La Motte Saint Euverte
BP 11
F-45800 St Jean de Braye
France
Tel: 33 388 660 49
Fax: 33 387 527 45
Geographical coverage: France

– 1042 – Svenska Solenergiföreningen

(Swedish Solar Energy Society)

c/o SERC
Box 10044
S-781 10 Borlänge
Sweden
Geographical coverage: Sweden

– 1043 – Waste Processing Association

Runnymede Malt House
Runnymede Road
Egham
Surrey TW20 9BO
United Kingdom
Tel: 0784 434377
Contact name: Peter King, Secretary
Geographical coverage: United Kingdom

– 1044 – World Fuel Cell Council

Stettenstrasse 2
D-6000 Frankfurt am Main
Germany
Tel: 49 69 597 5070
Geographical coverage: International

BOOKS

– 1045 – Energy Conscious Design: A Primer for European Architects

Geographical coverage: European
Publishers: Batsford, 1992

126 Energy (Non-Nuclear) *Alternative Energy*

Coverage: Intoduction to the theory and practice of designing energy efficient passive solar buildings
Price: £25.00
ISBN: 0713469196
Author: Lewis, O. and Goulding, J.

– 1046 – Energy Without End

Geographical coverage: United Kingdom
Publishers: Friends of the Earth (FoE), 1991
Coverage: Explains how modern systems like wind and water turbines, solar panels, photovoltaics and growing crops can harness the renewable natural energy flows
Price: £7.95
ISBN: 0905966872

– 1047 – Renewable Energy Sources for Fuels and Electricity

Geographical coverage: International
Publishers: Earthscan Publications
Editor: Johansson, T. B. et al
Language: English
Coverage: Contributions from fifty of the world's leading figures on satisfying future needs for power in ways which are compatible with sustainable progress
Price: £60.00

– 1048 – Solar Power via Satellite

Geographical coverage: United Kingdom
Publishers: Ellis Horwood, 1991
Coverage: Examines the technical, demographic, political, diplomatic and judicial aspects of solar power via satellite
Price: £35.00
ISBN: 0138248060
Author: Glaser, P. E. and Davidson, F. P.

DIRECTORIES/YEARBOOKS

– 1049 – Wind-Energie Jahrbuch

Gielsdorfer Strasse 16
D-5300 Bonn
Germany
Geographical coverage: Germany
Publishers: Wind-Energie Sekretariat
Editor: Schonbal, W.
Frequency: annual
Price: DM28
ISSN: 0720 3926

PERIODICALS

– 1050 – Alternative Times

35 Wedmore Street
London N19 4RU
United Kingdom
Geographical coverage: United Kingdom
Publishers: R Stevens
Frequency: quarterly
Coverage: Run down of alternative energy news and views plus items on nuclear power
Price: £3.00
ISSN: 0261 6033

– 1051 – Biomass and Bioenergy

Headington Hill Hall
Oxford OX3 0BW
United Kingdom
Tel: 0865 794141
Fax: 0865 743911
Geographical coverage: United Kingdom
Publishers: Pergamon Press plc
Frequency: monthly
Coverage: Presents research papers on all aspects of biomass and bioenergy
Annual subscription: £295.00
ISSN: 0961 9534

– 1052 – Energie Alternative: Habitat Territorio Energia

Via Fratelli Bressan 2
I-20126 Milan
Italy
Tel: 39 2 25 79 841
Fax: 39 2 25 52 779
Geographical coverage: Italy
Publishers: Editoriale P G SpA
Frequency: bi monthly
Coverage: Covers renewable energies and rational use of conventional ones. Also considers environmental problems
Annual subscription: Lit44000
ISSN: 0391 5360

– 1053 – Era Solar

Costa Rica 13
4 a-2
E-28016 Madrid
Spain
Tel: 34 1 350 58 85/250 62 16
Geographical coverage: Spain
Editor: Sentra Diaz de Cevallos, M.
Coverage: Aspects related to solar energy

– 1054 – Green Energy Matters

82 Rivington Street
London EC2A 3AY
United Kingdom
Tel: 071 613 0087
Fax: 071 613 0094
Geographical coverage: United Kingdom
Publishers: EconoMatters Ltd
Editor: Ouseley, R.
Coverage: Putting energy and the environment into perspective

– 1055 – International Journal of Ambient Energy

PO Box 25
Lutterworth
Leicestershire LE17 4FF
United Kingdom
Tel: 0455 202281
Geographical coverage: International
Publishers: Ambient Press Ltd
Editor: McVeigh, J. C.
Frequency: quarterly
Language: English
Coverage: Information on renewable energy
Annual subscription: £69.00
ISSN: 0143 0750

– 1056 – International Journal of Solar Energy

PO Box 90
Reading
Berks RG1 8JL
United Kingdom
Tel: 0734 560080
Fax: 0734 568211
Geographical coverage: International
Publishers: Harwood Academic Publishers
Editor: Palz, W.
Frequency: 6 issues per volume
Coverage: Science and engineering of solar energy
Annual subscription: £88.00
ISSN: 0142 5919

– 1057 – Practical Alternatives

Victoria House
Bridge Street
Rhayader
Wales LD6 5AG
United Kingdom
Tel: 0597 810929
Geographical coverage: United Kingdom
Publishers: David Stephens (editor and publisher)
Frequency: irregular
Coverage: Reports on building of a solar village with new survivor design of passive solar house as a means of reducing energy consumption

Price: £7.00
ISSN: 0262 4540

– 1058 – Safe Energy

11 Forth Street
Edinburgh
Scotland EH1 3LE
United Kingdom
Tel: 031 557 4283
Fax: 031 557 4284
Geographical coverage: United Kingdom
Publishers: SCRAM
Frequency: bi monthly
Coverage: Produced for the British anti-nuclear and safe energy movements
Price: £1.50
ISSN: 0140 7340

– 1059 – Solar Energy Materials and Solar Cells

PO Box 211
NL-1000 AE Amsterdam
Netherlands
Tel: 31 20 5803911
Fax: 31 20 5803598
Geographical coverage: International
Publishers: Elsevier Science Publishers BV
Editor: Lampert, C. M.
Frequency: monthly
Language: English
Annual subscription: Dfl903
ISSN: 0927 0248

– 1060 – Solar Energy Research and Development in the European Community: Series A–H

Spuiboulevard 50
PO Box 17
NL-3300 AA Dordrecht
Netherlands
Tel: 31 78 334911
Fax: 31 78 334254
Geographical coverage: European
Publishers: Kluwer Academic Publishers
Frequency: irregular
Language: English
Coverage: Covers all aspects of solar energy, including solar energy applications, photovoltaic power generation, energy from biomass, solar radiation data and solar energy in industry and agriculture

– 1061 – Solaria

Stapferstrasse 27
CH-5200 Brugg
Switzerland
Geographical coverage: Switzerland

Publishers: Keller und Co
Editor: Sabady, P.
Frequency: bi monthly
Price: Sfr15

– 1062 – Sonnenenergie und Wärmepumpe

Niederwall 53
D-4800 Bielefeld 1
Germany
Geographical coverage: Germany
Publishers: Bielefelder Verlagsanstalt GmbH
Editor: Urbanek, A.
Frequency: bi monthly
Annual subscription: DM54
ISSN: 0172 5912

– 1063 – Sunceva Energija

(Solar Energy)

Narodnog Ustanka 58
51000 Rijeka
Croatia
Tel: 38 51 322 10
Fax: 38 51 515 403
Geographical coverage: Croatia
Publishers: Croatian Solar Energy
Editor: Frankovic, B.
Language: Croatian text, summaries in English
Coverage: Studies solar energy and other renewable sources
Price: din300
ISSN: 0351 2797

– 1064 – Wind Engineering

107 High Street
Brentwood
Essex CM14 4RX
United Kingdom
Geographical coverage: United Kingdom
Publishers: Multi Science Publishing Co Ltd
Frequency: quarterly
Language: English
ISSN: 0309 524X

– 1065 – Windkraft Journal

D-2372 Brekendorf
Germany
Tel: 49 4353 551
Fax: 49 4353 796
Geographical coverage: Germany
Publishers: Verlag Natürliche Energie GmbH

– 1066 – Windpower Monthly

Torgny Moeller
Vrinners Hoved
DK-8420 Knebel
Denmark
Tel: 45 86 36 59 00
Fax: 45 86 36 56 26
Geographical coverage: International
Frequency: monthly
Language: English
Coverage: International wind energy news and views on environmental impact
ISSN: 0109 7318

REPORTS

– 1067 – Energy from Waste State-of-the-Art Report

Bremerholm 1
DK-1069 Copenhagen K
Denmark
Geographical coverage: International
Publishers: International Solid Wastes Association
Language: English
Coverage: ISWA's working group on waste incineration provides detailed national surveys
Price: £40.00

– 1068 – Wind Energy in Europe Report

c/o Ecotecnica
Demóstenes 6
E-08028 Barcelona
Spain
Tel: 34 3 330 78 60
Geographical coverage: European
Publishers: European Wind Energy Association

RESEARCH CENTRES

– 1069 – Centro de Estudios de la Energía Solar (CENSOLAR)

Avenida Republica Argentina 1
E-41011 Seville
Spain
Geographical coverage: Spain

– 1070 – Centro di Ecologia e Climatologia/Osservatorio Geofisico Sperimentale Macerata
(Centre of Ecology and Climatology/Geophysical Experimental Observatory)

Via Indipendenza 180
I-62100 Macerata
Italy
Tel: 39 733 30743
Geographical coverage: Italy
Publications: *Rendiconti Osservatorio Geofisico*, annual
Research activities: Atmospheric electricity; solar energy; wind energy; climatology

– 1071 – Comitato Nazionale per la Ricerca e per lo Sviluppo dell'Energia Nucleare e dell'Energie Alternativ
(Italian Commission for Nuclear and Alternative Energy Sources)

Via Regina Margherita 125
I-00198 Rome
Italy
Tel: 39 6 85281
Geographical coverage: Italy
Publications: *Notizario*, monthly; *Risparmio Energetico*, quarterly; *Nucleare*, quarterly and *Sicureza e Protezione*; quarterly
Research activities: Alternative energy source; fast breeder reactors

– 1072 – Deutsche Wetterdienst (DWD)
(German Meteorological Service)

Frankfurter Strasse 135
Postfach 10 04 65
D-6050 Offenbach am Main
Germany
Tel: 49 69 80620
Fax: 49 69 8062 339
Geographical coverage: Germany
Publications: Several reports issued on a regular basis; annual report
Research activities: Relations between climatology and technical problems e.g. utilisation of wind and solar energy, air conditioning

– 1073 – Energy Technology Support Unit (ETSU): *see entry 1097*

Geographical coverage: United Kingdom

– 1074 – Hahn-Meitner Institut Berlin GmbH (HMI)

Glienicker Strasse 100
D-1000 Berlin 39
Germany
Tel: 49 30 8009 1
Fax: 49 30 8009 2047
Contact name: Erich te Kaat, Scientific Director
Geographical coverage: Germany
Research activities: Solid state physics and solar energy

– 1075 – Institut für Energie- und Transportforschung Meissen
(Energy and Transport Research Institute)

Kynastweg 57
D-8250 Meissen
Germany
Geographical coverage: Germany
Research activities: Energy conservation and the introduction of alternative energy sources to agriculture

– 1076 – Institute of Meteorology and Physics of the Atmospheric Environment

National Observatory of Athens
PO Box 20048
GR-118 10 Athens
Greece
Tel: 30 1 3456 257
Fax: 30 1 3463 803
Geographical coverage: Greece
Research activities: Wind energy

– 1077 – Istituto per l'Edilizia ed il Risparmio Energetico: *see entry 1099*

Geographical coverage: Italy

– 1078 – Netherlands Energy Research Foundation ECN: *see entry 1140*

Geographical coverage: Netherlands

– 1079 – Schweizerische Meteorologische Anstalt
(Swiss Meteorological Institute)

Krabuhlstrasse 58
CH-8044 Zurich
Switzerland
Tel: 44 1 252 67 20
Geographical coverage: Switzerland
Publications: Various publications
Research activities: Solar radiation research

DATABASES

– 1080 – International Solar Energy Intelligence Report

Geographical coverage: International

Coverage: All aspects of alternative energy including solar energy, photovoltic and wind power
Database producer: Business Publishers Inc
Database host: DataStar; Dialog
Coverage period: 1988–

– 1081 – SESAME

Geographical coverage: European
Coverage: Energy technology projects including renewables
Database producer: CEC Energy, CEC Science Research and Development
Database host: Eurobases
Coverage period: 1975–
Update frequency: weekly

Energy Conservation and Efficiency

GOVERNMENT ORGANISATIONS

– 1082 – AFME

(Association for the Control and Saving of Energy)

27 rue Louis Vicat
F-75015 Paris
France
Tel: 33 1 476 524 85
Geographical coverage: France

– 1083 – Center for Energy Saving and Clean Technology

Oude Delft 180
NL-2611 HH Delft
Netherlands
Tel: 31 15 150 150
Geographical coverage: Netherlands

NON-GOVERNMENT ORGANISATIONS

– 1084 – Building Energy Efficiency Division

Building Research Establishment
Garston
Watford WD2 7JR
United Kingdom
Tel: 0923 664258
Geographical coverage: United Kingdom
Publications: Statistical information and International Energy Efficiency Project Papers

– 1085 – Combined Heat and Power Association (CHPA)

Grosvenor Gardens House
35–37 Grosvenor Gardens
Victoria
London SW1W 0BS
United Kingdom
Tel: 071 828 4077
Fax: 071 828 0310
Contact name: David I Green, Director
Geographical coverage: European
Aims and objectives: CHPA is a membership based association working to secure the wider use of combined heat and power and community heating, an energy efficient and environmentally friendly technology
Publications: *CHPA Bulletin*; newsletter quarterly, *CHPA Handbook*; annually. Various occasional technical and policy papers

– 1086 – European Association for Conservation of Energy (Euro ACE)

9 Sherlock Mews
London W1M 3RH
United Kingdom
Tel: 071 935 1495
Fax: 071 935 8346
Contact name: Mr Andrew Warren
Geographical coverage: European
Aims and objectives: Euro ACE campaigns for greater investment in energy efficient measures in order to reduce emissions in greenhouse gases
Publications: Various publications

– 1087 – Göncöl Alliance: *see entry 1036*

Geographical coverage: Hungary

DIRECTORIES/YEARBOOKS

– 1088 – European Directory of Renewable Energy Suppliers and Service

5 Castle Road
London NW1 8PR
United Kingdom
Geographical coverage: European
Publishers: James & James Science Publishers
Price: £24.95

PERIODICALS

– 1089 – Applied Energy

Crown House
Linton Road
Barking
Essex IG11 8JU
United Kingdom
Tel: 081 594 7272
Fax: 081 594 5942
Geographical coverage: United Kingdom
Publishers: Elsevier Science Publishers Ltd
Editor: Probert, S. D.
Frequency: monthly
Coverage: Covers energy conversion, conservation and the optimal management and use of energy and power sources
Annual subscription: £425.00
ISSN: 0306 2619

– 1090 – Economia Delle Fonti di Energia

Via Monza 106
I-20127 Milan
Italy
Tel: 39 2 28 27 651
Geographical coverage: Italy
Publishers: Franco Angeli Editore
Editor: Vacca, S.
Frequency: 3 issues a year
Annual subscription: Lit70000

– 1091 – Energie Alternative: Habitat Territorio Energia: *see entry 1052*

Geographical coverage: Italy

– 1092 – Fifth Fuel

9 Sherlock Mews
London W1M 3RH
United Kingdom
Tel: 071 935 1495
Geographical coverage: United Kingdom
Publishers: Association for the Conservation of Energy
Frequency: irregular
Coverage: Energy efficiency and conservation
Price: free

– 1093 – Institutul de Studii si Proiectari Energetice Bulletinul

Bd Lacul Tei nr 1
Sector 2
72301 Bucharest 30
Romania
Tel: 40 0 107080
Fax: 40 0 112502
Geographical coverage: Romania
Publishers: Institutul de Studii si Proiectan Energetice
Frequency: quarterly
Language: English, French, German and Spanish
ISSN: 1220 4145

– 1094 – International Journal of Global Energy Issues

110 avenue Louis Casai
CP 306
CH-1215 Geneva Airport
Switzerland
Fax: 41 22 7910885
Geographical coverage: International
Publishers: Inderscience Enterprises Ltd
Frequency: quarterly
Language: English
Coverage: Covers energy policy, management and conservation. Development of alternate energy and economic control
Annual subscription: $155
ISSN: 0954 7118

REPORTS

– 1095 – Economic Restructuring in Eastern Europe and Acid Rain Abatement Strategies 1991

IIASA
A-2361 Laxenberg
Austria
Geographical coverage: European
Publishers: IIASA, 1991
Coverage: Report presents an energy scenario for Eastern European countries in which a gradually improved energy efficiency is assumed
Author: Amann, M., Hordijk, L. et al

RESEARCH CENTRES

– 1096 – Centrum voor Energiebesparing en Schone Technologie
(Centre for Energy Conservation and Environmental Technologies)

Oude Delft 180
NL-2611 HH Delft
Netherlands
Tel: 31 15 150 150
Fax: 31 15 150 151
Contact name: Mr Hans Becht, Project Manager
Geographical coverage: Netherlands
Research activities: The role of hydrogen in the energy supply; sustainable mobility. Various projects on energy conservation

– 1097 – Energy Technology Support Unit (ETSU)

Building 156
Harwell
Oxfordshire OX11 0RA
United Kingdom
Tel: 0235 433517
Fax: 0235 432923
Contact name: Dr David Martin, Business Development Manager
Geographical coverage: United Kingdom
Aims and objectives: ETSU manages and promotes the results of government research, development and demonstration programmes in energy efficiency, renewables and clean coal technology
Publications: Various leaflets and reports

– 1098 – Fundacja na Rzecz Efelatywnego Wykorzystanos Energii
(Polish Foundation for Energy Efficiency)

ul Gorskiego 7
PL-00 330 Warsaw
Poland
Tel: 48 22 278454
Fax: 48 22 273271
Contact name: Mr Adam Gula
Geographical coverage: Poland
Aims and objectives: Promotes energy efficiency

– 1099 – Istituto per l'Edilizia ed il Risparmio Energetico
(Energy Conservation in Buildings Institute)

Istituto di Fisica Tecnica
Viale delle Scienze
I-90128 Palermo
Italy
Tel: 39 91 487401; 422511
Geographical coverage: Italy
Research activities: Passive solar energy; energy conservation in buildings

– 1100 – Specific Action for Vigorous Energy Efficiency (SAVE)

CEC
DGXVII-C/2-SAVE Program
Rue de la Loi 200
B-1049 Brussels
Belgium
Tel: 32 2 235 40 87
Fax: 32 2 236 42 54
Contact name: Armand Colling
Geographical coverage: European
Aims and objectives: The programme's objectives are to encourage energy efficiency throughout the EC, thereby contributing to better use of resources

DATABASES

– 1101 – BRIX/FLAIR

BRE
Garston
Watford WD2 7JR
United Kingdom
Tel: 0923 894040
Geographical coverage: International
Coverage: Energy conservation in building design
Database producer: Building Research Establishment, Fire Research Station
Database host: ESA–IRS
Coverage period: 1950–
Update frequency: monthly

– 1102 – Energy Conservation News

Geographical coverage: United Kingdom
Coverage: All aspects of energy conservation
Database producer: Business Communications Company Inc
Database host: DataStar; Dialog
Coverage period: 1976–

– 1103 – Energyline

Geographical coverage: International
Coverage: Energy conservation and development
Database producer: R R Bowker
Database host: DataStar; Dialog; ESA–IRS; Orbit
Coverage period: 1971–
Update frequency: 350 records monthly
Additional information: Available on CD–ROM

– 1104 – VANYTT

Halsingegatan 49
S-113 31 Stockholm
Sweden
Tel: 46 8 34 01 70
Geographical coverage: Sweden
Coverage: Covers aspects of indoor air pollution and energy conservation in buildings
Database producer: The Swedish Institute of Building Documentation (BYGGDOK)

Fossil Fuels

GOVERNMENT ORGANISATIONS

– 1105 – Bundesamt für Wirtschaft

(Federal Office for Trade and Industry Branch for Mineral Oil and Gas)

Frankfurter Strasse 29/31
D-6236 Eschborn 1
Germany
Geographical coverage: Germany

– 1106 – Ministerio de Industria Comercio y Turismo

Paseo de la Castellana 160
Planta 6
E-28046 Madrid
Spain
Tel: 34 1 3494545
Fax: 34 1 4578066
Geographical coverage: Spain
Aims and objectives: The Ministry is concerned with all matters relating to law and the exploitation of hydrocarbon reserves

– 1107 – Ministry of Industry

PO Box 8014
Dep N-0030 Oslo 1
Norway
Tel: 47 2 34 90 90
Fax: 47 2 34 95 25
Geographical coverage: Norway

– 1108 – Royal Ministry of Petroleum and Energy

Ploensgt 8
PO Box 8148 Dep
N-0030 Oslo
Norway
Tel: 47 2 34 90 90
Geographical coverage: Norway

– 1109 – State Pollution Control Authority (SFT): Oil Pollution Control Department

PO Box 125
N-3191 Horten
Norway
Tel: 47 33 44 161
Fax: 47 33 44 257
Geographical coverage: Norway
Aims and objectives: SFT is responsible for coordinating public and private emergency services to provide a national emergency response system

NON-GOVERNMENT ORGANISATIONS

– 1110 – Association Technique de l'Industrie du Gaz en France

62 rue de Courcelles
F-75008 Paris
France
Tel: 33 1 475 434 34
Fax: 33 1 422 749 43
Contact name: Pierre Henry, President
Geographical coverage: France
Aims and objectives: The Association covers all fuel activities concerning gases

– 1111 – Chemical and Oil Recycling Association

O'r Diwedd
Parc Lund
Kinmel Bay
Rhyl
Clwyd
Wales LL18 5JG
United Kingdom
Tel: 0745 332427
Contact name: Mr J Looker
Geographical coverage: United Kingdom
Aims and objectives: The Association aims to promote and represent the interest of members in the recovery and re-use of oils and solvents

134 Energy (Non-Nuclear) *Fossil Fuels*

– 1112 – Deutscherverein des Gas und Wasserfaches eV (DVGW)

Haupstrasse 71–79
D-8236 Eschborn 1
Germany
Geographical coverage: Germany

– 1113 – European Petroleum Industries Association (EUROPIA)

Avenue Louise 140
B-1050 Brussels
Belgium
Tel: 32 2 645 19 11
Fax: 32 2 646 49 17
Contact name: Gilbert Portal, Secretary General
Geographical coverage: European
Aims and objectives: Founded in 1989, the Association consists of 35 companies representing the petroleum industry across the EC. Its main objectives are to represent the view of the industry to the EC institutions, and to contribute to study and problem solving

– 1114 – Federazione Italiana Impresa Pubbliche Gas Acqua e Varie (FEDERGASACQUA)

Piazza Cola di Rienzo 80
I-00192 Rome
Italy
Geographical coverage: Italy

– 1115 – Geological Survey of the Netherlands

PO Box 157
NL-2000 AD Haarlem
Netherlands
Tel: 31 23 300300
Fax: 31 23 351614
Contact name: Dr H M van Montfrans, Head Subsurface Oil and Gas Department
Geographical coverage: Netherlands

– 1116 – International Oil Pollution Compensation Fund (IOPC)

4 Albert Embankment
London SE1 7SR
United Kingdom
Tel: 071 582 2606
Fax: 071 735 0326
Geographical coverage: International
Aims and objectives: IOPC is a compensation fund organised by the International Maritime Organisation (IMO) for victims of oil pollution damage
Publications: *Claims Manual*; annual report; statistics

– 1117 – International Petroleum Industry Environmental Conservation Association (IPIECA)

1 College Hill
1st Floor
London EC4R 2RA
United Kingdom
Tel: 071 248 3447
Fax: 071 489 9067
Geographical coverage: International
Aims and objectives: IPIECA was formed in 1974, by BP, Exxon, Mobil and Shell, to be the point of contact between the international petroleum industry and UNEP.
Publications: *IPIECA Newsletter*, 3 times a year

– 1118 – International Tanker Owners Pollution Federation Ltd (ITOPF)

Staple Hall
Stonehouse Court
87–90 Houndsditch
London EC3A 7AX
United Kingdom
Tel: 071 621 1255
Fax: 071 621 1783
Contact name: Dr I C White, Managing Director
Geographical coverage: International
Aims and objectives: ITOPF was established in 1968 as a service organisation. Devotes considerable efforts to the provision of technical services an the fields of response to marine oil spills, damage assessment, contingency planning, training and information
Publications: *Ocean Orbit Newsletter*; annually, technical information papers, training videos and others

– 1119 – Mineralöl Wirtschafts Verband eV (MWV)

(Association of Petroleum Industry)

Steindamm 71
D-2000 Hamburg 1
Germany
Tel: 49 40 28540
Fax: 49 40 2854/53
Contact name: Dr Peter Schluter, Executive Director
Geographical coverage: Germany

– 1120 – Oil Companies European Organisation for Environmental Health and Protection (CONCAWE)

Madouplein 1
B-1030 Brussels
Belgium
Tel: 32 2 220 31 11
Fax: 32 2 219 46 46
Contact name: Klaus Kohlhase, Chairman
Geographical coverage: European

Aims and objectives: Founded in 1963 by six oil companies and now has a membership of 36 companies. These represent 90% of all the refining capacity in Western Europe. The scope of its activities include air, land and water pollution control and health protection
Publications: Technical reports. Catalogues are available on request

– 1121 – Österreichisches Vereinigung für das Gas und Wasserfach

Schubertring 14
A-1010 Vienna
Austria
Geographical coverage: Austria

– 1122 – Société Suisse de l'Industrie de Gaz et des Eaux (SSIGE)

Grutlistrasse 44
CH-8027 Zurich
Switzerland
Geographical coverage: Switzerland

BOOKS

– 1123 – UK Marine Oil Pollution Legislation

Geographical coverage: United Kingdom
Publishers: Lloyds of London Press, 1987
Editor: Bates, J. H.
ISBN: 1850044109X

DIRECTORIES/YEARBOOKS

– 1124 – North Sea Oil and Gas Directory

Geographical coverage: United Kingdom
Publishers: Benn Business Information Services Ltd, 1992
Frequency: annual
Coverage: Lists organisations related to the North Sea oil and gas industry
Price: £62.00
ISSN: 0265 5039
ISBN: 0863821811

PERIODICALS

– 1125 – British Coal Research Association (BCRA) Quarterly

Mill Lane
Wingerworth
Chesterfield
Derbyshire S42 6NG
United Kingdom
Tel: 0246 209654
Fax: 0246 272247
Geographical coverage: United Kingdom
Publishers: BCRA Scientific and Technical Services Ltd
Editor: Edwards, D. G.
Frequency: quarterly
Coverage: Reviews published literature on coal, coke and allied topics
Annual subscription: £80.00

– 1126 – Coal Prospects and Policies in IEA Countries

2 rue André-Pascal
F-75775 Paris Cedex 16
France
Tel: 33 1 452 482 00
Geographical coverage: International
Publishers: OECD International Energy Agency
Frequency: irregular
Language: English
Price: varies

– 1127 – Coal Research Projects

IEA Coal Research
Gemini House
10–18 Putney Hill
London SW15 6AA
United Kingdom
Tel: 081 780 2111
Fax: 081 780 1746
Geographical coverage: International
Publishers: IEA Coal Research
Frequency: annual
Language: English
Coverage: Research projects in coal worldwide
Price: £500
Additional information: Available online

– 1128 – Colliery Guardian

Queensway House
2 Queensway
Redhill
Surrey RH1 1QS
United Kingdom
Tel: 0737 768611
Geographical coverage: United Kingdom
Publishers: F M J International Publications Ltd

Editor: Schwartz, M.
Frequency: 6 issues a year
Coverage: Explores coal mining and resources
Annual subscription: £72.00
ISSN: 0010 1281

– 1129 – Greenhouse Issues

IEA Coal Research
Gemini House
10–18 Putney Hill
London SW15 6AA
United Kingdom
Tel: 081 780 2111
Fax: 081 780 1746
Contact name: Deborah Norman
Geographical coverage: United Kingdom
Publishers: IEA Coal Research on behalf of IEA Greenhouse Gas R & D Programme
Coverage: Worldwide developments in the field of greenhouse gases and fossil fuels
Price: free

– 1130 – International Coal Letter

19 rue Capouillet
Box 1
B-1060 Brussels
Belgium
Tel: 32 2 536 86 11
Fax: 32 2 536 86 00
Geographical coverage: International
Publishers: International Coal Letter
Editor: Doerell, P. E.
Frequency: fortnightly
Language: English
Annual subscription: Bfr20500

– 1131 – Petroleum Review

61 New Cavendish Street
London W1M 8AR
United Kingdom
Tel: 071 636 1004
Fax: 071 255 1472
Geographical coverage: United Kingdom
Publishers: Institute of Petroleum
Editor: Reader, C
Frequency: monthly
Annual subscription: £60.00
ISSN: 0020 3076

REPORTS

– 1132 – Coal Gasification for IGCC Power Generation

Geographical coverage: International
Publishers: IEA Coal Research, 1991
Coverage: This report reviews the various gasification processes that have been developed and makes an assessment of their readiness for application in utility service
ISBN: 9290291907
Author: Takematsu, T. and Maude, C.

– 1133 – Global Status of Peatlands and their Role in Carbon Recycling, The

Geographical coverage: United Kingdom
Publishers: Friends of the Earth (FoE), 1992
Coverage: Assesses the role and significance of peatbogs in relation to the greenhouse effect
Price: £18.00
ISBN: 1857501055
Author: The Wetland Ecosystem Research Group, University of Exeter

– 1134 – NOx Control Installations on Coal Fired Plants

Geographical coverage: International
Publishers: IEA Coal Research, 1991
Coverage: This report is a compilation of data gathered from around the world on NOx installations on conventional coal fired units
ISBN: 9290291869
Author: Hjalmarson, A-K. and Soud, H. N.

– 1135 – Volatile Organic Compound Emissions in Western Europe: Report 6/87

Geographical coverage: European
Publishers: CONCAWE, Brussels
Coverage: Control options and their cost effectiveness for gasoline vehicles, distribution and refining

RESEARCH CENTRES

– 1136 – Coal Research Establishment (CRE) Technical Services

Stoke Orchard
Cheltenham GL52 4RZ
United Kingdom
Tel: 0242 673361
Fax: 0242 672429
Contact name: Peter Sage
Geographical coverage: United Kingdom
Aims and objectives: CRE is a consultancy operated by the British Coal Corporation, offering advice, testing and development facilities in many aspects of coal processing

– 1137 – Deutsche Wissenschaftliche Gesellschaft für Erdöl

(German Scientific Society for Oil Gas and Coal)

Steinstrasse 7
D-2000 Hamburg 1
Germany
Tel: 49 40 321512
Geographical coverage: Germany
Publications: Annual report
Research activities: Fuel development, inflammables and lubricants; coal chemistry

– 1138 – IEA Coal Research

IEA Coal Research
Gemini House
10–18 Putney Hill
London SW15 6AA
United Kingdom
Tel: 081 780 2111
Fax: 081 780 1746
Contact name: Ms Irene Smith, Head of Environment Group
Geographical coverage: International
Aims and objectives: IEA Coal research was established under the auspices of the IEA and is supported by 14 countries and the CEC. It provides information and analysis on: coal science, supply, transport and markets, utilisation technology and environmental aspects
Publications: *Coal Abstracts*, monthly; *Coal Calendar*, six times a year; reports and leaflets
Additional information: Databases available

– 1139 – IKU Continental Shelf and Petroleum Technology Research Institute

S P Andersen vei 15B
N-7034 Trondheim
Norway
Tel: 47 7 59 11 00
Fax: 47 7 59 11 02
Contact name: Per R Thomassen, Managing Director
Geographical coverage: Norway
Research activities: Offshore petroleum exploration, production and environmental technology for the Norwegian oil industry

– 1140 – Netherlands Energy Research Foundation ECN

PO Box 1
NL-1755 ZG Petten
Netherlands
Tel: 31 2246 49 49
Fax: 31 2246 4480
Contact name: H H van den Kroonenberg, General Managing Director
Geographical coverage: Netherlands
Research activities: Research and development in energy technology; nuclear fossil and renewable sources

DATABASES

– 1141 – Coal Database

IEA Coal Research
Gemini House
10–18 Putney Hill
London SW15 6AA
United Kingdom
Tel: 081 780 2111
Fax: 081 780 1746
Geographical coverage: International
Coverage: All aspects of the coal industry, including energy policy, reserves, processes, products, wastes, health and safety
Database producer: International Energy Agency Coal Research
Database host: via Energy Science and Technology Database on Dialog
Coverage period: 1978–
Update frequency: 1,100 records monthly
Additional information: Available on CD–ROM

– 1142 – Coal Research Projects

Geographical coverage: International
Coverage: All aspects of coal industry, including energy, pollution, organisations, monitoring projects and chemical processes
Database producer: International Energy Agency Coal Research
Database host: BELINDIS
Coverage period: 1989–
Update frequency: irregularly
Additional information: Available on CD–ROM

– 1143 – Oil Spill Intelligence Report

Geographical coverage: International
Coverage: Prevention, control and clean-up of oil spills
Database producer: Cutter Information Corporation
Database host: NewsNet Inc
Coverage period: 1990–

– 1144 – Tulsa

Geographical coverage: International
Coverage: All aspects of oil including spillages and pollution
Database producer: Petroleum Abstracts
Database host: Dialog; Orbit
Coverage period: 1965–

8
Energy (Nuclear)

Nuclear Power

GOVERNMENT ORGANISATIONS

– 1145 – Advisory Committee on Nuclear Energy, Ministry of Trade and Industry

Pohjoinen Makasiinikatu 6
SF-00130 Helsinki 13
Finland
Tel: 358 1 254 52
Fax: 358 1 60 26 95
Contact name: Sakari Immonen, General Secretary
Geographical coverage: Finland
Aims and objectives: Advisory body to Ministry of Trade and Industry on nuclear energy matters

– 1146 – Agence Internationale de l'Energie Atomique (IAEA): Organismo Internacional de Energia Atómica

(International Atomic Energy Agency)

Vienna International Centre
Wagramerstrasse 5
POB 100
A-1400 Vienna
Austria
Tel: 43 1 222 2360
Fax: 43 1 223 4564
Contact name: Dr Hans Blix, Director General
Geographical coverage: International
Aims and objectives: IAEA was established in 1957. The Agency seeks to accelerate and enlarge the contribution of atomic energy to peace, health and prosperity throughout the world
Publications: Publications catalogue available

– 1147 – Bundesministerium für Forschung und Technologie (BMFT)

Heinemannstrasse 2
D-5300 Bonn 2
Germany
Tel: 49 228 591
Fax: 49 593 601
Contact name: Christine Patermann, Direct of Cabinet Spokesperson
Geographical coverage: Germany
Aims and objectives: Federal ministry for research and technology

– 1148 – Commissariat à l'Energie Atomique (CEA)

31-33 rue de la Fédération
F-75752 Paris Cedex 15
France
Contact name: Jean Teillac, High Commissioner
Geographical coverage: France

– 1149 – Committee on the Use of Atomic Energy for Peaceful Purposes

55 A Chapaev St
1574 Sofia
Bulgaria
Tel: 359 2 02 17
Fax: 359 2 70 21 43
Contact name: Yanjo Yanev, Chairperson
Geographical coverage: Bulgaria
Aims and objectives: The Committee is responsible for state policy and control over the use of atomic energy

– 1150 – CSFR Atomic Energy Commission

Slezka 9
120 29 Prague 2
Czech Republic
Tel: 42 2 21 52 540
Fax: 42 2 21 52 467
Contact name: Karel Wagner
Geographical coverage: Czech Republic
Aims and objectives: Government body for regulation and development of civil nuclear energy

– 1151 – Direção Geral de Energia Departamento de Energia Nuclear

Avenida da Republica 45–50
P-1000 Lisbon
Portugal
Tel: 351 1 769753
Contact name: H Carreira Pich, Deputy Director General
Geographical coverage: Portugal

– 1152 – Federal Chancellory Division for Nuclear Energy Coordination and Nonproliferation

Renngasse 5
A-1014 Vienna
Austria
Tel: 43 222 53115 2924
Fax: 43 222 53115 2935
Contact name: Fritz W Schmidt, Director
Geographical coverage: Austria
Aims and objectives: National Authority for nuclear coordination, multilateral and bilateral relations and safety

– 1153 – Greek Atomic Energy Commission

GR-153 10 Aghia Paraskevi
Greece
Tel: 30 1 6515194
Fax: 30 1 6533939
Contact name: Charalambos Proukakis, President
Geographical coverage: Greece
Aims and objectives: Radiprotection, nuclear technology applications and nuclear energy

– 1154 – HM Nuclear Installations Inspectorate Health and Safety Executive

Baynards House
1 Chepstow Place
Westbourne Grove
London W2 4TF
United Kingdom
Tel: 071 243 6000
Fax: 071 727 4116
Contact name: S A Harrison, Chief Inspector
Geographical coverage: United Kingdom
Aims and objectives: Regulatory body for the UK civil nuclear power industry

– 1155 – Lithuania Ministry of Energy

36 Gedimino
2600 Vilnius
Lithuania
Tel: 7 0122 615140
Fax: 7 0122 626845
Contact name: Leonas Asmantas, Minister
Geographical coverage: Lithuania

– 1156 – Ministry of Atomic Energy and Industry

Ulitsa Kirova 18
Moscow
Russia
Tel: 7 095 925 86 70
Fax: 7 095 943 00 71
Geographical coverage: Russia

– 1157 – Ministry of Atomic Energy of the Russian Federation (MINATOM)

Staromonetny pereulok 26
109180 Moscow
Russia
Tel: 7 095 233 1718
Fax: 7 095 230 2420
Contact name: Viktor Mikhailov, Minister
Geographical coverage: Russia
Aims and objectives: MINATOM is a government body responsible for all aspects of utilisation of atomic energy

– 1158 – Ministry of Chernobyl

L Ukrainka Square 1
252196 Kiev
Ukraine
Tel: 7 044 296 8198
Fax: 7 044 294 7796
Geographical coverage: Ukraine

– 1159 – Ministère des Affaires Economiques Administration de l'Energie Service des Applications Nucléaires

Rue J A De Mot 30
B-1040 Brussels
Belgium
Tel: 32 2 233 61 61
Fax: 32 2 230 42 38
Contact name: T van Rentergem, Chief Engineer and Director
Geographical coverage: Belgium

Aims and objectives: Follow up of the activities related to nuclear energy for power production

– 1160 – National Atomic Energy Agency

Krucza 36
PL-00 921 Warsaw
Poland
Tel: 48 2 628 27 22
Fax: 48 2 229 01 64
Contact name: Andrzej Janikowski, President Ad Interim
Geographical coverage: Poland
Aims and objectives: The Agency is a coordinator of activities related to civil use of atomic energy

– 1161 – National Commission for the Control of Nuclear Activities

Ministry of the Environment
Bucharest
Romania
Tel: 40 0 81 24 07
Fax: 40 0 31 64 86
Contact name: Stefan Alexandru Olariu, Director, Nuclear Regulatory Division
Geographical coverage: Romania

– 1162 – Nuclear Free Local Authorities National Steering Committee

Nuclear Policy and Information Unit
Manchester Town Hall
Manchester M60 2LA
United Kingdom
Tel: 061 234 3222
Fax: 061 236 8864
Contact name: Stella Whittaker
Geographical coverage: United Kingdom
Aims and objectives: Statutory organisation which gives legal advice, policy guidance, research, promotion and information distribution on nuclear issues
Additional information: MP Database

– 1163 – Österreichische Kerntechnische Gesellschaft

(Austrian Nuclear Society)

Schuttelstrasse 115
A-1020 Vienna
Austria
Tel: 43 1 21701 268
Fax: 43 1 2189220
Contact name: Hans Grumm, President
Geographical coverage: Austria
Aims and objectives: Public information, advice to government organisations, promotion of nuclear techniques

NON-GOVERNMENT ORGANISATIONS

– 1164 – Associazione Nazionale di Ingegneria Nucleare (ANDIN)

Via Flavia 104
I-00187 Rome
Italy
Tel: 39 6 486415
Fax: 39 6 4744397
Contact name: Maurizio Cumo, President
Geographical coverage: Italy
Aims and objectives: ANDIN helps to promote the advancement of nuclear science, in particular fission reactors and their safety

– 1165 – British Nuclear Forum (BNF)

22 Buckingham Gate
London SW1E 6LB
United Kingdom
Tel: 071 828 0116
Fax: 071 828 0110
Contact name: J H Gittus, Director General
Geographical coverage: United Kingdom
Aims and objectives: The BNF is an association dedicated to the sound economic development of nuclear power in Britain
Publications: *BNF Bulletin*, magazine monthly; *Status Report on Nuclear Power in Europe*, annually

– 1166 – Danish Nuclear Society (DNS)

Vester Farimagsgade 29
DK-1780 Copenhagen V
Denmark
Tel: 45 33 15 65 65
Fax: 45 33 93 71 71
Contact name: Franz R Marcus, President
Geographical coverage: Denmark
Aims and objectives: DNS is one of the professional groups of the Danish engineering society

– 1167 – Deutsches Atomforum eV (DAIF)

Heussallee 10
D-5300 Bonn 1
Germany
Fax: 49 228 507 219
Contact name: Thomas Roser, Delegate General
Geographical coverage: Germany
Aims and objectives: DAIF aims to promote the civil use of nuclear energy

– 1168 – European Nuclear Society (ENS)

PO Box 5032
CH-3001 Berne
Switzerland
Tel: 44 031 216 111
Fax: 44 031 229 203
Contact name: Colette Lewiner, President
Geographical coverage: European
Aims and objectives: ENS ia a Federation of twenty-five nuclear societies from Europe

– 1169 – Foreningen Karnteknik

(Swedish Nuclear Society)

Box 1419
S-111 84 Stockholm
Sweden
Tel: 46 8 61380000
Fax: 46 8 796 71 02
Contact name: Lars Gustafsson, President
Geographical coverage: Sweden
Aims and objectives: Promote the development of nuclear technology and exchange of information

– 1170 – Forum Atómico Español

Boix y Morer 6
E-28003 Madrid
Spain
Tel: 34 1 553 63 03
Fax: 34 1 535 08 82
Contact name: Alfonso Alvarez Miranda, President
Geographical coverage: Spain
Aims and objectives: The Forum aims to improve public awareness of nuclear energy and its civil uses

– 1171 – Forum Atomique Européen (FORATOM)

22 Buckingham Gate
London SW1E 6LB
United Kingdom
Tel: 071 828 0116
Fax: 071 931 0646
Contact name: J T Corner, Secretary General
Geographical coverage: European

– 1172 – Forum Italiano dell'Energia Nucleare (FIEN)

Palazzo Taverna
Via di Monte Giordano 36
I-00186 Rome
Italy
Tel: 39 6 689 309
Contact name: Luigi Noe, President
Geographical coverage: Italy

– 1173 – Forum Nucléaire Belge (ASBL)

Avenue Lloyd George 7
B-1050 Brussels
Belgium
Tel: 32 2 647 22 92
Fax: 32 2 647 04 54
Contact name: Robert van den Damme, President
Geographical coverage: Belgium
Aims and objectives: ASBL promotes civil use of nuclear power

– 1174 – Hellenic Nuclear Society

NRCPS Demokritos
Ag Paraskevi 15310
Greece
Tel: 30 1 6513111
Fax: 30 1 6519180
Contact name: N Catsaros, Secretary General
Geographical coverage: Greece

– 1175 – International Industrial Association for Energy from Nuclear Fuel: Uranium Institute (UI)

Bowater House
68 Knightsbridge
London SW1X 7LT
United Kingdom
Tel: 071 225 0303
Fax: 071 225 0308
Geographical coverage: International
Aims and objectives: The UI is an association concerned with nuclear fuel. It researches into the peaceful uses of uranium and atomic energy, including supply and demand, international trade and public acceptance of nuclear energy
Publications: *Uranium and Nuclear Energy*; *Uranium Market Issues Report*, every two years; conference proceedings annually

– 1176 – International Nuclear Power and Waste (INLA)

22 square de Meeus
B-1040 Brussels
Belgium
Tel: 32 2 513 68 45
Fax: 32 2 518 38 33
Contact name: Frédéric Vandenabeele, General Secretary
Geographical coverage: International
Aims and objectives: INLA carries out international research into the legal problems relating to the civil use of nuclear power

– 1177 – Nederlands Atoomforum

PO Box 1
NL-1775 ZG Petten
Netherlands

Tel: 31 2246 4082
Fax: 31 2246 3490
Contact name: W N Nijs, President
Geographical coverage: Netherlands
Aims and objectives: Dutch participant in Foratom

– 1178 – Norsk Atomforum

c/o Institut for Energiteknikk
PO Box 40
N-2007 Kjeller
Norway
Tel: 47 6 80 60 00
Fax: 47 6 81 63 56
Contact name: R Lingjaerde, Secretary General
Geographical coverage: Norway
Aims and objectives: Norsk Atomforum provides information to members about nuclear power

– 1179 – Nuclear Electric plc

Public Relations
Barnett Way
Barnwood
Gloucester GL4 7RS
United Kingdom
Tel: 0452 652222
Geographical coverage: United Kingdom
Aims and objectives: The Company offers an information service on nuclear power and related activities
Publications: Information packs

– 1180 – Nuclear Energy Board

3 Clonskeagh Square
Clonskeagh Road
Dublin 4
Eire
Tel: 353 1 2697766
Fax: 353 1 2697437
Geographical coverage: Ireland

– 1181 – OECD Nuclear Energy Agency (NEA)

12 boulevard des Îles
F-92130 Issy-les-Moulineaux
France
Tel: 33 1 452 411 00
Fax: 33 1 452 411 10
Contact name: Kunihiko Uematsu, Director General
Geographical coverage: International
Aims and objectives: The OECD NEA was established in 1957. Its primary objective is to promote cooperation among the governments of its members in furthering the development of nuclear power as a safe, environmentally acceptable and economic energy source
Publications: Annual report and newsletter twice a year
Additional information: NEA Data Bank

– 1182 – Orszagos Atomenergia Bizottsag

(National Atomic Energy Commission)

PO Box 565
H-1374 Budapest
Hungary
Tel: 36 1 327 172
Fax: 36 1 142 7598
Contact name: Pal Tetenyi, President
Geographical coverage: Hungary
Aims and objectives: Consultative body to the government in matters relating to the civil use of nuclear energy

– 1183 – Österreichisches Atomforum

Brehmstrasse 16
A-1110 Vienna
Austria
Contact name: K Kirchner, President
Geographical coverage: Austria

– 1184 – Schweizerische Vereinigung für Atomenergie (SVA)

(Swiss Association for Atomic Energy)

Postfach 5032
CH-3001 Berne
Switzerland
Tel: 41 31 225882
Fax: 41 31 229203
Geographical coverage: Switzerland
Aims and objectives: Promote civil use of nuclear energy

– 1185 – Sociedad Nuclear Española (SNE)

Campoamor 17
E-28004 Madrid
Spain
Tel: 34 1 308 63 18
Fax: 34 1 308 63 44
Contact name: Francisco Vighi, President
Geographical coverage: Spain
Aims and objectives: SNE is an organisation of professionals working in the nuclear sector

– 1186 – Swedish Atomic Forum (SAFO)

c/o Energiforum
Box 94
S-18271 Stocksund
Sweden
Tel: 46 8 855740
Fax: 46 8 853366
Contact name: Carl Erik Wikdahl, Secretary General
Geographical coverage: Sweden
Aims and objectives: SAFO aims to promote the civil use of nuclear power

– 1187 – Swiss Nuclear Society

c/o P Scherrer Institute
Wurenlingen and Villigen
CH-5232 Villigen PSI
Switzerland
Tel: 41 56 992692
Fax: 41 56 982327
Contact name: G Yadigaroglu, President
Geographical coverage: Switzerland
Aims and objectives: The Society helps to promote the advancement of science and technology in the nuclear field

– 1188 – Türkiye Atom Enerjisi Kurumu

(Turkish Atomic Energy Authority)

Alacam Sokak no 9
Gankaya
Ankara
Turkey
Tel: 90 4 270958
Fax: 90 4 272834
Geographical coverage: Turkey

– 1189 – Türkiye Nukleer Enerji Kurumu

(Turkish Nuclear Energy Forum)

PK 167
Aksaray
Istanbul
Turkey
Tel: 90 1 2478972
Fax: 90 1 5676913
Contact name: Muammer Cetincelik, President
Geographical coverage: Turkey

PRESSURE GROUPS

– 1190 – International Physicians for the Prevention of Nuclear War (IPPNW)

126 Rogers Street
Cambridge
MA 02142
United States of America
Tel: 1 67 868 5050
Fax: 1 67 868 2560
Geographical coverage: International
Aims and objectives: Nuclear Disarmament Group

BOOKS

– 1191 – Nuclear Power and the Greenhouse Effect

Geographical coverage: United Kingdom
Publishers: UK Atomic Energy Authority on behalf of the Nuclear Industry, 1992
Coverage: Account of how nuclear power affects global warming
Author: Donaldson, D., Tolland, H. and Grimston, M.

– 1192 – States and Anti-Nuclear Movements

Geographical coverage: United Kingdom
Publishers: Edinburgh University Press, 1993
Coverage: Using case studies from many countries in Europe, this book compares government reactions to anti-nuclear movements in Western Europe
Price: £50.00
ISBN: 0748603964
Author: Flam, H.

DIRECTORIES/YEARBOOKS

– 1193 – Directory of the French Nuclear Industry

39–41 rue Louis Blanc
92400 Courbevoie Cedex 72
F-92038 Paris La Defense
France
Tel: 33 1 471 762 78
Fax: 33 1 471 762 82
Geographical coverage: France
Publishers: Groupe Intersyndical de l'Industrie Nucléaire (GIIN)
Frequency: irregular
Coverage: Reference guide to the French nuclear power industry
Price: Ffr150
ISSN: 0066 2593

– 1194 – World Energy and Nuclear Directory

Westgate House
The High
Harlow
Essex CM20 1YR
United Kingdom
Geographical coverage: International
Publishers: Longman Group
Frequency: irregular
Language: English
Coverage: Lists various sources concerned with the nuclear industry
Annual subscription: £250.00

– 1195 – World Nuclear Directory

Geographical coverage: International
Publishers: Longman, 1985
Editor: Wilson, C. W. J.
Coverage: A guide to organisations and research activities in atomic energy

– 1196 – World Nuclear Industry Handbook

Geographical coverage: International
Publishers: Nuclear Engineering International, 1993
Editor: Varley, J.
Frequency: annual
Coverage: Guide to the civil nuclear power industry

PERIODICALS

– 1197 – Allicht

Postbus 8107
NL-5004 GC Tilburg
Netherlands
Tel: 31 13 351 1535
Fax: 31 13 358 169
Geographical coverage: Netherlands
Publishers: Stichting Allicht
Frequency: bi monthly
Language: Dutch
Price: Dfl15
ISSN: 0168 3748

– 1198 – Alternative Times

35 Wedmore Street
London N19 4RU
United Kingdom
Geographical coverage: United Kingdom
Publishers: R Stevens
Frequency: quarterly
Coverage: Run down of alternative energy news and views plus items on nuclear power
Price: £3.00
ISSN: 0261 6033

– 1199 – Atom

Room G073
Building 329
Harwell Laboratory
Didcot
Oxfordshire OX11 0RA
United Kingdom
Tel: 0235 432520
Geographical coverage: United Kingdom
Publishers: United Kingdom Atomic Energy Authority (UKAEA) Communications Directorate
Editor: Cruickshank, Dr A.
Frequency: bi monthly
Coverage: All aspects of nuclear energy and UKAEA non-nuclear work
Price: free
ISSN: 0004 7015

– 1200 – Atom-Informationen

Heussallee 10
D-5300 Bonn 1
Germany
Fax: 49 228 507 219
Geographical coverage: Germany
Publishers: Deutsches Atomforum eV
Editor: Bretschneider, D.
Frequency: monthly
Annual subscription: DM180
ISSN: 0004 7031

– 1201 – Atomnaya Energiya

Ulitsa Kirova 18
Moscow
Russia
Tel: 7 095 925 86 70
Fax: 7 095 943 00 71
Geographical coverage: Russia
Publishers: Ministry of Atomic Energy and Industry
Frequency: monthly
Language: Russian, contents page in English
Annual subscription: $495
ISSN: 0004 7163

– 1202 – Atomo Petrolio Elettricita

Piazza Borghese 3
I-00186 Rome
Italy
Geographical coverage: Italy
Editor: Guarino, G.
Frequency: bi monthly
Coverage: Atomic energy
ISSN: 0004 718X

– 1203 – Atomwirtschaft-Atomtechnik

Kasernenstrasse 67
Postfach 10 27 17
D-4000 Dusseldorf 1
Germany
Tel: 49 211 8870
Geographical coverage: Germany
Publishers: Handelsblatt GmbH
Editor: Hossner, R.
Frequency: monthly
Language: German text, summaries in English and French
Annual subscription: DM228
ISSN: 0365 8414

– 1204 – ATW News

Kasernenstrasse 67
Postfach 10 27 17
D-4000 Dusseldorf 1
Germany
Tel: 49 211 8870
Geographical coverage: Germany

Publishers: Handelsblatt GmbH
Frequency: monthly
Language: English text
Coverage: Nuclear information from the Federal Ministry of Germany
ISSN: 0341 4213

– 1205 – Belgicatom

1 Heiken
B-2840 Haacht
Belgium
Geographical coverage: Belgium
Editor: Van Goethem, A.
Frequency: bi monthly
Language: Dutch, English and French text, summaries also in German
Coverage: Atomic space age and electronics
Annual subscription: Bfr200
ISSN: 0005 8408

– 1206 – Droit Nucléaire

38 boulevard Suchet
F-75016 Paris
France
Tel: 33 1 452 496 97
Geographical coverage: France
Publishers: OECD Nuclear Energy Agency
Language: French edition
Coverage: Covers legislative developments
ISSN: 1016 4995

– 1207 – Energia es Atomtechnika

Leanyvallalat
Kossuth Lajos ter 6–8
H-1073 Budapest 5
Hungary
Tel: 36 1 117 0011
Geographical coverage: Hungary
Publishers: Scientific Society for Energy Economics
Language: Hungarian text, summaries in English and German
Price: $34.50
ISSN: 0013 7316

– 1208 – Energiespectrum

PB 90053
NL-1006 BB Amsterdam
Netherlands
Tel: 31 20 5182882
Fax: 31 20 170350
Geographical coverage: Netherlands
Publishers: Energieonderzoek Centrum Nederland
Frequency: monthly
Language: Dutch
Price: Dfl77
ISSN: 0165 2117

– 1209 – France Commissariat à l'Energie Atomique Notes d'Information

29-33 rue de la Fédération
F-75015 Paris
France
Geographical coverage: France
Publishers: Commissariat à l'Energie Atomique
Frequency: 2 issues a year
Price: free
ISSN: 0029 3997

– 1210 – Iaderna Energiia

(Nuclear Energy)

Acad G Bonchev St
Bldg 6
113 Sofia
Bulgaria
Geographical coverage: Bulgaria
Publishers: Publishing House of the Bulgarian Academy of Science
Editor: Khristov, V.
Frequency: bi monthly
Language: Various languages, summaries in Bulgarian, English, Russian
Price: lv1.35
ISSN: 0204 6989

– 1211 – International Journal of Radioactive Materials Transport

PO Box 7
Ashford
Kent TN23 1YW
United Kingdom
Tel: 0233 641683
Fax: 0233 610021
Geographical coverage: International
Publishers: Nuclear Technology Publishing
Editor: Goldfinch, E. P.
Frequency: quarterly
Coverage: All aspects of transport of radioactive materials
Annual subscription: £85.00
ISSN: 0957 476X

– 1212 – Kernenergie

Leipziger Strasse 3-4
D-1086 Berlin
Germany
Geographical coverage: Germany
Publishers: Akademie Verlag Berlin
Editor: Schumann, G.
Frequency: monthly
Language: English and German
Annual subscription: DM320.40
ISSN: 0023 0642

– 1213 – NEA Newsletter

OECD Nuclear Energy Agency
38 boulevard Suchet
F-75016 Paris
France
Geographical coverage: International
Publishers: OECD Nuclear Energy Agency
Editor: Ferté, J. de la
Frequency: 2 issues a year
Language: English and French

– 1214 – Nuclear Energy

1 Heron Quay
London E14 4JD
United Kingdom
Tel: 071 987 6999
Fax: 071 538 4101
Geographical coverage: United Kingdom
Publishers: Thomas Telford Services Ltd
Editor: Crocker, V. S.
Frequency: bi monthly
Language: English
Coverage: Covers practical and theoretical aspects of nuclear energy
Annual subscription: £83.00
ISSN: 0140 4067

– 1215 – Nuclear Engineering International

Quadrant House
the Quadrant
Sutton
Surrey SM2 5AS
United Kingdom
Tel: 081 652 3356
Fax: 081 652 8986
Geographical coverage: International
Publishers: Reed Business Publishing Group
Editor: Varley, J.
Frequency: monthly
Language: English, summaries in French and German
Coverage: News, comments and feature articles on technical and commercial developments in the exploitation of nuclear energy
Annual subscription: £120.00
ISSN: 0029 5507

– 1216 – Nuclear Europe Worldscan

c/o ATAG Science and Technology
Postfach 5032
CH-3001 Berne
Switzerland
Fax: 41 31 229203
Geographical coverage: European
Publishers: European Nuclear Society
Editor: Holt, Dr P.
Frequency: bi monthly
Language: English
Annual subscription: $110

– 1217 – Nuclear Law Bulletin

38 boulevard Suchet
F-75016 Paris
France
Tel: 33 1 452 496 67
Geographical coverage: International
Publishers: OECD Nuclear Energy Agency
Language: English
Coverage: Covers legislative and regulatory developments, agreements and case law in the nuclear field throughout the world
Annual subscription: Ffr170
ISSN: 0304 341X

– 1218 – Revue Générale Nucléaire

48 rue de la Procession
F-75724 Paris Cedex 15
France
Tel: 33 1 456 707 70
Fax: 33 1 406 592 29
Geographical coverage: France
Publishers: Revue Générale de l'Electricité
Editor: Sorin, F.
Frequency: 6 issues a year
Language: French text, summaries in English and French
Coverage: Offers synthesis and analyses on all aspects of nuclear developments around the world
Annual subscription: Ffr640

– 1219 – Revue Générale Nucléaire: International Edition

48 rue de la Procession
F-75724 Paris Cedex 15
France
Tel: 33 1 456 707 70
Fax: 33 1 406 592 29
Geographical coverage: International
Publishers: Revue Générale de l'Electricité
Editor: Sorin, F.
Frequency: 2 issues a year
Language: English
Coverage: Covers nuclear energy field. Includes translations of some of the articles published in Revue Générale Nucléaire

– 1220 – Safe Energy

11 Forth Street
Edinburgh
Scotland EH1 3LE
United Kingdom
Tel: 031 557 4283
Fax: 031 557 4284
Geographical coverage: United Kingdom
Publishers: SCRAM

Frequency: bi monthly
Coverage: Produced for the British anti-nuclear and safe energy movements
Price: £1.50
ISSN: 0140 7340

– 1221 – Schweizerische Vereinigung für Atomenergie Bulletin

Postfach 5032
CH-3001 Berne
Switzerland
Tel: 41 31 225882
Fax: 41 31 229203
Geographical coverage: Switzerland
Publishers: Swiss Association for Atomic Energy
Editor: Zuehike, P. and Haehlen, P.
Price: Sfr290
ISSN: 0036 777X

– 1222 – Swedish Nuclear News

Simpevarp
S-570 93 Figeholm
Sweden
Fax: 46 491 86920
Geographical coverage: Sweden
Publishers: O K G Aktiebolag
Editor: Petrini, J.
Frequency: monthly
Language: English
Price: free

– 1223 – Turkish Journal of Nuclear Sciences

Karanfil Sokak
67 Bakanliklar
Ankara
Turkey
Tel: 90 4 170661
Fax: 90 4 181938
Geographical coverage: Turkey
Publishers: Turkish Atomic Energy Authority
Editor: Ozman, A.
Price: $10

REPORTS

– 1224 – Crisis in the French Nuclear Power Industry

Geographical coverage: France
Publishers: Greenpeace, 1991
Language: English and French
Coverage: Economic and industrial issues of the French nuclear power programme
Author: Nectoux, F.

RESEARCH CENTRES

– 1225 – All-Union Research Institute for Reactor Operators (VNIIAES)

Ferganskaya 25
Moscow 109507
Russia
Tel: 7 095 376 1550
Fax: 7 095 376 8333
Contact name: Armen Abagyan, Director General
Geographical coverage: Russia
Research activities: All areas of nuclear technology

– 1226 – Atomic Energy Technology

Harwell Laboratory
Didcot
Oxfordshire OX11 ORA
United Kingdom
Tel: 0235 821111
Fax: 0235 832591
Contact name: John Maltby, Chairperson
Geographical coverage: United Kingdom
Research activities: Fast reactors and breeders

– 1227 – Cekmece Nükleer Arastirma ve Egitim Merkezi

(Cekmece Nuclear Research and Training Centre)

PK 1 Havaalani
Istanbul
Turkey
Tel: 90 1 790734
Geographical coverage: Turkey
Research activities: Nuclear fuel technology; environmental pollution; nuclear engineering

– 1228 – Comitato Nazionale per la Ricerca e per lo Sviluppo dell'Energia Nucleare e dell' Energie Alternativ

Via Regina Margherita 125
I-00198 Rome
Italy
Tel: 39 6 85281
Contact name: U Colombo, President
Geographical coverage: Italy

– 1229 – Commission of the European Communities (CEC) – Joint Research Centre (JRC)

Rue de la Loi 200
B-1049 Brussels
Belgium
Tel: 32 2 235 8527
Fax: 32 2 235 0146
Contact name: J P Contzen, Director General

Geographical coverage: European
Aims and objectives: The JRC was founded in 1960 as a scientific and technical base for primarily nuclear research. Since then its activities have been broadened to encompass research on the environment and safety
Research activities: Nuclear safety for the European Community

– 1230 – European Institute for Transuranium Elements

Postfach 2340
D-7500 Karlsruhe
Germany
Tel: 49 7247 841
Fax: 49 7247 4046
Geographical coverage: European
Publications: *Nuclear Fuels and Actinide Research*; bi annual progress report; annual report
Research activities: Nuclear fuels and actinide research; reactor safety; radioactive waste management

– 1231 – Forschungszentrum Jülich GmbH (KFA)

PO Box 1913
D-5170 Jülich
Germany
Tel: 49 22461 613000
Fax: 49 22461 612525
Contact name: J Treusch, Chairperson
Geographical coverage: Germany
Research activities: National research centre providing basic and applied research

– 1232 – Institut for Energiiteknikk (IFE)

PO Box 40
N-2007 Kjeller
Norway
Tel: 47 6 8060 00
Fax: 47 6 8163 56
Contact name: N G Aamodt, Managing Director
Geographical coverage: Norway
Research activities: Nuclear and non-nuclear fields including petroleum technology

– 1233 – Netherlands Energy Research Foundation ECN

PO Box 1
NL-1755 ZG Petten
Netherlands
Tel: 31 2246 49 49
Fax: 31 2246 4480
Contact name: H H van den Kroonenberg, General Managing Director
Geographical coverage: Netherlands
Research activities: Research and development in energy technology; nuclear fossil and renewable sources

– 1234 – Nuclear Power and Plant Research Institute (VUJE)

Okruzna 5
918 64 Trnava
Slovakia
Tel: 42 805 412 55
Fax: 42 805 423 96
Contact name: Stanislav Novak, Director
Geographical coverage: Slovakia

– 1235 – Österreichisches Forschungszentrum Seibersdorg GmbH

A-2444 Seibersdorf
Austria
Tel: 43 2254 80 0
Fax: 43 2254 80 2118
Contact name: Peter Koss, Managing Director
Geographical coverage: Austria
Aims and objectives: Research and development

– 1236 – Stockholm International Peace and Research Institute (SIPRI)

Pipers Vag 28
S-17173 Solna
Sweden
Tel: 46 8 655 97 00
Fax: 46 8 655 97 33
Contact name: Adam Daniel Rotfeld, Director
Geographical coverage: International
Research activities: Examines questions of importance for peace and security

– 1237 – Studiecentrum voor Kernenergie – Centre d'Etude de l'Energie Nucléaire (CEN-SCK)

Rue Ch Lemairestraat 1
B-1160 Brussels
Belgium
Tel: 32 2 143 321 11
Fax: 32 2 143 150 21
Contact name: Carl M Malbrain, Managing Director
Geographical coverage: Belgium
Research activities: Applied research in nuclear technology

DATABASES

– 1238 – International Nuclear Information System (INIS)

Geographical coverage: International
Coverage: Nuclear industry and information on peaceful applications of nuclear science

Database producer: International Atomic Energy Authority
Database host: ESA–IRS

Coverage period: 1976–
Update frequency: 3,500 records fortnightly

Nuclear Safety

GOVERNMENT ORGANISATIONS

– 1239 – Bundesministerium für Umwelt Naturschutz und Reaktorsicherheit

Kennedyallee 10
D-5300 Bonn 1
Germany
Tel: 49 228 3052010
Contact name: Marlene Mühe, Public Relations
Geographical coverage: Germany

– 1240 – Consejo de Seguridad Nuclear

Justo Dorado 11
E-28040 Madrid
Spain
Tel: 34 1 5 34 91 40
Fax: 34 1 45 51 28 0
Contact name: Donato Fuejo Lago, President
Geographical coverage: Spain
Aims and objectives: Regulatory body for nuclear safety and radiological protection

– 1241 – Federal Chancellory Division for Nuclear Energy Coordination and Nonproliferation: *see entry 1152*

Geographical coverage: Austria

– 1242 – Greek Atomic Energy Commission: *see entry 1153*

Geographical coverage: Greece

– 1243 – HM Nuclear Installations Inspectorate Health and Safety Executive: *see entry 1154*

Geographical coverage: United Kingdom

– 1244 – Ministry for the Environment Nature Conservation and Nuclear Reactors Safety

Kennedyallee 5
Postfach 12 06 29
D-5300 Bonn 1
Germany
Tel: 49 228 3050
Fax: 49 228 305 3225
Geographical coverage: Germany

– 1245 – Nordic Committee for Nuclear Safety Research (NKS)

PO Box 49
DK-4000 Roskilde
Denmark
Tel: 45 42 37 12 12
Fax: 45 46 32 22 06
Contact name: Franz R Marcus, Secretary General
Geographical coverage: Denmark
Aims and objectives: NKS performs joint Nordic programmes related to nuclear safety

– 1246 – State Committee for Chernobyl Accident Problems

ul Krasnoarmayskaya 4/45
220020 Minsk
Byelorussia
Fax: 7 0172 26 59 42/29 34 39
Contact name: V M Burjak, First Deputy Chairperson
Geographical coverage: Byelorussia

– 1247 – State Committee for Nuclear and Radiation Safety

Kiev
Ukraine
Tel: 7 044 559 6962
Fax: 7 044 559 5344
Contact name: N Shteinberg, Chairperson
Geographical coverage: Ukraine

– 1248 – State Committee for the Supervision of Nuclear and Radiation Safety under the President of Russia

Taganskaya str 34
109147 Moscow
Russia
Tel: 7 095 272 47 10
Fax: 7 095 278 80 90
Contact name: Yuri Vishnevsky, Chairperson
Geographical coverage: Russia
Aims and objectives: Regulatory body

NON-GOVERNMENT ORGANISATIONS

– 1249 – Associazione Nazionale di Ingegneria Nucleare (ANDIN): see entry 1164

Geographical coverage: Italy

– 1250 – Commission of the European Communities Nuclear Safety Research Directorate

Rue de la Loi 200
B-1049 Brussels
Belgium
Tel: 32 2 591 77
Fax: 32 2 620 06
Contact name: Sergio Finzi, Director
Geographical coverage: European
Research activities: Matters related to nuclear safety

– 1251 – Kerntechnische Ausschuss (KTA)

(Nuclear Safety Standards Commission)

Albert Schweitzer Strasse 18
Postfach 1001 49
D-3320 Salzgitter 1
Germany
Tel: 49 5341 2205 0 21
Fax: 49 5341 2205 99
Contact name: I Kalinowski, Managing Director
Geographical coverage: Germany
Aims and objectives: KTA helps to develop safety standards in the area of nuclear technology and to promote their use

– 1252 – Sateilyturvakeskus (STUK)

(Finnish Centre for Radiation and Nuclear Safety)

PO Box 268
SF-00101 Helsinki
Finland
Tel: 358 1 708 21
Fax: 358 1 708 2210
Contact name: Antti P U Vuorinen, Director General
Geographical coverage: Finland
Aims and objectives: STUK is a regulatory body for nuclear power, plant safety and other uses of radiation

PERIODICALS

– 1253 – Übersetzungen- Kerntechnische Regeln

Schwertnergasse 1
D-5000 Cologne 1
Germany
Geographical coverage: Germany
Publishers: Gesellschaft für Reaktorsicherheit
Frequency: irregular
Language: English edition
Coverage: Safety codes and guides
Price: DM20

REPORTS

– 1254 – Safety Assessment of Radioactive Waste Repositories: Systematic Approaches to Scenario Development

Geographical coverage: European
Publishers: OECD Publications, 1992
Language: English
Coverage: Analysis of the long term safety of radioactive waste
Price: £21; Ff150; $38
ISBN: 9264136053

– 1255 – Working for Public Safety and Environmental Protection

Nuclear Policy Information Unit
Town Hall
Manchester M60 2LA
United Kingdom
Tel: 061 234 3244
Geographical coverage: United Kingdom
Publishers: Nuclear Free Local Authorities (NFLA)
Coverage: Review of safe nuclear waste management options for the future

RESEARCH CENTRES

– 1256 – Commission of the European Communities (CEC) – Joint Research Centre (JRC): *see entry 1228*

Geographical coverage: European

– 1257 – Eidgenössische Kommission für die Sicherheit von Kernanlagen
(Federal Commission for the Safety of Nuclear Installations)

c/o KSA Sekretariat
CH-5303 Würenlingen
Switzerland
Tel: 41 56 99 39 43
Geographical coverage: Switzerland

Research activities: Licensing of nuclear power plants, nuclear installations, research reactors and radioactive waste disposal

– 1258 – European Institute for Transuranium Elements: *see entry 1230*

Geographical coverage: European

– 1259 – Joint Research Centre Ispra Establishment

I-21020 Ispra (Varese)
Italy
Tel: 39 332 789111
Geographical coverage: Italy
Research activities: Reactor and Nuclear Plant Safety; treatment and storage of radioactive wastes

9
Health and Safety

General

GOVERNMENT ORGANISATIONS

– 1260 – Department of Health

Hawkins House
Dublin 2
Eire
Tel: 353 1 714711
Geographical coverage: Ireland

– 1261 – Health and Safety Executive

Baynards House
1 Chepstow Place
Westbourne Grove
London W2 4TF
United Kingdom
Tel: 071 243 6000
Fax: 071 727 2254
Geographical coverage: United Kingdom
Aims and objectives: The Health and Safety Executive is the Health and Safety Commission's major instrument. Through its inspectorates it enforces health and safety law in the majority of industrial premises, to protect both work people and the public

– 1262 – Ministry for Social Affairs and Health

Snellmanikatu 4–6
SF-00170 Helsinki 17
Finland
Geographical coverage: Finland

– 1263 – Ministry of Health

Deutschherrenstrasse 87
D-5300 Bonn 2
Germany
Tel: 49 228 9300
Fax: 49 228 9304978
Geographical coverage: Germany

– 1264 – Ministry of Health

20 Viale dell'Industria (EUR)
I-00144 Rome
Italy
Tel: 39 6 5994
Fax: 39 6 593 4774
Geographical coverage: Italy

– 1265 – Ministry of Health

Spitálska 6
813 05 Bratislava
Slovakia
Fax: 42 7 57508/57627
Geographical coverage: Slovakia

– 1266 – Ministry of Health and Consumer Affairs

Paseo del Prado 18–20
E-28071 Madrid
Spain
Tel: 34 1 420 0000
Fax: 34 1 429 3526
Geographical coverage: Spain

– 1267 – Ministry of Health and Humanitarian Policies

8 avenue de Segur
F-75700 Paris
France
Tel: 33 1 405 600 00
Geographical coverage: France

– 1268 – Ministry of Health and Social Security

Laugavegi 116
150 Reykjavik
Iceland
Tel: 354 1 60 97 00
Fax: 354 1 1 91 65
Geographical coverage: Iceland

– 1269 – Ministry of Health and Social Welfare

ul Miodowa 15
PL-00 246 Warsaw
Poland
Tel: 48 22 32 34 41
Geographical coverage: Poland

– 1270 – Ministry of Health Welfare and Social Security

Odos Aristoteloous 17
Athens
Greece
Tel: 30 1 523 2821
Geographical coverage: Greece

– 1271 – Ministry of Public Health

57–90 boulevard de la Pétrusse
Luxembourg
Luxembourg
Tel: 352 408 01
Fax: 352 48 49 03
Geographical coverage: Luxembourg

– 1272 – Ministry of Welfare Health and Cultural Affairs

Sir W Churchill-laan 362/366 (Boogaard Ctr)
POB 5406
NL-2280 HK Rijswijk
Netherlands
Tel: 31 70 340 7911
Geographical coverage: Netherlands

– 1273 – National Institute of Occupational Health

Lerso Parkalle 105
DK-2100 Copenhagen Ø
Denmark
Tel: 45 31 29 97 11
Contact name: IB Andersen, Director
Geographical coverage: Denmark

– 1274 – Rijksinstituut voor Volksgezundheid en Milieuhygiene (RIVM)

(National Institute of Public Health and Environmental Protection)

PO Box 1
NL-3720 BA Bilthoven
Netherlands
Tel: 31 30 749111
Contact name: Klaas van Egmond
Geographical coverage: Netherlands

– 1275 – Royal Ministry of Health and Social Affairs

Grubbegt 10
PO Box 8011 Dep
N-0030 Oslo
Norway
Tel: 47 2 34 90 90
Geographical coverage: Norway

– 1276 – Secretary of State for Public Health

Rue de la Loi 56
B-1040 Brussels
Belgium
Tel: 32 2 230 49 25
Geographical coverage: Belgium

NON-GOVERNMENT ORGANISATIONS

– 1277 – Associação Portuguesa para Estudos de Saneamento Basico (APESB)

Rua Antero sw Quental 44
P-1100 Lisbon
Portugal
Tel: 351 1 54 35 44
Geographical coverage: Portugal

154 Health and Safety *General*

– 1278 – Societatea Româna de Igiena si Sanatate Publica

(Romanian Society of Hygiene and Public Health)

1–3 Dr Leonte Str
76256 Bucharest
Romania
Tel: 40 1 6383970
Fax: 40 1 3123426
Contact name: Prof Manole Cucu, President
Geographical coverage: Romania

BOOKS

– 1279 – Health Cities

Geographical coverage: International
Publishers: Open University Press, 1991
Coverage: An introduction to the WHO Healthy Cities Project which describes its objectives and examines its performance and effects
Price: £32.50
ISBN: 0335094775
Author: Ashton, J.

DIRECTORIES/YEARBOOKS

– 1280 – Adresboek voor Milieuhygiene Veiligheids-Techniek en Recuperatie

(Directory of Environmental Health and Safety)

Tesselschadestraat 18
Amsterdam
Netherlands
Geographical coverage: Netherlands
Publishers: Uitgeversmij Diligentia
Coverage: Lists manufacturers of public health and safety equipment

– 1281 – Health and Safety Directory, 5th revised ed

Geographical coverage: United Kingdom
Publishers: Croner Publications Ltd, 1991
ISBN: 1870080432

PERIODICALS

– 1282 – Forum Städte-Hygiene

Königsallee 65
D-1000 Berlin 33
Germany
Geographical coverage: Germany
Publishers: Patzer Varlag GmbH und Co KG
Editor: Knoll, K. H. and Patzer, B.
Frequency: 6 issues a year
ISSN: 0342 202X

– 1283 – Fravar ved Anmeldte Arbejdsulykker

Landskrongade 33–35
DK-2100 Copenhagen
Denmark
Tel: 45 31 18 00 88
Geographical coverage: Denmark
Publishers: Direktoratet for Arbejdstilsynet
Frequency: 5 issues a year
Price: free
ISSN: 0109 5129

– 1284 – Health and Safety at Work

Tolley House
2 Addiscombe Road
Croydon
Surrey CR9 5AF
United Kingdom
Tel: 081 686 9141
Fax: 081 760 0588
Geographical coverage: United Kingdom
Publishers: Tolley Publishing Co Ltd
Editor: Allan, I.
Frequency: monthly
Coverage: Covers occupational health and safety
Annual subscription: £30.00

– 1285 – Human Environment in Sweden

Swedish Consulate General
825 Third Ave
New York
NY 100022
Sweden
Geographical coverage: Sweden
Publishers: Swedish Information Service
Frequency: quarterly
Language: English
Price: free

– 1286 – Informazione Innovativa

Via dell Palme 13
I-35137 Padua
Italy
Tel: 39 49 36435
Geographical coverage: Italy
Publishers: Centro Studi "l'Uomo e l'Ambiente"
Editor: Merlin, T.
Language: Text in Italian, summaries in English
Coverage: Clean technologies, energy sources, industrial health and safety
Annual subscription: Lit400000

– 1287 – Nova – Revista de Salud

Independencia 92
E-08902 L'Hospitalet
Barcelona
Spain
Geographical coverage: Spain
Publishers: Ibis SA
Frequency: 6 issues a year

– 1288 – Umwelttechnik

Spindelstrasse 2 Postfach
CH-8021 Zurich
Switzerland
Tel: 41 1 488 8400
Geographical coverage: Switzerland
Publishers: Cicero Verlag Ag
Editor: Schaetzle, P.
Frequency: monthly
Price: Sfr75

RESEARCH CENTRES

– 1289 – Arbetsmiljöinstitutet
(National Institute of Occupational Health)

S-171 84 Solna
Sweden
Tel: 46 8 730 91 00
Geographical coverage: Sweden
Research activities: Occupational health research

– 1290 – European Institute of Ecology and Cancer (EIEC)

Rue des Fripiers 24 bis
B-1000 Brussels
Belgium
Tel: 32 2 219 08 30
Contact name: Dr Emile–Gaston Peeters, Secretary General
Geographical coverage: European
Publications: *Medecine–Biology–Environnement*, twice a year
Research activities: Geocancerology

– 1291 – International Agency for Research on Cancer (IARC)

150 Cours Albert-Thomas
F-69372 Lyon 8
France
Tel: 33 72 73 84 85
Fax: 33 72 73 85 75
Contact name: Dr Lorenzo Tomatis, Director
Geographical coverage: International
Aims and objectives: Established in 1965 as a self-governing body within the framework of WHO, the Agency organises international research on cancer. It runs a programme of research on the environmental factors causing cancer
Research activities: Carries out research into the causes of human cancer with the aim of identifying possibilities for primary prevention

– 1292 – Kentron Yginis Keasfalias tis Ergassias (KYAE)
(Centre for Occupational Hygiene and Safety)

6 Dodekanissou Street
GR-174 56 Alimos
Greece
Tel: 30 1 9919–566
Geographical coverage: Greece
Research activities: Applied research in environmental and biological monitoring

DATABASES

– 1293 – EMBASE

Geographical coverage: United Kingdom
Coverage: All aspects of health and safety; chemical hazards and wastes
Database producer: Elsevier Science Publishers
Database host: BRS; DataStar; Dialog; DIMDI
Coverage period: 1974–
Additional information: Available on CD–ROM

– 1294 – HSELine

Geographical coverage: International
Coverage: Health
Database producer: Health and Safety Executive
Database host: ESA–IRS; Orbit
Coverage period: 1977–
Update frequency: 1,000 records monthly

Asbestos

NON-GOVERNMENT ORGANISATIONS

– 1295 – Asbestos Information Centre (AIC)

PO Box 69
Widnes
Cheshire WA8 9GW
United Kingdom
Tel: 051 420 5866
Fax: 051 420 5853
Contact name: J B Heron, Chairperson
Geographical coverage: United Kingdom
Aims and objectives: The Centre provides an information service on asbestos and its effects on health and the environment

– 1296 – Asbestos International Association (AIA)

10 rue de la Pepinidre
F-75008 Paris
France
Tel: 33 1 452 214 00
Fax: 33 1 429 498 86
Geographical coverage: International

– 1297 – Society for Prevention of Asbestosis and Industrial Diseases (SPAID)

38 Drapers Road
Enfield
Middlesex EN6 4HT
United Kingdom
Tel: 0707 873 025
Contact name: Nancy Tait, Secretary
Geographical coverage: United Kingdom
Aims and objectives: SPAID works towards promoting an asbestos free environment, legislation, research, protection for workers and research into electron microscopy
Publications: List available on request

REPORTS

– 1298 – Safety in the Use of Asbestos Report IV (2)

Geographical coverage: International
Publishers: ILO, 1986
ISBN: 9221051755

Biotechnology

GOVERNMENT ORGANISATIONS

– 1299 – Biotechnology Unit

Laboratory of the Government Chemist
Queens Rd
Teddington
Middlesex TW11 OLY
United Kingdom
Tel: 081 943 7354
Fax: 081 943 7304
Geographical coverage: United Kingdom
Aims and objectives: The Unit works towards fostering biotechnology

– 1300 – Concentration Unit for Biotechnology in Europe (CUBE)

DG XII Commission of the EC
Rue de la Loi 200
B-1049 Brussels
Belgium
Tel: 32 2 235 81 45
Fax: 32 2 235 53 65
Geographical coverage: European
Aims and objectives: CUBE aims to provide sources of information about Community policy with respect to biotechnology

NON-GOVERNMENT ORGANISATIONS

– 1301 – Asociación de Bioindustria

Calle Brue 72–74
6A planta
E-08009 Barcelona
Spain
Tel: 34 3 3183383
Fax: 34 3 3023568
Geographical coverage: Spain

– 1302 – Assobiotec

Via Academia 33
I-20131 Milan
Italy
Tel: 39 2 636 2306
Fax: 39 2 636 2310
Geographical coverage: Italy

– 1303 – Belgian Biotechnology Coordinating Group (BBCG)

Plant Genetic Systems
Jozef Plateaustraat 22
B-9000 Ghent
Belgium
Tel: 32 91 358 444
Fax: 32 91 233 855
Geographical coverage: Belgium
Aims and objectives: Participate as appropriate in the formulation of local, national and international policies

– 1304 – Bioindustry Association (BIA)

1 Queen Anne's Gate
London SW1H 9BT
United Kingdom
Tel: 071 222 2809
Fax: 071 222 8876
Geographical coverage: United Kingdom
Aims and objectives: BIA offers informational and other support to member companies; identifies key issues affecting the development of biotechnology; encourages positive public attitudes and promotes biotechnological legislation

– 1305 – Bioresearch Ireland (EOLAS)

Glasnevin
Dublin 9
Eire
Tel: 353 1 370177
Fax: 353 1 370176
Geographical coverage: Ireland
Aims and objectives: Irish science and technology agency
Additional information: Library Service

– 1306 – Federation of Biotechnology Industries of Denmark (FBID)

Novo Nordisk A/S
Nove Alle
DK-2880 Bagsvaerd
Denmark
Tel: 45 44 44 88 88
Fax: 45 42 98 46 47
Geographical coverage: Denmark
Aims and objectives: Offers informational and other support to member countries

– 1307 – ICSU Steering Committeee for Biotechnology (COBIOTECH): Centre of Bioengineering

Institute of Molecular Biology
ul Vavilova 32
Moscow B-312
Russia
Russia
Tel: 7 095 135 73 19
Contact name: Prof K G Skryabin, Secretary General
Geographical coverage: International
Aims and objectives: COBIOTECH publishes a journal, runs training courses, conferences and workshops and organises an International Biosciences Network

– 1308 – International Council of Scientific Unions (ICSU)

51 boulevard de Montmorency
F-75016 Paris
France
Tel: 33 1 452 503 29
Fax: 33 1 428 894 31
Geographical coverage: International
Aims and objectives: The Council was founded in 1931 as an international non-profit organisation to promote scientific activity in the different branches of science and their applications for the benefit of humanity
Publications: *Science International*, quarterly; yearbook

– 1309 – Kemian Keskusliitto

(Chemical Industry Federation of Finland)

PO Box 359
Hietaniemenkatu 2
SF-00101 Helsinki
Finland
Tel: 358 1 44 71 22
Fax: 358 1 44 79 60
Geographical coverage: Finland

– 1310 – Nederlandse Industriële en Agarische Biotechnologie Associatie (NIABA)

PO Box 185
NL-2260 AD Leidschendam
Netherlands
Tel: 31 70 3270464
Fax: 31 70 3203671
Geographical coverage: Netherlands
Aims and objectives: Offers informational and other support to member companies, promotes biotechnological education and encourages positive public attitudes

158 Health and Safety *Biotechnology*

– 1311 – Organbio

28 rue St Dominique
F-75007 Paris
France
Tel: 33 1 475 309 12
Fax: 33 1 475 598 62
Geographical coverage: France

– 1312 – Senior Advisory Group Biotechnology (SAGB): CEFIC

Avenue Louise 250
Box 71
B-1050 Brussels
Belgium
Tel: 32 2 640 20 95
Fax: 32 2 640 19 81
Geographical coverage: European
Aims and objectives: The purpose of SAGB is to promote a supportive climate in Europe

BOOKS

– 1313 – Assessing Biological Risks of Biotechnology

Geographical coverage: United Kingdom
Publishers: Butterworth Heinemann, 1991
Coverage: The emphasis is on ecological risks associated with the release of genetically engineered organisms in the environment
Price: £65.00
ISBN: 0409901997
Author: Ginsburg, L. R.

– 1314 – Biotechnology: The Science and the Business

Geographical coverage: United Kingdom
Publishers: Harwood Academic Publishers 1991
Coverage: Focuses on biotechnology in business
ISBN: 3718650940
Author: Moses, V. and Cape, E.

DIRECTORIES/YEARBOOKS

– 1315 – Information Sources in Biotechnology

Geographical coverage: United Kingdom
Publishers: Stockton Press, 2nd ed.,1986
Editor: Crafts-Lighty, A.
Coverage: A guide to information sources on biotechnology
ISBN: 0943818184

REPORTS

– 1316 – Biotechnology in Industry Healthcare and the Environment: Special Report no 2178

Geographical coverage: United Kingdom
Editor: Economic Intelligence Unit
Coverage: Report on biotechnology and its relation to the environment
Author: Moses, S. and Moses, V.

RESEARCH CENTRES

– 1317 – Centro Studi l'Uomo e l'Ambiente

(Centre for the Study of Man and the Environment)

Via delle Palme 13
I-35137 Padua
Italy
Tel: 39 49 38613
Geographical coverage: Italy
Research activities: Biotechnology

– 1318 – Cranfield Biotechnology Centre

Cranfield Institute of Technology
Cranfield
Bedfordshire MK43 OAL
United Kingdom
Tel: 0234 754339
Geographical coverage: United Kingdom
Aims and objectives: The Centre is involved in research contracts and postgraduate education
Publications: Research papers available
Research activities: Toxic gas monitoring

– 1319 – Inveresk Research International Ltd: *see entry 1429*

Geographical coverage: United Kingdom

– 1320 – Technological Institute – Royal Flemish Society of Engineers

Desuguinlei 214
B-2018 Antwerp
Belgium
Geographical coverage: Belgium

DATABASES

– 1321 – Biobusiness

Geographical coverage: International

Coverage: Aspects of biotechnology
Database producer: BIOSIS
Database host: BRS; DataStar; Dialog; DIMDI; ESA–IRS; STN
Coverage period: 1979–

– 1322 – Biorep

Geographical coverage: European
Coverage: Biotechnical research projects in the EC
Database producer: The Library, Royal Netherlands Academy of Arts and Sciences
Database host: ECHO
Coverage period: 1989–
Update frequency: annually

– 1323 – Biotechnology Abstracts

Geographical coverage: International
Coverage: Biotechnology
Database producer: Derwent Publications Ltd
Database host: Dialog; Orbit
Coverage period: 1982–
Update frequency: 1,250 records monthly

– 1324 – Current Biotechnology Abstracts

Geographical coverage: United Kingdom
Coverage: Biotechnology
Database producer: Royal Society of Chemistry
Database host: DataStar; ESA–IRS
Coverage period: Current

Environmental Health

NON-GOVERNMENT ORGANISATIONS

– 1325 – Centre d'Information du Plomb

1 boulevard de Vaugirard
F-75751 Paris Cedex 15
France
Tel: 33 1 538 52 33
Contact name: Mr Prevost
Geographical coverage: France
Aims and objectives: Health protection, pollutant effects on man and society

– 1326 – Consumentenbond

(Consumers Association)

Leeghwaterplein 26
NL-2521 CV 's-Gravenhage
Netherlands
Tel: 39 70 889377
Geographical coverage: Netherlands
Aims and objectives: Consumer affairs in relation to environment and health

– 1327 – Kansanterveyslaitos (NPHI)

(National Public Health Institute)

PO Box 95
SF-70701 Kuopio
Finland
Tel: 358 71 20 12 11
Fax: 358 71 20 12 65
Contact name: Prof Dr Juoko Tuomisto, Director of Department
Geographical coverage: Finland
Aims and objectives: NPHI aims to promote public health in Finland

– 1328 – Oil Companies European Organisation for Environmental Health and Protection (CONCAWE)

Madouplein 1
B-1030 Brussels
Belgium
Tel: 32 2 220 31 11
Fax: 32 2 219 46 46
Contact name: Klaus Kohlhase, Chairman
Geographical coverage: European
Aims and objectives: Founded in 1963 by six oil companies and now has a membership of 36 companies. These represent 90% of all the refining capacity in Western Europe. The scope of its activities include air, land and water pollution control and health protection
Publications: Technical reports. Catalogues are available on request

– 1329 – Swiss Association for Health Techniques

Postfach 305
CH-8035 Zürich
Switzerland
Tel: 41 1 734 1096
Geographical coverage: Switzerland
Aims and objectives: The Association helps to promote environmental health

– 1330 – World Health Organisation Division of Environmental Health (WHO/EHE)

Avenue Appia
CH-1211 Geneva 27
Switzerland
Tel: 41 22 7912111
Fax: 41 22 7910746
Geographical coverage: International
Aims and objectives: WHO was established in 1948 as the central agency directing international health work. EHE aims to promote and protect human health by preventing and controlling conditions in the environment that adversely affect human health
Publications: Catalogue of publications supplied free on request

BOOKS

– 1331 – Environmental Respiratory Diseases

Geographical coverage: United Kingdom
Publishers: Van Nostrand Reinhold International, 1992
Coverage: Covers diseases of the respiratory system caused by environmental contaminants and pollution problems
ISBN: 0442008473
Author: Cordasco, E. M.

PERIODICALS

– 1332 – Ambio: A Journal of the Human Environment Research and Management

Box 5005
S-104 05 Stockholm
Sweden
Tel: 46 8 150430
Geographical coverage: Sweden
Publishers: Royal Swedish Academy of Sciences
Editor: Rosemarin, A.
Frequency: 18 issues a year (6 elsewhere)
Language: English
Coverage: Environmental issues pertaining to health

– 1333 – Environmental Health

Chadwick House
Rushworth Street
London SE1 0OT
United Kingdom
Tel: 071 928 6006
Fax: 071 261 1960
Geographical coverage: United Kingdom
Publishers: The Institution of Environmental Health Officers
Editor: King, H.
Frequency: monthly
Coverage: Health and Safety
Annual subscription: £64.00

– 1334 – Environmental Health News

Black Prince Road
London SE1 7SJ
United Kingdom
Tel: 071 582 3661
Fax: 071 582 9405
Geographical coverage: United Kingdom
Publishers: EHN Ltd
Editor: Randall, B.
Frequency: weekly
Coverage: Provides an overview of the major events and developments in local authorities and the commercial sector, new legislation, new techniques
Price: 70p
Annual subscription: £35.00

– 1335 – European Bulletin of Environment and Health

WHO Regional Office for Europe
Scherfigsvej 8
DK-2100 Copenhagen
Denmark
Geographical coverage: European
Publishers: ISIS Publishing Corporation
Editor: Racco, M.
Frequency: quarterly
Language: English and Italian
Coverage: Health issues in relation to the environment

– 1336 – Excerpta Medica Section 46: Environmental Health and Pollution Control

PO Box 548
NL-1000 AM Amsterdam
Netherlands
Tel: 31 20 5803911
Fax: 31 20 5803222
Geographical coverage: Netherlands
Publishers: Excerpta Medica
Frequency: 10 issues a year
Language: English
Annual subscription: Dfl909
ISSN: 0300 5194

– 1337 – GSF Mensch und Umwelt

Ingolstädter Landstrasse 1
D-8042 Neuherberg
Germany

Tel: 49 89 3187 2711
Fax: 49 89 3187 3324
Geographical coverage: Germany
Publishers: GSF Forschungszentrum fuer Umwelt und Gesundheit
Editor: Klemm, C.
Frequency: annual
Coverage: Scientific information for the public on environmental and health research topics
Price: free
ISSN: 0175 4521

RESEARCH CENTRES

– 1338 – Batelle Institute Ltd

15 Hanover Square
London W1R 9AJ
United Kingdom
Tel: 071 493 0184
Fax: 071 629 9705
Geographical coverage: International
Publications: *Batelle Today*, quarterly; President's report annually
Research activities: Health and environment

– 1339 – Ecole de Santé Publique – Université Libre de Bruxelles

Laboratoire de Médecine du Travail et d'Hygiène du Milieu
808 Rte de Lennick
B-1070 Brussels
Belgium
Tel: 32 2 569 56 00
Contact name: Mr H Bastenier
Geographical coverage: Belgium
Aims and objectives: Environmental and occupational hygiene

– 1340 – Hollustuvernd Rikisins

(National Centre for Hygiene Food Control and Environmental Protection)

Sidumúli 13
Box 8953
128 Reykjavik
Iceland
Tel: 354 1 91 81844
Geographical coverage: Iceland
Research activities: Environmental and public health matters in Iceland

– 1341 – Medizinisches Institut für Umwelthygiene

(Medical Institute for Environmental Hygiene)

Auf'm Hennekamp 50
D-4000 Düsseldorf
Germany
Tel: 49 211 33890
Geographical coverage: Germany
Research activities: Injurious effects of air and noise pollution

DATABASES

– 1342 – Environmental Bibliography

Geographical coverage: International
Coverage: Pollution; waste management; health hazards
Database producer: Environmental Studies Institute, Int. Academy, Santa Barbara
Database host: Dialog; Mead Data Central
Coverage period: 1973–
Update frequency: 4,000 records every two months
Additional information: Available on CD–ROM

Food Quality and Safety

GOVERNMENT ORGANISATIONS

– 1343 – Ministry of Agriculture Fisheries and Food (MAFF)

Whitehall Place
London SW1A 2HH
United Kingdom
Tel: 071 270 8080
Geographical coverage: United Kingdom
Aims and objectives: MAFF administers the Government's agriculture, horticulture and fisheries policies in England and has responsibilities for food, trade and animal health throughout the UK
Publications: *Food Additives: A Balanced Approach*

NON-GOVERNMENT ORGANISATIONS

– 1344 – Food and Agriculture Organisation of the United Nations (FAO)

Via delle Terme di Caracalla
I-00100 Rome
Italy

162 Health and Safety *Food Quality and Safety*

Tel: 39 6 57971
Fax: 39 6 57973152/5782610
Contact name: Edouard Saouma, Director General
Geographical coverage: International
Aims and objectives: FAO was established in 1945. The Organisation aims to increase production of agriculture, forestries and fisheries. It has an environment and energy programme. Environmental concerns are a major component of the organisation's field work
Publications: FAO publishes a number of yearbooks. It also publishes a quarterly, *Plant Protection Bulletin* and *Food Outlook Monthly*

– 1345 – Food Commission: Formerly Food Additives Campaign Team (FACT)

3rd Floor
Viking House
5–11 Worship Street
London EC2A 2BH
United Kingdom
Tel: 071 628 7774
Fax: 071 628 0817
Geographical coverage: United Kingdom
Aims and objectives: The Food Commission provides information on all aspects of food

– 1346 – Internationale Scharrelvees Controle (ISC)

(International Eco Meat Control)

PO Box 649
NL-3500 AP Utrecht
Netherlands
Tel: 31 30 340811
Geographical coverage: European
Aims and objectives: ISC is an independent organisation which controls member farmers and butchers on the observance of the eco regulations

– 1347 – Linking Environment and Farming (LEAF)

National Agricultural Centre
Stoneleigh
Warwickshire CV8 2LZ
United Kingdom
Tel: 0203 696969
Fax: 0203 696900
Contact name: Caroline Drummond, Project Coordinator
Geographical coverage: United Kingdom
Aims and objectives: LEAF's aims are to develop and promote farm practices which combine care and concern for the environment, with the responsible use of modern methods to produce safe and wholesome food
Publications: *Leafletter*, three times yearly; brochures and information booklets

– 1348 – Soil Association, The

86–88 Colston Street
Bristol BS1 5BB
United Kingdom
Tel: 0272 290661
Contact name: Sue Stolton, Publicity Administrator
Geographical coverage: United Kingdom
Aims and objectives: The Soil Association campaigns to maintain and improve the quality of food, health and the environment
Publications: List available on request

BOOKS

– 1349 – E is for Additives

Geographical coverage: United Kingdom
Coverage: Provides you with facts to make informed decisions about what you buy
Price: £4.50
Author: Hanssen, M. and Marsden, J.

RESEARCH CENTRES

– 1350 – British Industrial Biological Research Association: *see entry 1423*

Geographical coverage: United Kingdom

– 1351 – NIPH Department of Toxicology: *see entry 1430*

Geographical coverage: Norway

DATABASES

– 1352 – Food Safety Briefing (FSB)

The Lansbury Estate
Knaphill
Woking
Surrey GU21 2EP
United Kingdom
Geographical coverage: International
Coverage: Environmental health; reviews of new publications concerning food and safety
Database producer: Environmental Health Advisory Service (EHAS)
Database host: Direct dial to EHAS via Telecom Gold
Coverage period: 1988–
Update frequency: weekly

– 1353 – **Food Science and Technology Abstracts**

Geographical coverage: International
Coverage: All aspects of food sciences
Database producer: International Food Information Service
Database host: DataStar; Dialog; DIMDI; ESA–IRS; Orbit; STN
Coverage period: 1969–
Update frequency: 1,700 records monthly

Radiation Protection

GOVERNMENT ORGANISATIONS

– 1354 – **Bestuur van de Volksgezondheid – Ministerie van Volksgezondheid**

Rijksadministratief Centrum/Vesalius Gebouw
B-1010 Brussels
Belgium
Tel: 32 2 564 11 55
Geographical coverage: Belgium
Aims and objectives: Aims to collect the maximum number of documents concerning air pollution noise and radioactivity

– 1355 – **Consejo de Seguridad Nuclear**

Justo Dorado 11
E-28040 Madrid
Spain
Tel: 34 1 5 34 91 40
Fax: 34 1 45 51 28 0
Contact name: Donato Fuejo Lago, President
Geographical coverage: Spain
Aims and objectives: Regulatory body for nuclear safety and radiological protection

– 1356 – **Gabinete de Protecção e Segurança Nuclear**

Avenida da Republica 45–6
P-1000 Lisbon
Portugal
Tel: 351 1 793615
Fax: 351 1 7973482
Contact name: Marques de Cavalho, General Director
Geographical coverage: Portugal
Aims and objectives: Environmental radioactivity surveillance, safety assessment and emergency advice

– 1357 – **Greek Atomic Energy Commission**

GR-153 10 Aghia Paraskevi
Greece
Tel: 30 1 6515194
Fax: 30 1 6533939
Contact name: Charalambos Proukakis, President
Geographical coverage: Greece
Aims and objectives: Radiprotection, nuclear technology applications and nuclear energy

– 1358 – **Sociedad Española de Protección Radiológica (SEPR)**
(Spanish Society for Radiological Protection)

Avenida Compultense 22
E-28040 Madrid
Spain
Tel: 34 1 244 12 00
Contact name: Rafael Saenz-Gancedo, President
Geographical coverage: Spain
Aims and objectives: Scientific and organisational problems in the field of protection against ionising and non-ionising radiation

– 1359 – **State Committee for Nuclear and Radiation Safety**

Kiev
Ukraine
Tel: 7 044 559 6962
Fax: 7 044 559 5344
Contact name: N Shteinberg, Chairperson
Geographical coverage: Ukraine

– 1360 – **State Committee for the Supervision of Nuclear and Radiation Safety under the President of Russia**

Taganskaya str 34
109147 Moscow
Russia
Tel: 7 095 272 47 10
Fax: 7 095 278 80 90
Contact name: Yuri Vishnevsky, Chairperson
Geographical coverage: Russia
Aims and objectives: Regulatory body

NON-GOVERNMENT ORGANISATIONS

– 1361 – Associazione Italiana di Protezione Contro le Radiazioni (AIRP)
(Italian Radiation Protection Association)

c/o ENEA
Via Mazzini 2
I-40138 Bologna
Italy
Tel: 39 51 498259
Contact name: Dr G Busuoli, President
Geographical coverage: Italy
Aims and objectives: The Association organises conferences and panels on topics in radiation protection and helps in the solution of problems for workers in the field
Publications: *Annali Di Radioprotezione*; *Bolletino*; bi monthly

– 1362 – International Commission on Radiological Protection (ICRP)

Clifton Avenue
Sutton
Surrey SM2 5PU
United Kingdom
Tel: 081 642 4680
Contact name: Prof B Lindell, Chairperson
Geographical coverage: International
Aims and objectives: The Commission was established in 1928. It has been looked to as the appropriate body to give general guidance on the more widespread use of radiation sources caused by rapid developments in the field of nuclear energy
Publications: Reports; Annals of the ICRP, quarterly

– 1363 – International Radiation Protection Association (IRPA)

PO Box 662
NL-5600 A R Eindhoven
Netherlands
Tel: 31 40 473355
Fax: 31 40 435020
Contact name: Chris J Huyskens, Executive Officer
Geographical coverage: International
Aims and objectives: IRPA is the umbrella organisation for international communications and cooperation in radiation protection

– 1364 – National Radiological Protection Board (NRPB)

Chilton
Didcot
Oxfordshire OX11 0RQ
United Kingdom
Tel: 0235 831600
Fax: 0235 833891
Contact name: Matthew Gaines, Information Officer
Geographical coverage: United Kingdom
Aims and objectives: NRPB is an independent statutory body set up by the Radiological Protection Act 1970. It aims to provide advice and services on radiation standards to persons including government departments, with responsibilities in the UK
Publications: *Radiological Protection Bulletin*; Documents of the NRPB; technical reports and memoranda, booklets and brochures

– 1365 – Radiological Protection, Institute of Ireland

3 Clonskeagh Square
Dublin 14
Eire
Tel: 353 1 2697766
Fax: 353 1 2697766
Contact name: Mary Upton, Chairperson
Geographical coverage: Ireland
Aims and objectives: National regulatory authority for activities involving ionising radiation

– 1366 – Sateilyturvakeskus (STUK)
(Finnish Centre for Radiation and Nuclear Safety)

PO Box 268
SF-00101 Helsinki
Finland
Tel: 358 1 708 21
Fax: 358 1 708 2210
Contact name: Antti P U Vuorinen, Director General
Geographical coverage: Finland
Aims and objectives: STUK is a regulatory body for nuclear power, plant safety and other uses of radiation

– 1367 – World Radiation Data Centre

Voeikov Main Geo. Observatory
ul Karbysheva 7
St Petersburg 197018
Russia
Geographical coverage: International

BOOKS

– 1368 – Living with Radiation: National Radiological Protection Board

Geographical coverage: United Kingdom
Publishers: HMSO Publications, 1991
ISBN: 0859513203

– 1369 – Natural Sources of Ionising Radiation in Europe: Radiation Atlas

Geographical coverage: European
Publishers: Commission of the European Communities
Coverage: An atlas, of the levels of natural radiation in Europe. It contains maps for cosmic rays, gamma rays outdoors and indoors – the principal causes of human exposure
Author: Green, B. M. R. Hughes, J. S. and Lomas, P. R.

– 1370 – Radon in Dwellings

Geographical coverage: United Kingdom
Publishers: National Radiological Protection Board (NRPB), 1992
Price: £10.00
ISBN: 0859513491
Author: Green, B. M. R. et al

PERIODICALS

– 1371 – Journal of Radiological Protection

Techno House
Redcliffe Way
Bristol BS1 6NX
United Kingdom
Tel: 0272 297481
Fax: 0272 294318
Geographical coverage: United Kingdom
Publishers: IOP Publishing
Editor: Charles, Dr M. W.
Frequency: quarterly
Coverage: Scientific papers on radiation protection
Annual subscription: £23.80
ISSN: 0952 4746

– 1372 – Radiprotection

Z I de Courtaboeuf
BP 112
F-91922 les Ulis Cedex A
France
Geographical coverage: France
Publishers: Société Française de Radiprotection
Frequency: quarterly
Language: Text and summaries in English and French
Annual subscription: Ffr235
ISSN: 0033 8451

REPORTS

– 1373 – Risks from Radon in Homes: Report of a Working Group of the Institute of Radiation Protection

Geographical coverage: United Kingdom
Publishers: H & H Scientific Consultants Ltd, 1990
Coverage: Provides information on the behaviour of Radon, its measurement in UK homes, remedial action and quantification of risks
ISBN: 0948234704X
Author: Duggan, M. J. et al

RESEARCH CENTRES

– 1374 – Commission of the European Communities (CEC) – Joint Research Centre (JRC)

Rue de la Loi 200
B-1049 Brussels
Belgium
Tel: 32 2 235 8527
Fax: 32 2 235 0146
Contact name: J P Contzen, Director General
Geographical coverage: European
Aims and objectives: The JRC was founded in 1960 as a scientific and technical base for primarily nuclear research. Since then its activities have been broadened to encompass research on the environment and safety
Research activities: Nuclear safety for the European Community

– 1375 – Eidgenössische Kommission zur Überwachung der Radioaktivität

(Swiss Federal Commission of Radioactivity Surveillance)

c/o Physics Department
University of Fribourg
CH-1700 Fribourg/Perolles
Switzerland
Tel: 41 37 82 62 36
Geographical coverage: Switzerland
Publications: Annual reports in German and French
Research activities: Radioacitivity monitoring; environmental monitoring around nuclear facilities; industries using radionuclides

– 1376 – Gesellschaft für Strahlen und Umweltforschung GmbH (GSF)

(Radiation and Environmental Research Centre)

Ingolstädter Landstrasse 1
D-8042 Neuherberg
Germany
Tel: 49 89 31870
Contact name: Prof Dr H W Levi, Director

– 1377 – Institut für Immission- Arbeits- und Strahlenschutz

(Institute for Emission Labour and Radiation Protection)

Hertzstrasse 173
Baden-Württemberg
D-7500 Karlsruhe 21
Germany
Tel: 49 721 75031
Fax: 49 721 758758
Geographical coverage: Germany
Research activities: Concerned with the emission of substances, radiation and noises

– 1378 – Institut für Strahlenhygiene (ISH)

(Institute for Radiation Hygiene)

Ingolstädter Landstrasse 1
D-8042 Neuherberg
Germany
Tel: 49 89 3187555
Contact name: Prof Dr Alexander Kaul, Head
Geographical coverage: Germany
Publications: *ISH–Berichte*; *BGA–Schriften*
Research activities: Interests include radiation effects and protection: genetic effects; somatic effects; radiochemistry

– 1379 – Laboratório Nacional de Engenharia e Tecnologia Industrial (LNETI)

(National Laboratory for Engineering and Industrial Technology)

Estrado Paco do Lumiar 22
P-1600 Lisbon
Portugal
Tel: 351 1 7586141/7589181
Contact name: A Coelho Carvalho, President
Geographical coverage: Portugal
Publications: *Corrosão e Proteção de Materials*; annual reports
Research activities: LNETI has been given the necessary facilities to modernise national industry and to assure its development

Toxicology and Chemicals

GOVERNMENT ORGANISATIONS

– 1380 – Department of Health

Richmond House
79 Whitehall
London SW1A 2NS
United Kingdom
Tel: 071 210 3000
Geographical coverage: United Kingdom
Aims and objectives: The Department is responsible in England for the administration of the National Health Service. Representing the UK in the World Health Organisation and other international forums

– 1381 – Medical Research Council, Toxicology Unit

Woodmansterne Road
Carshalton
Surrey SM5 4EF
United Kingdom
Tel: 081 643 8000
Geographical coverage: United Kingdom

– 1382 – Miljoministeriet Miljostyrelsen Kemikaliekontrollen

(The State Chemical Supervision Service)

12 Skovbrynet
DK-2800 Lyngby
Denmark
Tel: 45 02 87 70 66
Contact name: Hedegaard Poulsen, Director
Geographical coverage: Denmark
Aims and objectives: Controls and observes pesticide and toxic substance legislation

– 1383 – Ministry of Agriculture Fisheries and Food, Fisheries Laboratory: Fisheries Radiological Laboratory

Pakefield Road
Lowestoft
Suffolk NR33 0HT
United Kingdom
Tel: 0502 562244
Geographical coverage: United Kingdom

NON-GOVERNMENT ORGANISATIONS

– 1384 – Danish Toxicology Centre (DTC)

Agern Allee 15
DK-2970 Hoersholm
Denmark
Tel: 45 42 57 00 55
Fax: 45 42 57 19 19
Contact name: Ms Lisbeth Valentin Hansen
Geographical coverage: Denmark
Aims and objectives: DTC aims to procure, assess and communicate toxicological data and aspects regarding chemical legislation

– 1385 – Environmental Detergent Manufacturers Association (EDMA)

Mouse Lane
Steyning BN44 3DG
United Kingdom
Tel: 0903 879077
Fax: 0903 879052
Contact name: Robin Bines, Chairperson
Geographical coverage: European
Aims and objectives: The Association debates environmental consequences of detergents and lobbies for high standards on EC legislation. Also provides information to consumers, manufacturers, the media and national regulatory bodies

– 1386 – European Chemical Industry Ecology and Toxicology Centre (ECETOC)

Avenue E Van Nieuwenhuyse 4
Box 6
B-1160 Brussels
Belgium
Tel: 32 2 675 36 00
Fax: 32 2 675 36 25
Contact name: Derek Stringer, Director
Geographical coverage: European
Aims and objectives: ECETOC's main objective is to coordinate information and on going research into products and processes developed in the chemical industry, specifically in areas of toxicology and ecotoxicology

– 1387 – European Chlorinated Solvent Association (ECSA)

Michelangelo
Avenue E Van Nieuwenhuyse 4
B-1160 Brussels
Belgium
Tel: 32 2 640 20 95
Fax: 32 2 640 62 498
Geographical coverage: European
Aims and objectives: ECSA is a sector group of CEFIC. The members promote environmentally safe and economically viable use of chlorinated solvents. They also sponsor research on environmental and toxicological issues
Publications: *Solvents Digest*; *ECSA Newsletter*

– 1388 – European Transport Law

Maria Henriettalei 1
B-2018 Antwerp
Belgium
Tel: 32 3 2313655
Fax: 32 3 23442380
Geographical coverage: European
Aims and objectives: Concerned with transport of dangerous goods and transfer of dangerous waste

– 1389 – International Commission for Protection Against Environmental Mutagens and Carcinogens

Medical Bio Lab TNO
Lange Kleiweg 139 PB 45
NL-2280 AA Rijswijk
Netherlands
Tel: 31 15138777
Contact name: Dr J D Jansen, Secretary
Geographical coverage: International
Aims and objectives: The Commission helps to identify and promote scientific principles and to determine guidelines and regulations for the prevention of the effects of chemicals on human genetic material

– 1390 – International Programme on Chemical Safety (IPCS)

World Health Organisation
CH-1211 Geneva 27
Switzerland
Tel: 42 22 791 21 11
Fax: 42 22 791 07 46
Contact name: Dr Michel Mercier, Manager, Division of Environmental Health
Geographical coverage: International
Aims and objectives: IPCS aims to carry out and disseminate evaluations of the risk to human health and the environment from exposure to chemicals
Publications: Environmental Health Criteria Documents; Health and Safety Guides; International Chemical Safety Cards; Technical Report Series

– 1391 – International Society for Environmental Toxicology and Cancer

PO Box 134
Park Forest
IL 60466
United States of America
Tel: 1 312 755 2080
Geographical coverage: International

– 1392 – Society for Environmental Toxicology and Chemistry (SETAC)

Avenue Prekelinden 149
B-1200 Brussels
Belgium
Fax: 32 2 462 28 84
Geographical coverage: European
Aims and objectives: SETAC has formed a Life Cycle Analysis group

– 1393 – UNEP'S International Register of Potentially Toxic Chemicals (IRPTC)

Palais des Nations
CH-1211 Geneva 10
Switzerland
Tel: 41 22 985850/988400
Geographical coverage: International
Aims and objectives: UN information service. Collects information on toxic and environmental effects of chemicals
Publications: *IRPTC Bulletin*; *IRPTC Legal File*; *Treatment and Disposal Methods for Waste Chemicals*
Database producer: IRPTC
Coverage period: 1980–
Update frequency: continuously

PRESSURE GROUPS

– 1394 – Communities Against Toxics

c/o 129 Edison Way
Hemlington
Cleveland TS8 9ES
United Kingdom
Tel: 0642 595740
Geographical coverage: United Kingdom

BOOKS

– 1395 – C is for Chemicals

Geographical coverage: United Kingdom
Publishers: Soil Association
Coverage: Chemical hazards and how to avoid them
Price: £4.99
Author: Mansfield, P. and Monro, J.

– 1396 – Dioxins in the Environment Pollution Paper 27

Geographical coverage: United Kingdom
Publishers: HMSO, 1989
Price: £8.60

– 1397 – Ecogenetics: Genetic Predisposition to the Toxic Effects of Chemicals

Geographical coverage: European
Publishers: Chapman and Hall, 1991, UK
Editor: Grandjean, P.
Language: English
Price: £35.00
ISBN: 0412392909

– 1398 – Environmental and Health Controls on Lead

Geographical coverage: International
Publishers: International Lead and Zinc Study Group, 1989
Language: English
Coverage: A review of current and proposed regulations controlling lead in works, lead in the atmosphere and lead in water in 25 countries
Price: £20.00

– 1399 – Eutrophication of Freshwaters

Geographical coverage: International
Publishers: Chapman and Hall, 1991
Coverage: Explains causes and effects of eutrophication worldwide
Price: £35.00
ISBN: 0412329700
Author: Harper, D.

– 1400 – Metals and their Compounds in the Environment

Geographical coverage: United Kingdom
Publishers: VCH, 1990
Coverage: Contains detailed information on the environmental metals which influence the health of plants, animals and humans
Price: £127.00
ISBN: 352726521X
Author: Merian, E.

– 1401 – Neurotoxicity

Geographical coverage: United Kingdom
Publishers: Van Nostrand Reinhold International, 1992
Coverage: This text aims to clarify what neurotoxins are, how they affect the brain, how government policies regulate chemicals, how to test, monitor and assess for risks
Price: £47.00
ISBN: 0442010478

– 1402 – Occupational Exposure Limits for Airborne Toxic Substances, Third Edition

Geographical coverage: International
Publishers: ILO, 1991, (Occupational Health and Safety Series No 37)
Coverage: The latest available values either prescribed or recommended in 15 countries
ISBN: 9221072932

– 1403 – Waste Not Want Not: The Production and Dumping of Toxic Waste

Geographical coverage: United Kingdom
Publishers: Earthscan Publications Ltd, 1992
Coverage: Campaigning book on the production and dumping of toxic waste in Britain and Ireland
Price: £9.95
ISBN: 185383095X
Author: Allen, R.

– 1404 – Water Pollution by Fertilisers and Pesticides

Geographical coverage: International
Publishers: OECD, 1986
ISBN: 9264128565

PERIODICALS

– 1405 – Acta Toxicologica et Therapeutica

Via Trento 53
I-43100 Parma
Italy
Fax: 39 521 771268
Geographical coverage: International
Publishers: C E M Casa Editoriale Maccari
Editor: Cerrati, A. and Debemarm, M.
Frequency: quarterly
Language: English, French and Italian
Coverage: International journal of toxicology, pharmacology and therapy
Annual subscription: Lit60000
ISSN: 0393 635X

– 1406 – Aquatic Toxicology

PO Box 211
NL-1000 AE Amsterdam
Netherlands
Tel: 31 20 5803911
Fax: 31 20 5803598
Geographical coverage: Netherlands
Publishers: Elsevier Science Publishers BV
Editor: Malins, D. C. and Jensen, A.
Frequency: monthly
Language: English
Coverage: Environmental toxicology
Annual subscription: Dfl1203
ISSN: 0166 445X

– 1407 – Arhiv za Higijenu Rada i Toksikologiju

(Archives of Industrial Hygiene and Toxicology)

University of Zagreb
PO Box 291
41001 Zagreb
Croatia
Fax: 38 41 274 572
Geographical coverage: Croatia
Publishers: Institute for Medical Research and Occupational Health
Editor: Plestina, R.
Frequency: quarterly
Language: Croatian and English
Price: $40
ISSN: 0004 1254

– 1408 – Chemico–Biological Interactions

PO Box 85
Limerick
Eire
Tel: 353 61 61944
Fax: 353 61 62144
Geographical coverage: Ireland
Publishers: Elsevier Scientific Publishers, Ireland
Frequency: 15 issues a year
Language: English
Annual subscription: $960

– 1409 – Ecotoxicology

2–6 Boundary Row
London SE1 8HN
United Kingdom
Tel: 071 865 0066
Fax: 071 522 9623
Geographical coverage: United Kingdom
Publishers: Chapman and Hall
Frequency: quarterly
Annual subscription: £50.00
ISSN: 0963 9292

– 1410 – Environmental Toxicology and Chemistry

Headington Hill Hall
Oxford
Oxon OX3 0BW
United Kingdom
Tel: 0865 794141
Fax: 0865 743592
Geographical coverage: United Kingdom

Publishers: Pergamon Press plc
Editor: Ward, H.
Frequency: 12 issues a year
Coverage: Physical sciences
Annual subscription: DM732

– 1411 – Environmental Toxicology and Water Quality

Baffins Lane
Chichester
West Sussex PO19 1UD
United Kingdom
Tel: 0243 770351
Fax: 0243 775878
Geographical coverage: United Kingdom
Publishers: John Wiley and Sons Ltd
Editor: Dutka, B.
Frequency: quarterly
Coverage: Covers all aspects of environmental toxicology and water quality
Annual subscription: $179

– 1412 – European Journal of Pharmacology: Environmental Toxicology and Pharmacology Section

PO Box 211
NL-1000 AE Amsterdam
Netherlands
Tel: 31 20 5803911
Fax: 31 20 5803705
Geographical coverage: European
Publishers: Elsevier Science Publishers BV
Editor: Koeman, J. H. and Nijkamp, F. P.
Frequency: 6 issues a year
Language: English
Coverage: Publishes research in mechanistic studies concerning the toxic effects in man and vertebrate animals of drugs and environmental contaminants
Annual subscription: Dfl371
ISSN: 0926 6917

– 1413 – Haznews

140 Battersea Park Road
London SW11 4NB
United Kingdom
Tel: 071 498 2511
Fax: 071 498 2343
Geographical coverage: International
Publishers: Profitastral Ltd
Editor: Coleman, D.
Frequency: monthly
Language: English
Coverage: Provides news and features on technology, research, legislation, company developments on hazardous waste management
Annual subscription: £176.00
ISSN: 0953 5357

– 1414 – Journal de Toxicologie Clinique et Experimentale

120 boulevard St Germain
F-75280 Paris Cedex 06
France
Tel: 33 1 463 421 60
Fax: 33 1 458 729 99
Geographical coverage: France
Publishers: Masson
Editor: Roche, I.
Frequency: 8 issues a year
Price: Ecus 158
ISSN: 0753 2830

– 1415 – Journal of Biochemical Toxicology

Postfach 10 11 61
D-6940 Weinheim
Germany
Tel: 49 6201 602 0
Geographical coverage: Germany
Publishers: V C H Verlagsgesellschaft mbH
Editor: Hodgson, E.
Frequency: 4 issues a year
Language: English
Annual subscription: DM370
ISSN: 0887 2082

– 1416 – Journal of Hazardous Materials

PO Box 211
NL-1000 AE Amsterdam
Netherlands
Tel: 31 20 5803911
Geographical coverage: International
Publishers: Elsevier Scientific Publishing Co
Frequency: quarterly
ISSN: 0304 3894

– 1417 – Toxicology

PO Box 85
Limerick
Eire
Tel: 353 61 61944
Fax: 353 61 62144
Geographical coverage: International
Publishers: Elsevier Scientific Publishers Ireland
Editor: Witschi, H. P. and Netter, K. J.
Frequency: 24 issues a year
Language: English
Coverage: Publishes original scientific papers on the biological effects arising from the administration of chemical compounds
Annual subscription: $1514
ISSN: 0300 483X

REPORTS

– 1418 – Chlorine Production and Use and their Environmental Risks

Brussels Office
Rue de la Concorde 51
B-1040 Brussels
Belgium
Belgium
Tel: 32 2 514 01 24
Fax: 32 2 512 32 65
Geographical coverage: European
Publishers: Institute for European Environmental Policy, 1991

– 1419 – Death in Small Doses – The Effects of Organochlorines on Aquatic Ecosystems

Geographical coverage: International
Publishers: Greenpeace International, 1992
Editor: Johnston, P.
Coverage: The report draws on a range of scientific findings to demostrate the urgent need for a complete phase out of all releases of organochlorines
ISBN: 1871532612

– 1420 – Poisonous Fish

Geographical coverage: United Kingdom
Publishers: Friends of the Earth (FoE), 1991
Coverage: A survey of available data on the contamination of freshwater fish flesh with persistent toxic chemicals
Price: £13.00
ISBN: 1857500350

– 1421 – Product is the Poison, – The Case for a Chlorine Phase-Out

Geographical coverage: International
Publishers: Greenpeace International, 1991
Author: Thornton, J.

– 1422 – Single European Dump, The: Free Trade in Hazardous and Nuclear Wastes in the New Europe

Greenpeace EC Unit
Avenue de Tervuren 36
B-1040 Brussels
Belgium
Tel: 32 2 736 99 27
Fax: 32 2 736 44 60
Geographical coverage: European
Publishers: Greenpeace, December 1991
Language: English
Author: Carroll, S.

RESEARCH CENTRES

– 1423 – British Industrial Biological Research Association

Woodmansterne Road
Carshalton
Surrey SM5 4DS
United Kingdom
Tel: 081 643 4411
Geographical coverage: United Kingdom
Research activities: Toxicology of environmental chemicals, food additives etc

– 1424 – Department of Toxicology

Faculty of Agricultural Sciences
Biotechnion
De Dreijen 12
NL-6703 BC Wagen
Netherlands
Tel: 31 8370 82137
Geographical coverage: Netherlands
Research activities: Cell and genetic toxicology; environmental effects of pesticides

– 1425 – Fraunhofer-Institut für Toxicologie und Aerosolforschung

(Fraunhofer Institute for Toxicology and Aerosol Research)

Nikolai-Fuchs-Strasse 1
D-3000 Hannover 61
Germany
Tel: 49 511 5350 301
Fax: 49 511 5350 155
Geographical coverage: Germany
Research activities: Protecting and promoting human health standards and the environment

– 1426 – Fraunhofer-Institut für Umweltchemie und Ökotoxikologie (IUCT)

(Fraunhofer Institute for Environmental Chemistry and Toxicology)

Postfach 1260
Grafschaft
D-5948 Schmallenberg
Germany
Tel: 49 29 72 30 2 0
Fax: 49 29 72 30 23 19
Geographical coverage: Germany
Research activities: Exposure to and ecotoxicological effects of environmental chemicals including pesticides and other chemical products

– 1427 – Institut National de la Santé et de la Recherche Medicale

Université de Paris
11 rue J B Clement
F-92290 Chatenay-Malabry
France
Tel: 33 1 6604518
Contact name: Mr Boudene
Geographical coverage: France
Research activities: Toxicology and air pollution

– 1428 – Institut Pasteur de Lille

Domaine du Certia BP 15
F-59650 Villeneuve d'Ascq
France
Tel: 33 20561100
Geographical coverage: France

– 1429 – Inveresk Research International Ltd

Inveresk Gate
Musselburgh
Edinburgh EH21 7UB
United Kingdom
Tel: 031 665 6881
Fax: 031 665 9976
Geographical coverage: United Kingdom
Publications: *Regulatory Affairs Bulletin*, quarterly
Research activities: Toxicology, mutagenesis; ecotoxicology; biotechnology

– 1430 – NIPH Department of Toxicology

Geitmyrsveien 75
N-0462 Oslo 4
Norway
Tel: 47 2 35 6020
Geographical coverage: Norway
Research activities: Toxicological evaluation and consultation related to contaminants in food and drinking water

– 1431 – Zentralstelle für Gesamtverteidigung

Rebhaldenstrasse 2
CH-9450 Altstätten
Switzerland
Tel: 41 71 75 40 08
Contact name: Dr P Bützer
Geographical coverage: Switzerland
Research activities: Modelling toxic gas effects

DATABASES

– 1432 – Chemical Exposure

Geographical coverage: International
Coverage: Toxicology and biomedical effects of chemicals such as pesticides and drugs
Database producer: Science Applications International Corporation
Database host: Dialog
Coverage period: 1974–
Update frequency: 2,000 records annually

– 1433 – Chemical Safety Newsbase

Geographical coverage: International
Coverage: Occupational hazards in the chemical industry; hazardous reactions, fires, explosions, waste management, legislation and regulations
Database producer: Royal Society of Chemistry
Database host: DataStar; Dialog; ESA–IRS; Orbit; STN
Coverage period: 1981–
Update frequency: monthly

– 1434 – Environmental Chemicals Data and Information Network (ECDIN)

ECDIN Group
I-21020 Ispra (Varese)
Italy
Tel: 39 332 789720
Geographical coverage: European
Language: English, Danish, Dutch, French, German and Italian
Coverage: Contains information on 63,000 chemical compounds, acute toxicity data on approximately 20,000 compounds and additional data on 2,000 substances
Database producer: CEC Joint Research Centre
Database host: Datacentralen; CEC Joint Research Council
Coverage period: 1970–
Update frequency: Periodically as new data is available

– 1435 – Hazardous Substances Databank (HSDB)

Geographical coverage: International
Coverage: Toxicology, chemistry and hazardous substances
Database producer: National Library of Medicine
Database host: DataStar; DIMDI; TOXNET
Coverage period: Current
Update frequency: constant
Additional information: Available on CD–ROM

– 1436 – Major Hazard Incident Data Service (MHIDAS)

Geographical coverage: International

Coverage: Major Hazard Incident Data Service; including health and safety, all incidents involving hazardous materials in the UK and internationallly
Database producer: Health & Safety Executive, IAEA
Database host: ESA–IRS
Coverage period: 1964–
Update frequency: quarterly

– 1437 – Medline

Geographical coverage: International
Coverage: Includes toxicology
Database producer: US National Library of Medicine
Database host: DataStar; Dialog; DIMDI; Blaise; Telesystems Questel
Coverage period: 1966–
Update frequency: 24,000 records monthly

– 1438 – Registry of Toxic Effects of Chemicals on Substances (RTECS)

Geographical coverage: International
Coverage: Registry of Toxic Effects of Chemical Substances, including toxicology measurements of 100,000+ substances
Database producer: US National Institute for Occupational Safety and Health
Database host: Blaise; Dialog; DIMDI
Update frequency: quarterly

– 1439 – Toxic Materials Transport

Geographical coverage: International
Coverage: Transport of hazardous materials
Database producer: Business Publishers Inc
Database host: DataStar; Dialog
Coverage period: 1985–

– 1440 – Toxline

Geographical coverage: International
Coverage: Adverse effects of chemicals and related products on humans and animals. Includes data from Pesticide Abstracts
Database producer: National Library of Medicine and Food Science
Database host: Blaise; DataStar; Dialog; DIMDI; National Library of Medicine
Coverage period: 1940–
Update frequency: 12,000 records monthly
Additional information: Available on CD–ROM

10
Pollution

General

GOVERNMENT ORGANISATIONS

– 1441 – Her Majesty's Inspectorate of Pollution (HMIP)

Romney House
43 Marsham Street
London SW1P 3PY
United Kingdom
Tel: 071 276 8061
Geographical coverage: United Kingdom
Aims and objectives: Inspectorate of the Department of the Environment

– 1442 – Royal Commission on Environmental Pollution (RCEP)

Church House
Great Smith Street
London SW1P 3BZ
United Kingdom
Tel: 071 276 2080
Fax: 071 276 2098
Contact name: B Glicksman, Secretary
Geographical coverage: United Kingdom
Aims and objectives: The Commission was established by Royal Warrant in 1970. It advises on matters both national and international, concerning the pollution of the environment
Publications: Reports on various aspects of environmental pollution. List available on request

– 1443 – Statens Forurensningstilsyn (SFT)

(The Norwegian State Pollution Control Authority)
PO Box 8100 Dep
N-0032 Oslo 1
Norway
Tel: 22 57 34 00
Fax: 22 67 67 06
Geographical coverage: Norway
Aims and objectives: SFT carries out professional surveys and evaluations for the Ministry of Environment and is generally responsible for monitoring the status of pollution in air and water
Publications: Publications catalogue available

NON-GOVERNMENT ORGANISATIONS

– 1444 – Association Vincotte ASBL

125 rue de Rhode
B-1630 Linkebeek
Brabant
Belgium
Tel: 32 2 358 3580
Contact name: Mr M Warzee
Geographical coverage: Belgium
Aims and objectives: The Association carries our control measurements and studies on air and water pollutants, noise and solid wastes

– 1445 – Centre d'Information du Plomb

1 boulevard de Vaugirard
F-75751 Paris Cedex 15
France

Tel: 33 1 538 52 33
Contact name: Mr Prevost
Geographical coverage: France
Aims and objectives: Health protection, pollutant effects on man and society

– 1446 – Hellenic Association on Environmental Pollution

14 Xenofontos Street
GR-105 57 Athens
Greece
Tel: 30 1 32 43 534
Contact name: Mr John G Panaioannou, President
Geographical coverage: Greece

– 1447 – Inter Environnement Bond Beter Leefmilieu Centre d'Inform et de Documentation sur l'Environnement

49 rue d'Arlon
B-1040 Brussels
Belgium
Tel: 32 2 230 31 44
Contact name: Mr Y Schillebeeck
Geographical coverage: Belgium
Aims and objectives: Environment documentation centre concerning air water and soil pollution

– 1448 – Oil Companies European Organisation for Environmental Health and Protection (CONCAWE)

Madouplein 1
B-1030 Brussels
Belgium
Tel: 32 2 220 31 11
Fax: 32 2 219 46 46
Contact name: Klaus Kohlhase, Chairman
Geographical coverage: European
Aims and objectives: Founded in 1963 by six oil companies and now has a membership of 36 companies. These represent 90% of all the refining capacity in western Europe. The scope of its activities include air, land and water pollution control and health protection
Publications: Technical reports. Catalogues are available on request

BOOKS

– 1449 – Environmental Respiratory Diseases

Geographical coverage: United Kingdom
Publishers: Van Nostrand Reinhold International, 1992
Coverage: Covers diseases of the respiratory system caused by environmental contaminants and pollution problems
ISBN: 0442008473
Author: Cordasco, E. M.

– 1450 – Future is Now, The

Geographical coverage: Norway
Publishers: State Pollution Control Authority
Editor: Sivertsen, S. and Lundgaard, H.
Language: English
Coverage: A summary of SFT's long term planning
ISBN: 8290031343

– 1451 – Green PC, The

Geographical coverage: United Kingdom
Publishers: Tab Books, 1993
Coverage: Takes a look at how computing affects our environment and what computer users can do to reverse the growing problem of computer generated pollution
Price: £7.95
ISBN: 0830643117
Author: Anzovin, S.

– 1452 – Integrated Pollution Control in Europe and North America

Geographical coverage: International
Publishers: The Conservation Foundation/IEEP, 1990
Editor: Haigh, N. and Irwin, F.
Price: £15.00
ISBN: 0891641173

– 1453 – Pollution Control Policy of the European Communities

Geographical coverage: European
Publishers: Graham and Trotman, 1983
ISBN: 0860104117
Author: Johnson, S. P.

DIRECTORIES/YEARBOOKS

– 1454 – Achema Handbook Pollution Control

Geographical coverage: Germany
Publishers: Dechema Deutsche Gesellschaft für Chemisches Apparatewesen Chemische Technik eV
Frequency: every 3 years
Language: English, French and German
Coverage: Manufacturers of technical equipment and products for environment engineering

– 1455 – Eco Technics: International Pollution Control Directory

Geographical coverage: International
Publishers: ECO Verlags AG

176 Pollution *General*

Frequency: every 2 years
Coverage: Lists 35,000 manufacturers and service organisations worldwide concerned with pollution control

– 1456 – Forurensning i Norge 1991
(Pollution in Norway)

Geographical coverage: Norway
Publishers: Utslipp Miljotilstand Perpektiv 1991
Frequency: annual

– 1457 – Guide de l'Environnement et des Techniques Antipollution
(Environmental Protection and Antipollution Directory)

Geographical coverage: France
Publishers: Compagnie Française d'Editions
Frequency: every 2 years
Coverage: Industrial units, government departments, associations, institutions concerned with pollution control

– 1458 – POLMARK – The European Pollution Control and Waste Management Industry Directory

Geographical coverage: European
Publishers: ECOTEC Research and Consulting Ltd, 1992
Frequency: annual
Language: English
Coverage: Comprehensive directory of European pollution control – includes some market analysis information
Price: $425
ISBN: 0863545890

PERIODICALS

– 1459 – Code Permanent Environnement et Nuisances

80 avenue de la Marne
F-92546 Montrouge Cedex
France
Tel: 33 1 409 268 68
Fax: 33 1 465 600 15
Geographical coverage: France
Publishers: Editions Legislatives et Administratives
Frequency: monthly
Coverage: Focuses on the effects on the environment of commerce and industry, water, air and noise pollution

– 1460 – Ecological Engineering

PO Box 211
NL-1000 AE Amsterdam
Netherlands
Tel: 31 20 5803911
Fax: 31 20 5803705
Geographical coverage: Netherlands
Publishers: Elsevier Science Publishers BV
Editor: Costanza, R.
Frequency: quarterly
Language: English
Coverage: Publishes contributions in ecotechnology including bioengineering, pollution control and sustainable agriculture
Annual subscription: Dfl290
ISSN: 0925 8574

– 1461 – ENDS

Unit 24
Finsbury Business Centre
40 Bowling Green Lane
London EC1R 0NE
United Kingdom
Tel: 071 278 4745
Fax: 071 837 7612
Geographical coverage: United Kingdom
Publishers: Environmental Data Services Ltd
Editor: Mayer, M.
Frequency: monthly
Coverage: Environmental policy in the UK and EEC; pollution control, waste management and the profiles of key environmental organisations. This is a key environmental information resource
ISSN: 0260 1249

– 1462 – Enviro

SNV
S-171 85 Solna
Sweden
Geographical coverage: Sweden
Publishers: Swedish Environment Protection Agency
Editor: Hanneberg, P.
Frequency: monthly
Language: English
Coverage: Disseminates views on air and sea pollution and acidification of the environment
Price: free
ISSN: 1101 7341

– 1463 – Environmental Policy and Practice

52 Kings Road
Richmond
Surrey TW10 6EP
United Kingdom
Tel: 081 995 0877
Fax: 081 747 9663
Geographical coverage: United Kingdom
Publishers: EDD Publications
Editor: Aston, G.
Frequency: quarterly

Coverage: Pollution control and waste management environmental law
Price: £20.00
Annual subscription: £66.00

– 1464 – Environmental Pollution

Crown House
Linton Road
Barking
Essex IG11 8JU
United Kingdom
Tel: 081 594 7272
Fax: 081 594 5942
Geographical coverage: International
Publishers: Elsevier Science Publications Ltd
Editor: Dempster, J. P. and Manning, J. W.
Frequency: monthly
Coverage: The biological chemical and physical aspects of environmental pollution and pollution control
Annual subscription: £645.00
ISSN: 0269 7491

– 1465 – Excerpta Medica Section 46: Environmental Health and Pollution Control

PO Box 548
NL-1000 AM Amsterdam
Netherlands
Tel: 31 20 5803911
Fax: 31 20 5803222
Geographical coverage: Netherlands
Publishers: Excerpta Medica (Subsidiary of Elsevier Science Publishers)
Frequency: 10 issues a year
Language: English
Coverage: Covers all aspects of air soil and water pollution
Annual subscription: Dfl909
ISSN: 0300 5194

– 1466 – Fundamental Aspects of Pollution Control and Environmental Science

PO Box 211
NL-1000 AE Amsterdam
Netherlands
Tel: 31 20 5803911
Fax: 31 20 5803705
Geographical coverage: Netherlands
Publishers: Elsevier Science Publishers BV
Frequency: irregular
Language: English
Price: varies

– 1467 – Inquinamento

Via Mecenate 91
I-20138 Milan
Italy
Tel: 39 2 5808 4302
Fax: 39 2 5801 2592
Geographical coverage: Italy
Publishers: Etas srl
Editor: Berbenni, P.
Frequency: monthly
Coverage: Technical publication on water, air, soil and noise pollution, waste, recycling and clean energy treatment techniques
Annual subscription: Lit165000
ISSN: 0001 4982

– 1468 – International Journal of Environment and Pollution

110 avenue Louis Casai
Case Postale 309
CH-1215 Geneva Airport
Switzerland
Fax: 41 22 7910885
Geographical coverage: International
Publishers: Inderscience Enterprises Ltd
Frequency: quarterly
Language: English
Coverage: Examines environmental policy, pollution control and sustainable development in these fields
Annual subscription: $155
ISSN: 0957 4352

– 1469 – Literaturberichte über Wasser, Abwasser, Luft und Feste Abfallstoffe

Wollgrasweg 49
Postfach 72 01 43
D-7000 Stuttgart 70
Germany
Tel: 49 711 455038
Geographical coverage: Germany
Publishers: Gustav Fischer Verlag
Frequency: irregular
Language: German

– 1470 – Ministerio da Qualidade de Vida Comissão Nacional do Ambiente Boletim

Praça Duque de Saldanha 31
P-1096 Lisbon Codex
Portugal
Geographical coverage: Portugal
Publishers: Ministerio da Qualidade de Vida Comissão Nacional do Ambiente
Frequency: 12 issues a year

178 Pollution *General*

– 1471 – Pollustop

4 rue Leneveux
F-75014 Paris
France
Geographical coverage: France
Editor: Forel, J.
Frequency: 12 issues a year

– 1472 – Pollution

Cross Street Court (Ground Floor)
Cross Street
Peterborough PE1 1XA
United Kingdom
Tel: 0733 52454
Geographical coverage: United Kingdom
Publishers: Springfield Information Services
Editor: Franks, J. W. F.
Frequency: monthly
Coverage: Ecology and environment
Annual subscription: £75.00
ISSN: 0048 4700

– 1473 – Pollution Prevention

Rococo House
281 City Road
London EC1V 1LA
United Kingdom
Tel: 071 250 1234
Fax: 071 253 6664
Geographical coverage: United Kingdom
Publishers: Wirelink Communications Ltd
Editor: Wolfe, P.
Frequency: 7 issues a year
Coverage: Environmental/waste management
Annual subscription: £12.40

– 1474 – Polska Akademia Nauk Instytut Podstaw Inzynierii Srodowiska Prace i Studia

Rynek 9
Wroclaw
Poland
Geographical coverage: Poland
Publishers: Ossolineum Publishing House of the Polish Academy of Sciences
Editor: Godzik, S.
Frequency: irregular
Language: Text in Polish, summaries in English and Russian
Price: varies

– 1475 – Reinwater

Vossiusstraat 20
NL-1071 AD Amsterdam
Netherlands
Tel: 31 20 719322
Fax: 31 20 753806
Geographical coverage: Netherlands
Publishers: Stichting Reinwater
Frequency: quarterly
Language: Dutch
Coverage: Covers problems and solutions of pollution in the Netherlands by focusing on the main European rivers
Annual subscription: Dfl47.50

– 1476 – Wasser Luft und Boden

Lise-Meitner Strasse 2
Postfach 2760
D-6500 Mainz 1
Germany
Tel: 49 6131 992 01
Fax: 49 6131 992 100
Geographical coverage: Germany
Publishers: Vereinigte Fachverlage GmbH
Frequency: 9 issues a year
Coverage: Technical journal concerned with industrial water economics, purification of air and waste disposal
Annual subscription: DM220
ISSN: 0938 8303

– 1477 – Water Air and Soil Pollution

PO Box 17
NL-3300 AA Dordrecht
Netherlands
Tel: 31 78 334911
Fax: 31 78 334254
Geographical coverage: International
Publishers: Kluwer Academic Publishers
Editor: Mc Cormack, B.
Frequency: 24 issues a year
Language: English
Annual subscription: Dfl2064
ISSN: 0049 6979

RESEARCH CENTRES

– 1478 – Bundesamt für Umweltschutz
(Federal Office of Environmental Protection)

Hallwylstrasse 4
CH-3003 Berne
Switzerland
Tel: 41 31 61 93 11
Geographical coverage: Switzerland
Research activities: Water protection; air pollution; noise control; soil pollution; control of chemicals and waste

- 1479 - Cekmece Nükleer Arastirma ve Egitim Merkezi

(Cekmece Nuclear Research and Training Centre)

PK 1 Havaalani
Istanbul
Turkey
Tel: 90 1 790734
Geographical coverage: Turkey
Research activities: Nuclear fuel technology; environmental pollution; nuclear engineering

- 1480 - Centrum voor de Studie Van Water Bodem en Lucht Becewa Vzw

271 Krijgslaan
B-9000 Ghent
Belgium
Tel: 32 91 22 77 59
Contact name: Mr Vercruysse
Geographical coverage: Belgium
Research activities: Soil, water and air pollution and waste management

- 1481 - Environmental Pollution Control Project - Athens

Patission 147
GR-11251 Athens
Greece
Tel: 30 1 86 50 111/86 50 476
Geographical coverage: Greece
Publications: Interim technical report (4 volumes)
Research activities: Pollution problems in greater Athens and Greece

- 1482 - Instytut Podstaw Inzynierii Srodowiska PAN

(Environmental Engineering Institute)

Ulica M Curie Sklodowskiej 34
PL-41 800 Zabrze
Poland
Tel: 48 32 716481
Geographical coverage: Poland
Publications: *Archives of Environmental Protection*, *Prace i Studia*
Research activities: Atmospheric pollution protection; water pollution protection; reclamation of soil; Biological reclamation of dumps

- 1483 - International Institute for Applied Systems Analysis (IIASA)

A-2361 Laxenberg
Austria
Tel: 43 2236 715210
Fax: 43 2236 73149
Contact name: Elisabeth Krippl, Office of Communications
Geographical coverage: International
Aims and objectives: IIASA is part of an international network of scientific institutions working together to study global change. IIASA is supported by scientific organisations in 15 countries. Founded in 1972 on the initiative of the USA and ex-USSR
Publications: Catalogue available on request
Research activities: Conducts environmental studies and projects in acid rain, air pollution and other aspects of the environment

- 1484 - Paint Research Association

Waldegrave Road
Teddington
Middlesex TW11 8LD
United Kingdom
Tel: 081 977 4427
Fax: 081 943 4705
Geographical coverage: United Kingdom
Publications: Various publications
Research activities: Industrial microbiology; air sampling and control of environmental pollution

- 1485 - Studiecentrum voor Ecologie en Bosbouw

(Ecology and Forestry Research Council)

Craenevenne 140
B-3600 Genk (Bokrijk)
Belgium
Tel: 32 9 236 27 91
Fax: 32 9 235 58 05
Geographical coverage: Belgium
Research activities: Air, noise, surface water, groundwater, wastewater, solid waste, soil, biological environment and other topics

- 1486 - Studies en Consultancy voor Milieu en Omgeving - TNO

(TNO Study and Information Centre for Environment)

PO Box 186
NL-2600 AD Delft
Netherlands
Tel: 31 15 696900
Fax: 31 15 613186
Geographical coverage: Netherlands
Publications: *Environmental Research in the Netherlands*
Research activities: Air and water pollution prevention; noise; soil pollution; wastes; ecology

DATABASES

- 1487 - Environmental Bibliography

Geographical coverage: International
Coverage: Pollution; waste management; health hazards

Database producer: Environmental Studies Institute, Int.Academy, Santa Barbara
Database host: Dialog; Mead Data Central
Coverage period: 1973–
Update frequency: 4,000 records every two months
Additional information: Available on CD–ROM

– 1488 – Pollution Abstracts

Geographical coverage: International

Coverage: All aspects of pollution including source and control; air, land and water pollution, toxicology, health, noise, radiation, pesticides, waste, environmental quality
Database producer: Cambridge Scientific Abstracts
Database host: BRS; DataStar; Dialog; ESA–IRS; NewsNet Inc
Coverage period: 1970–
Update frequency: 157,000 records monthly

Air Pollution

GOVERNMENT ORGANISATIONS

– 1489 – Bestuur van de Volksgezondheid – Ministerie van Volksgezondheid

Rijksadministratief Centrum/Vesalius Gebouw
B-1010 Brussels
Belgium
Tel: 32 2 564 11 55
Geographical coverage: Belgium
Aims and objectives: Aims to collect the maximum number of documents concerning air pollution noise and radioactivity

– 1490 – Bundesantstalt für Arbeitsschutz und Unfallforschung

Vogelpothsweg 50–52
D-4600 Dortmund
Dorstfeld
Germany
Tel: 49 231 17631
Geographical coverage: Germany
Aims and objectives: Air quality control

– 1491 – Département de la Pollution Atmosphérique de l'Institut National de Recherche Chimique Appliquée

(Air Pollution Dept of the National Institute of Applied Chemical Research)

BP 1
F-91710 Vert-le-Petit
Essonnée
France
Tel: 33 49 82 475
Geographical coverage: France
Aims and objectives: The Institute's concerns are air pollution and industrial hygiene

– 1492 – Division for Air Monitoring and Industrial Compliance

Telemark County Statens Hus
N-3708 Skien
Norway
Tel: 47 3 58 61 20
Fax: 47 3 53 00 20
Geographical coverage: Norway
Aims and objectives: The Division is responsible for monitoring atmospheric pollution

NON-GOVERNMENT ORGANISATIONS

– 1493 – Association pour la Prevention de la Pollution Atmosphérique (APPA)

(Association for the Prevention of Atmospheric Pollution)

58 rue de Rocher
F-5008 Paris
France
Tel: 33 1 429 369 30
Fax: 33 1 429 341 99
Contact name: M Pierre Gaussens
Geographical coverage: France
Aims and objectives: APPA aims to promote clean air through the reduction of pollution

– 1494 – Atmospheric Research and Information Centre (ARIC)

The Manchester Metropolitan University
Chester Street
Manchester M1 5GD
United Kingdom
Tel: 061 247 1590/93
Fax: 061 247 6318
Contact name: Mrs Sue Hare, Information Officer
Geographical coverage: United Kingdom

Aims and objectives: National and international information Centre on air pollution, particularly acid rain and global climate change. Also research into the chemistry of acid deposition

– 1495 – Ceská Asociace IUAPPA (CA IUAPPA)
(Czech Association of IUAPPA)

Doubravcická 10
Prague 10 100 00
Czech Republic
Tel: 42 2 78 11 674
Fax: 42 2 78 11 674
Geographical coverage: Czech Republic

– 1496 – Comitato di Studio perl'Inquinamento Atmosferico (CSIA)
(Air Pollution Study Committee)

Via Assarotti 15/8 SC B
I-16122 Genoa
Italy
Tel: 39 10 893922
Fax: 39 10 887766
Geographical coverage: Italy
Aims and objectives: CSIA deals with scientific, technical, legislative and economic aspects of air pollution control

– 1497 – Cooperative Programme for Monitoring and Evaluation of Long Range Transboundary Air Pollution (EMEP)

EMEP West
Norwegian Meteorological Institute
PO Box 43 Blindern
N-0313 Oslo
Norway
Geographical coverage: European
Aims and objectives: EMEP presents information on emissions, transports, concentrations and depositions of sulphur and nitrogen pollutants

– 1498 – Coordinadora de Organizaciones de Defensa Ambiental (CODA)

Plaza de Santo Domingo 7
E-2013 Madrid
Spain
Tel: 34 1 559 60 25
Fax: 34 1 559 78 97
Geographical coverage: Spain

– 1499 – European Association for the Science of Air Pollution (EURASAP)

Air Pollution Group
Imperial College
London SW7 2AZ
United Kingdom
Tel: 071 589 5111
Fax: 071 584 7596
Geographical coverage: European
Aims and objectives: EURASAP was formed in 1986 to advance education and knowledge in air pollution science and its applications. It brings together in conference scientists throughout Europe working in the relevant disciplines
Publications: Regular newsletter

– 1500 – Fundación Para la Ecología y la Protección del Medio Ambiente (FEPMA)

Castellana 8
E-28046 Madrid
Spain
Tel: 34 1 575 41 68
Geographical coverage: Spain

– 1501 – Hravtsko Drutvo za Zastitu Zraka
(Croatian Air Pollution Prevention Association) (CAPPA)

University of Zagreb
CAPPA
Ksaverska c 2
41000 Zagreb
Croatia
Tel: 38 41 434 188
Fax: 38 41 274 572
Geographical coverage: Croatia

– 1502 – Ilmansuojeluyhdistys ry ISY
(Finnish Air Pollution Prevention Society) (FAPPS)

PO Box 335
SF-00131 Helsinki
Finland
Tel: 358 1 45 66 159
Fax: 358 1 45 52 408
Contact name: Mr Kari Larjava, President
Geographical coverage: Finland

– 1503 – Information Centre for Air Protection, Polish Ecological Club

ul Armii Czerwonej
Al Korfantego 2/220
PL-40 960 Katowice
Poland
Tel: 48 32 586071
Geographical coverage: Poland

– 1504 – International Union of Air Pollution Prevention Associations (IUAPPA)

136 North Street
Brighton BN1 1RG
United Kingdom
Tel: 0273 326313
Fax: 0273 735802
Contact name: John Langston, Director General
Geographical coverage: United Kingdom
Aims and objectives: The Union was formed in 1964 and now has 28 members. It aims to promote public education worldwide in all matters relating to the value and importance of clean air and methods and consequences of air pollution control
Publications: *Clean Air Around the World*; newsletter, quarterly; members handbook

– 1505 – Kommission Reinhaltung der Luft (KRdL) in VDI und DIN

(Commission on Air Pollution Prevention in VDI and DIN)

Robert-Stolz-Strasse 5
Postfach 10 11 39
D-4000 Düsseldorf 1
Germany
Tel: 49 211 6214 532
Fax: 49 211 6214 575
Contact name: Dr Ing Klaus Grefen, Secretary
Geographical coverage: Germany

– 1506 – National Society for Clean Air and Environmental Protection (NSCA)

136 North Street
Brighton BN1 1RG
United Kingdom
Tel: 0273 326313
Fax: 0273 735802
Contact name: Tim Brown/Mary Stevens, Information Department
Geographical coverage: United Kingdom
Aims and objectives: Founded in 1899 as the Smoke Abatement Society, NCSA's objectives are to secure environmental improvement by promoting clean air through the reduction of air pollution, noise and other contaminants
Publications: *Clean Air*, quarterly; *Pollution Handbook*, annually; annual report and other leaflets and reports

– 1507 – Polish Ecological Club/Information Centre for Air Protection

Grunwaldski 8-10
PL-40 950 Katowice
Poland
Tel: 48 32 594 315
Contact name: Piotr Poborski
Geographical coverage: Poland
Aims and objectives: Concerned with all aspects of air pollution

– 1508 – Ren Luft Foreningen (RLF)

(Norwegian Clean Air Association)

PO Box 2312 Solli
N-0201 Oslo
Norway
Tel: 47 2 838330
Contact name: Vigdis Ekeberg, Secretary
Geographical coverage: Norway

– 1509 – Vereniging der Lucht

(Society for Clean Air in the Netherlands – CLAN)

PO Box 6012
NL-2600 JA Delft
Netherlands
Tel: 31 15 69 68 84
Fax: 31 15 61 31 86
Contact name: Dr J van Ham, Secretary
Geographical coverage: Netherlands
Aims and objectives: CLAN's concerns are air, climate and acid rain

BOOKS

– 1510 – Air Pollution Control in the European Community

Geographical coverage: European
Publishers: Graham and Trotman Ltd, 1991
Editor: Bennett, G.
Language: English
Coverage: Examines the implication of EC Directives on air pollution and control in the member states
ISBN: 1853335673

– 1511 – Air Pollution's Toll on Forests and Crops

Geographical coverage: International
Publishers: Yale University Press, 1990
Coverage: Examines the effects of air pollution on forestry and crops
ISBN: 0300045697
Author: MacKenzie, J. J. and El-Ashry, M. T.

– 1512 – Ambient Air Pollutants from Industrial Sources

Geographical coverage: Netherlands
Publishers: Elsevier Science Publishers, 1985, Netherlands
Editor: Suess, M. J., Grefen, K. and Reinisch, D
Language: English

Price: Dfl150
ISBN: 0444806059

– 1513 – Clean Air Around the World

Geographical coverage: International
Publishers: IUAPPA, 1991
Editor: Loveday Murley
Language: English
Coverage: A reference book for which specialist organisations from many countries have compiled up to date information concerning their national air pollution problems
Price: £32.00
ISBN: 1871688019

– 1514 – Continuous Emission Monitoring

Geographical coverage: United Kingdom
Publishers: Van Nostrand Reinhold International, 1992
Coverage: A guide to the field of continuous emission monitoring (CEM) addressing technologies and practices in the monitoring of pollutants emitted from industrial stacks
Price: £43.50
ISBN: 0442007248
Author: Jahnke, J. A.

– 1515 – Occupational Exposure Limits for Airborne Toxic Substances, Third Edition

Geographical coverage: International
Publishers: ILO, 1991 (Occupational Health and Safety Series No 37)
Coverage: The latest available values either prescribed or recommended in 15 countries
ISBN: 9221072932

– 1516 – Skogsskjötsel i en Utslippstid

(Forest Management in a time of emissions)

Geographical coverage: Norway
Publishers: Det Norske Skogselskap, 1992
Coverage: Describes the effects of air pollution on forests and what can be done to mitigate the effects
Price: Nkr80
Author: Braadland, B. and Rognerud, P. A.

– 1517 – Skogsutsikter

Göteborgs Universitet
Medicinaregatan 20B
S-413 90 Gothenburg
Sweden
Geographical coverage: Sweden
Publishers: Institutionen för Miljövard, 1992
Language: Swedish
Coverage: Reviews state of European forests and the problems of air pollution
Author: Elvingson, P.

CONFERENCE PAPERS

– 1518 – Desulphurisation

Geographical coverage: United Kingdom
Publishers: Taylor and Francis, 1991
Coverage: Covers current technologies and strategies for reducing sulphur dioxide emissions from both large utility plant and small industrial systems
Price: £25.50
ISBN: 1560322322
Author: Kyte, W. S., Institution of Chemical Engineers

– 1519 – North Atlantic Treaty Organisation Expert Panel on Air Pollution Modelling Proceedings

Information and Press Service
B-1110 Brussels
Belgium
Geographical coverage: International
Publishers: North Atlantic Treaty Organisation (NATO)
ISSN: 0377 7669

PERIODICALS

– 1520 – Air Pollution and Noise Bulletin

Birmingham Central Libraries
Chamberlain Square
Birmingham B3 3HQ
United Kingdom
Tel: 021 235 4392
Geographical coverage: United Kingdom
Publishers: APNB Science Technology and Management Services
Editor: Pratt, D.
Frequency: 6 issues a year
Coverage: The bulletin is a current awareness service for specialists in the field. Its purpose is to alert people to recent publications in English
Price: free

– 1521 – Annals of Occupational Hygiene

Headington Hill Hall
Oxford OX3 0BW
United Kingdom
Tel: 0865 794141
Fax: 0865 743952
Geographical coverage: United Kingdom
Publishers: Pergamon Press plc

Editor: McKellinson, J.
Frequency: 6 issues a year
Coverage: Life sciences including air pollution monitoring
Annual subscription: £260.00
ISSN: 0003 4878

– 1522 – Clean Air

136 North Street
Brighton BN1 1RG
United Kingdom
Tel: 0273 326313
Fax: 0273 735802
Geographical coverage: United Kingdom
Publishers: National Society for Clean Air and Environmental Protection
Frequency: quarterly
Annual subscription: £18.00
ISSN: 0300 5143

– 1523 – Difesa Ambientale

Via A Capecelatro 5
I-20148 Milan
Italy
Geographical coverage: Italy
Publishers: Centro Informazioni Studi Ambientali
Editor: Serbelloni, L. C.
Frequency: 12 issues a year

– 1524 – Etudes de Pollution Atmosphérique a Paris et dans le Départements Periphériques

Prefecture de Police
Laboratoire Ctrl
39 bis rue de Dantzig
F-75015 Paris
France
Geographical coverage: France
Frequency: annual
ISSN: 0071 1942

– 1525 – Finishing

Turrett House
171 High Street
Rickmansworth
Hertfordshire WD3 1SN
United Kingdom
Tel: 0923 777000
Fax: 0923 771297
Geographical coverage: United Kingdom
Publishers: Turrett Group plc
Editor: Tomkins, G.
Frequency: monthly
Annual subscription: $118
ISSN: 0264 2506

– 1526 – Journal of Atmospheric Chemistry

PO Box 17
NL-3300 AA Dordrecht
Netherlands
Tel: 31 78 334911
Fax: 31 78 334254
Geographical coverage: Netherlands
Publishers: D Reidel Publishing Co
Editor: Crutzen, P. J. and Ehhalt, D. H.
Frequency: quarterly
Language: English

– 1527 – Lufthygienischer Monatsbericht

Unter den Eichen 7
D-6200 Wiesbaden
Germany
Tel: 49 611 5810
Fax: 49 611 581221
Geographical coverage: Germany
Publishers: Landesanstalt für Umwelt
Frequency: monthly
Annual subscription: DM30

– 1528 – MST Luft

Forsoeganlaeg Risoe
DK-4000 Roskilde
Denmark
Geographical coverage: Denmark
Publishers: Miljoestyrelsen, Luftforureningslaboratorium
Editor: Fenger, J.
Frequency: irregular
Language: Danish
Price: free

– 1529 – Pollution Atmosphérique

58 rue de Rocher
F-75008 Paris
France
Tel: 33 1 429 369 30
Fax: 33 1 429 341 99
Geographical coverage: France
Publishers: Association pour la Prévention de la Pollution Atmosphérique
Editor: Sommer, M.
Frequency: quarterly
Language: French with English summaries
Annual subscription: Ffr394
ISSN: 0032 3632

– 1530 – Refrigeration and Air Conditioning

Maclaren House
19 Scarbrook Road
Croydon
Surrey CR9 1QH
United Kingdom
Tel: 081 760 9690
Fax: 081 681 1672
Geographical coverage: United Kingdom
Publishers: EMAP Vision Ltd
Editor: Bailey, A.
Frequency: monthly
Annual subscription: £34.50
ISSN: 0263 5739

– 1531 – Smog

Belvedere Golfo Paradiso 21
I-16036 Recco
Genoa
Italy
Geographical coverage: Italy
Publishers: Lega Italiana Contro Fumi e Rumori
Frequency: 3 issues a year

– 1532 – Warren Spring Laboratory UK Smoke and Sulphur Dioxide Monitoring Networks

Gunnels Wood Road
Stevenage
Hertfordshire SG1 2BX
United Kingdom
Tel: 0438 741122
Fax: 0438 360858
Geographical coverage: United Kingdom
Publishers: Warren Spring Laboratory
Frequency: irregular
Price: varies

REPORTS

– 1533 – Acid Deposition and Vehicle Emissions: European Environmental Pressures on Britain

Geographical coverage: United Kingdom
Publishers: Royal Institute of International Affairs, Policy Studies Unit
Editor: Brackley, P.
ISBN: 0566051257

– 1534 – Air Pollution from Vehicles:
see entry 1614

Geographical coverage: United Kingdom

– 1535 – Aircraft Pollution: Environmental Impacts and Future Solutions

Geographical coverage: International
Publishers: WWF, August 1991
Language: English
Coverage: Issues of aircraft emissions pollution
ISBN: 2880850827
Author: Barrett, Dr. M.

– 1536 – Bikes Not Fumes

Cotterell House
69 Meadrow
Godalming
Surrey GU7 3HS
United Kingdom
Tel: 0483 417217
Fax: 0483 426994
Geographical coverage: United Kingdom
Publishers: Cycling Tourists Club (CTC)
Price: £8.00
Author: Earth Resources Centre

– 1537 – Environment in Czechoslovakia, The

Slezka 9
120 29 Prague 2
Czech Republic
Geographical coverage: Czech Republic
Publishers: Department of the Environment, 1990
Coverage: Report on the state of the country's environment
Author: Federal Committee for the Environment

– 1538 – Protecting the Earth: A Status Report with Recommendations for a New Energy Policy

Geographical coverage: Germany
Publishers: Deutscher Bundestag, 1991
Coverage: Provides an account of the magnitude of the threat of the greenhouse effect. Recommendations are offered for national actions to reduce energy related emissions of radioactive trace gases
ISBN: 3924521719
Author: Enquête Commission, Germany

– 1539 – Volatile Organic Compound Emissions in Western Europe: Report 6/87

Geographical coverage: European
Publishers: CONCAWE, Brussels
Coverage: Control options and their cost effectiveness for gasoline vehicles, distribution and refining

– 1540 – WATT Committee on Energy: Air Pollution Acid Rain and the Environment

Geographical coverage: United Kingdom
Publishers: Elsevier, 1988
ISBN: 1851662227
Author: Watt Committee

RESEARCH CENTRES

– 1541 – Centre Interprofessionnel Technique d'Etudes de la Pollution Atmosphérique (CITEPA)

(Interprofessional Technical Centre for Studies in Atmospheric Pollution)

3 rue Henri Heine
F-75016 Paris
France
Tel: 33 1 452 712 88
Fax: 33 1 405 007 45
Contact name: Mr Remy Bouscaren, Director
Geographical coverage: France
Aims and objectives: CITEPA do studies of air pollution from industrial sources and traffic
Research activities: Air pollution

– 1542 – Department of Air Pollution

Faculty of Agricultural Sciences
De Dreijen 12
NL-6703 BC Wageningen
Netherlands
Tel: 31 8370 82106
Geographical coverage: Netherlands
Research activities: Air pollution; mutagenicity of air pollution; aerosol research; transport and chemical transformation of air pollutants

– 1543 – Département Pollution des Eaux de l'Institut National de Recherche Chimique Appliquée: *see entry 1639*

Geographical coverage: France

– 1544 – Environmental Research Unit

St Martins House
Waterloo Road
Dublin 4
Eire
Tel: 353 1 602511
Fax: 353 1 680009
Geographical coverage: Ireland
Research activities: Air and water quality
Additional information: Library Service

– 1545 – Hava Kirlenmesi Arastirmalari ve Denetimi Türk Milli Komitesi

(Turkish National Committee for Air Pollution Research and Control) (TUNCAP)

Dokuz Eylül University
Department of Environmental Engineering
Bornova 35100 Izmir
Turkey
Tel: 90 51 882108
Fax: 90 51 887864
Contact name: Prof Dr Aysen Müezzinoglu, President
Geographical coverage: Turkey

– 1546 – Institut de Recherche des Transports Centre de Documentation d'Evaluation et de Recherches Nuisances

109 avenue Salvador Allendre
F-69500 Bron
France
Tel: 33 78269093
Geographical coverage: France
Research activities: Transport; air pollution; noise pollution

– 1547 – Institut für Immission- Arbeits- und Strahlenschutz

(Institute for Emission Labour and Radiation Protection)

Hertzstrasse 173
Baden-Württemberg
D-7500 Karlsruhe 21
Germany
Tel: 49 721 75031
Fax: 49 721 758758
Geographical coverage: Germany
Research activities: Concerned with the emission of substances, radiation and noises

– 1548 – Institut National de la Santé et de la Recherche Medicale

Université de Paris
11 rue J B Clement
F-92290 Chatenay-Malabry
France
Tel: 33 1 6604518
Contact name: Mr Boudene
Geographical coverage: France
Research activities: Toxicology and air pollution

– 1549 – Istituto Sull'Inquinamento Atmosferico

(Institute of Atmospheric Pollution)

Via Salaria Km 29 300
CP 10
I-00016 Monterotondo Stazione
Rome
Italy

Tel: 39 6 900 53 49
Geographical coverage: Italy
Research activities: Atmospheric acidity, photochemical smog and toxic emissions

– 1550 – Landesanstalt für Immissionsschutz Nordrhein-Westfalen
(North Rhine Westphalia State Centre of Air Quality, Noise and Vibration Control)

Wallneyer Strasse 6
D-4300 Essen-Bredeney 1
Germany
Tel: 49 201 799 50
Fax: 49 201 799 54 46
Geographical coverage: Germany
Publications: Various publications
Research activities: Air quality surveillance; abatement of harmful emissions; protection against noise and vibration

– 1551 – Medizinisches Institut für Umwelthygiene
(Medical Institute for Environmental Hygiene)

Auf'm Hennekamp 50
D-4000 Düsseldorf
Germany
Tel: 49 211 33890
Geographical coverage: Germany
Research activities: Injurious effects of air and noise pollution

– 1552 – Norsk Institut for Luftforskning (NILU)
(Norwegian Institute for Air Research)

PO Box 64
N-2001 Lillestrom
Norway
Tel: 47 6 814170
Fax: 47 6 819247
Geographical coverage: Norway
Publications: Reports
Research activities: Air quality measurements; air pollution modelling; atmospheric corrosion studies and other activities

DATABASES

– 1553 – Air/Water Pollution Report
Geographical coverage: United Kingdom
Coverage: All aspects of air and water pollution in the UK
Database producer: Business Publishers Inc
Database host: DataStar; Dialog
Coverage period: 1988–

– 1554 – IBSEDEX
Geographical coverage: International
Coverage: Indoor air pollution; heating; ventilation; energy
Database producer: Building Services Research and Information Association
Database host: ESA–IRS
Coverage period: 1960–
Update frequency: weekly

– 1555 – VANYTT
Halsingegatan 49
S-113 31 Stockholm
Sweden
Tel: 46 8 34 01 70
Geographical coverage: Sweden
Coverage: Covers aspects of indoor air pollution and energy conservation in buildings
Database producer: The Swedish Institute of Building Documentation (BYGGDOK)

Air Quality and Pollution

NON-GOVERNMENT ORGANISATIONS

– 1556 – Norwegian Clean Air Campaign
Postboks 94
N-1364 Hvalstad
Norway
Tel: 47 2 78 38 60
Fax: 47 2 90 15 87
Geographical coverage: Norway
Aims and objectives: Works towards promoting clean air through the reduction of air pollution and other contaminants

Land Pollution

BOOKS

– 1557 – Acid Deposition: Volume 1 Sources Effects and Controls

Geographical coverage: United Kingdom
Publishers: The British Library/Technical Communications, 1989
Editor: Longhurst, J. W. S.
Coverage: A collection of papers, with a European perspective on acid deposition. Monitoring freshwater acidification, soils and forest systems, structural materials and control technologies
ISBN: 0946655332

– 1558 – Council of Europe: Farming and Wildlife

Geographical coverage: European
Publishers: Council of Europe, 1989
Coverage: Includes agricultural pollution
ISBN: 9287116857

– 1559 – Recovery and Rehabilitation of Contaminated Land

Geographical coverage: United Kingdom
Publishers: Ellis Horwood, 1992
Coverage: This book looks at the geological, physical, chemical and biological problems posed in reclamation of the land
Price: £85.00
ISBN: 0137687710
Author: Bewley, R. J. F. and Sharp, D. H.

REPORTS

– 1560 – Contaminated Land: Counting the Cost of Our Past

NatWest Bank plc
Environmental Management Unit
1/2 Broadgate
London ECM2 2AD
United Kingdom
Geographical coverage: United Kingdom
Coverage: A report which looks at section 143 of the Environmental Protection Act which requires the creation and maintenance of public registers of land which has been subjected to a contaminative use
Price: on application

– 1561 – Mercury in Soil – Distribution Speciation and Biological Effects

Allmänna Förlaget
S-106 47 Stockholm
Sweden
Geographical coverage: International
Publishers: Nordic Council of Ministers, 1992
Coverage: Review of the international literature on mercury in soils and on the effects on microbial and biochemical processes as well as on species diversity of biota soil

RESEARCH CENTRES

– 1562 – Geological Survey of Sweden, Department of Forest Soils

Swedish University of Agricultural Sciences
Box 7001
S-750 07 Uppsala
Sweden
Tel: 46 18 672212
Geographical coverage: Sweden

– 1563 – Istituto Sperimentale per lo Studio e la Difesa del Suolo

(Research Institute for the Study and Conservation of the Soil)

Piazza Massimo d'Azeglio 30
I-50121 Florence
Italy
Tel: 39 55 2477242
Geographical coverage: Italy
Publications: Annual report
Research activities: Soil chemistry; soil physics; soil conservation techniques

– 1564 – Soil Survey and Land Research Centre (SSLRC)

Cranfield Institute of Technology
Silsoe Campus
Silsoe
Bedfordshire MK45 4DT
United Kingdom
Tel: 0525 60428
Fax: 0525 61147
Geographical coverage: United Kingdom
Aims and objectives: The Centre's work relates to sustainable soil and water management, pollution prevention and the monitoring of soil quality
Publications: Large number of soil maps and books

– 1565 – Statens Planteavis Laboratorium

(Danish Laboratory for Soil and Crop Research)

Lottenborgvej 24
DK-2800 Lyngby
Denmark
Tel: 45 42 93 09 99
Geographical coverage: Denmark
Publications: Reports in *Tidsskrift for Planteavl*
Research activities: Soil chemistry; plant nutrition; pollution problems and soil and plant analysis

DATABASES

– 1566 – GeoArchive

Geographical coverage: International
Coverage: Geology and related disciplines including land pollution and reclamation
Database producer: Geosystems
Database host: Dialog
Coverage period: 1974–
Update frequency: monthly

Marine Pollution

GOVERNMENT ORGANISATIONS

– 1567 – State Pollution Control Authority (SFT): Oil Pollution Control Department

PO Box 125
N-3191 Horten
Norway
Tel: 47 33 44 161
Fax: 47 33 44 257
Geographical coverage: Norway
Aims and objectives: SFT is responsible for coordinating public and private emergency services to provide a national emergency response system

NON-GOVERNMENT ORGANISATIONS

– 1568 – Arbeitsgemeinschaft Information Meeresforschung Meerestechnik

Stilleweg 2
D-3000 Hannover 51
Germany
Tel: 49 511 6468655
Geographical coverage: Germany
Aims and objectives: Marine pollution and sea water quality

– 1569 – Baltic Marine Environment Protection Committee Helsinki Commission (HELCOM)

Mannerheimintie 12 A
SF-00100
Helsinki 10
Finland
Tel: 358 1 90 602
Fax: 358 1 90 644 577
Geographical coverage: European
Aims and objectives: The Commission was established in 1974 to protect the marine environment of the Baltic Sea from all types of pollution. Its members are the governments of Denmark, Finland, Germany, Poland, Sweden and the former USSR

– 1570 – Bonn Commission

New Court
48 Carey Street
London WC2A 2JE
United Kingdom
Geographical coverage: European
Aims and objectives: The Bonn Commission aims to protect the North Sea from pollution

– 1571 – Centre National pour l'Exploitation des Océans

Boîte Postale 337
F-29273 Brest
Finistère
France
Tel: 33 98804650
Geographical coverage: France
Aims and objectives: Concerns marine pollution

– 1572 – Institut Atlantique

120 rue de Longchamp
F-75116 Paris
France
Tel: 33 1 7272436
Geographical coverage: France
Aims and objectives: Pollution control and environmental information

– 1573 – International Oil Pollution Compensation Fund (IOPC)

4 Albert Embankment
London SE1 7SR
United Kingdom
Tel: 071 582 2606
Fax: 071 735 0326
Geographical coverage: International
Aims and objectives: IOPC is a compensation fund organised by the International Maritime Organisation (IMO) for victims of oil pollution damage
Publications: *Claims Manual*; annual report; statistics

– 1574 – International Tanker Owners Pollution Federation Ltd (ITOPF)

Staple Hall
Stonehouse Court
87–90 Houndsditch
London EC3A 7AX
United Kingdom
Tel: 071 621 1255
Fax: 071 621 1783
Contact name: Dr I C White, Managing Director
Geographical coverage: International
Aims and objectives: ITOPF was established in 1968 as a service organisation. Devotes considerable efforts to the provision of technical services an the fields of response to marine oil spills, damage assessment, contingency planning, training and information
Publications: *Ocean Orbit Newsletter*, annual; technical information papers, training videos and others

– 1575 – Oslo Commission (OSCOM)

New Court
48 Carey Street
London WC2 2JE
United Kingdom
Tel: 071 242 9927
Fax: 071 831 7427
Geographical coverage: European
Aims and objectives: The Commission was set up under the Convention for the Prevention of Marine Pollution by Dumping from Ships and Aircraft in Oslo 1974, and now works jointly with the Paris Commission set up under the Convention for the Prevention of Marine Pollution

– 1576 – Secretariat for the Protection of the Mediterranean Sea

Placa Lesseps 1
E-08023 Barcelona
Spain
Tel: 34 3 217 16 95
Geographical coverage: European
Aims and objectives: The Secretariat, established in 1983, seeks to protect the sea from pollution by information exchange and action amongst its members

BOOKS

– 1577 – Manual of Methods in Aquatic Environment Research

Geographical coverage: International
Publishers: Food and Agriculture Organisation (FAO), 1992
Language: English
Coverage: Biological assessment of marine pollution with particular reference to benthos
Price: $8.00
ISBN: 9251031363

– 1578 – Marine Pollution

Geographical coverage: United Kingdom
Publishers: Hemisphere Publications, 1989
Editor: Albaiges, J.
ISBN: 0891168621

– 1579 – Recreational Water Quality Management: Vol 1 Coastal Waters

Geographical coverage: European
Publishers: Ellis Horwood, 1992
Coverage: This study describes management strategies for the coastal water environment and its waste disposal and pollution problems
Price: £40.00
ISBN: 0137700253
Author: Kay, D.

– 1580 – UK Marine Oil Pollution Legislation

Geographical coverage: United Kingdom
Publishers: Lloyds of London Press, 1987
Editor: Bates, J. H.
ISBN: 185004410ttps9X

CONFERENCE PAPERS

– 1581 – Pollution of the Mediterranean Sea

Geographical coverage: European
Publishers: Pergamon Books, 1987
Editor: Miloradov, M.
ISBN: 0080355781

PERIODICALS

– 1582 – Marine Pollution Bulletin

Headington Hill Hall
Oxford OX3 0BW
United Kingdom
Tel: 0865 794141
Fax: 0865 743952
Geographical coverage: United Kingdom
Publishers: Pergamon Press plc
Editor: Clark, R. B.
Frequency: 24 issues a year
Annual subscription: £235.00
ISSN: 0025 326X

– 1583 – Marine Pollution Research Titles

Plymouth Marine Laboratory
Citadel Hill
Plymouth PL1 2PB
United Kingdom
Tel: 0752 222772
Fax: 0752 226865
Geographical coverage: United Kingdom
Editor: Moulder, D. S.
Frequency: monthly
Annual subscription: £66.00
ISSN: 0264 8059

REPORTS

– 1584 – Baltic Environment Proceedings no 39: Baltic Marine Environment Protection Commission

Mannerheimintie 12 A
SF-00100
Helsinki 10
Finland
Tel: 358 1 90 602
Fax: 358 1 90 644
Geographical coverage: Finland

– 1585 – International Maritime Dangerous Goods (IMDG)

IMO
4 Albert Embankment
London SE1 7SR
United Kingdom
Geographical coverage: International
Aims and objectives: IMDG Code includes regulations for the prevention of pollution by harmful substances carried by sea in packaged form

RESEARCH CENTRES

– 1586 – Bundesforschungs Anstalt für Fischerei

(Federal Research Centre for Fisheries)

Palmaille 9
D-2000 Hamburg 50
Germany
Tel: 49 40 389 05113
Geographical coverage: Germany
Research activities: All aspects of fisheries including marine pollution and environmental protection analysis

– 1587 – Deutsches Hydrographisches Institut

(German Hydrographic Institute)

Transport
Bernhard-Nocht-Strasse 78
D-2000 Hamburg 36
Germany
Tel: 49 40 31901
Fax: 49 40 31905150
Geographical coverage: Germany
Research activities: Monitoring seawater for radioactivity and other noxious substances

– 1588 – Group of Experts on the Scientific Aspects of Marine Pollution (GESAMP)

4 Albert Embankment
London SE1 7SR
United Kingdom
Tel: 071 735 7611
Fax: 071 587 3210
Geographical coverage: United Kingdom
Aims and objectives: GESAMP was formed in 1969 to serve as a mechanism for encouraging coordination, collaboration and harmonisation of activities related to marine pollution of common interest to the sponsoring bodies
Publications: Reports and studies
Research activities: Oceanographic research aspects of marine pollution including monitoring

– 1589 – **Institutl Roman de Cercetari Marine**
(Marine Research Institute of Romania)

vul Lenin 300
8700 Constanta
Romania
Tel: 40 16 43288
Geographical coverage: Romania
Publications: Annual report
Research activities: Studies in the Black Sea and Atlantic Ocean on water pollution; marine technology and other aspects

– 1590 – **International Laboratory of Marine Radioactivity**

2 avenue Prince Hereditaire Albert
MC-9800 Monaco
Monaco
Tel: 33 93 50 44 88
Fax: 33 93 25 73 46
Geographical coverage: International
Research activities: Occurrence and behaviour of radioactive substances and other forms of pollution in the marine environment

DATABASES

– 1591 – **Baltic**

Geographical coverage: Sweden
Coverage: Aspects of pollution in the Baltic
Database producer: Swedish National Environment Protection Agency
Database host: DIMDI
Coverage period: 1980–

– 1592 – **Oceanic Abstracts**

Geographical coverage: International
Coverage: Marine biology; pollution of oceans and estuaries
Database producer: Cambridge Scientific Abstracts
Database host: BRS; Dialog; ESA–IRS
Coverage period: 1964–
Update frequency: 1,500 records monthly

– 1593 – **Oil Spill Intelligence Report**

Geographical coverage: International
Coverage: Prevention, control and clean-up of oil spills
Database producer: Cutter Information Corporation
Database host: NewsNet Inc
Coverage period: 1990–

– 1594 – **Tulsa**

Geographical coverage: International
Coverage: All aspects of oil including spillages and pollution
Database producer: Petroleum Abstracts
Database host: Dialog; Orbit
Coverage period: 1965–

Noise Pollution and Abatement

GOVERNMENT ORGANISATIONS

– 1595 – **Bestuur van de Volksgezondheid – Ministerie van Volksgezondheid:** see entry 1489

Geographical coverage: Belgium

NON-GOVERNMENT ORGANISATIONS

– 1596 – **Association Internationale Contre le Bruit (AICB)**
(International Association Against Noise)

Hirschenplatz 7
CH-6004 Lucerne
Switzerland
Tel: 41 41 51 30 13
Fax: 41 41 52 80 15
Geographical coverage: International
Aims and objectives: AICB aims to promote noise control at the international level

– 1597 – **Deutsche Verkehrswissenschaftliche Gesellschaft eV:** see entry 1613

Geographical coverage: Germany

– 1598 – **National Society for Clean Air and Environmental Protection (NSCA):** see entry 1506

Geographical coverage: United Kingdom

– 1599 – Nederlandse Stichting Geluidshinder

(Noise Abatement Society)

PO Box 381
NL-1600 AJ Delft
Netherlands
Tel: 31 15 56723
Geographical coverage: Netherlands
Aims and objectives: The Society is concerned with noise pollution and abatement

– 1600 – Österreichischer Arbeitsring für Lärmbekämpfung (OAL)

(Austrian Noise Abatement Society)

Wexstrasse 19-23
A-1200 Vienna
Austria
Tel: 43 1 222 33 92 36
Fax: 43 1 222 35 35 11 204
Geographical coverage: Austria
Aims and objectives: OAL works towards noise control and abatement
Publications: *OAL–Richtlinien*

PERIODICALS

– 1601 – Applied Acoustics

Rippleside Commercial Estate
Barking IG11 0SA
United Kingdom
Geographical coverage: United Kingdom
Publishers: Elsevier Science Publishers Ltd
Editor: Lord, P.
Frequency: bi monthly
Coverage: Design of buildings, measurements and control of industrial noise and vibration, transportation noise and hearing
ISSN: 0003 682X

– 1602 – Geluid en Omgeving

Louiszalaan 485
B-1050 Brussels
Belgium
Tel: 32 2 723 11 11
Fax: 32 2 649 84 80
Geographical coverage: Belgium
Publishers: C E D Samson (Subsidiary of Wolters Samson Belge nv)
Frequency: quarterly
Language: Flemish
Coverage: Follows latest developments in noise control
Annual subscription: Bfr1365

– 1603 – Noise and Vibration Bulletin

107 High Street
Brentwood
Essex CM14 4RX
United Kingdom
Tel: 0277 224632
Fax: 0277 224632
Geographical coverage: United Kingdom
Publishers: Multi Science Publishing Co Ltd
Editor: Hughes, B.
Frequency: monthly
Coverage: Focuses on noise pollution and noise control
Annual subscription: £92.00
ISSN: 0029 0947

– 1604 – Noise and Vibration Worldwide

Crown House
Linton Road
Barking
Essex IG11 8JU
United Kingdom
Tel: 081 594 7272
Fax: 081 594 5942
Geographical coverage: International
Publishers: Elsevier Science Publishers Ltd
Editor: Barrett, S.
Frequency: 11 issues a year
Language: English
Coverage: Devoted to the engineering discipline of noise control and vibration reduction
Annual subscription: £85.00
ISSN: 0957 4565

REPORTS

– 1605 – Railway Noise Standards: Let's Get Them Right

Lynton House
7–12 Tavistock Square
London WC1H 9LT
United Kingdom
Tel: 071 388 2684
Geographical coverage: United Kingdom
Publishers: Technica, 1990
Coverage: Technical report on noise impact from railways

RESEARCH CENTRES

– 1606 – Institut de Recherche des Transports Centre de Documentation d'Evaluation et de Recherches Nuisances: *see entry 1546*

Geographical coverage: France

– 1607 – Institut für Immission- Arbeits- und Strahlenschutz: see entry 1547

Geographical coverage: Germany

– 1608 – Institute of Sound and Vibration Research

The University
Southampton SO9 5NH
United Kingdom
Tel: 0703 592310
Fax: 0703 593033
Geographical coverage: United Kingdom
Research activities: Noise control

– 1609 – Landesanstalt für Immissionsschutz Nordrhein-Westfalen: see entry 1550

Geographical coverage: Germany

– 1610 – Lydteknisk Institut (LI)

(Danish Acoustical Institute)

Building 356
Akademivej
DK-2800 Lyngby
Denmark
Tel: 45 45 93 12 11
Fax: 45 45 93 19 90
Geographical coverage: Denmark
Publications: Research reports
Research activities: Acoustics noise and vibration

– 1611 – Medizinisches Institut für Umwelthygiene: see entry 1551

Geographical coverage: Germany

– 1612 – Physikalisch Technische Versuchsanstalt für Wärme und Schalltechnik

(Testing Institute for Heat and Sound Technology)

Wexstrasse 19–23
A-1200 Vienna
Austria
Tel: 43 222 33 92 36
Geographical coverage: Austria
Research activities: Noise control and sound insulation in buildings

Traffic Pollution

NON-GOVERNMENT ORGANISATIONS

– 1613 – Deutsche Verkehrswissenschaftliche Gesellschaft eV

Apostelnstrasse 9
D-5000 Cologne 1
Germany
Tel: 49 221 241193
Geographical coverage: Germany
Aims and objectives: Concerned with traffic, traffic noise, water quality and noise reduction

REPORTS

– 1614 – Air Pollution from Vehicles

Geographical coverage: United Kingdom
Publishers: HMSO
ISBN: 0115510001
Author: Transport and Road Research Laboratory

– 1615 – Bikes not Fumes: see entry 1536

Geographical coverage: United Kingdom

– 1616 – Emissions for Heavy Duty Vehicles

Geographical coverage: United Kingdom
Publishers: HMSO, 1991
Coverage: Looks at the effect of diesel emissions, considering the scope for further reductions, beyond those recently agreed for new vehicles within the EC
Price: £12.00
Author: Royal Commission on Environmental Pollution

– 1617 – Procedures for Enhancing the Use of Environmentally Friendly Vehicles

Birger Jarlstorg 5
S-111 28 Stockholm
Sweden
Geographical coverage: Sweden
Publishers: Swedish Transport Research Board, 1992

Language: Full report in Swedish with English summary
Coverage: Describes the possibilities of decreasing the environmental effects of road traffic by introducing alternative fuels
Author: Eriksson, G.

RESEARCH CENTRES

- 1618 - Office National d'Etudes Recherches Aerospatiales

29 avenue de la Division Leclerc
F-92320 Chatillon
France
Tel: 33 1 7352111
Geographical coverage: France
Research activities: Air transport and air pollution

Water Pollution

GOVERNMENT ORGANISATIONS

- 1619 - Ministère de l'Environnement: Direction de la Prévention des Pollutions de l'Eau

14 boulevard du Général Leclerc
F-92512 Neuilly-sur-Seine
France
Tel: 33 1 475 812 12
Geographical coverage: France

- 1620 - National Rivers Authority (NRA)

30–34 Albert Embankment
London SE1 7TL
United Kingdom
Tel: 071 820 0101
Fax: 071 820 1603
Geographical coverage: United Kingdom
Aims and objectives: The NRA was established by the Water Act 1989. It represents something quite new in protecting the water environment; an independent watchdog with clearly defined duties and the resources to put them into effect
Publications: Various publications

NON-GOVERNMENT ORGANISATIONS

- 1621 - European Water Pollution Control Association (EWPCA)

Markt 71
D-5205 St Augustin 1
Germany
Tel: 49 2241 2320
Fax: 49 2241 232-35
Contact name: Dr Sigurd Van Reissen
Geographical coverage: European
Aims and objectives: The Association was set up in 1981. It aims to promote the science and practice of water pollution control in Europe. Members are composed of organisations in 20 countries

- 1622 - International Association on Water Pollution Research and Control (IAWPRC)

1 Queen Anne's Gate
London SW1H 9BT
United Kingdom
Tel: 071 222 38 48
Fax: 071 233 1197
Contact name: Anthony Milburn, Executive Director
Geographical coverage: International
Aims and objectives: Founded in 1965, IAWPRC works to encourage international communication, cooperative efforts and exchange of information on water pollution control research and control and water quality management
Publications: *Water Research*, monthly; *Water Science and Technology*, monthly; *Water Quality International*, quarterly

- 1623 - International Water and Sanitation Centre (IRC)

PO Box 93190
NL-2509 AD The Hague
Netherlands
Tel: 31 70 331 4133
Fax: 31 70 381 4034
Geographical coverage: International
Aims and objectives: IRC is concerned with knowledge generation and transfer and technical information exchange for water supply and sanitation improvement in developing countries

196 Pollution *Water Pollution*

– 1624 – Norwegian Institute of Technology

Division of Hydraulic and Sanitary Engineering
N-7034 Trondheim
Norway
Tel: 47 7 59 47 59
Contact name: Prof Hallvard Odegaard
Geographical coverage: Norway

BOOKS

– 1625 – Eutrophication of Freshwaters

Geographical coverage: International
Publishers: Chapman and Hall, 1991
Coverage: Explains causes and effects of eutrophication worldwide
Price: £35.00
ISBN: 0412329700
Author: Harper, D.

– 1626 – Examination of Water for Pollution Control: A Reference Handbook

Geographical coverage: European
Publishers: Pergamon Press, 1982, UK
Editor: Seuss, M. J.
Price: $130
ISBN: 0080252559

– 1627 – Waste, Wastewater, Air Laws and Technology

Geographical coverage: Germany
Publishers: Bohman Druck und Verlag Gesellschaft mbH & Co KG
Language: German
Author: List, W. and Kuntscher, H.

– 1628 – Water Pollution by Fertilisers and Pesticides

Geographical coverage: International
Publishers: OECD, 1986
ISBN: 9264128565

DIRECTORIES/YEARBOOKS

– 1629 – International Association on Water Pollution Research and Control Yearbook

Geographical coverage: International
Publishers: Kogan Page, 1989
ISBN: 1850919623

PERIODICALS

– 1630 – A E S Ambiente e Sicurezza: Rivista Dell'Antiquinamento

Piazza della Repubblica 26
I-20124 Milan
Italy
Geographical coverage: Italy
Publishers: Eris SpA
Editor: Meinardi, S.
Frequency: 12 issues a year
Language: Italian with summaries in English
Coverage: Environmental studies
Annual subscription: Lit43000
ISSN: 0391 7339

– 1631 – Abwassertechnik (AWT)

Postfach 1460
D-6200 Wiesbaden
Germany
Tel: 49 611 791 0
Fax: 49 611 791 285
Geographical coverage: Germany
Publishers: Bauverlag GmbH
Frequency: 6 issues a year
Coverage: Devoted to treatment of sewerage, recycling and water pollution
Annual subscription: DM114
ISSN: 0932 3708

– 1632 – Aquatic Toxicology

PO Box 211
NL-1000 AE Amsterdam
Netherlands
Tel: 31 20 5803911
Fax: 31 20 5803598
Geographical coverage: Netherlands
Publishers: Elsevier Science Publishers BV
Editor: Malins, D. C. and Jensen, A.
Frequency: monthly
Language: English
Coverage: Environmental toxicology
Annual subscription: Dfl1203
ISSN: 0166 445X

– 1633 – European Water Pollution Control

PO Box 211
NL-1000 AE Amsterdam
Netherlands
Tel: 31 20 5803911
Fax: 31 20 5803598
Geographical coverage: European
Publishers: Elsevier Science Publishers BV
Frequency: 6 issues a year

Language: English
ISSN: 0925 5060

– 1634 – Wasser Abwasser GWF

Postfach 80 13 60
Rosenheimerstrasse 145
D-8000 Munich 80
Germany
Tel: 49 89 411 20
Fax: 49 89 411 22 07
Geographical coverage: Germany
Publishers: R Oldenbourg Verlag GmbH
Frequency: monthly
Coverage: Covers information and chemical research on water pollution
Price: DM32.00
Annual subscription: DM308

– 1635 – Water Quality International

Headington Hill Hall
Oxford OX3 OBW
United Kingdom
Tel: 0865 794141
Fax: 0865 743911
Geographical coverage: International
Publishers: Pergamon Press plc
Editor: Horobin, W. A.
Frequency: 5 issues a year
Language: English
Coverage: Worldwide developments in the scientific and technical aspects of water pollution control
Annual subscription: £75.00
ISSN: 0892 211X

REPORTS

– 1636 – Trace Element Occurrence in British Groundwaters

Geographical coverage: United Kingdom
Publishers: British Geological Survey, 1989
Editor: Edmunds, W. M. Cook, J. M. Miles, D. G.
Coverage: The BGS reveals zones of acidified groundwater with abnormal concentrations, mostly of aluminium, but also of metals such as zinc, copper and nickel
ISBN: 0852721897

– 1637 – Water Pollution Incidents in England and Wales 1990

Geographical coverage: United Kingdom
Publishers: National Rivers Authority, 1992
Frequency: annual
Coverage: Analysis of water pollution incident statistics in England and Wales
Price: £3.50
ISBN: 1873160143

RESEARCH CENTRES

– 1638 – Centre de Recherche et de Contrôle Lainier et Chimique Celac

69H avenue du Parc
B-4655 Chaineux
Liege
Belgium
Tel: 32 87 33 01 47
Contact name: Mr L Rousseau
Geographical coverage: Belgium
Research activities: Water pollution, sludges dust and filtration

– 1639 – Département Pollution des Eaux de l'Institut National de Recherche Chimique Appliquée

(Water Pollution Dept of the National Institute of Applied Chemical Research)

BP 1
F-91710 Vert-le-Petit
Essonnée
France
Tel: 33 49 82 475
Geographical coverage: France
Research activities: Air pollution; liquid wastes; water pollution

– 1640 – Directorate of Fisheries Research

Fisheries Laboratory
Pakefield Road
Lowestoft
Suffolk NR33 62244
United Kingdom
Tel: 0502 62244
Geographical coverage: United Kingdom
Publications: Fisheries research technical reports; laboratory leaflets; aquatic environment monitoring reports
Research activities: Distribution and effects of radionuclides, metals, pesticides and other pollutants on the aquatic environment

– 1641 – Eidgenössische Anstalt für Wasserversorgung Abwasserreiningung und Gewässerschutz

(Swiss Federal Institute for Water Resources and Water Pollution Control)

Überlandstrasse 133
CH-8600 Dübendorf
Switzerland
Tel: 41 1 823 55 11
Geographical coverage: Switzerland
Publications: Various publications

Research activities: Quantitative evaluation of chemical biological and physical processes of natural waters

– 1642 – Greek National Centre for Marine Research

Aghios Kosmas
GR-166 04 Hellinikon
Athens
Greece
Tel: 30 1 9820 214
Geographical coverage: Greece
Publications: *Thalassographica*, annual
Research activities: Carries out research in the fields of oceanography, fisheries, inland waters, aquaculture and water pollution

– 1643 – Institut für Wasser- Boden- und Lufthygiene
(Institute for Water Soil and Air Hygiene)

Corrensplatz 1
Bundesgesundheitsamt Postfach
D-1000 Berlin 33
Germany
Tel: 49 30 83082313
Geographical coverage: Germany
Research activities: Drinking water quality and treatment, wastewater and environmental hygiene, water pollution control

– 1644 – Istituto di Ricerca sulle Acque (IRSA)
(Water Research Institute)

Via Reno 1
I-00198 Rome
Italy
Tel: 39 6 8841451
Geographical coverage: Italy
Publications: *Quaderni dell'Istituto di Ricerca sulle Acque*, serial; *Notizario Metodi Analittici per le Acque*, quarterly
Research activities: Water supply and water pollution

– 1645 – Vesien ja ympäristöntutkimuslaitos
(Water and Environment Research Institute)

PO Box 436
SF-00101 Helsinki
Finland
Tel: 358 1 19 291
Geographical coverage: Finland
Publications: Scientific papers, instructions, recommendations and annual reports
Research activities: Water pollution control; water supply; wastewater treatment

DATABASES

– 1646 – Air/Water Pollution Report: *see entry 1553*

Geographical coverage: United Kingdom

– 1647 – Aquatic Sciences and Fisheries Abstracts

Geographical coverage: International
Coverage: Aquatic pollution and environmental quality; marine and freshwater science technology and management; ecology and ecosytems
Database producer: ASFIS/Cambridge Scientific Abstracts
Database host: BRS; Dialog; DIMDI; ESA–IRS;
Coverage period: 1975–
Update frequency: 1,000 records monthly

– 1648 – Water Resources Abstracts

Geographical coverage: International
Coverage: Water quality, pollution, waste treatment
Database producer: US Department of the Interior Geological Survey
Database host: Dialog
Coverage period: 1971–
Update frequency: every two months
Additional information: Available on CD–ROM

11
Businesses and Services

Eco Labelling

GOVERNMENT ORGANISATIONS

– 1649 – Bundesministerium für Umwelt Naturschutz und Reaktorsicherheit

Postfach 12 06 29
D-5300 Bonn 1
Germany
Tel: 49 228 305 2350
Fax: 49 228 305 3524
Contact name: Dr Edda Muller
Geographical coverage: Germany
Aims and objectives: Coordinates information requests regarding eco labelling

– 1650 – Department of Consumer Affairs: Ministry of Family and Consumer Affairs

PO Box 8036
Dep N-0030 Oslo 1
Norway
Tel: 47 2 34 90 90
Fax: 47 2 34 27 17
Contact name: Inger Skoglund
Geographical coverage: Norway
Aims and objectives: Coordinates information requests regarding eco labelling

– 1651 – Department of the Environment (DoE)

Room A118
Romney House
43 Marsham Street
London SW1P 3PY
United Kingdom
Tel: 071 276 8218
Fax: 071 276 8600
Contact name: Mr Peter Walton
Geographical coverage: United Kingdom
Aims and objectives: Coordinates information requests regarding eco labelling

– 1652 – Environment Policy Section

Department of the Environment
Customs House
Dublin 1
Eire
Tel: 353 1 679 33 77
Fax: 353 1 742 710
Contact name: Mr Brendan Linehan
Geographical coverage: Ireland
Aims and objectives: Coordinates information requests regarding eco labelling

– 1653 – Federal Office of Environment, Forests and Landscape

Hallwylstrasse 4
CH-3003 Berne
Switzerland
Tel: 41 31 61 64 93
Fax: 41 31 61 99 81
Contact name: Dr Eduard Back
Geographical coverage: Switzerland
Aims and objectives: Coordinates information requests regarding eco labelling

– 1654 – Ministère de l'Environnement

14 boulevard du Général Leclerc
F-92524 Neuilly-sur-Seine
France

Tel: 33 1 475 812 12
Fax: 33 1 474 504 74
Contact name: M. Jean-Paul Ventre
Geographical coverage: France
Aims and objectives: Coordinates information requests regarding labelling programmes

– 1655 – Ministère de la Santé Publique, Chef du Service des Nuisances

Cité Administrative
Quartier Vesala
B-1010 Brussels
Belgium
Tel: 32 2 10 48 74
Contact name: Inspecteur–Directeur
Geographical coverage: Belgium
Aims and objectives: Coordinates information requests regarding labelling programmmes

– 1656 – Ministry for Health Sports and Consumer Protection

Radetzkystrasse 2
A-1031 Vienna
Austria
Tel: 43 1 711 58
Fax: 43 1 711 58
Contact name: Dr Gerard Schuster
Geographical coverage: Austria
Aims and objectives: Coordinates information requests regarding labelling programmes

– 1657 – Ministry of Economic Affairs

Laan van Nieuw Oost Indie 123
K503
PB 20101
NL-2500 EC 's-Gravenhage
Netherlands
Tel: 31 70 37 97 994
Fax: 31 70 37 97 340
Contact name: Mark Hoevers
Geographical coverage: Netherlands
Aims and objectives: Coordinates information requests regarding eco labelling

– 1658 – Ministry of Environment and Natural Resources

Rua do Secuolo no 51
P-1200 Lisbon
Portugal
Tel: 351 1 364755/363730
Fax: 351 1 3460150
Contact name: Prof Dr Fernando Real
Geographical coverage: Portugal
Aims and objectives: Coordinates information requests regarding eco labelling

– 1659 – Ministry of Housing Physical Planning and Environment

PO Box 450
NL-2260 MB Leidschendam
Netherlands
Contact name: Franz van Buul
Geographical coverage: Netherlands
Aims and objectives: Coordinates information requests regarding eco labelling

– 1660 – Ministry of the Environment

Fredsgatan 8
S-103 33 Stockholm
Sweden
Tel: 46 8 763 10 00
Fax: 46 8 24 16 29
Contact name: Ms Anna Elzvik
Geographical coverage: Sweden
Aims and objectives: Coordinates information requests regarding eco labelling

---

NON-GOVERNMENT ORGANISATIONS

– 1661 – Comité Européen de Normalisation (CEN): *see entry 1707*

Geographical coverage: European

– 1662 – Environmental Detergent Manufacturers Association (EDMA)

Mouse Lane
Steyning BN44 3DG
United Kingdom
Tel: 0903 879077
Fax: 0903 879052
Contact name: Robin Bines, Chairperson
Geographical coverage: European
Aims and objectives: The Association debates environmental consequences of detergents and lobbies for high standards on EC legislation. Also provides information to consumers, manufacturers, the media and national regulatory bodies

– 1663 – Finnish Standards Association

PL 205
SF-00121 Helsinki
Finland
Tel: 358 0 64 56 01
Fax: 358 0 64 31 47
Contact name: Ms Eeva-Liisa Arponen
Geographical coverage: Finland
Aims and objectives: Coordinates information requests regarding labelling programmes

– 1664 – Stiftelsen Miljomerking i Norge
(Norwegian Foundation for Environmental Labelling)

Kr. Augusts gt.5
N-0030 Oslo 1
Norway
Tel: 47 2 36 07 10
Fax: 47 2 36 07 29
Contact name: Mr Geir-Olav Fjeldheim
Geographical coverage: Norway
Aims and objectives: Coordinates information requests regarding eco labelling

– 1665 – Swedish Standards Institution Environmental Labelling Programme

Box 3295
S-103 66 Stockholm
Sweden
Tel: 46 8 613 53 27
Fax: 46 8 214 835
Contact name: Mr Bo Assarson
Geographical coverage: Sweden
Aims and objectives: Coordinates information requests regarding eco labelling

REPORTS

– 1666 – Eco Labelling

Geographical coverage: United Kingdom
Publishers: HMSO
Coverage: Eight report on eco labelling in two volumes
Price: £8.90 and £22.0
Author: House of Commons Environment Committee

– 1667 – Eco-Labelling of Paper Products

Miljoministeriet Miljostyrelsen
Strandgade 29
DK-1401 Copenhagen
Denmark
Tel: 45 31 57 83 10
Fax: 45 31 57 24 49
Geographical coverage: Denmark
Publishers: Danish Environmental Protection Agency
Coverage: Guidelines for the awarding of eco-labels for paper products
Price: Dkr95.00

– 1668 – Environment Labelling in OECD Countries

Geographical coverage: International
Publishers: OECD, 1991
Language: English
Coverage: Development of environmental labelling in the OECD
ISBN: 9264135383

RESEARCH CENTRES

– 1669 – Business and Environment Research Unit

University of Bradford
Bradford BD7 1DP
United Kingdom
Tel: 0274 733466
Geographical coverage: United Kingdom

Environment and Industry

GOVERNMENT ORGANISATIONS

– 1670 – Bundesministerium für Öffentliche Wirtschaft und Verkehr
(Federal Ministry for Nationalised Industries and Transport)

Radetzkystrasse 2
A-1031 Vienna
Austria
Geographical coverage: Austria

– 1671 – Department of Trade and Industry Environment Unit

151 Buckingham Palace Road
London SW1W 9SS
United Kingdom
Tel: 0800 585 794
Fax: 0438 360 858
Geographical coverage: United Kingdom
Aims and objectives: DTI's environmental objectives are to encourage firms to respond to environmental challenges and market opportunities at home and abroad; to encourage research and development and marketing of environmental technologies and products
Publications: Various publications

– 1672 – Department of Trade and Industry Environmental Enquiry Point

Warren Spring Laboratory
Gunnels Wood Road
Stevenage SG1 2BX
United Kingdom
Tel: 0800 585794
Fax: 0438 360858
Contact name: Sue Plum, Manager
Geographical coverage: United Kingdom
Aims and objectives: The environmental helpline aims to help British business to find answers to environmental problems

NON-GOVERNMENT ORGANISATIONS

– 1673 – Association of Chemical Industries (VCI)

Karl Strasse 21
D-6000 Frankfurt 1
Germany
Tel: 49 69 255 64 71
Geographical coverage: Germany

– 1674 – British Non-Ferrous Metals Association

10 Greenfield Crescent
Birmingham B15 3AU
United Kingdom
Tel: 021 456 3322
Fax: 021 456 1394
Contact name: J Heaton, Legislative Affairs Executive
Geographical coverage: United Kingdom
Aims and objectives: The Association deals with all potential UK and EC legislation on environmental issues affecting the non-ferrous metal industry

– 1675 – Business and Industry Advisory Committee (BIAC)

13–15 chaussée de la Muette
F-75016 Paris
France
Tel: 33 1 452 448 38
Fax: 33 1 428 878 38
Contact name: Marc Patten, Assistant to the Secretary General
Geographical coverage: International
Aims and objectives: The role of BIAC is to provide OECD and its members with the viewpoint of international business and industry on all aspects of OECD policies. Recent policies cover environment and waste management

– 1676 – Chemical Industries Association

King's Building
Smiths Square
London SW1 3JJ
United Kingdom
Tel: 071 834 3399
Geographical coverage: United Kingdom

– 1677 – Cotton Council International (CCI)

239 Old Marylebone Road
London NW1 5QT
United Kingdom
Tel: 071 402 0029
Fax: 071 724 8979
Geographical coverage: International

– 1678 – European Federation of Chemical and General Workers' Unions (FESCID)

Avenue Emile de Béco 109
B-1050 Brussels
Belgium
Tel: 32 2 648 24 97
Fax: 32 2 646 06 85
Contact name: Franco Bisegna, Secretary General
Geographical coverage: European
Aims and objectives: FESCID represents the views of EC Member State chemical and general workers' unions to the EC institutions

– 1679 – European Federation of Pharmaceutical Industries (EFPIA)

Avenue Louise 250
Box 91
B-1050 Brussels
Belgium
Tel: 32 2 640 68 15
Fax: 32 2 647 60 49
Contact name: Nelly Baudrihaye, Director General
Geographical coverage: European
Aims and objectives: EFPIA represents the views of pharmaceutical associations in sixteen countries

– 1680 – Industry and Environment Programme Activity Centre (IE/PAC)

UNEP
39/43 quai André Citroën
F-75739 Paris Cedex 15
France
Tel: 33 1 405 888 50
Fax: 33 1 405 888 74
Geographical coverage: International

Aims and objectives: Works towards environmentally sound forms of industrial development

– 1681 – International Environment Bureau (IEB): *see entry 1727*

Geographical coverage: International

– 1682 – International Lead and Zinc Study Group

Metro House
58 St James Street
London SW1A 1LD
United Kingdom
Tel: 071 499 9373
Fax: 071 493 3725
Contact name: Bejoy Das Gupta, Economist
Geographical coverage: International
Aims and objectives: The Group was formed in 1959, by the UN as an independent organisation to provide information on the supply of lead and demand position of lead and zinc and its probable development. Makes special studies of the world situation in lead and zinc
Publications: Full catalogue of books available from the Study Group

– 1683 – International Wool Secretariat (IWS)

6–7 Carlton Gardens
London SW1Y 5AE
United Kingdom
Tel: 071 930 7300
Fax: 071 930 8884
Geographical coverage: International

– 1684 – Irish Industry Confederation – Environmental Policy Committee

Confederation House
Kildare Street
Dublin 2
Eire
Tel: 353 1 779801
Fax: 353 1 777823
Geographical coverage: Ireland

BOOKS

– 1685 – Proposal for a Council Regulation (EC): *see entry 1723*

Geographical coverage: European

CONFERENCE PAPERS

– 1686 – Initiatives for the Environment: A Publication from General Motors Europe

PO Box
Stelzenstrasse 4
CH-8152 Glattbrugg (Zürich)
Switzerland
Geographical coverage: European
Publishers: General Motors Europe AG Public Affairs, October 1991
Language: English
Coverage: Brochure detailing how General Motors has sought to tackle environmental issues affecting its plants and products

DIRECTORIES/YEARBOOKS

– 1687 – Environment Industry Yearbook

Geographical coverage: United Kingdom
Publishers: The Environment Press, 1992
Coverage: Arranged under headings such as Equipment & Services, Environmental Consultants etc, it contains detailed profiles of over 2,000 UK organisations
Price: £54.00
ISBN: 0951909606

– 1688 – Industrial Environmental Services Directory

Geographical coverage: United Kingdom
Publishers: Information for Industry Ltd
Language: annual
ISSN: 0964 2390
ISBN: 0951837508

– 1689 – Informationskaller Bygg and Miljo

(Information Sources for Construction and the Environment)

Haslingegatan 49
S-11331 Stockholm
Sweden
Tel: 46 8 34 01 70
Geographical coverage: Sweden
Publishers: BYGGDOK (Swedish Institute of Building Documentation)

– 1690 – Packaging Industry Directory 1993

Geographical coverage: United Kingdom
Publishers: Benn Business Information Services Ltd
Frequency: annual

Coverage: Wide coverage of the packaging industry in the UK. Also lists associations worldwide
Price: £68.00
ISSN: 0269 9834
ISBN: 0863821626

PERIODICALS

- 1691 - Economia Industrial

Dr Fleming 7 2
E-28036 Madrid
Spain
Geographical coverage: Spain
Publishers: Ministerio de Industria y Energia, Secretaria General Tecnica Centro de Publicaciones
Frequency: bi monthly

- 1692 - Energieanwendung: see entry 1000

Geographical coverage: Germany

- 1693 - Entsorgungspraxis: see entry 2189

Geographical coverage: Germany

- 1694 - Environment and Industry Digest: see entry 2190

Geographical coverage: United Kingdom

- 1695 - Environmental Protection Bulletin

Davis Building
165–171 Railway Terrace
Rugby
Warwickshire CV21 3HQ
United Kingdom
Tel: 0788 578214
Fax: 0788 560833
Geographical coverage: United Kingdom
Publishers: Institution of Chemical Engineers
Editor: Gardner, D.
Frequency: bi monthly
Coverage: An exchange of information and experience on environmental matters related to the chemical and process industries
Annual subscription: £105.00
ISSN: 0957 9052

- 1696 - Industrie et Environnement

5 rue de Dovai
F-75009 Paris
France
Tel: 33 1 429 315 43
Fax: 33 1 429 397 92
Geographical coverage: France
Editor: Lavernhe, C.

- 1697 - Industry and the Environment

Tour Mirabeau
39–43 quai André Citroën
F-75739 Paris Cedex 15
France
Tel: 33 1 405 888 50
Fax: 33 1 405 888 74
Geographical coverage: European
Publishers: UNEP Industry and Environment Office
Editor: Aloisi de Larderel, J.
Frequency: quarterly
Language: English
Coverage: Focuses on a main theme. Articles by government officials, industry managers, scientists and representatives of NGOs share the results of research and experience
ISSN: 0378 9993

- 1698 - Modern Plastics International

1221 Avenue of the Americas
New York
NY 10020
United States of America
Tel: 1 212 512 6267
Fax: 1 212 512 6111
Geographical coverage: International
Publishers: McGraw Hill Inc
Editor: Brownbill, D.
Frequency: monthly
Annual subscription: $157
ISSN: 0026 8283

- 1699 - Umwelt

Heinrichstrasse 24
Postfach 10 10 54
D-4000 Düsseldorf 1
Germany
Tel: 49 211 61880 0
Fax: 49 211 6188 112
Geographical coverage: Germany
Publishers: Verein Deutscher Ingenieure
Editor: Firnhaber, H.
Frequency: 10 issues a year
Coverage: Specialists and engineers in charge of environmental protection in industry, administrative bodies, communal authorities, service companies, engineering consultancies and technical schools
Annual subscription: DM216
ISSN: 0041 6355

REPORTS

– 1700 – German Chamber of Industry and Commerce: German Packaging Laws

Mecklenburg House
16 Buckingham Gate
London SW1E 6LB
United Kingdom
Tel: 071 233 5656
Geographical coverage: Germany

– 1701 – Tourism Industry and the Environment, The

Geographical coverage: United Kingdom
Publishers: Economist Intelligence Unit, 1992
ISBN: 0850585910
Author: Jenner, P. and Smith, C.

RESEARCH CENTRES

– 1702 – Espoo Research Centre

Box 44
SF-02271 Espoo
Finland
Tel: 358 90 804 71
Geographical coverage: Finland
Research activities: Fertilisers; agrochemicals; speciality chemicals for pulp and paper; chemical and biological plant protection

– 1703 – International Tin Research Institute

Kingston Lane
Uxbridge
Middlesex UB8 3PJ
United Kingdom
Tel: 0895 272406
Fax: 0895 251841
Contact name: Dr L A Hobbs
Geographical coverage: International
Aims and objectives: The Institute aims to develop and promote the use of tin. Scientific and technical study of the metal, its alloys and compounds, and of industrial processes that use tin

DATABASES

– 1704 – Base Relacional de la Industria y Servicios Ambientales (BRISA)

Jorge Juan 47
E-28001 Madrid
Spain
Fax: 34 1 577 09 10
Geographical coverage: Spain
Coverage: Covers environmental industries and services in Spain

Environmental Assessment

GOVERNMENT ORGANISATIONS

– 1705 – Nordic Council of Ministers: *see entry 2049*

Geographical coverage: European

NON-GOVERNMENT ORGANISATIONS

– 1706 – ASH Partnership

Unit 1/7
St Mary's Workshops
67 Giles Street
Edinburgh EH6 6DD
United Kingdom
Tel: 031 554 7619
Geographical coverage: United Kingdom

– 1707 – Comité Européen de Normalisation (CEN)

Central Secretariat
Rue de Stassart 36
B-1050 Brussels
Belgium
Geographical coverage: European
Aims and objectives: CEN has a working group looking at terminology, symbols and criteria for life cycle assessment of packaging

– 1708 – Fraunhofer-Institut für Lebensmitteltechnologie und Verpackung

(Food Technology and Packaging)

Schragenhofstrasse 35
D-8000 Munich 50
Germany
Tel: 49 89 149 0090
Fax: 49 89 149 00980
Geographical coverage: Germany

– 1709 – Natural Resources Research (NRR)

21 Church Lane
Loughton
Milton Keynes MK5 8AS
United Kingdom
Tel: 0908 666275
Fax: 0908 666275
Contact name: Mike Flood
Geographical coverage: United Kingdom

– 1710 – Society for Environmental Toxicology and Chemistry (SETAC)

Avenue Prekelinden 149
B-1200 Brussels
Belgium
Fax: 32 2 462 28 84
Geographical coverage: European
Aims and objectives: SETAC has formed a Life Cycle Analysis group

– 1711 – Society for the Promotion of Life Cycle Development (SPOLD)

Avenue Prekelinden 149
B-1200 Brussels
Belgium
Tel: 32 2 772 90 80
Fax: 32 2 772 90 80
Geographical coverage: European
Aims and objectives: SPOLD aims to improve life cycle analysis techniques

BOOKS

– 1712 – Environmental Assessment: A Guide to the Procedures

Geographical coverage: United Kingdom
Publishers: HMSO, 1992
ISBN: 0117522449
Author: Department of the Environment, Welsh Office

– 1713 – Man and the Environment: Product Life Cycle Analysis – What does it mean?

Batelle Institute eV
Forschung Entwicklung Innovation
Frankfurt am Main
Germany
Tel: 49 69 7908 0
Geographical coverage: Germany
Publishers: Batelle Information, 1991
Author: Klöpffer, W. and Rippen, G.

– 1714 – Role of Environmental Assessment in the Planning Process, The

Geographical coverage: United Kingdom
Publishers: Mansell, 1988
Author: Clark, M. and Hetherington, J.

– 1715 – Vital Signs 1992/93: The Trends that are Shaping our Future

Geographical coverage: United Kingdom
Publishers: Earthscan, 1992
Coverage: This book is a collection of information from around the world on economic social and environmental health
Price: £9.95
ISBN: 1853831417

PERIODICALS

– 1716 – Environmental Assessment Report

Holbeck Manor
Horncastle
Lincolnshire LN9 6PU
United Kingdom
Tel: 0507 533444
Geographical coverage: United Kingdom
Publishers: The Institute of Environmental Assessment
Editor: Tarling, J.
Language: English

Environmental Auditing

NON-GOVERNMENT ORGANISATIONS

– 1717 – Environmental Management and Auditing Services Ltd

34–38 Chapel Street
Little Germany
Bradford BD1 5DN
United Kingdom
Tel: 0274 742196
Geographical coverage: United Kingdom

– 1718 – Environmental Management and Auditing Services Ltd

Westland House
1 Westland Square
Dublin 2
Eire
Tel: 353 1 779199
Geographical coverage: Ireland

– 1719 – Environmental Management and Auditing Services Ltd

KUB Consultancy
Hogeschoollaan 225
Postbus 1046
NL-5004 BA Tilburg
Netherlands
Tel: 31 13 662010
Geographical coverage: Netherlands

BOOKS

– 1720 – Annotated Bibliography on Environmental Auditing

Office of Policy Planning and Evaluation
401 M Street SW
Washington DC 20460
United States of America
Geographical coverage: International
Publishers: USA Environment Protection Agency
Coverage: Guide to sources on environmental auditing

– 1721 – Environmental Auditing

Geographical coverage: United Kingdom
Publishers: Technical Communications and British Library, 1992
Coverage: A guide to best practice in the UK and Europe
ISBN: 0946655588
Author: Grayson, L.

– 1722 – Environmental Auditing Handbook: A Guide to Corporate and Environmental Risk Management

Geographical coverage: United Kingdom
Publishers: McGraw Hill
ISBN: 0070268592
Author: Harrison, L. L.

– 1723 – Proposal for a Council Regulation (EC)

Geographical coverage: European
Coverage: Proposal allowing voluntary participation by companies in the industrial sector in a Community ECO-audit scheme
ISBN: 927741782X

Environmental Business

NON-GOVERNMENT ORGANISATIONS

– 1724 – Ecological Studies Institute (ESI): see entry 1935

Geographical coverage: European

– 1725 – European Investment Bank (EIB)

Blvd Konrad Adenauer 100
L-2950 Luxembourg
Luxembourg
Tel: 352 43 7 91
Contact name: Cesare Matteuzzi

Geographical coverage: European
Aims and objectives: EIB is the European Community's financial institution. In its on-going task to encourage and cultivate the objectives of the Community, the EIB coordinates numerous financing projects, including projects in the environment sector.
Publications: EIB Information Bulletin; annual report; press releases and EIB Papers

– 1726 – Informat Green Alert

VO Tec Centre
Hambridge Lane
Newbury
Berkshire RG14 5TN
United Kingdom
Tel: 0635 34867
Fax: 0635 40212
Geographical coverage: European
Aims and objectives: Green Alert is a part of Infomat's commercial information service. For an annual subscription customers will receive a weekly digest of European environmental issues aimed at business use

– 1727 – International Chamber of Commerce (ICC)

38 Cours Albert 1
F-75008 Paris
France
Tel: 33 1 456 234 56
Fax: 33 1 422 586 63
Geographical coverage: International
Aims and objectives: The ICC is a non-governmental organisation serving world business. It set up in 1978 a Commission of the Environment to promulgate sound environmental policies for industry and to encourage and help their input to projects and organisations
Publications: Environmental factsheets, guidelines, updates and many pamphlets

– 1728 – International Environment Bureau (IEB)

61 route de Chêne
CH-1208 Geneva
Switzerland
Tel: 41 22 786 5111
Fax: 41 22 736 0336
Contact name: D M Roderick, Chairperson
Geographical coverage: International
Aims and objectives: The Bureau was founded in 1986, as a specialist division of the International Chamber of Commerce. It publishes a bimonthly newsletter and reports and supplies information, mainly relating to industrial matters, on demand to companies and agencies

– 1729 – Union of Industrial and Employers' Confederation of Europe (UNICE)

Rue Joseph II 40
Box 4
B-1040 Brussels
Belgium
Tel: 32 2 237 65 11
Fax: 32 2 231 14 45
Contact name: Daniel Cloquet, Director Industrial Affairs
Geographical coverage: European
Aims and objectives: UNICE is the officially recognised representative of business interests with the institutions of the EC. It focuses primarily on aspects of EC environment policy, looking at the various media and policy instruments

BOOKS

– 1730 – BS 7750: What the New Environmental Management Standards Mean for Your Business: see entry 2167

Geographical coverage: United Kingdom

– 1731 – Corporate Environmental Responsibility: Law and Practice: see entry 1943

Geographical coverage: United Kingdom

– 1732 – Corporate Responsibility Europe

Geographical coverage: European
Publishers: Chesterford Publishing, 1993
Coverage: Practical editorial data on both specific issues relating to environmental and social policy, and the wider context of best practice for companies operating within the EC
Price: £47.50

– 1733 – Director's Guide to Environmental Issues, The

Geographical coverage: United Kingdom
Publishers: Director Books, 1992
Coverage: Discusses the implications for the business community of the different environmental issues
Price: £29.95
ISBN: 1870555554
Author: Salter, J.

– 1734 – Environmental Guidelines to World Industry

Geographical coverage: International
Publishers: International Chamber of Commerce, 1990
Coverage: Also has supplementary guidelines on waste

– 1735 – Greener Marketing

Geographical coverage: United Kingdom
Publishers: Greenleaf Publishing, 1992
Editor: Charter, M.
Coverage: This book has drawn from a range of expert contributors. Covers strategic issues and practical implications of greener marketing
ISBN: 1874719004

– 1736 – Greening of Business, The

Geographical coverage: United Kingdom
Publishers: Gower, 1991
Editor: Rhys, A. D.
ISBN: 0566072815

– 1737 – In the Company of Green: Corporate Communications for the New Environment

Geographical coverage: United Kingdom
Publishers: ISBA Publications
Coverage: Broad outline of the issues which are affecting companies and their products
Price: £16.00
ISBN: 0906241235
Author: Bernstein, D.

– 1738 – Management for a Small Planet

Geographical coverage: United Kingdom
Publishers: Sage Publications
Coverage: Provides an overview of the social, scientific and economic concepts which should underlie environmentally sound decision making at the business level
Price: £12.95
ISBN: 0803946341

– 1739 – Managing the Environment: The Greening of European Business

Geographical coverage: United Kingdom
Publishers: Business International, 1990
Author: Robins, N.

DIRECTORIES/YEARBOOKS

– 1740 – ENTEC Directory of Environmental Technology: *see entry 2184*

Geographical coverage: United Kingdom

– 1741 – Environment Contacts: A Guide for Business (Who does what in Government Departments)

Geographical coverage: United Kingdom
Publishers: Department of Trade and Industry
Frequency: every 2 years
Coverage: Environment contacts within the Government Departments

– 1742 – Institute of Materials Management Members' Reference Book and Buyers Guide 1993, The

Geographical coverage: United Kingdom
Publishers: Guardian Communications Ltd (part of Argus Business Publications Ltd), 1993
Editor: Craig, P.
Frequency: annual
Coverage: Covers, trade organisations worldwide, buyers and suppliers in the UK and associated trade journals
ISSN: 0960 3832
ISBN: 0861088750

– 1743 – Spirit of Versailles: The Business of Environmental Management

38 Cours Albert
F-75008 Paris
France
Tel: 33 1 456 234 56
Geographical coverage: France
Publishers: International Chamber of Commerce

– 1744 – World Environmental Business Handbook

Geographical coverage: International
Publishers: Euromonitor, 1993
Language: English
Coverage: Authoritative analysis covering all main environmental issues and considerations by region
Price: £95/US $190
ISBN: 0836684528

PERIODICALS

– 1745 – Business Strategy and the Environment

34–38 Chapel Street
Little Germany
Bradford BD1 5DN
United Kingdom
Tel: 0274 729315
Fax: 0274 306981
Geographical coverage: European
Publishers: European Research Press Ltd

Editor: Smith, D.
Frequency: quarterly
Language: English
Coverage: Research and development in environmental issues pertinent to industry
Annual subscription: £50.00
ISSN: 0964 4733

– 1746 – Confederation of British Industry News

Centre Point
103 New Oxford Street
London WC1A 1DU
United Kingdom
Tel: 071 379 7400
Fax: 071 240 2651
Geographical coverage: United Kingdom
Publishers: ABC Business Press
Editor: Dembinski, M.
Coverage: Covers many aspects of business including environmental affairs
Annual subscription: £25.00
ISSN: 0261 6661

– 1747 – Environment Bulletin

World Bank
Room S-5055
1818 H Street NW
Washington DC 20433
United States of America
Geographical coverage: International
Publishers: World Bank, Environment Department
Editor: Snyder, A.
Frequency: quarterly
Language: English
Coverage: A newsletter of the World Bank environment committee

– 1748 – Environment Business

521 Old York Road
London SW18 1TG
United Kingdom
Tel: 081 877 9130
Fax: 081 877 9938
Geographical coverage: United Kingdom
Publishers: Information for Industry Ltd
Editor: Newham, M.
Frequency: fortnightly
Coverage: Key information for senior managers working in pollution control, environmental section and sustainable resource management
Annual subscription: £227.00
ISSN: 0959 7042

– 1749 – Environment Risk

Nestor House
Playhouse Yard
London EC4 5EX
United Kingdom
Tel: 071 779 8888
Fax: 071 779 8617
Geographical coverage: International
Publishers: Euromoney Publications plc
Editor: Carr, J.
Frequency: 10 issues a year
Coverage: Approaches environmental issues from the perspective of multinational companies
Annual subscription: £220.00
ISSN: 0965 3813

– 1750 – Environmental Law Brief: see entry 1960

Geographical coverage: United Kingdom

– 1751 – European Environmental Business News

15th Floor
149 avenue Louise
B-1050 Brussels
Belgium
Tel: 32 2 533 16 49
Fax: 32 2 534 18 45
Geographical coverage: European
Publishers: Centurion Corporation
Editor: Jones, T. S.
Frequency: 6 issues a year
Language: English
Coverage: European newspaper for the environmental industry
ISSN: 1060 3573

– 1752 – Marketing

22 Lancaster Gate
London W2 3LY
United Kingdom
Geographical coverage: United Kingdom
Publishers: Marketing Publications Ltd
Editor: Lester, T.
Frequency: weekly
Coverage: Includes features on environment related products and services
Price: £40.00
ISSN: 0025 3650

RESEARCH CENTRES

– 1753 – Institute of Materials Management (IMM)

Cranfield Institute of Technology
Cranfield
Bedford MK43 0AL
United Kingdom
Tel: 0234 750662
Fax: 0234 750875
Contact name: Christine Rowar, Executive Officer
Geographical coverage: United Kingdom
Aims and objectives: To promote materials management in manufacturing and service companies. To stimulate and promote research

DATABASES

– 1754 – Base Relacional de la Industria y Servicios Ambientales (BRISA): see entry 1704

Geographical coverage: Spain

– 1755 – INFOMAT

Geographical coverage: European
Coverage: International business information, covering Europe and developing countries from periodical sources
Database producer: Information Access Co./Predicasts
Database host: DataStar; Dialog; ESA–IRS; Orbit
Coverage period: current

– 1756 – PTS PROMT

Geographical coverage: International
Coverage: Business products and markets; including initiatives and response to environmental issues
Database producer: Information Access Co./Predicasts
Database host: DataStar; Dialog; FT Profile and other hosts
Coverage period: 1972–
Update frequency: daily

Environmental Consultants

NON-GOVERNMENT ORGANISATIONS

– 1757 – Environmental Resources Ltd

106 Gloucester Place
London W1H 3DB
United Kingdom
Tel: 071 465 7200
Fax: 071 935 8355
Contact name: Verina Ingram, Commercial Services
Geographical coverage: United Kingdom
Aims and objectives: ERL is a multidisciplinary environmental management consultancy. It provides business and governments with specialist advice on all aspects of the environment

– 1758 – ERL Brussels

20 avenue des Celtes
B-1040 Brussels
Belgium
Tel: 32 2 280 04 70
Fax: 32 2 230 28 62
Geographical coverage: Belgium

Aims and objectives: ERL provides business and governments with specialist advice on all aspects of the environment

– 1759 – ERL España

Calle Reyes 7
1 izq
E-28015 Madrid
Spain
Tel: 34 1 522 2750
Fax: 34 1 523 3991
Geographical coverage: Spain
Aims and objectives: ERL provides business and governments with specialist advice on all aspects of the environment

– 1760 – ERL Italia

Via G B Morgagni n 4
I-20129 Milan
Italy
Tel: 39 2 295 22275
Fax: 39 2 295 22592
Geographical coverage: Italy

Businesses and Services *Environmental Consultants*

Aims and objectives: ERL provides business and governments with specialist advice on all aspects of the environment

– 1761 – ERL Nederland

PO Box 710
NL-2700 HG Zoetermeer
Netherlands
Tel: 31 79 522 777
Fax: 31 79 512 127
Geographical coverage: Netherlands
Aims and objectives: ERL provides business and governments with specialist advice on all aspects of the environment

– 1762 – ERL Umwelt Consult: Rhein–Main–Neckar GmbH

Darmstädter Strasse 190
D-6140 Bensheim 3
Germany
Tel: 49 6251 76057
Fax: 49 6251 787699
Geographical coverage: Germany
Aims and objectives: ERL provides business and governments with specialist advice on all aspects of the environment

– 1763 – International Institute for Environment and Development (IIED)

3 Endsleigh Street
London WC1H 0DD
United Kingdom
Tel: 071 388 2117
Fax: 071 388 2826
Geographical coverage: International
Aims and objectives: The Institute assists and advises governmental and private organisations concerned with the links between environment and development. It is concerned to promote the use of natural resources through sustainable development

DIRECTORIES/YEARBOOKS

– 1764 – 1992/93 ENDS Directory and Market Analysis

Geographical coverage: United Kingdom
Publishers: Environmental Data Services, 1992
Frequency: annual
Coverage: Comprehensive guide to UK environmental consultancies
Price: £96.00
ISBN: 090767304X

– 1765 – Adressbuch Umwelt-Experten

(Addressbook of Environmental Experts)

Geographical coverage: Germany
Publishers: Eberhard Blottner Verlag, Taunusstein 1991
Coverage: Directory of environmental experts
Price: DM76

– 1766 – INFOTERRA World Directory of Environmental Expertise

Geographical coverage: International
Publishers: United Nations Environment Programme (UNEP), 1987
ISBN: 9280711520

– 1767 – Vachers European Companion and Consultants Register

Geographical coverage: European
Publishers: Vachers Publications, 1993
Editor: Gunn, Mrs E.
Frequency: quarterly
Language: English
Coverage: Diplomatic, political and commercial reference book
Price: £11.00
ISSN: 0958 0336

PERIODICALS

– 1768 – UK Centre for Economic and Environmental Development Bulletin

3e King's Parade
Cambridge CB2 1SJ
United Kingdom
Tel: 0223 67799
Fax: 0223 67794
Geographical coverage: United Kingdom
Publishers: UK Centre for Economic and Environmental Development
Editor: Harbinson, J.
Frequency: bi monthly
Coverage: Economic analyses of environmental issues
Annual subscription: £15.00
ISSN: 0268 7402

RESEARCH CENTRES

– 1769 – East European Environmental Research (ISTER)

Vaci u 62–64
H-1056 Budapest
Hungary

Tel: 36 1 136 70
Fax: 36 1 136 70
Contact name: Mr Janos Vargha
Geographical coverage: Hungary
Research activities: Research and consultancy on activities in the interests of sustainable development and new environmental policies

LIBRARIES

– 1770 – Environmental Information Service: The British Library

25 Southampton Buildings
London WC2A 1AW
United Kingdom
Tel: 071 323 7955
Fax: 071 323 7954
Contact name: Helen Woolston
Geographical coverage: United Kingdom
Aims and objectives: The Library offers a comprehensive range of information services to the general public, companies and organisations who are concerned with legal, technical and business questions on the environment
Publications: Various publications

Environmental Education

GOVERNMENT ORGANISATIONS

– 1771 – Foundation for Environmental Education

PO Box 130030
NL-3507 LA Utrecht
Netherlands
Tel: 31 30 71 3734
Geographical coverage: Netherlands

NON-GOVERNMENT ORGANISATIONS

– 1772 – Council for Occupational Standards and Qualifications in Environmental Conservation (COSQUEC): *see entry 2164*

Geographical coverage: United Kingdom

– 1773 – Foundation for Environmental Education in Europe (FEEE): European Office Friluftsradet

(Open Air Council)
Olof Palmes Gade 10
DK-2100 Copenhagen Ø
Denmark
Contact name: Mr Ole Lovig Simonsen, President
Geographical coverage: European
Aims and objectives: FEEE is a network of organisations in Europe promoting environmental education, both by carrying out campaigns and creating an awareness of the concept of environmental education
Publications: *Blue Flag Campaign*

– 1774 – Hungarian Institute for Materials Handling and Packaging

Rigo utca 3
PO Box 189
H-1431 Budapest
Hungary
Tel: 36 1 1137460
Fax: 36 1 1338170
Contact name: Gyozo Polhammer
Geographical coverage: Hungary
Aims and objectives: The Institute provides information and reference service, postgraduate courses, laboratory and quality tests

– 1775 – Instituut voor Natuurbeschermingseducatie (IVN)

Plantage Middenlaan 41
NL-1081 DC Amsterdam
Netherlands
Tel: 31 20 228115
Fax: 31 20 266091
Geographical coverage: Netherlands

– 1776 – Institution of Mechanical Engineers: *see entry 1792*

Geographical coverage: United Kingdom

214 Businesses and Services *Environmental Education*

– 1777 – Institution of Water and Environmental Management (IWEM): *see entry 2007*

Geographical coverage: United Kingdom

– 1778 – International Environment Bureau: *see entry 2174*

Geographical coverage: International

– 1779 – International Youth Federation for Environmental Studies and Conservation (IYF)

Klostermolle
Klostermollevej 48
DK-8660 Skanderborg
Denmark
Geographical coverage: International
Aims and objectives: IYF was founded in 1956 under the auspices of the IUCN. It is a federation of regional national and local youth organisations concerned with the study and conservation of the environment and has some 130 member organisations in 54 countries
Publications: *Youth in Environmental Action in 1987*; reports and booklets

– 1780 – National Association for Environmental Education (NAEE)

University of Wolverhampton
Walsall Campus
Gorway
Walsall WS1 3BD
United Kingdom
Tel: 0922 31200
Contact name: Philip Neal, General Secretary
Geographical coverage: United Kingdom
Aims and objectives: The NAEE is the organisation for all those involved with environmental education in schools and colleges
Publications: *Environmental Education*; magazine three times a year; a wide range of practical guides; occasional papers

– 1781 – Natur og Ungdom

(Nature and Youth)

Torggt. 34
N-0183 Oslo 1
Norway
Tel: 47 2 364218
Geographical coverage: Norway

DICTIONARIES/ENCYCLOPEDIAS

– 1782 – Dictionary of Environment and Development

Geographical coverage: United Kingdom
Publishers: WWF UK in association with Earthscan
Coverage: An A–Z of environment and development
Price: £15.00
ISBN: 185383078X

– 1783 – Dictionary of the Environment

Geographical coverage: United Kingdom
Publishers: Paladin, 1990
Coverage: Provides an overview of the the major areas of environmental conern
Price: £6.99
ISBN: 0586085432
Author: Elsworth, S.

DIRECTORIES/YEARBOOKS

– 1784 – Directory of Environmental Journals and Media Contacts

Geographical coverage: United Kingdom
Publishers: Council for Environmental Conservation
Editor: Cairns, T. et al
Frequency: irregular
Coverage: Lists environment journals mainly in the UK
ISBN: 0903158329

– 1785 – Guide to Resources in Environmental Education

Geographical coverage: United Kingdom
Publishers: The Conservation Trust, 1991
Editor: Berry, P. and Lydford, C.
ISBN: 0907153372

PERIODICALS

– 1786 – Environmental Education – Journal of the NAEE

University of Wolverhampton
Walsall Campus
Walsall WS1 3BD
United Kingdom
Tel: 0922 31200
Geographical coverage: United Kingdom
Publishers: National Association for Environmental Education (NAEE)
Editor: Armstrong, S.
Frequency: 3 issues a year

Price: £3.00
Annual subscription: £8.50

– 1787 – International Journal of Environmental Studies

42 William IV Street
London WC2
United Kingdom
Geographical coverage: International
Publishers: Gordon and Breach Science Publishers
Frequency: bi monthly
ISSN: 0020 7233

– 1788 – Streetwise

University of Brighton
68 Grand Parade
Brighton BN2 2JY
United Kingdom
Tel: 0273 673416
Fax: 0273 679179
Geographical coverage: United Kingdom
Publishers: National Association for Urban Studies
Editor: Welsh, R.
Frequency: quarterly
Coverage: Urban policy issues and curriculum development in environmental education in UK and worldwide
Annual subscription: £15.50
ISSN: 0957 6517

RESEARCH CENTRES

– 1789 – Cranfield Biotechnology Centre

Cranfield Institute of Technology
Cranfield
Bedfordshire MK43 0AL
United Kingdom
Tel: 0234 754339
Geographical coverage: United Kingdom
Aims and objectives: The Centre is involved in research contracts and postgraduate education
Publications: Research papers available
Research activities: Toxic gas monitoring

– 1790 – ECO Environmental Education Trust

10–12 Picton Street
Montpelier
Bristol BS6 5QA
United Kingdom
Tel: 0272 420162
Contact name: Monica Barlow
Geographical coverage: United Kingdom
Aims and objectives: The Trust carries out research and improvement in the provision of environmental information

Environmental Engineering

NON-GOVERNMENT ORGANISATIONS

– 1791 – Institution of Engineers in Ireland

22 Clyde Road
Ballsbridge
Dublin 4
Eire
Tel: 353 1 684341
Fax: 353 1 685508
Contact name: Prof Thomas Casey, Chairperson
Geographical coverage: Ireland
Aims and objectives: The Institution lectures on water and environmental engineering topics

– 1792 – Institution of Mechanical Engineers

1 Birdcage Walk
London SW1H 9JJ
United Kingdom
Tel: 071 222 7899
Fax: 071 222 4557
Contact name: Alan Knowles, Public Relations Manager
Geographical coverage: United Kingdom
Aims and objectives: IMechE is among the world's foremost professional development centres in the field of mechanical engineering. It is involved in the education, training and professional development of engineers, and acts as an international centre for technology
Publications: Full catalogue of books available from Mechanical Engineering Publications Ltd

216 Businesses and Services *Environmental Engineering*

BOOKS

– 1793 – Environmental Engineering in the Process Plant

Geographical coverage: International
Publishers: McGraw Hill Book Company, 1992
Coverage: A collection of 36 articles taken from Chemical Enginering Magazine
Price: £38.50
ISBN: 0070110352
Author: Chopey, N. P.

– 1794 – Introduction to Environmental Engineering

Geographical coverage: United Kingdom
Publishers: McGraw Hill Book Company, 1991
Coverage: Emphasises fundamental concepts, definitions and problem solving in its presentation of environmental engineering
Price: £28.50
ISBN: 0070159114
Author: Davis, M. L. and Cornwell, D. A.

DICTIONARIES/ENCYCLOPEDIAS

– 1795 – Encyclopedia of Environmental Science and Engineering: *see entry 2134*

Geographical coverage: United Kingdom

PERIODICALS

– 1796 – Ecological Engineering

PO Box 211
NL-1000 AE Amsterdam
Netherlands
Tel: 31 20 5803911
Fax: 31 20 5803705
Geographical coverage: Netherlands
Publishers: Elsevier Science Publishers BV
Editor: Costanza, R.
Frequency: quarterly
Language: English
Coverage: Publishes contributions in ecotechnology including bioengineering, pollution control and sustainable agriculture
Annual subscription: Dfl290
ISSN: 0925 8574

– 1797 – Ecological Modelling

PO Box 211
NL-1000 AE Amsterdam
Netherlands
Tel: 31 20 5803911
Fax: 31 20 5803598
Geographical coverage: Netherlands
Publishers: Elsevier Science Publishers BV
Editor: Joergensen, S. E.
Frequency: 20 issues a year
Language: English
Coverage: Combines mathematical modelling systems analysis and computer techniques with ecology and environmental management
Annual subscription: Dfl1365
ISSN: 0304 3800

– 1798 – Environment Protection Engineering

Wybrzeze Wyspianskiego 27
PL-50 370 Wroclaw
Poland
Tel: 48 71 20 23 04
Fax: 48 71 22 33 64
Geographical coverage: Poland
Publishers: Politechnika Wroclawska
Editor: Winnicki, T. and Pawlowski, L.
Frequency: quarterly
Language: Text in English, summaries in Polish and Russian
Coverage: Papers dealing with water purification, wastewater treatment, solid waste disposal, neutralisation and sterilisation
Price: $80
ISSN: 0324 8828

– 1799 – Environmental Engineering

Northgate Avenue
Bury St Edmunds
Suffolk IP32 6BW
United Kingdom
Tel: 0284 763277
Fax: 0284 704006
Geographical coverage: United Kingdom
Publishers: Mechanical Engineering Publications
Editor: Whiteley, F. A.
Frequency: quarterly
Coverage: Articles and papers in noise, shock, vibration, contamination control and packaging
Annual subscription: £31.00
ISSN: 0954 5824

– 1800 – Green Engineering

Northgate Avenue
Bury St Edmunds
Suffolk
IP32 6BW
United Kingdom
Tel: 0284 763277
Fax: 0284 704006
Geographical coverage: United Kingdom
Publishers: MEP Ltd

Editor: Caird-Daley, P.
Frequency: monthly
Coverage: Engineering environmental issues
Annual subscription: £99.00

– 1801 – Green Engineering – A Current Awareness Bulletin

Geographical coverage: United Kingdom
Publishers: Mechanical Engineering Publications Ltd
Editor: Wright, L.
Frequency: monthly
Coverage: Focuses on environmental issues relevant to the engineering profession
ISSN: 0960 8796

– 1802 – Manufacturing Engineering:
see entry 1964
Geographical coverage: International

RESEARCH CENTRES

– 1803 – Building Research Establishment (BRE)

Garston
Watford WD2 7JR
United Kingdom
Tel: 0923 894040
Fax: 0923 664010
Contact name: Christine Cornwell
Geographical coverage: United Kingdom
Aims and objectives: The BRE carries out research for the Department of the Environment, provides technical consultancy on construction problems for the public and private sector and disseminates research results via publications, seminars etc.
Publications: *BRE Update*; List available on request
Research activities: Research and development into structural and civil engineering and environmental design and engineering

– 1804 – Instytut Podstaw Inzynierii Srodowiska PAN
(Environmental Engineering Institute)

Ulica M Curie Sklodowskiej 34
PL-41 800 Zabrze
Poland
Tel: 48 32 716481
Geographical coverage: Poland
Publications: *Archives of Environmental Protection*; *Prace i Studia*
Research activities: Atmospheric pollution protection; Water pollution protection; Reclamation of soil; Biological reclamation of dumps

DATABASES

– 1805 – Compendex Plus

Geographical coverage: International
Coverage: Engineering and Technology
Database producer: Engineering Information Inc
Database host: BRS; DataStar; Dialog; ESA–IRS; Orbit; STN;
Coverage period: 1970–
Update frequency: 12,000 records monthly
Additional information: Available on CD–ROM

– 1806 – ICONDA

Geographical coverage: International
Coverage: Large scale engineering projects such as bridges, dams roads and tunnels
Database producer: Info Centre for Regional Planning and Building Construction
Database host: Orbit
Coverage period: 1976–
Update frequency: monthly

Environmental Groups and Movements

GOVERNMENT ORGANISATIONS

– 1807 – Anders Gaan Leven (AGALEV)

Tweekerkenstraat 78
B-1040 Brussels
Belgium
Tel: 32 2 230 66 66
Fax: 32 2 230 47 86
Geographical coverage: Belgium
Aims and objectives: Green party of the Flemish part of Belgium

– 1808 – ARC

Bundestag HT 108
D-5300 Bonn 1
Germany
Tel: 49 228 16 91 98
Fax: 49 228 16 78 78/16
Geographical coverage: Germany
Aims and objectives: Rainbow Group

– 1809 – Comhaontas Glas

(The Green Party)

5A Upper Fownes Street
Dublin 2
Eire
Tel: 353 1 771436
Geographical coverage: Ireland
Aims and objectives: Green party
Publications: *Nuacht Glas*, six a year

– 1810 – Council of Ministers

Secretariat
Rue de la Loi 170
B-1048 Brussels
Belgium
Tel: 32 2 234 61 11
Fax: 32 2 234 73 81
Contact name: Frederik Moys, Head of Environment Division
Geographical coverage: European
Aims and objectives: COM is the institution which directly represents the governments of the member states of the European Community. Environmental affairs are the responsibility of Directorate-General DXI

– 1811 – Council of Ministers

Rue Belliard 62
B-1040 Brussels
Belgium
Tel: 32 2 233 21 11
Fax: 32 2 231 10 75
Contact name: Chris van den Bilke
Geographical coverage: Belgium
Aims and objectives: Contact for the environment

– 1812 – Council of Ministers

Rue d'Arlon 73
B-1040 Brussels
Belgium
Tel: 32 2 233 08 11
Fax: 32 2 230 93 84
Contact name: Axel Kristiansen
Geographical coverage: Denmark
Aims and objectives: Contact for the environment

– 1813 – Council of Ministers

Rue Ducale 67–71
B-1000 Brussels
Belgium
Tel: 32 2 511 49 55
Fax: 32 2 514 53 09
Contact name: Laurence Auer
Geographical coverage: France
Aims and objectives: Contact for the environment

– 1814 – Council of Ministers

Rue J De Lalaing 19–21
B-1040 Brussels
Belgium
Tel: 32 2 238 18 11
Fax: 32 2 238 19 78
Contact name: Reinhard Krapp
Geographical coverage: Germany
Aims and objectives: Contact for the Environment

– 1815 – Council of Ministers

Avenue de Cortenberg 71
B-1040 Brussels
Belgium
Tel: 32 2 739 56 11
Fax: 32 2 735 59 79
Contact name: Miltiadis Vassilopoulos
Geographical coverage: Greece
Aims and objectives: Contact for the environment

– 1816 – Council of Ministers

Avenue Galilée 5
Bte 22
B-1030 Brussels
Belgium
Tel: 32 2 218 06 05
Fax: 32 2 218 13 47
Contact name: Jim Humphreys
Geographical coverage: Ireland
Aims and objectives: Contact for the environment

– 1817 – Council of Ministers

Rue du Marteau 9
B-1040 Brussels
Belgium
Tel: 32 2 220 04 11
Fax: 32 2 219 34 49
Contact name: Julio Tonini
Geographical coverage: Italy
Aims and objectives: Contact for the environment

– 1818 – Council of Ministers

Rue du Noyer 211
B-1040 Brussels
Belgium

Tel: 32 2 733 99 77
Fax: 32 2 736 14 29
Contact name: Jean-Paul Munchen
Geographical coverage: Luxembourg
Aims and objectives: Contact for the environment

– 1819 – Council of Ministers

Avenue des Arts 66
B-1040 Brussels
Belgium
Tel: 32 2 513 77 75
Fax: 32 2 513 08 29
Contact name: Edward Kronenburg
Geographical coverage: Netherlands
Aims and objectives: Contact for the environment

– 1820 – Council of Ministers

Rue Marie-Thérèse 11–13
B-1040 Brussels
Belgium
Tel: 32 2 211 12 11
Fax: 32 2 218 15 42
Contact name: Fernando Almiro do Vale
Geographical coverage: Portugal
Aims and objectives: Contact for the environment

– 1821 – Council of Ministers

Boulevard du Régent 52
B-1000 Brussels
Belgium
Tel: 32 2 509 86 11
Fax: 32 2 511 26 30
Contact name: Hilario Dominguez Hernandez
Geographical coverage: Spain
Aims and objectives: Contact for the environment

– 1822 – Council of Ministers

Ron Point Schuman 6
B-1040 Brussels
Belgium
Tel: 32 2 287 82 11
Fax: 32 2 287 83 98
Contact name: Simon Featherstone
Geographical coverage: United Kingdom
Aims and objectives: Contact for the environment

– 1823 – Déi Gréne Alternativ

(Green Alternative Party)

BP 454
L-2014 Luxembourg
Luxembourg
Tel: 352 46 37 40
Fax: 352 46 37 43
Geographical coverage: Luxembourg
Aims and objectives: Green party
Publications: *Grénge Spoun*, weekly

– 1824 – Ecolo

28 rue Basse-Marcelle
B-5000 Namur
Belgium
Tel: 32 81 22 78 71
Fax: 32 81 23 06 03
Geographical coverage: Belgium
Aims and objectives: Green party of the French part of Belgium

– 1825 – Estonian Green Party

Narva Mnt 27
202400 Tartu
Estonia
Tel: 7 0142 681319
Fax: 7 0142 529579
Geographical coverage: Estonia
Aims and objectives: Green party

– 1826 – Green Party

10 Station Parade
Balham High Road
London SW12 9AL
United Kingdom
Tel: 081 673 0045
Fax: 081 675 4434
Geographical coverage: United Kingdom
Publications: *Econews*, quarterly

– 1827 – Green Party of Bulgaria

39 Dondukov Boulevard
1594 Sofia
Bulgaria
Tel: 359 2 390 039
Geographical coverage: Bulgaria
Aims and objectives: Green party

– 1828 – Groenen, De

(Green Party)

PO Box 3244
DK-1001 AA Amsterdam
Netherlands
Tel: 31 20 996418
Fax: 31 20 5523426
Geographical coverage: Netherlands
Aims and objectives: Green party

– 1829 – Gronne, De

Slugten 10
DK-3300 Frederiksvaerk
Denmark
Tel: 45 42 34 89 19
Geographical coverage: Denmark
Aims and objectives: Green environmentalist party

– 1830 – Grüne Alternative, Die

Stiftgasse 6
A-1070 Vienna
Austria
Tel: 43 1 521 250
Fax: 43 1 521 2540
Geographical coverage: Austria
Aims and objectives: Green party
Publications: *Impuls Grün*, monthly

– 1831 – Grüne Liga

Friedrichstrasse 165
D-01080 Berlin
Germany
Tel: 49 30 392 2316
Geographical coverage: Germany
Aims and objectives: Network of eastern eco-groups
Publications: *Grüne Liga–Rundbrief*, monthly

– 1832 – Grüne Partei – Ost

Haus der Demokratie
Friedrichstrasse 165
D-01080 Berlin
Germany
Tel: 49 30 229 1657
Geographical coverage: Germany
Aims and objectives: Green party

– 1833 – Grüne Partei der Schweiz/Parti Ecologiste Suisse (GPS)
(Green Party of Switzerland)

Marienstrasse 11
CH-3005 Berne
Switzerland
Tel: 41 31 441441
Geographical coverage: Switzerland
Aims and objectives: Green party
Publications: *GPS Info/Verts*, quarterly in German and French

– 1834 – Izmir Yesiller Partisi
(Green Party of Izmir)

Kore Sht Cad 96 K7/701
Alsancak
35220 Izmir
Turkey
Tel: 90 51 635023
Geographical coverage: Turkey
Aims and objectives: Also SOS Akdeniz Grubu/SOS Mediterranean Group
Publications: *Agackakan*, monthly

– 1835 – Izquierda de los Pueblos

San Martin 13-4
E-20005 San Sebastian
Spain
Tel: 34 43 422470/425839
Fax: 34 43 290988
Contact name: Juan Maria Bandres Molet
Geographical coverage: Spain

– 1836 – Lithuanian Green Party

Pylimo 4
2001 Vilnius
Lithuania
Geographical coverage: Lithuania
Aims and objectives: Green party

– 1837 – MDP/CDE–Green Party

Rua Febo Moniz 13 r/c
P-1100 Lisbon
Portugal
Tel: 351 1 561788
Fax: 351 1 524790
Geographical coverage: Portugal
Aims and objectives: Green party

– 1838 – Miljopartiet de Gröna (MP)
(Green Party)

Riksdagen
S-10012 Stockholm
Sweden
Tel: 46 8 786 5684
Fax: 46 8 215316
Geographical coverage: Sweden
Aims and objectives: Green party

– 1839 – Miljopartiet de Gronne

Grensen 8
N-0159 Oslo
Norway
Tel: 47 2 429758
Geographical coverage: Norway
Aims and objectives: Green party
Publications: *Gronn Kontakt*, monthly

– 1840 – Polska Partia Zielonych (PPZ)
(Polish Party of Greens)

PO Box 783
PL-30 960 Krakow 1
Poland
Tel: 48 12 551098
Fax: 48 12 377731
Geographical coverage: Poland
Aims and objectives: Pacifist environmental group
Publications: *Swiat*, weekly

– 1841 – Socialistik Folkeparti

Skaering Hedevej 190
K-8250 EGA
Denmark
Tel: 45 86 22 65 67
Contact name: John Iversen
Geographical coverage: Denmark
Aims and objectives: Green party

– 1842 – Verdi

c/o Gruppo Verde/Fraktion Die Grünen
Via Crispi 9
I-39100 Bolzano/Bozen
Italy
Tel: 39 471 99 30 15
Fax: 39 471 97 84 44
Contact name: Alexander Langer
Geographical coverage: Italy
Aims and objectives: Green party

– 1843 – Verdi Arcobaleno

Via Giovanni Rasoni 15
I-20145 Milan
Italy
Tel: 39 2 469 21 01
Fax: 39 2 39 43 41
Contact name: Virginio Bettini
Geographical coverage: Italy
Aims and objectives: Green party

– 1844 – Verts

5 rue de Platière
F-69001 Lyon
France
Tel: 33 78 27 44 03
Contact name: Mme Djida Tazdait
Geographical coverage: France
Aims and objectives: Green party

– 1845 – Vihreä Litto

(Green League)

Eerikinkatu 24A
SF-00100 Helsinki
Finland
Tel: 358 0 693 3877
Fax: 358 0 693 3799
Geographical coverage: Finland
Aims and objectives: Green party
Publications: *Vihreä Lanka*, weekly

– 1846 – Zeleni Slovenije

(Greens of Slovenia)

Komenskoga 7
61000 Ljubljana
Slovenia
Tel: 38 61 312368
Geographical coverage: Slovenia
Aims and objectives: Anti militarist/pacifist environmental group

NON-GOVERNMENT ORGANISATIONS

– 1847 – Bon Beter Leefmilieu (BBL)

Overwinningsstrasse 26
B-1060 Brussels
Belgium
Tel: 32 2 539 22 17
Fax: 32 2 539 09 21
Contact name: Ms Ingrid Pauwels
Geographical coverage: Belgium
Aims and objectives: BBL is an umbrella organisation of groups working in the field of environment and nature protection

– 1848 – Cyprus Association for the Protection of the Environment

Chanteclare Building
PO Box 3810
Nicosia
Cyprus
Geographical coverage: Cyprus

– 1849 – Deutsche Naturschutzring (DNR)

Am Michaelshof 8–10
D-5300 Bonn 2
Germany
Tel: 49 228 359005
Fax: 49 228 359096
Contact name: Sascha Müller-Kraenner, EC-Coordination
Geographical coverage: Germany
Aims and objectives: DNR is a league of 105 German environmental NGOs

– 1850 – Environment Council

80 York Way
London N1 9AG
United Kingdom
Tel: 071 278 4736
Fax: 071 837 9688
Contact name: Rachel Adatia, Information Programme
Geographical coverage: United Kingdom
Aims and objectives: The Council was formed in 1969 and is the umbrella body for UK organisations concerned with the environment.
Publications: *News from the Environment Council*; *Habitat*, magazine ten times a year; *Business and Environment Handbook*

– 1851 – Environmental Politics in Bulgaria: Ecological Studies Institute

60 Chandos Place
London
United Kingdom
Tel: 071 379 4324
Geographical coverage: Bulgaria

– 1852 – Europäische Ökologische Aktion (ECOROPA)

(European Group for Ecological Action)

Görresstrasse 33
D-8000 Munich 40
Germany
Tel: 49 89 52 97 70
Geographical coverage: European

– 1853 – European Environment Bureau (EEB)

Rue de la Victoire 26
B-1047 Brussels
Belgium
Tel: 32 2 539 00 37
Fax: 32 2 539 09 21
Contact name: Raymond van Ermen, Secretary General
Geographical coverage: European
Aims and objectives: The Bureau, founded in 1974, is a federation of 148 environmental NGO's which liaises with EC institutions on environmental affairs
Publications: *Metamorphosis*, newsletter, published in nine languages

– 1854 – Federation of Ecologist Alternative Organisations

37 Themistokleous Street
GR-10677 Athens
Greece
Tel: 30 1 3602 644
Fax: 30 1 3639 930
Geographical coverage: Greece

– 1855 – International Organisation of Consumer Unions (IOCU)

Emmastraat 9
NL-2595 EG The Hague
Netherlands
Tel: 31 70 347 63 31
Fax: 31 70 383 49 76
Contact name: Marjolijn Peters, Coordinator Environmental Programmes
Geographical coverage: International
Aims and objectives: Established in 1960, IOCU aims to link the activities of consumer organisations and to promote the expansion of the consumer movement worldwide. Concerns include hazardous products and technologies, wastes and the environmental aspects of consumption

– 1856 – International Professional Association for Environment Affairs (IPRE)

31 rue Montoyer
Box 1
B-1040 Brussels
Belgium
Tel: 32 2 513 60 83
Fax: 32 2 514 33 86
Contact name: Mark Dubrulle, Executive Vice President
Geographical coverage: International
Aims and objectives: Founded in 1976, the IPRE aims to provide a forum to discuss environmental issues through workshops and symposia
Publications: *Proceedings*, semi annual; newsletter

– 1857 – Media Natura (MN)

21 Tower Street
London WC2H 9NS
United Kingdom
Tel: 071 240 49 36
Fax: 071 240 22 91
Contact name: Michael Keating, Director
Geographical coverage: United Kingdom
Aims and objectives: A UK charity, established in 1988, MN utilises the support of professionals in the media industry to help environment and development groups communicate their messages more effectively. It runs the annual British Environment and Media Awards (BEMAs)
Publications: Bulletin twice a year; media reports; *Commercial Communications Charter for the Environment*

– 1858 – Scientists for Global Responsibility (SGR)

Unit 3
Down House
Broomhill Road
London SW18 4QJ
United Kingdom
Tel: 081 871 5175
Fax: 081 877 1940
Geographical coverage: United Kingdom
Aims and objectives: SGR, incorporating Scientists against Nuclear Arms and Psychologists for Peace, is an independent organisation, formed to combat the irresponsible use of science and technology. Its objectives are to promote and coordinate research and education

– 1859 – Skakavac

Dobrovoljacka 3
71000 Sarajevo
Bosnia Hercegovina

Tel: 38 71 51 27 98
Fax: 38 71 21 62 38
Geographical coverage: Bosnia Hercegovina
Aims and objectives: Environmental group
Publications: *Eko Oko*

– 1860 – Svensk Polska Miljoforeningen (SPM): Szwedsko Polskie Towarzystwo Ochrony Srodowiska

(The Swedish Polish Association for Environment Protection)

SNV
S-170 11 Drottingholm
Sweden
Contact name: Zofia Kukulska
Geographical coverage: Sweden
Aims and objectives: SPM helps to stimulate cooperation between organisations, enterprises, authorities and individuals in Sweden and Poland interested in the protection of the environment

– 1861 – United Nations Environment Programme (UNEP)

PO Box 30552
Nairobi
Kenya
Kenya
Tel: 254 2 333930
Fax: 254 2 520711
Contact name: Dr Mostafa K Tolba, Executive Director
Geographical coverage: International
Aims and objectives: UNEP was established in 1972. Its role is to coordinate, catalyse and stimulate environmental action primarily within the UN organisation. It works to identify gaps where little or nothing is being done, and to stimulate the necessary action

– 1862 – Youth Ecological Movement of Byelorussia

c/o Sergei Dorozhko
40 Marxa
SU-220600 Minsk GSP
Byelorussia
Tel: 7 0172 39 91 29
Fax: 7 0172 29 35 72
Geographical coverage: Byelorussia

PRESSURE GROUPS

– 1863 – Amici della Terre

Via del Sudario 35
I-00186 Rome
Italy
Tel: 39 6 6875308/6868289
Fax: 39 6 683 08610
Contact name: Laura Radiconcini
Geographical coverage: Italy
Aims and objectives: FoE environmental pressure group

– 1864 – Amigos da Terra

Travessa Marques de Sampaio 44-Rc
P-1200 Lisbon
Portugal
Tel: 351 1 347 9599
Fax: 351 1 347 3586
Contact name: Mário Alves
Geographical coverage: Portugal
Aims and objectives: FoE environmental pressure group

– 1865 – Amigos de la Tierra

Juan Pradillo 26-1
E-28039 Madrid
Spain
Tel: 34 1 311 2186/311 2437
Fax: 34 1 311 4874
Contact name: Sandy Hemingway
Geographical coverage: Spain
Aims and objectives: FoE environmental pressure group

– 1866 – Amis de la Terre, Les

Place de la Vingeanne
B-5100 Dave
Belgium
Tel: 32 81 401 478
Fax: 32 81 402 354
Contact name: Alain Hanssen
Geographical coverage: Belgium
Aims and objectives: FoE environmental pressure group

– 1867 – Amis de la Terre, Les

38 rue Mesley
F-75003 Paris
France
Tel: 33 1 488 733 44
Fax: 33 1 488 728 23
Contact name: Guy Aznar/Pierre Samuel
Geographical coverage: France
Aims and objectives: FoE environmental pressure group

– 1868 – Bund

Postfach 30 02 20
D-5300 Bonn 3
Germany
Tel: 49 228 400970
Fax: 49 228 4009740
Contact name: Arno Behlau
Geographical coverage: Germany
Aims and objectives: Environmental pressure group

– 1869 – Earthwatch

Harbour View
Bantry
Cork
Eire
Tel: 353 27 50968/51283
Fax: 353 27 50545
Contact name: Jeremy Wates
Geographical coverage: Ireland
Aims and objectives: Environmental pressure group

– 1870 – Ecoglasnost

Dondukov Blv 39
1000 Sofia
Bulgaria
Tel: 359 2 80 23 23
Fax: 359 2 88 15 30
Contact name: Paula Rakowska
Geographical coverage: Bulgaria
Aims and objectives: Environmental pressure group

– 1871 – Estonian Green Movement

Box 318
EE2400 Tartu
Estonia
Tel: 7 014 34 30198/73517
Fax: 7 014 34 35440
Contact name: Tõnu Oja
Geographical coverage: Estonia
Aims and objectives: Environmental pressure group

– 1872 – Freunde der Erde

Mariahiferstrasse 105/2/1/13
A-1060 Vienna
Austria
Tel: 43 1 222 597 1443
Fax: 43 1 222 597 3743
Contact name: Sonia Arnecke
Geographical coverage: Austria
Aims and objectives: FoE environmental pressure group

– 1873 – Friends of the Earth (FoE)

26–28 Underwood Street
London N1 7JQ
United Kingdom
Tel: 071 490 2665
Fax: 071 490 0881
Contact name: Andrew Dilworth
Geographical coverage: United Kingdom
Aims and objectives: FoE believes in informing and empowering the public over environmental issues, and encourages people to take action through its network of over 300 local groups and its youth section
Publications: *Earth Matters*, magazine quarterly. Free publications list available on request

– 1874 – Friends of the Earth (FoE)

PO Box 3411
Limassol
Cyprus
Tel: 357 5 34 70 42
Fax: 357 5 34 70 43
Contact name: Solomon A Ioannou
Geographical coverage: Cyprus

– 1875 – Friends of the Earth International (FoEI)

PO Box 19199
NL-1000 GD Amsterdam
Netherlands
Tel: 31 20 622 1369
Fax: 31 20 627 5287
Contact name: Eka Morgan, Information Officer
Geographical coverage: International
Aims and objectives: FoEI was founded in 1971 and now has member groups in 33 countries around the world. Each country member group is autonomous and each is bound in a common cause – the conservation, restoration and rational use of the Earth's resources
Publications: List available on request

– 1876 – Friends of the Earth International (FoEI): European Coordination – CEAT

Rue Blanche 29
B-1050 Brussels
Belgium
Tel: 32 2 537 72 28
Fax: 32 2 537 55 96
Contact name: Inge Nalbach, Coordinator
Geographical coverage: International
Aims and objectives: The Brussels office is the regional one linking members of FoE in Europe. FoE works to promote conservation, restoration and rational use of the environment and the earth's natural resources
Publications: *FoE Link*, semi annual

– 1877 – Green Alternative European Link (GRAEL)

European Parliament Office
Rue Belliard 97–113
B-1047 Brussels
Belgium
Tel: 32 2 284 30 44
Fax: 32 2 230 78 37
Contact name: Maurizeo Cancelmo
Geographical coverage: European
Aims and objectives: Created in 1984, GRAEL has 11 member countries. It is part of the Rainbow Group and works towards complete disarmament, a no compromise environment protection policy, equal rights for all areas of society and free exercise of civil rights

– 1878 – Greenpeace Austria

Auenbruggergasse 2
A-1030 Vienna
Austria
Tel: 43 1 713 00 31
Fax: 43 1 713 00 30
Geographical coverage: Austria
Aims and objectives: Environmental pressure group

– 1879 – Greenpeace Belgium

Vooruitgangstraat 317
B-1210 Brussels
Belgium
Tel: 32 2 215 19 44
Fax: 32 2 215 19 50
Geographical coverage: Belgium
Aims and objectives: Environmental pressure group

– 1880 – Greenpeace Czechoslovakia

U Prasne Brany 3
110 00 Prague 1
Czech Republic
Tel: 42 2 26 95 16
Fax: 42 2 22 17 89
Geographical coverage: Czech Republic
Aims and objectives: Environmental pressure group

– 1881 – Greenpeace Denmark

Linnesgade 25
DK-1361 Copenhagen K
Denmark
Tel: 45 33 93 53 44
Fax: 45 33 93 53 99
Geographical coverage: Denmark
Aims and objectives: Environmental pressure group

– 1882 – Greenpeace Finland

Kirkokatu 28
Helsinki
Finland
Tel: 358 1 66 19 92
Fax: 358 1 66 18 99
Geographical coverage: Finland
Aims and objectives: Environmental pressure group

– 1883 – Greenpeace France

28 rue des Petites Ecuries
F-75010 Paris
France
Tel: 33 1 477 046 89
Fax: 33 1 477 046 91
Geographical coverage: France
Aims and objectives: Environmental pressure group

– 1884 – Greenpeace Germany

Vorsetzen 53
D-2000 Hamburg 11
Germany
Tel: 49 40 31186–0
Fax: 49 40 31186–141
Geographical coverage: Germany
Aims and objectives: Environmental pressure group

– 1885 – Greenpeace Greece

44 Kalidromiou Street
GR-114 73 Athens
Greece
Tel: 30 1 3640774
Fax: 30 1 3604008
Geographical coverage: Greece
Aims and objectives: Environmental pressure group

– 1886 – Greenpeace International

Stichting Greenpeace Council
Keizergracht 176
NL-1016 DW Amsterdam
Netherlands
Tel: 31 20 523 65 55
Fax: 31 20 523 65 00
Geographical coverage: International
Aims and objectives: Greenpeace is an international, independent environmental pressure group which acts against abuse in the natural world. It has offices in 20 countries and a worldwide membership of over 3,000,000 supporters

– 1887 – Greenpeace International EC Unit

Avenue de Tervuren 36
B-1040 Brussels
Belgium
Tel: 32 2 736 99 27
Fax: 32 2 736 44 60
Contact name: Frank Schwalba-Hoth, Coordinator
Geographical coverage: International
Aims and objectives: Set up in
1971, Greenpeace believes in using direct action in its campaigns to protect wildlife and the environment against destructive processes. The policy followed is one of complete non-violence, and its actions cause no injury

– 1888 – Greenpeace Ireland

44 Upper Mount Street
Dublin 2
Eire
Tel: 353 1 619836 .
Fax: 353 1 605258
Geographical coverage: Ireland
Aims and objectives: Environmental pressure group

– 1889 – Greenpeace Italy

28 Via Manlio Gelsomini
I-00153 Rome
Italy
Tel: 39 6 578 11 73
Fax: 39 6 578 35 31
Geographical coverage: Italy
Aims and objectives: Environmental pressure group

– 1890 – Greenpeace Luxembourg

22 rue Dicks
L-4081 Esch/Alzette
Luxembourg
Tel: 352 546 252
Fax: 352 545 405
Geographical coverage: Luxembourg
Aims and objectives: Environmental pressure group

– 1891 – Greenpeace Netherlands

Keizergracht 174
NL-1016 DW Amsterdam
Netherlands
Tel: 31 20 626 18 77
Fax: 31 20 622 12 72
Geographical coverage: Netherlands
Aims and objectives: Environmental pressure group

– 1892 – Greenpeace Norway

St Olavsgt 11
PO Box 6803
St Olavs Plass
N-0130 Oslo 1
Norway
Tel: 47 2 205101
Fax: 47 2 205114
Geographical coverage: Norway
Aims and objectives: Environmental pressure group

– 1893 – Greenpeace Russia

Ulitsa Kalyayevskaya 21
103006 Moscow
Russia
Tel: 7 095 251 9073
Geographical coverage: Russia
Aims and objectives: Environmental pressure group

– 1894 – Greenpeace Spain

Rodriquez San Pedro 58
4 Pisa
E-28015 Madrid
Spain
Tel: 34 1 543 47 04
Fax: 34 1 543 97 79
Geographical coverage: Spain
Aims and objectives: Environmental pressure group

– 1895 – Greenpeace Sweden

Box 8913
S-402 73 Gothenburg
Sweden
Tel: 46 31 222255
Fax: 46 31 232429
Geographical coverage: Sweden
Aims and objectives: Environmental pressure group

– 1896 – Greenpeace Switzerland

Muellerstrasse 37
Postfach 276
CH-8026 Zurich
Switzerland
Tel: 41 1 2413441
Fax: 41 1 2413821
Geographical coverage: Switzerland
Aims and objectives: Environmental pressure group

– 1897 – Greenpeace UK

Canonbury Villas
London N1 2PN
United Kingdom
Tel: 071 354 5100/359 7396
Fax: 071 359 4372/359 4062
Geographical coverage: United Kingdom
Aims and objectives: Greenpeace is an international, independent environmental pressure group which acts against abuse to the natural world
Publications: *Campaign Report*, newsletter; campaigning brief for adults; environmental factsheets for children. List available on request

– 1898 – Greenpeace Ukraine

Hotel Kiev
Room 707
26-1 Grushevsky Str
252021 Kiev
Ukraine
Tel: 7 044 293 3261
Geographical coverage: Ukraine
Aims and objectives: Environmental pressure group

– 1899 – International Physicians for the Prevention of Nuclear War (IPPNW)

126 Rogers Street
Cambridge
MA 02142
United States of America
Tel: 1 67 868 5050
Fax: 1 67 868 2560
Geographical coverage: International
Aims and objectives: Nuclear Disarmament Group

– 1900 – Irish Women's Environmental Network

4 Naffau Street
Dublin 2
Eire
Tel: 353 1 6795123
Geographical coverage: Ireland
Aims and objectives: Environmental pressure group

– 1901 – Jordens Vanner

Fjällgatan 23A
S-116 28 Stockholm
Sweden
Tel: 46 8 7022017
Fax: 46 8 6426261
Contact name: Charles Berkow
Geographical coverage: Sweden
Aims and objectives: Environmental pressure group

– 1902 – Mouvement Ecologique Luxembourg (MECO)

(Luxembourg Ecological Movement)

5 rue St Nicolas
L-9263 Diekirch
Luxembourg
Tel: 352 808 935
Fax: 352 808 935
Contact name: Mr Claude Turmes
Geographical coverage: Luxembourg
Aims and objectives: MECO works towards lobbying the government and providing information on many aspects of the environment

– 1903 – NOAH International

Norrebrogade 39 1
DK-2200 Copenhagen N
Denmark
Tel: 45 35 36 12 12
Fax: 45 35 36 12 17
Contact name: Mads Teisen
Geographical coverage: Denmark
Aims and objectives: Environmental pressure group

– 1904 – Norges Naturvernforbund (NNV)

PO Box 2113
Grünerlokka
N-0505 Oslo 5
Norway
Tel: 47 2 715520
Fax: 47 2 715640
Contact name: Dag Hareide/Gunnar Album
Geographical coverage: Norway
Aims and objectives: Environmental pressure group

– 1905 – Polski Klub Ekologiczny

Krzystof Kamieniecki
Miedzynardowa 32/34A m167
PL-03 922 Warsaw
Poland
Geographical coverage: Poland
Aims and objectives: Environmental pressure group

– 1906 – Scottish Campaign to Resist the Atomic Menace (SCRAM)

11 Forth Street
Edinburgh
Scotland EH1 3LE
United Kingdom
Tel: 031 557 4283
Fax: 031 557 4284
Geographical coverage: United Kingdom
Aims and objectives: SCRAM aims to oppose nuclear power and advocate a sane, environmentally sound energy policy
Publications: *Safe Energy Journal*, monthly

– 1907 – Slovak Union of Nature and Landscape Protectors (SZOPK)

Gorkeho 6
811 01 Bratislava
Slovakia
Tel: 42 7 506 65
Fax: 42 7 506 65
Contact name: Jozef Gregor/Mikulas Huba
Geographical coverage: Slovakia
Aims and objectives: Environmental pressure group

– 1908 – Vereniging Milieudefensie

Damrak 26
NL-7102 LJ Amsterdam
Netherlands
Tel: 31 20 622 13 66
Fax: 31 20 627 52 87
Contact name: Teo Wams
Geographical coverage: Netherlands
Aims and objectives: Environmental pressure group

– 1909 – Women's Environmental Network (WEN)

Schitwc q 1c
CH-3645 Gwatt
Switzerland
Geographical coverage: Switzerland
Aims and objectives: Environmental pressure group

– 1910 – Women's Environmental Network (WEN)

Aberdeen Studios
22 Highbury Grove
London N2 2EA
United Kingdom
Tel: 071 354 8823
Fax: 071 354 0464
Geographical coverage: United Kingdom
Aims and objectives: The Women's Environmental Network is one of UK's leading environment pressure groups. It is a registered charity, educating, informing and empowering women who care about the environment
Publications: Various

– 1911 – Zelenyi Svit

Kontaktova 4
Kiev 70
Ukraine
Tel: 7 044 416 5218/417 0283
Fax: 7 044 417 4383/440 3017
Geographical coverage: Ukraine
Aims and objectives: Environmental pressure group

BOOKS

– 1912 – Environmental Action in Eastern Europe: *see entry 2073*

Geographical coverage: United Kingdom

– 1913 – Realism in Green Politics

Geographical coverage: Germany
Publishers: Manchester University Press, 1992
Coverage: This study provides a critique of ecological fundamentalism based on the experience of West German Green politics
Price: £35.00
ISBN: 0719037018
Author: Wiesenthal, H.

– 1914 – States and Anti-Nuclear Movements

Geographical coverage: United Kingdom
Publishers: Edinburgh University Press, 1993
Coverage: Using case studies from many countries in Europe, this book compares government reactions to anti-nuclear movements in Western Europe
Price: £50.00
ISBN: 0748603964
Author: Flam, H.

DIRECTORIES/YEARBOOKS

– 1915 – Anuario Profesional del Medio Ambiente 1993

Tel: 34 1 541 6768
Geographical coverage: Spain
Publishers: SPA S L
Frequency: annual
Coverage: Main Spanish source covering the environmental sector. Includes some 1,700 addresses of organisations related to the environment
Price: £30.00
ISBN: 8460447111

– 1916 – Directory of European Environmental Organisations

Geographical coverage: European
Publishers: Blackwell Publishers, 1991
Language: English
Coverage: Lists organisations in Europe that are concerned with the environment
ISBN: 0631183868
Author: Deziron, M. and Bailey, L.

– 1917 – EC Environmental Guide

Avenue des Arts 50
Bte 5
B-1040 Brussels
Belgium
Tel: 32 2 513 68 92
Geographical coverage: European
Publishers: EC Committee of the American Chamber of Commerce
Frequency: annual
Language: English
Coverage: A guide to organisations and contacts on the environment in the European Community. Also lists some main organisations country by country
Price: £32.00

– 1918 – Environnement: Le Guide des Sources d'Information

Geographical coverage: France
Publishers: Publié par a Jour, 1991
Coverage: Environmental institutions and organisations
ISBN: 2903685371
Author: Clicquot de Mentque, C. and Fauchet, J.

– 1919 – NGO Directory for Central and Eastern Europe 1992

Aloys-Schultestrasse 6
D-5300 Bonn 1
Germany
Geographical coverage: European
Publishers: Institut für Europäische Umweltpolitik

Editor: Juras, A.
Language: English
Coverage: Lists NGO's in Central and Eastern Europe which are concerned with the environment
Price: DM40
ISBN: 3893470077

– 1920 – Who's Who in the Environment – England

Geographical coverage: United Kingdom
Publishers: The Environment Council, 1992
Coverage: Organisations in England which are concerned with the environment at regional or national level
ISBN: 0903158388
Additional information: Available on CD–ROM

– 1921 – Who's Who in the Environment – Scotland

Geographical coverage: United Kingdom
Publishers: The Environment Council, 1989
Coverage: Organisations in Scotland which are concerned with the environment at regional or national level
ISBN: 1853970903

– 1922 – Who's Who in the Environment – Wales

Geographical coverage: United Kingdom
Publishers: The Environment Council, 1991
Coverage: Organisations in Wales which are concerned with the environment at regional or national level
ISBN: 0903115836

– 1923 – World Directory of Environmental Organisations

Geographical coverage: International
Publishers: Institute of Public Affairs, 1990
Coverage: Lists organisations related to the environment
ISBN: 091210287X

– 1924 – World Guide to Environmental Issues and Organisations

Geographical coverage: International
Publishers: Longmans, 1990
Editor: Brackley, P.
Language: English
Coverage: Lists major organisations concerned with the environment
ISBN: 0582062705

PERIODICALS

– 1925 – Econews

10 Station Parade
Balham High Road
London SW12 9AL
United Kingdom
Tel: 081 673 0045
Geographical coverage: United Kingdom
Publishers: The Green Party
Frequency: 6 issues a year
Coverage: UK Green politics and environmental issues
Annual subscription: £4.00

– 1926 – Ökolgiepolitik

Kaiserplatz 17
D-5300 Bonn 1
Germany
Tel: 49 228 261633
Fax: 49 228 211188
Geographical coverage: Germany
Publishers: Ökologisch Demokratische Partei
Editor: Guhde, E.
Frequency: bi monthly
Coverage: Political ecology and party related information for party members and others interested in environmental politics
Price: DM20

Environmental Laws and Legislation

GOVERNMENT ORGANISATIONS

– 1927 – Direção Geral de Energia
(General Directorate for Energy)

241 Rua da Beneficencia
P-1093 Lisbon
Portugal
Tel: 351 1 771091
Geographical coverage: Portugal
Aims and objectives: Energy planning; legislation; alternative sources of energy; energy policy; energy conservation; energy statistics

– 1928 – Miljoministeriet Miljostyrelsen Kemikaliekontrollen
(The State Chemical Supervision Service)

12 Skovbrynet
DK-2800 Lyngby
Denmark
Tel: 45 02 87 70 66
Contact name: Hedegaard Poulsen, Director
Geographical coverage: Denmark
Aims and objectives: Controls and observes pesticide and toxic substance legislation

– 1929 – Ministerio de Industria Comercio y Turismo

Paseo de la Castellana 160
Planta 6
E-28046 Madrid
Spain
Tel: 34 1 3494545
Fax: 34 1 4578066
Geographical coverage: Spain
Aims and objectives: The Ministry is concerned with all matters relating to law and the exploitation of hydrocarbon reserves

NON-GOVERNMENT ORGANISATIONS

– 1930 – British Non-Ferrous Metals Association: see entry 1674

Geographical coverage: United Kingdom

– 1931 – British Scrap Federation

16 High Street
Brampton
Huntingdon
Cambridgeshire PE18 8TU
United Kingdom
Tel: 0480 455249
Fax: 0480 453680
Contact name: J A Clubb, Executive Director
Geographical coverage: United Kingdom
Aims and objectives: The Federation advises on health and safety and new Government legislation
Publications: *News Review Newsletter*

– 1932 – Centre for International Environmental Law (CIEL)

Kings College London
Manresa Road
London SW3 6LX
United Kingdom
Tel: 071 352 8123
Fax: 071 351 6453
Contact name: Mr Philip Sands, Director
Geographical coverage: International
Aims and objectives: CIEL was established in order to promote international law as a means of protecting the global environment through teaching, research and legal advice
Publications: *CIEL Discussion Papers*

– 1933 – Conseil Européen du Droit de l'Environnement (CEDE)
(European Council on Environmental Law)

Politiques et Sociales
Place d'Athènes
F-67084 Strasbourg
France
Geographical coverage: European

– 1934 – Danish Toxicology Centre (DTC)

Agern Allee 15
DK-2970 Hoersholm
Denmark
Tel: 45 42 57 00 55
Fax: 45 42 57 19 19
Contact name: Ms Lisbeth Valentin Hansen
Geographical coverage: Denmark
Aims and objectives: DTC aims to procure, assess and communicate toxicological data and aspects regarding chemical legislation

– 1935 – Ecological Studies Institute (ESI)

49 Wellington Street
London WC2E 7BN
United Kingdom
Tel: 071 836 0341
Fax: 071 240 9205
Geographical coverage: European
Aims and objectives: ESI monitors environmental legislation, policy changes and business investment which affects the environment in Eastern Europe
Publications: *East West Environment Newsletter*, monthly

– 1936 – Environmental Law Association

17 Marni Street
GR-14562 Kifissia
Athens
Greece
Geographical coverage: Greece

– 1937 – European Transport Law

Maria Henriettalei 1
B-2018 Antwerp
Belgium
Tel: 32 3 2313655
Fax: 32 3 23442380
Geographical coverage: European
Aims and objectives: Concerned with transport of dangerous goods and transfer of dangerous waste

– 1938 – Global Legislators Organisation for a Balanced Environment (GLOBE): GLOBE International/EC

Rue du Taciturne 50
B-1040 Brussels
Belgium
Tel: 32 2 230 65 89
Fax: 32 2 230 95 30
Contact name: François Roelants du Vivier, Director
Geographical coverage: European
Aims and objectives: GLOBE was created in 1989, with the aim of working towards improving the state of the global environment through the legislative process and other means

– 1939 – International Council of Environmental Law (ICEL): Environmental Law Center

Adenauerallee 214
D-5300 Bonn 1
Germany
Tel: 49 228 269 22 40
Fax: 49 228 269 22 50
Contact name: Eric Howard, Information Officer
Geographical coverage: European
Aims and objectives: Founded in 1969, ICEL endeavours to foster the exchange and dissemination of information about environmental law, policy and administration and encourage advice and assistance among its membership
Publications: *Environmental Policy and Law*, semi annual

– 1940 – Internationale Scharrelvees Controle (ISC)

(International Eco Meat Control)

PO Box 649
NL-3500 AP Utrecht
Netherlands
Tel: 31 30 340811
Geographical coverage: European
Aims and objectives: ISC is an independent organisation which controls member farmers and butchers on the observance of the eco regulations

– 1941 – National Westminster Bank

41 Lothbury
London EC2P 2BP
United Kingdom
Tel: 071 726 1631
Contact name: Tim Blythe, Deputy Head of Group Media Relations
Geographical coverage: United Kingdom
Aims and objectives: The Bank offers a computer advisory service on environmental legislation which businesses can consult when compiling their environmental plans

– 1942 – Stichting Natuur en Milieu (SNM)

(Netherlands Society for Nature and the Environment)

Donkerstraat 17
NL-3511 KB Utrecht
Netherlands
Tel: 31 30 33 13 28
Fax: 31 30 33 13 11
Contact name: Mr Ralph Hallo, Coordinator International Affairs
Geographical coverage: Netherlands
Aims and objectives: SNM's principal work is the analysis of policy and legislation and advocacy in the interests of nature and environmental protection

BOOKS

– 1943 – Corporate Environmental Responsibility: Law and Practice

Geographical coverage: United Kingdom
Publishers: Butterworth Law Publishers Ltd
Coverage: Introduction to corporate environmental law

Price: £65.00
ISBN: 0406001383
Author: Salter, J. R.

– 1944 – Environmental and Health Controls on Lead

Geographical coverage: International
Publishers: International Lead and Zinc Study Group, 1989
Language: English
Coverage: A review of current and proposed regulations controlling lead in works, lead in the atmosphere and lead in water in 25 countries
Price: £20.00

– 1945 – Environmental Dispute Handbook

Geographical coverage: International
Publishers: John Wiley & Sons, 1991
Coverage: This text contains over 50 US Federal Statutes and the relevant passages from the Code of Federal Regulation, related to various aspects of environmental protection
Price: £147.20
ISBN: 0471525863
Author: Carpenter, D. A. and Cushman, R. F.

– 1946 – Environmental Law

Geographical coverage: United Kingdom
Publishers: Blackstone Press, 1991
ISBN: 1854311174
Author: Ball, S. and Bell, S.

– 1947 – Environmental Law in UNEP

Geographical coverage: International
Publishers: UNEP Environmental Law and Institution Unit, 1991
ISBN: 9280712969

– 1948 – Europe and the Environment:
see entry 2077

Geographical coverage: European

– 1949 – European Community Environmental Legislation 1967–1987

Geographical coverage: European
Publishers: Office for Official Publications of the European Communities (OPOCE)
Coverage: Available free in all languages except English from the Directorate-General XI, Commission of the European Communities, Brussels, 1987. These volumes are being updated and will be published by OPOCE

– 1950 – Garners Environmental Law

Geographical coverage: United Kingdom
Publishers: Butterworths
Editor: Garner, J. F.
Price: £130.00
ISBN: 0406205620

– 1951 – International Environmental Law

Geographical coverage: International
Publishers: Clarendon Press, 1992
Coverage: Deals with international environmental law
Price: £45.00
ISBN: 0198762828
Author: Birnie, P. and Boyle, A.

– 1952 – Legal 500, The, 5th ed

Geographical coverage: United Kingdom
Publishers: Legalease Ltd, 1992
ISBN: 187085408X
Author: Pritchard, J.

– 1953 – Neurotoxicity

Geographical coverage: United Kingdom
Publishers: Van Nostrand Reinhold International, 1992
Coverage: This text aims to clarify what neurotoxins are, how they affect the brain, how government policies regulate chemicals, how to test, monitor and assess for risks
Price: £47.00
ISBN: 0442010478

– 1954 – Transnational Environmental Liability and Insurance

Geographical coverage: International
Publishers: Graham & Trotman Publishers, 1992
Coverage: A review and analysis of the critical issues as brought to a head in several recent pollution disasters (such as Exxon Valdez, Chernobyl and the Gulf War) which face the international community
Price: £78.00
ISBN: 1853337781
Author: Kroner, R. R.

– 1955 – UK Marine Oil Pollution Legislation

Geographical coverage: United Kingdom
Publishers: Lloyds of London Press, 1987
Editor: Bates, J. H.
ISBN: 1850044109X

– 1956 – UK Waste Law

Geographical coverage: European
Publishers: Sweet and Maxwell, 1992

Coverage: Provides an overview of waste management law in the EC and the UK
Price: £35.00
ISBN: 0421430109
Author: Bates, J. H.

– 1957 – Umwelthaftung und Umweltstrafrecht

(Environmental Liability and Environmental Criminal Law)

Geographical coverage: Germany
Publishers: Heinz Hillermeier, Regierungsdirektor Landratsamt Neustadt/Aisch–Bad Windsheim
Coverage: Reference book on legislation, jurisdiction and administration
Price: DM73

– 1958 – Waste, Wastewater, Air Laws and Technology

Geographical coverage: Germany
Publishers: Bohman Druck und Verlag Gesellschaft mbH & Co KG
Language: German
Author: List, W. and Kuntscher, H.

– 1959 – Water and Drainage Law

Geographical coverage: United Kingdom
Publishers: Sweet and Maxwell, 1990
ISBN: 0421387009
Author: Bates, J. H.

PERIODICALS

– 1960 – Environmental Law Brief

Ludgate House
107 Fleet Street
London EC4A 2AB
United Kingdom
Tel: 071 936 2016
Fax: 071 936 2303
Geographical coverage: United Kingdom
Publishers: Legal Studies and Services (Publishing) Ltd
Editor: Battersby, S.
Frequency: 10 issues a year
Coverage: A concise update for companies on the impact of environmental regulation and law
Annual subscription: £95.00
ISSN: 0959 0617

– 1961 – European Environmental Law Review

Sterling House
66 Wilton Road
London SW1V 1DE
United Kingdom
Tel: 071 821 1123
Fax: 071 630 5229
Geographical coverage: European
Publishers: Graham and Trotman
Editor: Bramwell, Mrs E.
Frequency: monthly
Coverage: Environmental law throughout Europe
Annual subscription: £97.50
ISSN: 0966 1646

– 1962 – Journal of Environmental Law

Pinkhill House
Southfield Road
Eynsham
Oxford OX8 1JJ
United Kingdom
Tel: 0865 882283
Fax: 0865 882890
Geographical coverage: United Kingdom
Publishers: Oxford University Press
Editor: Macrory, R.
Frequency: 2 issues a year
Coverage: Environmental law in pollution control, wastes management, habitat protection, biotechnology and regimes for common natural resources
Annual subscription: £45.00
ISSN: 0952 8873

– 1963 – Journal of Planning and Environmental Law

11 New Fetter Lane
London EC4P 4EE
United Kingdom
Tel: 071 583 9855
Geographical coverage: United Kingdom
Publishers: Sweet and Maxwell Ltd
Frequency: monthly
Coverage: Planning and environment law
ISSN: 0307 4870

– 1964 – Manufacturing Engineering

Box 930
Dearborn
MI 48121-0930
United States of America
Tel: 1 313 271 1500
Fax: 1 313 271 2861
Geographical coverage: International
Publishers: Society of Manufacturing Engineers
Editor: Coleman, J.
Frequency: monthly

Language: English
Coverage: Explains what recent legislation including the Environmental Protection Act, means for factory waste
Price: $60
ISSN: 0361 0853

– 1965 – Natura e Societa

Borgo Felino 54
Parma
Italy
Tel: 39 521 481871
Geographical coverage: Italy
Publishers: Federazione Nazionale pro Natura
Editor: de Marchi, A.
Frequency: quarterly
Coverage: Presents articles on the protection of nature, naturalistic education, current events in nature, conservative and legislative laws for nature
Annual subscription: Lit5000
ISSN: 0393 8875

– 1966 – Natural Environment Research Council News: see entry 2141

Geographical coverage: United Kingdom

– 1967 – Nuclear Law Bulletin

38 boulevard Suchet
F-75016 Paris
France
Tel: 33 1 452 496 67
Geographical coverage: International
Publishers: OECD Nuclear Energy Agency
Language: English
Coverage: Covers legislative and regulatory developments, agreements and case law in the nuclear field throughout the world
Annual subscription: Ffr170
ISSN: 0304 341X

– 1968 – Revue Juridique de l'Environnement

11 rue d'Algerie
F-69001 Lyon
France
Geographical coverage: France
Publishers: Publications Périodiques Spécialisées
Editor: Prieur, M.
Frequency: quarterly
Annual subscription: Ffr280

REPORTS

– 1969 – Climate Change: Designing a Tradeable Permit System

2 rue André Pascal
F-75775 Paris Cedex
France
Geographical coverage: International
Publishers: OECD, 1992
Language: English
Coverage: This report contains technical papers presented at a workshop on tradeable permits organised by OECD. A second volume discussing the use of taxes will also be published

– 1970 – Managing the Environment: see entry 2017

Geographical coverage: United Kingdom

DATABASES

– 1971 – ABEL

Geographical coverage: European
Coverage: Official Journal of the EC, L series
Database producer: Office for Official Publications of the EC
Database host: EUR–OP
Update frequency: daily

– 1972 – Celex

Geographical coverage: European
Coverage: All EC legislation including treaties and agreements, between member states of the EC
Database host: Eurobases Context; FT Profile; Mead Data Central (via Nexis)
Coverage period: 1951–
Update frequency: monthly

– 1973 – European Parliament Online Query Service (EPOQUE)

Geographical coverage: European
Coverage: EPOQUE is the European Parliament Library catalogue
Database producer: European Parliament
Database host: European Parliament
Coverage period: 1979–
Update frequency: daily

– 1974 – European Update: see entry 2117

Geographical coverage: European

– 1975 – LEXIS

Geographical coverage: International
Coverage: LEXIS incorporates UK Law library and US Environmental Law Library databases. Includes data from European and Commonwealth law libraries, English General Library and English Local Government Library
Database host: Mead Data Central Inc
Coverage period: c.1970–
Update frequency: monthly

– 1976 – SPEARHEAD

Geographical coverage: European
Coverage: 1992 related legislation summarised, including proposals, research and development
Database producer: UK DTI
Database host: Context; DataStar; FT Profile
Coverage period: 1985–

Environmental Management

GOVERNMENT ORGANISATIONS

– 1977 – Bundesministerium für Umwelt Jugend and Familie

(Ministry of the Environment Youth and Family)

Radetzkystrasse 2
A-1030 Vienna
Austria
Tel: 43 1 711 580
Fax: 43 1 711 584221
Geographical coverage: Austria

– 1978 – Department of the Environment

Customs House
Dublin 1
Eire
Tel: 353 1 6793377
Fax: 353 1 779278
Geographical coverage: Ireland
Aims and objectives: The Department develops policies and legislation to promote, oversee and assist the efficient and effective provision of inter-related services; including environmental protection

– 1979 – Department of the Environment (DoE)

2 Marsham Street
London SW1P 3EB
United Kingdom
Tel: 071 276 3000
Geographical coverage: United Kingdom
Aims and objectives: The Department's responsibilities include local government structure and finance; land use planning; housing and construction; conservation of the built and natural heritage; environmental protection and water; and sport and recreation
Publications: Various publications

Additional information: Library Services Headquarters; also focal point for UNEP's INFOTERRA

– 1980 – Federal Ministry for the Environment Federal Republic of Germany

Kennedyallee 5
D-5300 Bonn 2
Germany
Tel: 49 228 3053227
Fax: 49 228 3052044
Geographical coverage: Germany

– 1981 – Miljoministeriet Miljostyrelsen

(Ministry of the Environment)

Strandgade 29
DK-1401 Copenhagen K
Denmark
Tel: 45 31 57 83 10
Fax: 45 31 57 24 49
Geographical coverage: Denmark

– 1982 – Miljoverndepartementet

(Ministry of Environment)

Myntgaten 2
PO Box 8013
Department N-0030 Oslo 1
Norway
Tel: 47 2 234909
Fax: 47 2 234956
Geographical coverage: Norway
Additional information: Library Service

– 1983 – Ministère de l'Environnement et des Eaux et Forêts

51 rue de Prague
L-2918 Luxembourg-Ville
Luxembourg
Tel: 352 478 870
Geographical coverage: Luxembourg

– 1984 – Ministère de l'Environnement et du Cadre de Vie

14 boulevard du Général Leclerc
F-92524 Neuilly-sur-Seine
France
Geographical coverage: France

– 1985 – Ministerio de Ambiente e Recursos Naturais

Rua de Seculo
P-1200 Lisbon
Portugal
Tel: 351 1 356 2751
Geographical coverage: Portugal

– 1986 – Ministerio de Obras Públicas y Urbanismo – Dirección General de Medio Ambiente (DGMA)

Nuevos Ministerios
Paseo de la Castellana 67
E-28071 Madrid
Spain
Tel: 34 1 553 16 00
Fax: 34 1 543 65 77
Geographical coverage: Spain

– 1987 – Ministero per i Beni Culturali e Ambientali

Via del Collejio Romano 27
I-00186 Rome
Italy
Tel: 39 6 6723
Geographical coverage: Italy

– 1988 – Ministerstwo Ochrony Srodowiska i Zasobow Naturalriych

ul Wawelska 52–54
PL-02 067 Warsaw
Poland
Tel: 48 22 253355
Geographical coverage: Poland

– 1989 – Ministry for Environment and Water Management

Fo utca 46–50
Pf 351
H-1011 Budapest
Hungary
Tel: 36 1 201 3843
Fax: 36 1 201 2846
Geographical coverage: Hungary

– 1990 – Ministry of Energy and Natural Resources

Ataturk Blvd
143 Daekanluklar
Turkey
Tel: 90 4 1174455
Fax: 90 4 1177971
Geographical coverage: Turkey

– 1991 – Ministry of Housing Physical Planning and Environment: Central Dep for Information and Interior Relations

PO Box 20951
NL-2500 EJ The Hague
Netherlands
Tel: 31 70 3264201
Geographical coverage: Netherlands
Publications: *Environmental News from the Netherlands*, quarterly
Additional information: Library Service

– 1992 – Ministry of the Environment

Wetstraat 56
B-1040 Brussels
Belgium
Tel: 32 2 238 28 11
Geographical coverage: Belgium

– 1993 – Ministry of the Environment

ul William Gladstone 67
1000 Sofia
Bulgaria
Tel: 359 2 876 151
Geographical coverage: Bulgaria

– 1994 – Ministry of the Environment

Slotsholmsgade 12
DK-1216 Copenhagen K
Denmark
Tel: 45 33 92 33 88
Geographical coverage: Denmark

– 1995 – Ministry of the Environment

Ratakatv 3
PO Box 399
SF-00531 Helsinki
Finland
Tel: 358 1 19911
Fax: 358 1 1991499
Geographical coverage: Finland

– 1996 – Ministry of the Environment

Solvholsgotu 4
150 Reykjavik
Iceland
Tel: 354 1 609650
Fax: 354 1 624566
Geographical coverage: Iceland

– 1997 – Ministry of the Environment

Piazza Venezia 11
I-00187 Rome
Italy
Tel: 39 6 67593204
Geographical coverage: Italy

– 1998 – Ministry of the Environment

Bd Libertatis 12
70005 Bucharest
Romania
Tel: 40 0 316104
Geographical coverage: Romania

– 1999 – Ministry of the Environment and Natural Resources

Tegelbacken 2
S-103 33 Stockholm
Sweden
Tel: 46 8 7631000
Fax: 46 8 241629
Geographical coverage: Sweden

– 2000 – Ministry of the Environment Housing and Public Works

17 Amaliados Street
Ambelokipi
GR-115 23 Athens
Greece
Tel: 30 1 6431461/9
Geographical coverage: Greece

– 2001 – Ministry of the Environment of the Slovak Republic

Hloboka 2
812 35 Bratislava
Slovakia
Tel: 42 7 424 515
Fax: 42 7 311 368
Contact name: Jozef Skultety, Director
Geographical coverage: Slovakia

– 2002 – Office of Public Works

St Stephens Green 51
Dublin 2
Eire
Tel: 353 1 613111
Fax: 353 1 610747
Geographical coverage: Ireland

NON-GOVERNMENT ORGANISATIONS

– 2003 – Advisory Committee for the Coordination of Information Systems (ACCIS)

Palais des Nations
CH-1211 Geneva 10
Switzerland
Tel: 41 22 798 8591
Geographical coverage: International
Aims and objectives: ACCIS was established in 1983 to facilitate access by member states to UN information. It is the primary UN information agency. ACCIS maintains its own database of computerised databases. Information is supplied by 39 constituent UN agencies
Publications: *Directory of UN Databases and Information Services*; and *ACCIS Guide to UN Information Sources on the Environment*

– 2004 – Bundesdeutscher Arbeitskreis für Umweltbewusstes Management (BAUM)

Christian Forsterstrasse 19
D-2000 Hamburg 20
Germany
Tel: 0049 40 407721
Contact name: Georg Wintr, Chairperson of the Board of Management
Geographical coverage: Germany
Aims and objectives: BAUM helps to work out joint solutions to environmental problems

– 2005 – Centre for Environmental Management and Planning (CEMP)

Auris Business Centre
23 St Machar Drive
Aberdeen AB2 1RY
United Kingdom
Tel: 0224 272482
Fax: 0224 487658
Geographical coverage: International
Aims and objectives: Provides a service, based in Aberdeen University (and a part of the Aberdeen University Research and Industrial Services – AURIS) encompassing research, consultancy, training, conferences and information on environmental issues worldwide

– 2006 – Economic and Social Committee (ESC): Section for Protection of the Environment

Rue Ravenstein 2
B-1000 Brussels
Belgium
Tel: 32 2 519 90 11
Fax: 32 2 513 48 93
Contact name: Ceballo Herrero Francisco
Geographical coverage: European
Aims and objectives: The ESC is an advisory and consultative body intended to ensure the involvement of all economic and social groups in the development of the EC. The environment, public health and consumer affairs section is one of nine sections in the ESC

– 2007 – Institution of Water and Environmental Management (IWEM)

15 John Street
London WC1N 2EB
United Kingdom
Tel: 071 831 3110
Fax: 071 405 4967
Contact name: Stephen Myers
Geographical coverage: United Kingdom
Aims and objectives: The objects for which the Institution was established are to advance the science and practice of water and environmental management for the public benefit and to promote education, training, research and study in the said science
Publications: *Water and Environment Management Journal*, six times a year. List available on request

BOOKS

– 2008 – Croners Environmental Management

Geographical coverage: United Kingdom
Publishers: Croner Publications Ltd
Price: £99.20
ISBN: 1855241153

– 2009 – Environmental Management in the Soviet Union

Geographical coverage: European
Publishers: Cambridge University Press, 1991
Coverage: A study of Soviet environmental problems
Price: £30.00
ISBN: 052136079X
Author: Pryde, P. R.

DIRECTORIES/YEARBOOKS

– 2010 – CD ROM Directory, 5th ed

Geographical coverage: International
Publishers: TFPL Ltd, 1991
Frequency: annual
Coverage: Includes descriptions of a number of environmental databases
ISBN: 1870889231

– 2011 – Directory of the Environment

Geographical coverage: United Kingdom
Publishers: Green Print
Frequency: Will be updated
Price: £13.99
ISBN: 1854250361
Author: Frisch, M.

– 2012 – ECO Directory of Environmental Databases in the UK 1992

Geographical coverage: United Kingdom
Publishers: ECO Environmental Education Trust, 1992
Editor: Barlow, M. et al
Coverage: Covers a selection of databases from research, statutory and commercial organisations where their information is felt to be of particular relevance to environmental groups
ISBN: 1874666008

– 2013 – Green Globe Yearbook

Geographical coverage: International
Publishers: Oxford University Press, 1992
Coverage: This is the first edition of an annual publication on the international politics of environmental management
Price: £30.00
ISBN: 0198233221
Author: Bergesen, H. O. and Norderhaug, M.

PERIODICALS

– 2014 – Ambio: A Journal of the Human Environment Research and Management

Box 5005
S-104 05 Stockholm
Sweden
Tel: 46 8 150430
Geographical coverage: Sweden
Publishers: Royal Swedish Academy of Sciences
Editor: Kessler, E.
Frequency: 6 issues a year (18 elsewhere)
Language: English
Coverage: The human environment, research and management
ISSN: 0044 7447

– 2015 – Journal of Environmental Management

24–28 Oval Road
London NW1 7DX
United Kingdom
Tel: 071 267 4466
Fax: 071 482 2293
Geographical coverage: United Kingdom
Publishers: Academic Press Ltd
Editor: Jeffers, J. N. R.
Frequency: 12 issues a year (2 volumes)
Coverage: Presents papers on all aspects of management and use of the environment, both natural and man made
Annual subscription: $350
ISSN: 0301 4797

– 2016 – Ministerio da Qualidade de Vida Comissão Nacional do Ambiente Boletim

Praça Duque de Saldanha 31
P-1096 Lisbon Codex
Portugal
Geographical coverage: Portugal
Publishers: Ministerio da Qualidade de Vida Comissão Nacional do Ambiente
Frequency: monthly
Price: free

REPORTS

– 2017 – Managing the Environment

Geographical coverage: United Kingdom
Publishers: British Institute of Management, 1991
Coverage: This report summarises the key environmental issues facing organisations and individual managers from a management perspective. Also summarises the main provisions of UK and EC law
Price: £30.00
ISBN: 0859462102
Author: Coe, T.

RESEARCH CENTRES

– 2018 – Centre d'Enseignement et de Recherche pour la Gestion des Resources Naturelles et l' Environnement

(Management of Natural Resources and Environment Study Centre)

Le Palatino
17 avenue de Choisy
F-75013 Paris
France
Tel: 33 1 458 412 55
Geographical coverage: France
Research activities: Urban hydrology; surface hydrology; environmental management; urban ecology

DATABASES

– 2019 – Acompline/Urbaline

Geographical coverage: International
Coverage: Two databases combined into a single online file, covering urban and regional planning, transport, leisure, recreation and urban environmental issues
Database producer: London Research Centre
Database host: ESA-IRS; London Research Centre
Coverage period: 1974–
Update frequency: Acompline monthly; Urbaline daily
Additional information: Available on CD-ROM

– 2020 – Centro de Documentación del Medio Ambiente: Escuela de Organización Industrial

Gregorio del Amo 6
E-28040 Madrid
Spain
Tel: 34 1 349 56 00
Fax: 34 1 554 23 94
Contact name: Antonio Vilches Guerra
Geographical coverage: Spain
Coverage: Environment-related topics

Environmental Monitoring

GOVERNMENT ORGANISATIONS

– 2021 – Division for Air Monitoring and Industrial Compliance

Telemark County Statens Hus
N-3708 Skien
Norway
Tel: 47 3 58 61 20
Fax: 47 3 53 00 20
Geographical coverage: Norway
Aims and objectives: The Division is responsible for monitoring atmospheric pollution

NON-GOVERNMENT ORGANISATIONS

– 2022 – Earthwatch

PO Box 403
680 Mt Auburn Street
Watertown
MA 02272
United States of America
Tel: 1 617 926 8200
Fax: 1 617 926 8532
Geographical coverage: International
Aims and objectives: Founded in 1971, Earthwatch is an international association of citizens and scientists working to sustain the world's environment, to monitor global change, to conserve endangered species and to foster world health and international cooperation
Publications: *Earthwatch*, magazine six times a year; annual report

– 2023 – European Geographical Information Systems Foundation (EGIS)

University of Utrecht
Heidelberglaan 8
NL-3508 TC Utrecht
Netherlands
Tel: 31 30 539111
Fax: 31 30 521818
Geographical coverage: European
Aims and objectives: Geographical Information Systems application to environmental management. The technological core of much environmental monitoring is GIS, which provides on-screen map displays of data generated from many sources

– 2024 – Global Environmental Monitoring System

GEMS/PAC UNEP
PO Box 30552
Nairobi
Kenya
Kenya
Tel: 254 2 333930
Geographical coverage: International
Aims and objectives: GEMS is a collaboration of about 20 monitoring networks throughout the world, comprised of UN agencies (such as WHO, FAO, UNESCO, WMO), intergovernmental organisations (such as IUCN), and national governments

BOOKS

– 2025 – Environmental Monitoring and Analytical Specimen Banking

Geographical coverage: United Kingdom
Publishers: Springer Verlag, 1992
Coverage: This text represents state-of-the-art in sampling, storing and characterising specimens from environmental, biological and human origin for environmental monitoring
Price: £60.00
ISBN: 3540550011
Author: Rossbach, M. and Schladot, J. D.

CONFERENCE PAPERS

– 2026 – Environmental Monitoring: Meeting the Technical Challenge

Geographical coverage: International
Publishers: The Institute of Physics
Coverage: Contributors examine the current state of technology and methodology employed by the industry and discuss future developments
Price: £24.00
ISBN: 0854985298
Author: Cashell, E. M.

DIRECTORIES/YEARBOOKS

– 2027 – Directory of Emissions Monitoring Equipment Suppliers 1992

University of Leeds
Leeds LS2 9JT
United Kingdom
Tel: 0532 332511
Geographical coverage: United Kingdom
Publishers: University of Leeds, Department of Fuel and Energy, 1992

PERIODICALS

– 2028 – Annals of Occupational Hygiene

Headington Hill Hall
Oxford OX3 0BW
United Kingdom
Tel: 0865 794141
Fax: 0865 743952
Geographical coverage: United Kingdom
Publishers: Pergamon Press plc
Editor: McKellinson, J.
Frequency: 6 issues a year
Coverage: Life sciences including air pollution monitoring
Annual subscription: £260.00
ISSN: 0003 4878

– 2029 – Warren Spring Laboratory UK Smoke and Sulphur Dioxide Monitoring Networks

Gunnels Wood Road
Stevenage
Hertfordshire SG1 2BX
United Kingdom
Tel: 0438 741122
Fax: 0438 360858
Geographical coverage: United Kingdom
Publishers: Warren Spring Laboratory
Frequency: irregular
Price: varies

RESEARCH CENTRES

– 2030 – British Textile Technology Group

Wira House
West Park Ring Road
Leeds LS16 6QL
United Kingdom
Tel: 0532 781381
Geographical coverage: United Kingdom
Research activities: Energy audits; water economy and recovery; effluent treatment; environmental monitoring in textile industries

– 2031 – Cranfield Biotechnology Centre: *see entry 1789*

Geographical coverage: United Kingdom

– 2032 – Department of Environmental Science

University of Stirling
Stirling FK9 4LA
Scotland
United Kingdom
Tel: 0786 473171
Fax: 0786 463000
Contact name: Donald A Davidson
Geographical coverage: United Kingdom
Research activities: Geographical Information systems application

– 2033 – Eidgenössische Kommission zur Überwachung der Radioaktivität

(Swiss Federal Commission of Radioactivity Surveillance)

c/o Physics Department
University of Fribourg
CH-1700 Fribourg/Perolles
Switzerland
Tel: 41 37 82 62 36
Geographical coverage: Switzerland
Publications: Annual reports in German and French
Research activities: Radioacitivity monitoring; environmental monitoring around nuclear facilities; industries using radionuclides

– 2034 – Instytut Ksztaltowania Srodowiska

(Environmental Protection Institute, Laboratory of Environmental Monitoring)

Krucza 5/11
PL-00 548 Warsaw
Poland
Tel: 48 22 295263
Geographical coverage: Poland
Research activities: Nature conservation; ecology; atmospheric monitoring; soil conservation; monitoring equipment

Businesses and Services *Environmental Policy*

– 2035 – Monitoring and Assessment Research Centre (MARC)

The Old Coach House
Campden Hill
London W8 7AD
United Kingdom
Tel: 071 376 1577
Fax: 071 937 5396
Contact name: Professor Peter Peterson, Director
Geographical coverage: United Kingdom
Aims and objectives: MARC undertakes work on various environmental issues such as water quality monitoring, urban air pollution. Run training workshops in developing countries and compile the UNEP Environmental Data Report
Publications: *UNEP Environmental Data Report*, every two years

– 2036 – Warren Spring Laboratory: *see entry 2207*

Geographical coverage: United Kingdom

– 2037 – World Conservation Monitoring Centre (WCMC)

219 Huntingdon Road
Cambridge CB3 0DL
United Kingdom
Tel: 0223 277314
Fax: 0223 277136
Contact name: Jo Taylor, Information Officer
Geographical coverage: International
Aims and objectives: The WCMC is a joint venture between IUCN, UNEP and WWF. Its mission is to support conservation and sustainable development through the provision of information on the world's biological diversity
Publications: Various
Research activities: Global Forest Mapping Project – A Geographical Information System which is the core of much environmental monitoring

DATABASES

– 2038 – Global Resource Information Database (GRID)

Geographical coverage: International
Coverage: Specialises in climate related activities; comparable data are collected from GEMS sources and combined into the GRID, allowing assessment of environmental trends and change
Database producer: GEMS/PAC
Database host: UN Agencies
Coverage period: 1960–
Update frequency: continuously

Environmental Policy

GOVERNMENT ORGANISATIONS

– 2039 – Center for Information on the Environment

Rue du Luxembourg 20
B-1040 Brussels
Belgium
Tel: 32 2 514 01 70
Geographical coverage: Belgium
Aims and objectives: Ensuring maximum environmental protection by developing strategies and legislation

– 2040 – Committee on the Use of Atomic Energy for Peaceful Purposes

55 A Chapaev St
1574 Sofia
Bulgaria
Tel: 359 2 02 17
Fax: 359 2 70 21 43
Contact name: Yanjo Yanev, Chairperson
Geographical coverage: Bulgaria
Aims and objectives: The Committee is responsible for state policy and control over the use of atomic energy

– 2041 – Concentration Unit for Biotechnology in Europe (CUBE)

DG XII
Commission of the EC
Rue de la Loi 200
B-1049 Brussels
Belgium
Tel: 32 2 235 81 45
Fax: 32 2 235 53 65
Geographical coverage: European
Aims and objectives: CUBE aims to provide sources of information about Community policy with respect to biotechnology

– 2042 – Department of the Environment: see entry 1978

Geographical coverage: Ireland

– 2043 – Environment Policy Committee: Confederation of Irish Industry

Kildare Street
Dublin 2
Eire
Tel: 353 1 779801
Fax: 353 1 777823
Geographical coverage: Ireland

– 2044 – Environment Policy Section: see entry 1652

Geographical coverage: Ireland

– 2045 – Ministerium für Soziale Gesundheit und Energie des Landes Schleswig-Holstein

Kronshagener Weg 130a
D-2300 Kiel 1
Germany
Tel: 49 431 1695 0
Fax: 49 431 15169
Contact name: Dr Klaus Rave
Geographical coverage: Germany
Aims and objectives: The Ministry is responsible for energy policy, energy technology and price regulations

– 2046 – Ministry of Agriculture Fisheries and Food: Conservation Policy and Environmental Protection Division

Nobel/Ergon House
17 Smith's Square
London SW1P 3HX
United Kingdom
Tel: 071 236 6563/5654
Geographical coverage: United Kingdom

– 2047 – Ministry of Health and Humanitarian Policies

8 avenue de Segur
F-75700 Paris
France
Tel: 33 1 405 600 00
Geographical coverage: France

– 2048 – National Environmental Policy Plan Plus (NEPP): Ministry of Housing Planning and the Environment

Rijnstraat 8
NL-2515 XP The Hague
Netherlands
Tel: 31 70 339 3741
Geographical coverage: Netherlands

– 2049 – Nordic Council of Ministers

Store Strandstraede 18
DK-1255 Copenhagen K
Denmark
Tel: 45 33 11 47 11
Fax: 45 33 11 47 11
Contact name: Fridtjov Clemet, Secretary General
Geographical coverage: European
Aims and objectives: The governments of Denmark, Finland, Iceland, Norway and Sweden cooperate through the Nordic Council of Ministers. The member states undertake to harmonise regulations for protecting the environment and to assess measures affecting their neighbours
Publications: Books and pamphlets on Nordic cooperation

– 2050 – Nuclear Free Local Authorities National Steering Committee

Nuclear Policy and Information Unit
Manchester Town Hall
Manchester M60 2LA
United Kingdom
Tel: 061 234 3222
Fax: 061 236 8864
Contact name: Stella Whittaker
Geographical coverage: United Kingdom
Aims and objectives: Statutory organisation which gives legal advice, policy guidance, research, promotion and information distribution on nuclear issues
Additional information: MP Database

– 2051 – OECD Environment Committee

2 rue André Pascal
F-75016 Paris
France
Tel: 33 1 452 482 00
Fax: 33 1 452 478 76
Geographical coverage: International
Aims and objectives: The Environmental Committee was formed in 1970, and has made a number of specific recommendations and decisions on the economic, legal and scientific aspects of environmental management. The "polluter pays principle" was originated by the OECD
Publications: Books, pamphlets, codes etc. Full details are available on request

NON-GOVERNMENT ORGANISATIONS

– 2052 – Belgian Biotechnology Coordinating Group (BBCG)

Plant Genetic Systems
Jozef Plateaustraat 22
B-9000 Ghent
Belgium
Tel: 32 91 358 444
Fax: 32 91 233 855
Geographical coverage: Belgium
Aims and objectives: Participate as appropriate in the formulation of local, national and international policies

– 2053 – Business and Industry Advisory Committee (BIAC): see entry 1674

Geographical coverage: International

– 2054 – Centre for European Policy Studies (CEPS)

Rue Ducale 33
B-1000 Brussels
Belgium
Tel: 32 2 513 40 88
Fax: 32 2 511 59 60
Contact name: Peter Ludlow, Director
Geographical coverage: European
Aims and objectives: CEPS is a European think tank whose activities cover economic, political and security issues. The centre organises a large number of conferences, on a range of topics, including environmental affairs

– 2055 – Centre for Our Common Future (COCF)

Palais Wilson
52 rue de Paquis
CH-1201 Geneva
Switzerland
Tel: 41 22 732 71 17
Fax: 41 22 738 50 46
Geographical coverage: International
Aims and objectives: The Centre is an independent organisation set up in 1988 following the publication of the Brundtland Report. It serves as an information exchange on activities taking place globally in keeping with the concepts of sustainable development
Publications: *Brundtland Review*, quarterly newsletter, and other publications

– 2056 – Commission of the European Communities

8 Storey's Gate
London SW1P 3AT
United Kingdom
Tel: 071 973 1992
Fax: 071 222 0900
Geographical coverage: European
Aims and objectives: The Commission is the administrative institution of the European Communities, maintaining the sole right of initiative of Community legislation and acting as guardian of the Treaties. Its executive body is made up of 17 members, chosen by agreement

– 2057 – European Environmental Alliance (EEA)

European Parliament Office
Rue Belliard 97–113
B-1047 Brussels
Belgium
Tel: 32 2 284 52 45
Fax: 32 2 284 92 45
Contact name: Ken Collins, President
Geographical coverage: European
Aims and objectives: Created in 1988, EEA aims to implement strict norms for the protection of the environment within the EC, to integrate the environment in all European policies

– 2058 – European Federation for Transport and Environment (T & E)

17 George Street
Croydon CR0 1LA
United Kingdom
Tel: 081 681 71 85
Fax: 081 666 04 22
Contact name: Chris Bowers, Coordinator
Geographical coverage: European
Aims and objectives: The T & E is an umbrella organisation for the various groups across Europe which campaign for a more environmental approach to transport. Its goal is to promote a transportation policy which has the lowest impact on the environment

– 2059 – Institute for European Environmental Policy (IEEP)

Aloys-Schulte-Strasse 6
D-5300 Bonn 1
Germany
Tel: 49 228 21 38 10
Fax: 49 228 22 19 82
Contact name: Dr Jan Bongaerts, Director
Geographical coverage: European
Aims and objectives: The IEEP was founded in Bonn in 1976 to provide an independent

body for the analysis of environmental policies in Europe. It seeks to increase the awareness of the European dimension of environmental protection and to advance European policy making
Publications: List available on request

– 2060 – International Council of Environmental Law (ICEL): *see entry 1939*

Geographical coverage: International

– 2061 – International Professional Association for Environment Affairs (IPRE): *see entry 1855*

Geographical coverage: International

– 2062 – Stichting Natuur en Milieu (SNM): *see entry 1943*

Geographical coverage: Netherlands

– 2063 – Union of Industrial and Employers' Confederation of Europe (UNICE): *see entry 1729*

Geographical coverage: European

– 2064 – Wissenschaftszentrum Berlin (WZB)

(International Institute for Environment and Safety)

Reichpietschufer 50
D-1000 Berlin 30
Germany
Tel: 49 30 25 49 1 0
Geographical coverage: Germany
Aims and objectives: WZB focuses on evaluation of selected policy areas from environmental perspectives

– 2065 – World Resources Institute (WRI)

1709 New York Avenue
NW Washington
DC 20006
United States of America
Tel: 1 202 638 6300
Fax: 1 202 638 0036
Contact name: James Gustav Speth, President
Geographical coverage: International
Aims and objectives: WRI, founded in 1982, is an independent research and policy institute. It currently focuses on forests and biological diversity; energy, climate and pollution; economics and resource and environmental information

PRESSURE GROUPS

– 2066 – Scottish Campaign to Resist the Atomic Menace (SCRAM): *see entry 1906*

Geographical coverage: United Kingdom

– 2067 – Transport 2000

Walkden House
10 Melton Street
London NW1 2EJ
United Kingdom
Tel: 071 388 8386
Fax: 071 388 2481
Contact name: Mr Stephen Joseph, Executive Director
Geographical coverage: United Kingdom
Aims and objectives: The aim of Transport 2000 is to promote environmentally sound and socially responsible transport policies that put the interests of pedestrians, public transport users and vulnerable road users to the fore

BOOKS

– 2068 – Acid Politics

Geographical coverage: United Kingdom
Publishers: Pinter Publishers, 1990
Coverage: A comparison of different policy approaches to acid rain recognition and control in the UK and West Germany
Price: £30.00
ISBN: 1852931167

– 2069 – Corporate Responsibility Europe: *see entry 1732*

Geographical coverage: European

– 2070 – Economic Policy Towards the Environment

Geographical coverage: International
Publishers: Blackwell Publishers, 1991
Editor: Helm, D.
Language: English
Coverage: Particular topics covered are the valuation of environmental damage, global warming, environmental taxes and tradeable permits
ISBN: 0631182020

– 2071 – EEC Environmental Policy and Britain

Geographical coverage: European
Publishers: Longman, second edition, 1990 update
Price: £38.50

ISBN: 0582059593
Author: Haigh, N.

– 2072 – Energy Policies and the Greenhouse Effect Volume 1: Policy Appraisal

Geographical coverage: United Kingdom
Publishers: Royal Institute of International Affairs, 1991
Coverage: Concentrates on policy issues arising from attempts to reduce the emissions of greenhouse gases from the energy sector
Author: Grubb, M.

– 2073 – Environmental Action in Eastern Europe

Geographical coverage: United Kingdom
Publishers: M. E. Sharpe, 1993
Coverage: Examines the efforts to develop strategies for dealing with the environmental crisis both by governments and at the grassroots level by newly emerging green movements
Price: £44.95
ISBN: 156324036X
Author: Jancar-Webster, B.

– 2074 – Environmental Policies for Cities in the 1990s

Geographical coverage: International
Publishers: OECD, 1990
ISBN: 9264134352

– 2075 – Environmental Policy in the European Community

Geographical coverage: European
Publishers: Office for Official Publications of the EC, 1990
Language: English
Coverage: Explains the EC's projects on environmental protection
ISBN: 928261414X

– 2076 – Environmental Practice in Local Government

Geographical coverage: United Kingdom
Publishers: ACC; ADC; AMA, 1990
Coverage: A guide by the local authority associations
ISBN: 090178365X

– 2077 – Europe and the Environment

Geographical coverage: European
Publishers: Industry Society Press, 1992
Coverage: A guide which examines the environmental policies of the European Commission and how these policies translate into legislation
Price: £5.00
ISBN: 0852904711
Author: Taylor, C. and Press, A.

– 2078 – European Community, The

Geographical coverage: European
Publishers: Fourmat Publishing, 1992
Coverage: This book discusses agriculture and the environment as well as the European social programme
Price: £22.50
ISBN: 1851901485
Author: Ishani, M. G. K.

– 2079 – European Integration and Environmental Policy

Geographical coverage: European
Publishers: Belhaven Press, 1993
Coverage: An overview of European policy making in relation to environmental issues at a national and Community wide scale
Price: £42.00
ISBN: 1852932821
Author: Liefferink, J. D. and Mol, A. P. J.

– 2080 – Greening of Urban Transport, The: Planning for Walking and Cycling in Western Cities

Geographical coverage: European
Publishers: Belhaven Press, 1990
Coverage: Outlines the principles and illustrates the practice of motor traffic restraint and exclusion in urban transport planning
Price: £27.50
ISBN: 1852930926
Author: Tolley, R. D.

– 2081 – Milieubeleid in de Europese Gemeenschap, Het

Geographical coverage: Netherlands
Publishers: Office for Official Publications of the EC, 1990
Language: Dutch
Coverage: Explains EC's projects on environmental protection
ISBN: 9282614174

– 2082 – Miljopolitik i Det Europaeiske Faellesskab

Geographical coverage: Denmark
Publishers: Office for Offical Publications of the EC, 1990
Language: Danish

Coverage: Explains the EC's projects on environmental protection
ISBN: 9282614115

– 2083 – Politica Ambientale nella Comunità Europea, La

Geographical coverage: Italy
Publishers: Office for Official Publications of the EC, 1990
Language: Italian
Coverage: Explains EC's projects on environmental protection
ISBN: 9282614166

– 2084 – Política de Ambiente na Comunidade Europeia, A

Geographical coverage: Portugal
Publishers: Office for Official Publications of the EC, 1990
Language: Portuguese
Coverage: Explains EC's projects on environmental protection

– 2085 – Política de medio ambiente en la Communidad Europea

Geographical coverage: Spain
Publishers: Office for Official Publications of the EC, 1990
Language: Spanish
Coverage: Explains EC's projects on environmental protection
ISBN: 9282614107

– 2086 – Politique de l'Environnement dans la Communauté Européenne, La

Geographical coverage: France
Publishers: Office for Official Publications of the EC, 1990
Language: French
Coverage: Explains EC's projects on environmental protection
ISBN: 9282614158

– 2087 – Politique Européenne de l' Environnement, La

Geographical coverage: European
Publishers: Romillat, Paris 1990
Author: Engref

– 2088 – This Common Inheritance: Britain's Environmental Strategy

Geographical coverage: United Kingdom
Publishers: HMSO, 1990
Price: £24.50
ISBN: 0101120028

– 2089 – Transport Policy and the Environment

Geographical coverage: European
Publishers: European Conference of Ministers of Transport (ECMT), 1990, Paris
Language: English and French
Price: Ffr225
ISBN: 9282111474
Author: Prepared in cooperation with OECD

– 2090 – Transport Policy and the Environment

Geographical coverage: International
Publishers: E & F N Spon, 1992
Coverage: Presents the current thinking from leading authorities worldwide on transport and the environment
Price: £35.00
ISBN: 0419178708
Author: Banister, D. and Button, K.

– 2091 – Umweltpolitik in der Europäischen Gemeinschaft, Die

Geographical coverage: Germany
Publishers: Office for Official Publications of the EC, 1990
Language: German
Coverage: Explains EC's projects on environmental protection
ISBN: 9282614123

DIRECTORIES/YEARBOOKS

– 2092 – Directory of Corporate Environmental Policy

Geographical coverage: United Kingdom
Publishers: Industry and Environment Associates, 1992
Frequency: annual
Coverage: Contemporary comparison of the environmental policies and organisations operating in the UK

– 2093 – Manual of Environmental Policy: The EC and Britain

Geographical coverage: International
Publishers: Longman, 1991 (looseleaf)
Frequency: 2 updates a year
ISBN: 0582087155
Author: Haigh, N.

PERIODICALS

– 2094 – ENDS
Unit 24
Finsbury Business Centre
40 Bowling Green Lane
London EC1R 0NE
United Kingdom
Tel: 071 278 4745
Fax: 071 837 7612
Geographical coverage: United Kingdom
Publishers: Environmental Data Services Ltd
Editor: Mayer, M.
Frequency: monthly
Coverage: Environmental policy in the UK and EEC; pollution control, waste management and the profiles of key environmental organisations. This is a key environmental information resource
ISSN: 0260 1249

– 2095 – Environmental News from the Netherlands
PO Box 20951
NL-2500 EZ The Hague
Netherlands
Geographical coverage: Netherlands
Publishers: VROM
Editor: Maurits Groen Milieu & Communicatie
Frequency: quarterly
Language: English
Coverage: Articles on environmental policy and research in the Netherlands
Price: free
ISSN: 0925 2940

– 2096 – Environmental Policy and Practice
52 Kings Road
Richmond
Surrey TW10 6EP
United Kingdom
Tel: 081 995 0877
Fax: 081 747 9663
Geographical coverage: United Kingdom
Publishers: EDD Publications
Editor: Aston, G.
Frequency: quarterly
Coverage: Pollution control and waste management environmental law
Price: £20.00
Annual subscription: £66.00

– 2097 – European Environment
Rue de Genève 6
B-1140 Brussels
Belgium
Tel: 32 2 242 60 20
Fax: 32 2 242 94 10
Geographical coverage: European
Publishers: Europe Information Service
Editor: Paoloni, M.
Frequency: fortnightly
Language: English
Coverage: Latest events in the field of environment and consumer policy, with features on activities of the EC and public or private organisations

– 2098 – Freight
Hermes House
St Johns Road
Tunbridge Wells TN4 9UZ
United Kingdom
Tel: 0892 26171
Geographical coverage: United Kingdom
Publishers: Freight Transport Association
Frequency: monthly
Coverage: Transport policy, developments and products relating to freight
ISSN: 1016 0849

– 2099 – Journal of Transport Economics and Policy
University of Bath
Claverton Down
Bath BA2 7AY
United Kingdom
Tel: 0225 826302
Fax: 0225 826767
Geographical coverage: United Kingdom
Editor: Glaister, S.
Frequency: 3 issues a year
Annual subscription: £41.00
ISSN: 0022 5258

– 2100 – Nordic Journal of Environmental Economics Protection
National Agency of Environment
Strandgade 29
DK-1401 Copenhagen K
Denmark
Tel: 45 31 57 83 10
Geographical coverage: Denmark

– 2101 – Streetwise: *see entry 1788*
Geographical coverage: United Kingdom

– 2102 – Studi Parlamentari e di Politica
Via Massa Fiscalgia 1
I-00127 Rome
Italy

Fax: 39 6 6071824
Geographical coverage: Italy
Publishers: Edistudio
Editor: Greco, N.
Frequency: quarterly
Language: Text in Italian, summaries in English
Annual subscription: Lit115000
ISSN: 0303 9714

REPORTS

– 2103 – Controlling Pollution in the Round: Change and Choice in Environmental Regulation in UK and Germany

Geographical coverage: Germany
Publishers: Anglo-German Foundation, 1991
Language: English
Coverage: Provides and analysis of the experience of coordinating and strengthening environmental regulations in Britain and Germany
Price: £12.50
ISBN: 0905492692
Author: Weale, A., O'Riordan, T. and Kramme, L.

– 2104 – Corporate Environmental Policy Statements

Geographical coverage: United Kingdom
Publishers: Confederation British Industry (CBI), 1992
Author: Confederation of British Industry

– 2105 – Energi och Miljö i Norden 1991

Store Strandstraede 18
DK-1255 Copenhagen K
Denmark
Geographical coverage: Denmark
Publishers: Nordic Council of Ministers, 1991
Language: Danish
Coverage: Report on the way taxes and charges can be used as levers to control developments in the energy sector

– 2106 – Energy Taxation and Environmental Policy in EFTA Countries 1991

Rue de Varembé 9–11
CH-1211 Geneva 20
Switzerland
Geographical coverage: European
Publishers: EFTA, 1991
Language: English
Coverage: Summarises the experience of energy taxation in EFTA countries

– 2107 – Environment Policy: Standards and Opportunities for Development

Geographical coverage: United Kingdom
Publishers: National Materials Handling Centre, 1991
Coverage: Environmental performance of the road freight industry

– 2108 – Future for UK Environmental Policy Special Report no 2182, The

Geographical coverage: United Kingdom
Publishers: Economist Intelligence Unit, 1991
ISBN: 0850585627
Author: Webb, A.

– 2109 – Green Paper on the Environment – A Community Strategy for Sustainable Mobility

Geographical coverage: European
Publishers: Commission of the EC, 1992
Coverage: The Green Paper proposes introducing a framework for a Community strategy for sustainable mobility which would set down broad guidelines to contain the impact of transport on the environment

– 2110 – Implications for the EC's Environmental Policy of the Treaty on European Union, The

Geographical coverage: European
Publishers: IEEP, London 1992
Coverage: Maastricht and the Environment
Author: Wilkinson, D.

– 2111 – Monthly Environmental Policy

Ecological Studies Institute
49 Wellington Street
London WC2E 7BN
United Kingdom
Tel: 071 836 0341
Fax: 071 240 9205
Geographical coverage: European
Coverage: Reports on Poland, the Czech Republic, Slovakia and Hungary

– 2112 – Paradise Deferred: Environmental Policymaking in Central and Eastern Europe

10 St James Square
London SW1Y 4LE
United Kingdom
Geographical coverage: European
Publishers: EEP/RIIA
Coverage: Reviews environmental policy developments in the region

Price: £10.00
ISBN: 0905031474
Author: Fisher, D.

– 2113 – Protecting the Earth: A Status Report with Recommendations for a New Energy Policy

Geographical coverage: Germany
Publishers: Deutscher Bundestag, 1991
Coverage: Provides an account of the magnitude of the threat of the greenhouse effect. Recommendations are offered for national actions to reduce energy related emissions of radioactive trace gases
ISBN: 3924521719
Author: Enquête Commission, Germany

RESEARCH CENTRES

– 2114 – Department of Political and Social Sciences

European Univ Inst
Via dei Roccettini 9
I-50016 San Domenico di Fiesole
Italy
Geographical coverage: Italy
Research activities: Transport policy

– 2115 – East European Environmental Research (ISTER): *see entry 1769*

Geographical coverage: Hungary

DATABASES

– 2116 – Agra Europe

Geographical coverage: European
Coverage: EC Common Agricultural Policy, food trade
Database producer: Agra Europe
Database host: Agra Europe via Telecom Gold
Coverage period: 1963–
Update frequency: weekly

– 2117 – European Update

Geographical coverage: European
Coverage: EC policy analysis, including industry, commerce, tax, legislation and Eastern Europe
Database producer: DRT Europe Services
Database host: DRT Europe Services
Update frequency: daily

– 2118 – Greenhouse Effect Report

Geographical coverage: International
Coverage: Global warming; scientific assessments, implications and policy responses relating to business, technology, economics, international action
Database producer: Business Publishers Inc
Database host: via PTS Newsletter on DataStar or Dialog
Coverage period: 1988–
Update frequency: monthly

– 2119 – UN Conference on Environment and Development (UNCED)

Geographical coverage: International
Coverage: Official documentation relating to the UN Conference on Environment and Development, in Rio, 1992
Database producer: UNCED
Database host: Manchester Host; GeoNet; Poptel
Coverage period: 1989–
Update frequency: monthly

Environmental Science

NON-GOVERNMENT ORGANISATIONS

– 2120 – Association des Régions des Alpes Centrales (ARGE ALP)

(Association of the Central Alps)

Amt der Tiroler Landesregierung
A-6010 Innsbruck
Austria
Tel: 43 512 53 2222/28701
Geographical coverage: Austria

– 2121 – Centro Ecologico

(Ecological Centre)

Apartado 4045
P-1501 Lisbon Cedex
Portugal
Geographical coverage: Portugal

– 2122 – Centro Ecológico Nacional SA

Francisco Giralte 2
E-28002 Madrid
Spain
Tel: 34 1 411 41 77
Fax: 34 1 261 56 26
Geographical coverage: Spain

– 2123 – Független Ökological Központ (FÖK)

(Independent Ecological Centre)

Miklos ter 1
H-1035 Budapest
Hungary
Tel: 36 1 16 86 229
Fax: 36 1 17 54 774
Contact name: Dr Judit Vasarhelyi
Geographical coverage: Hungary
Aims and objectives: The Centre acts as a resource centre for other NGOs

– 2124 – International Association for Ecology (INTECOL)

c/o Institute of Ecology
University of Georgia
Athens GA 20602
Greece
Tel: 1 404 542 2968
Geographical coverage: International
Aims and objectives: Established in 1967, INTECOL is mainly an umbrella organisation for over thirty national, regional and international societies concerned with various disciplines of ecology
Publications: *INTECOL Newsletter*, bi monthly; and various bulletins

– 2125 – International Organisation for Human Ecology

Karlsplatz 13
A-1040 Vienna
Austria
Tel: 43 222 56014599
Contact name: Dr Helmut Knotig, Secretary General
Geographical coverage: International
Aims and objectives: The Organisation's aims are to support and conduct scientific research and to organise scientific events in the field

– 2126 – Programme on Man and the Biosphere (MAB)

Divison of Ecological Sciences
UNESCO
7 place de Fontenoy
F-75700 Paris
France
Tel: 33 1 456 840 60
Fax: 33 1 456 716 90
Geographical coverage: International
Aims and objectives: The MAB programme, formed in 1971, is an international programme of research, training, demonstration and information diffusion. It has field projects and training activities in over 100 countries
Publications: Technical reports and a series of books. A Catalogue is available in English and French
Additional information: MABIS Database

– 2127 – Scientific Committee on Problems of the Environment (SCOPE)

51 boulevard de Montmorency
F-75016 Paris
France
Tel: 33 1 452 504 98
Fax: 33 1 428 894 31
Geographical coverage: International
Aims and objectives: SCOPE, one of the inter-disciplinary committees of the ICSU, is an international non-governmental, non-profit organisation established in 1969 with the objectives of advancing knowledge of the influence of humans on their environment

– 2128 – Scientists for Global Responsibility (SGR): *see entry 1858*

Geographical coverage: United Kingdom

BOOKS

– 2129 – Essential Ecology

Geographical coverage: International
Publishers: Blackwell Scientific Publications, 1993
Coverage: The authors illustrate principles of ecology with examples, drawing on work done in a broad spectrum of countries
Price: £25.01
ISBN: 0632021780
Author: Greenwood, J. J. D. and Hubbard, S. F.

– 2130 – UNEP 1991 United Nations Environment Programme Environmental Data Report

Geographical coverage: International
Publishers: Basil Blackwell, 1991
Editor: UNEP
Frequency: biennial
Coverage: Statistical review of the global environment. Scientific information is gathered from worldwide sources
ISSN: 0956 9324
ISBN: 0631180834

DICTIONARIES/ENCYCLOPEDIAS

– 2131 – Concise Oxford Dictionary of Earth Sciences, The

Geographical coverage: United Kingdom
Publishers: Oxford Paperbacks, 1991
Coverage: Containing over 6,000 entries, this dictionary defines words and terms pertaining to the earth sciences
Author: Allaby, A. and Allaby, M.

– 2132 – Dictionary of Ecology and Environment

Geographical coverage: United Kingdom
Publishers: Peter Collin Publishing, 1992
Language: English and French
Coverage: About 3,000 main words are listed in each language
Price: £19,95
ISBN: 0948549297
Author: Collin, P. H. and Schuwer, M.

– 2133 – Dictionary of Environmental Science and Technology 1991

Geographical coverage: United Kingdom
Publishers: Open University Press, 1991
Price: £12.99
ISBN: 0335092306
Author: Porteous, A.

– 2134 – Encyclopedia of Environmental Science and Engineering

Geographical coverage: United Kingdom
Publishers: McGraw Hill Book Company, 1993
Coverage: Includes articles on all major environmental subjects
Price: £73.95
ISBN: 0070513961
Author: Parker, S. P. and Corbitt, R.

– 2135 – Wörterbuch Ökologie
(Dictionary of Ecology)

Geographical coverage: Germany
Publishers: VCH, 1991
Coverage: German/English and English/German dictionary
Price: £44.00
ISBN: 3527281754
Author: Ohrbach, K. H.

PERIODICALS

– 2136 – Ecosystems of the World

PO Box 211
NL-1000 AE Amsterdam
Netherlands
Tel: 31 20 5803911
Geographical coverage: International
Publishers: Elsevier Science Publishers BV
Frequency: irregular
Language: English

– 2137 – Energetica

Piaţa Presei Libere 1
71341 Bucharest
Romania
Tel: 40 0 18 06 30
Fax: 40 0 18 48 03
Geographical coverage: Romania
Publishers: Editura Technica
Editor: Pavel, E.
Frequency: monthly
Language: Romanian, occasionally in English, French or German
Coverage: Covers scientific and related economic and ecological problems stemming from the utilisation of all forms of energy
Price: lei800
ISSN: 0421 1715

– 2138 – Environmental Software

Crown House
Linton Road
Barking
Essex IG11 8JU
United Kingdom
Tel: 081 594 7272
Fax: 081 594 5942
Geographical coverage: United Kingdom
Publishers: Elsevier Science Publishers Ltd
Editor: Zannetti, P.
Frequency: quarterly
Coverage: Covers development of analytical and numerical techniques in environmental sciences and engineering and their computer software implementation
Annual subscription: £133.00
ISSN: 0266 9838

– 2139 – Environmetrics

Baffins Lane
Chichester
West Sussex PO19 1UD
United Kingdom
Tel: 0243 779777
Fax: 0243 775878
Geographical coverage: United Kingdom
Publishers: John Wiley and Sons Ltd

Environmental Science Businesses and Services 253

Editor: El Shaarawi, A H.
Frequency: quarterly
Coverage: Concerned with the development and application of statistical methodology in the environmental sciences
Annual subscription: $225
ISSN: 1180 4009

– 2140 – Irish Journal of Environmental Science

St Martins House
Waterloo Road
Dublin 4
Eire
Geographical coverage: Ireland
Publishers: Foras Forbartha
Frequency: six monthly
ISSN: 0332 1665

– 2141 – Natural Environment Research Council News

Polaris House
North Star Avenue
Swindon
Wiltshire SN2 1EU
United Kingdom
Tel: 0793 411 500
Fax: 0793 411 610
Contact name: Jeremy Baldwin, Head of Public Relations
Geographical coverage: United Kingdom
Publishers: Natural Environment Research Council (NERC)
Editor: Jones, L.
Frequency: quarterly
Coverage: Information items and research articles on current political, legislative, technological and scientific developments in the environmental sciences
Price: free
ISSN: 0951 5305

– 2142 – Nature

4 Little Essex Street
London WC2R 3LF
United Kingdom
Tel: 071 836 6633
Geographical coverage: International
Publishers: Macmillan Magazines Ltd
Editor: Maddox, J.
Frequency: weekly
Coverage: International weekly journal of science
Annual subscription: £160.00
ISSN: 0256 842084

– 2143 – New Scientist

Kings Reach Tower
Stamford Street
London SE1 9LS
United Kingdom
Tel: 071 261 7307
Geographical coverage: International
Publishers: Holborn Publishing Group
Editor: Dickson, D.
Frequency: weekly
Language: English
Coverage: Covers science related fields and topics including environmental studies
Annual subscription: £67.00
ISSN: 0028 6664

– 2144 – Science of the Total Environment

PO Box 211
NL-1000 AE Amsterdam
Netherlands
Tel: 31 20 5803911
Fax: 31 20 5803598
Geographical coverage: International
Publishers: Elsevier Science Publishers BV
Editor: Hamilton, E. I. and Nriagu, J. O.
Frequency: 42 issues a year
Language: English
Coverage: Publishes research into man made changes in the environment. Emphasis on applications of environmental chemistry
Annual subscription: Dfl3514
ISSN: 0048 9697

– 2145 – Studies in Environmental Science

PO Box 211
NL-1000 AE Amsterdam
Netherlands
Tel: 31 20 5803911
Fax: 31 20 5803705
Geographical coverage: Netherlands
Publishers: Elsevier Science Publishers BV
Frequency: irregular
Language: English
Price: varies

REPORTS

– 2146 – IIASA Research Plan 1993

Geographical coverage: European
Publishers: IIASA, 1992
Coverage: The Research Plan underscores IIASA's current scientific contributions and its potential importance in an array of global change research topics

RESEARCH CENTRES

– 2147 – Gruppi Ricerca Ecologica (GRE)

(Ecology Research Groups)

Via dell Frasche
5 Rome
Italy
Tel: 39 6 4746701
Geographical coverage: Italy

– 2148 – Institut für Umwelt Informatik

(Environmental Information Science Institute)

Roseggerstrasse 17
A-8700 Leoben
Austria
Tel: 43 3842 43053 14
Fax: 43 3842 43053 55
Geographical coverage: Austria
Research activities: Gathering, processing and evaluating information in the field of earth sciences

– 2149 – Institute of Terrestrial Ecology (ITE)

Monks Wood Experimental Station
Abbots Ripton
Huntingdon PE17 2LS
United Kingdom
Tel: 04873 381
Fax: 04873 467
Geographical coverage: United Kingdom
Aims and objectives: The ITE is part of the Natural Environment Research Council (NERC). It undertakes specialist ecological research in all aspects of terrestrial environment and seeks to understand the ecology of species and of natural and man-made communities
Publications: Annual report. List of publications available on request
Research activities: Studies the factors determining the structure, composition and processes of land and freshwater systems

– 2150 – Instituut voor Oecologisch Onderzoek

(Institute for Ecological Research)

Boterhoeksestraat 22
NL-6666 GA Heteren
Netherlands
Tel: 31 8306 91111
Geographical coverage: Netherlands
Publications: Annual progress report

– 2151 – International Institute for Applied Systems Analysis (IIASA)

A-2361 Laxenberg
Austria
Tel: 43 2236 715210
Fax: 43 2236 73149
Contact name: Elisabeth Krippl, Office of Communications
Geographical coverage: International
Aims and objectives: IIASA is part of an international network of scientific institutions working together to study global change. IIASA is supported by scientific organisations in 15 countries. Founded in 1972 on the initiative of the USA and ex-USSR
Publications: Catalogue available on request
Research activities: Conducts environmental studies and projects in acid rain, air pollution and other aspects of the environment

– 2152 – Irish Science and Technology Agency (EOLAS)

Glasnevin
Dublin 9
Eire
Tel: 353 1 370101
Fax: 353 1 369823
Contact name: Mr Martin Reilly
Geographical coverage: Ireland
Aims and objectives: The State Agency responsible for the development, application and promotion of science and technology
Additional information: Library Service

– 2153 – Natural Environment Research Council (NERC)

Polaris House
North Star Avenue
Swindon
Wiltshire SN2 1EU
United Kingdom
Tel: 0793 411500
Fax: 0793 411501
Geographical coverage: United Kingdom
Aims and objectives: NERC established in 1965, has the responsibility for planning, encouraging and carrying out research in the physical and biological sciences which explain the natural processes of the environment. The council is divided into four main areas of study
Publications: Catalogue available on request
Research activities: Marine and atmospheric sciences; terrestrial and freshwater sciences; earth sciences and polar sciences

– 2154 – Norske Videnskaps-Akademi, Det
(The Norwegian Science Academy)

Drammensveien 78
N-0271 Oslo 2
Norway
Geographical coverage: Norway

– 2155 – Schweizer Arbeitsgemeinschaft für Umweltforschung (SAGUF)
(Swiss Association for Environmental Research)

President
Institute ETH
Zürichbergstrasse 38
CH-8044 Zürich
Switzerland
Tel: 41 1 2564588
Contact name: Prof Dr Frank Kloetzli, President
Geographical coverage: Switzerland
Aims and objectives: The Association's aims include the coordination and dissemination of information on environmental research
Publications: Symposia proceedings

– 2156 – Swedish Environmental Research Institute (IVL)

Hälsingegatan 43
PO Box 210 60
S-100 31 Stockholm
Sweden
Tel: 46 8 729 1500
Fax: 46 8 318 516
Geographical coverage: Sweden
Aims and objectives: IVL is an independent research organisation with the aim to carry out environmental research and provide industry and governmental agencies with an impartial basis for action

DATABASES

– 2157 – BIOSIS Previews

Geographical coverage: International
Coverage: Life sciences; agriculture; ecosystems; water; land use
Database producer: BIOSIS; Silver Platter
Database host: BRS; DataStar; Dialog; DIMDI; ESA–IRS; STN International
Coverage period: 1969–
Update frequency: fortnightly
Additional information: Available on CD–ROM

– 2158 – Enviroline (ENVN)

Geographical coverage: International
Coverage: Broad coverage of scientific, engineering, political and socio–economic aspects of environmental research, issues and awareness
Database producer: R R Bowker
Database host: DataStar; Dialog; DIMDI; ESA–IRS; Orbit
Coverage period: 1971–
Update frequency: monthly
Additional information: Available on CD–ROM

– 2159 – Life Sciences Collection

Geographical coverage: International
Coverage: Specialist life sciences such as ecology
Database producer: Cambridge Scientific Abstracts
Database host: BRS; Dialog;
Coverage period: 1978–
Update frequency: monthly
Additional information: Available on CD–ROM

– 2160 – New Scientist

Geographical coverage: International
Coverage: Scientific news and research magazine. Includes regular features on environmental issues
Database producer: New Scientist, IPC Magazines
Database host: FT Profile
Coverage period: 1985–
Update frequency: weekly
Additional information: Disc available

– 2161 – SciSearch

Geographical coverage: International
Coverage: Multidisciplinary database covering journal literature in all aspects of science and technology
Database producer: Institute for Scientific Information
Database host: DataStar; Dialog; DIMDI
Coverage period: 1974–
Update frequency: 28,000 records fortnightly

– 2162 – Umweltforschungsdatenbank (UFORDAT)

Umweltbundesamt
Fachgebiet Z 2.4
Bismarckplatz 1
D-1000 Berlin 33
Germany
Tel: 49 30 8903 2423
Fax: 49 30 8903 2285
Geographical coverage: Germany
Coverage: Covers projects and institutions engaged in environmental research and development
Database producer: Umweltbundesamt UMPLIS
Database host: Data Star, FIZ Technik and STN International
Coverage period: 1974–

Environmental Standards

NON-GOVERNMENT ORGANISATIONS

– 2163 – British Standards Institution (BSI)

Linford Wood
Milton Keynes MK14 6LE
United Kingdom
Tel: 0908 221166
Geographical coverage: United Kingdom
Aims and objectives: The BSI is the UK standards producing body, a Royal Charter organisation responsible for technical documents (standards) and the licensing and fixing of marks
Publications: BSI's catalogue

– 2164 – Council for Occupational Standards and Qualifications in Environmental Conservation (COSQUEC)

The Red House
Pillows Green
Staunton
Gloucestershire GL19 3NU
United Kingdom
Tel: 0452 84825
Fax: 0452 840824
Contact name: Keith Turner, Executive Coordinator
Geographical coverage: United Kingdom
Aims and objectives: The objectives of COSQUEC are to promote education and training in environmental conservation
Publications: *Green Side Up*; newsletter two to three times a year; *Occupational Standards in Environmental Conservation*

– 2165 – International Organisation for Standardisation (ISO)

CP 56
CH-1211 Geneva 20
Switzerland
Tel: 41 22 749 01 11
Fax: 41 22 733 34 30
Contact name: Klaus G Lingner, Coordinator of Environmental Standards
Geographical coverage: International
Aims and objectives: Founded in 1947, the Organisation promotes the development of standardisation to facilitate international exchange of goods and services. ISO works through a network of technical committees, many of which relate directly to the environment

– 2166 – Swedish Standards Institution Environmental Labelling Programme:
see entry 1665
Geographical coverage: Sweden

BOOKS

– 2167 – BS 7750: What the New Environmental Management Standards Mean for Your Business

Geographical coverage: United Kingdom
Publishers: Technical Communications Publishing Ltd, 1992
Coverage: This helps explain what BS 7750 is and what it means. The first part covers the background and its development. The second part is a detailed review of papers presented at a conference on BS 7750
Price: £25.00
ISBN: 094665560X
Author: Grayson, L.

– 2168 – Directions for Internationally Compatible Environmental Data

Geographical coverage: International
Publishers: Hemisphere Publishing Corporation, 1990
Coverage: Assesses the current status and projected needs for measurement standards, data banks and global monitoring
Price: £43.00
ISBN: 1560320591

REPORTS

– 2169 – Construction Materials and the Environment

Geographical coverage: United Kingdom
Publishers: Economic Intelligence Unit, 1990
Coverage: Preparing for stricter building product standards
ISBN: 0850583373
Author: Lorch, R.

Environmental Technology

NON-GOVERNMENT ORGANISATIONS

– 2170 – Centre for Alternative Technology (CAT)

Machynlleth
Powys
Wales SY20 9AZ
United Kingdom
Tel: 0654 702400
Contact name: Roger Kelly, Director
Geographical coverage: United Kingdom
Aims and objectives: The Centre founded in 1973, demonstrates and promotes sustainable technologies and ways of living, including renewable energy sources, energy conservation and organic growing. It provides working displays of a wide range of alternative technologies
Publications: *Clean Slate* newsletter, many pamphlets, information sheets, resource booklets and technical papers

– 2171 – Centre for Exploitation of Science and Technology (CEST)

5 Berners Road
London N1 0PW
United Kingdom
Tel: 071 354 9942
Fax: 071 354 4301
Contact name: Judith Crowe, Information Scientist, Researcher
Geographical coverage: United Kingdom
Aims and objectives: CEST helps industrial groups to identify opportunities for applying technologies to meet requirements imposed by change drivers. Demands for improved environmental quality are a major driver of change
Publications: Annual Review; plus periodic project reports

– 2172 – European Network of Environmental Research Organisations (ENERO)

Parc Technologique ALATA b p n 2
F-60660 Verneuil-en-Halatte
France
Tel: 33 1 445 566 77
Fax: 33 1 445 566 99
Contact name: Christine Heuraux, General Secretary
Geographical coverage: European
Aims and objectives: ENERO is an association created to exchange, coordinate and advance environmental technology

– 2173 – Global Environmental Technology Network (GET)

Division of Environmental Health
WHO
CH-1211 Geneva 27
Switzerland
Tel: 41 22 791 37 60
Fax: 41 22 791 07 46
Geographical coverage: International
Aims and objectives: GET is a network comprised of institutions and individuals actively involved with research and training in environmental technology

– 2174 – International Environment Bureau

PO Box 301
N-1324 Lysaker
Norway
Tel: 47 2 581 800
Fax: 47 2 581 875
Contact name: Jan-Olaf Willums, Executive Director
Geographical coverage: International
Aims and objectives: The IEB is a specialised division of the ICC. The Institute organises programmes on technology transfer, industry action plan for a sustainable energy and global climate, environmental education and improving communication between governments

– 2175 – International Environmental Service Centre (IESC)

ul. Czackiego 3/5
PL-00 049 Warsaw
Poland
Tel: 48 22 267 461
Geographical coverage: Poland
Aims and objectives: IESC works for the promotion of environmentally sound technologies for Poland

– 2176 – Network for Environmental Technology Transfer (NETT)

Avenue Louise 207
Bte 10
B-1050 Brussels
Belgium
Tel: 32 2 645 09 40
Fax: 32 2 646 42 66
Contact name: Annez De Tadoada
Geographical coverage: European
Aims and objectives: NETT aims to promote cooperation between industrial enterprises and users in the field of clean technologies

258 Businesses and Services *Environmental Technology*

– 2177 – RAPRA Technology

Shawbury
Shrewsbury
Shropshire SY4 4NR
United Kingdom
Tel: 0939 250383
Fax: 0939 251118
Contact name: Dr M Copley, Chief Executive
Geographical coverage: United Kingdom
Aims and objectives: Technology transfer in the polymer and associated industries. Research testing, production facilities and analytical laboratories
Additional information: RAPRA Abstracts Database

– 2178 – Swiss Centre for Appropriate Technology (SKAT)

Varnbüelstrasse 14
CH-9000 St Gallen
Switzerland
Tel: 41 71 233481
Geographical coverage: Switzerland

– 2179 – Umweltbundesamt

(German Environmental Agency)

Bismarckplatz 1
D-1000 Berlin 33
Germany
Tel: 49 30 8030
Fax: 49 30 8903 2285
Geographical coverage: Germany
Aims and objectives: The Umweltbundesamt was established in 1974 as a scientific federal authority responsible for questions relating to environmental protection. It provides data and services in the field of environmental research
Publications: Publications of the Umweltbundesamt are also available in English

BOOKS

– 2180 – Environmental Technology Transfer to Eastern Europe: A Selected Bibliography 1991

Geographical coverage: European
Publishers: Stockholm Environment Institute, 1991
Language: All western languages
Coverage: Covers material from 1985 onwards
Author: Johansson, E.

– 2181 – Waste, Wastewater, Air Laws and Technology: *see entry 1959*

Geographical coverage: Germany

DICTIONARIES/ENCYCLOPEDIAS

– 2182 – Dictionary of Environmental Science and Technology 1991: *see entry 2133*

Geographical coverage: United Kingdom

DIRECTORIES/YEARBOOKS

– 2183 – Directory of Environmental Technology

Geographical coverage: United Kingdom
Publishers: University of York Enterprise
Editor: Final, S. and Taylor, C.
Coverage: This publication supplements the addresses with a range of short articles on all aspects of environmental technology and its application
Price: £20.00
ISBN: 0953003104

– 2184 – ENTEC Directory of Environmental Technology

Geographical coverage: United Kingdom
Publishers: Kogan Page Ltd, 1993
Editor: Larson, J. E. G.
Language: English, French, German and Spanish
Coverage: European overview of all environmental products and services and the major companies and organisations who provide them
Price: £125.00
ISBN: 0749408537

– 2185 – NAMAS Concise Directory of Accredited Laboratories 1991

Tel: 081 943 6311
Geographical coverage: United Kingdom
Publishers: Executive of the National Measurement and Accreditation Service
Price: £100 for 3 yrs

PERIODICALS

– 2186 – Aviation and Space Technology

1221 Ave of the Americas
New York NY 10020
United States of America
Tel: 1 609 426 5526
Fax: 1 609 426 6068
Geographical coverage: International
Publishers: McGraw Hill
Editor: Fink, D. E.

Frequency: weekly
Language: English
Coverage: Provides current news and information related to the aerospace industry
Price: $105
ISSN: 0005 2175

– 2187 – Energie & Milieutechnologie
(Energy and Environmental Technology)

Postbus 235
NL-2280 AE Rijswijk
Netherlands
Tel: 31 70 3988100
Fax: 31 70 3988276
Geographical coverage: Netherlands
Publishers: Stam Tijdschriften BV
Frequency: 8 issues a year
Language: Dutch
Price: Dfl94
ISSN: 0925 2924

– 2188 – Energieanwendung

Karl-Heinestrasse 27
D-7031 Leipzig
Germany
Tel: 37 41 4081011
Fax: 37 41 4012571
Geographical coverage: Germany
Publishers: Deutscher Verlag für Grundstoffindustrie
Frequency: monthly
Coverage: Covers the technology of energy and energy economics in industry agriculture and transportation
Annual subscription: DM198
ISSN: 0013 7405

– 2189 – Entsorgungspraxis

Postfach 6666
D-4830 Gütersloh 100
Germany
Tel: 49 5241 802241
Fax: 49 5241 73055
Geographical coverage: Germany
Publishers: Bertelsmann Fachzeitschriften
Editor: Wedig, R.
Frequency: monthly
Coverage: Deals with environmental technology in industry and municipal economy
Annual subscription: DM105
ISSN: 0724 6870

– 2190 – Environment and Industry Digest

258 Banbury Road
Oxford OX2 7DH
United Kingdom
Tel: 0865 512242
Fax: 0865 310981
Geographical coverage: United Kingdom
Publishers: Elsevier Advanced Technology
Frequency: monthly
Coverage: Covers technology of environmental protection across a broad range of industries, including manufacturing, power generation, water and the process industries
Annual subscription: £165.00
ISSN: 0958 2126

– 2191 – Environmental Technology

79 Rusthall Ave
Chiswick
London W4 1BN
United Kingdom
Tel: 081 995 4160
Geographical coverage: United Kingdom
Publishers: Selper Ltd
Editor: Lester, J. N. and Harrison, R M.
Frequency: monthly
Annual subscription: £180.00

– 2192 – Informazione Innovativa

Via dell Palme 13
I-35137 Padua
Italy
Tel: 39 49 36435
Geographical coverage: Italy
Publishers: Centro Studi "l'Uomo e l'Ambiente"
Editor: Merlin, T.
Language: Text in Italian, summaries in English
Coverage: Clean technologies, energy sources, industrial health and safety
Annual subscription: Lit400000

– 2193 – Miljoe og Teknologi

A-S Hartmannsvej 47–49
DK-2920 Gentofte
Denmark
Tel: 45 39 40 80 00
Fax: 45 39 40 82 80
Geographical coverage: Denmark
Publishers: Forlaget John Vaboe
Frequency: bi monthly
ISSN: 0901 747X

– 2194 – Options

IIASA
A-2361 Laxenburg
Austria
Tel: 43 2236 71521 0
Fax: 43 2236 71313
Geographical coverage: European
Publishers: IIASA
Editor: Clark, M.

– 2195 – Umwelt Technologie Aktuell

Roesslerstrasse 90
Postfach 11 05 64
D-6100 Darmstadt 11
Germany
Tel: 49 6151 8090 0
Fax: 49 6151 809045
Geographical coverage: Germany
Publishers: G I T Verlag GmbH
Editor: Giebeler, E.
Annual subscription: DM48

Frequency: quarterly
Language: English
Coverage: Includes articles on environmental technology

RESEARCH CENTRES

– 2196 – Centre for Environmental Technology (ICCET)

Imperial College of Science and Technology
Exhibition Road
London SW7 2AX
United Kingdom
Tel: 071 589 5111
Fax: 071 584 7596
Geographical coverage: United Kingdom
Aims and objectives: ICCET is an interdisciplinary centre set up within Imperial College to coordinate and promote research and training in environmental technology
Publications: *MSc Environmental Technology Abstracts*; occasional research reports and papers
Research activities: Environmental technology

– 2197 – Centro de Investigación Energética Medioambientale y Tecnólogica (CIEMAT)

Avenida Compultense 22
E-28040 Madrid
Spain
Tel: 34 1 34 66 555/34 66 676
Fax: 34 1 34 66 005
Contact name: Mrs Pilar Garcia-Santesmases
Geographical coverage: Spain
Research activities: Covering all energy sources and environmentally related technologies, including conventional and radiological issues

– 2198 – Forschungsinstitut für Internationale Technische und Wirtschaftliche Zusammenarbeit der RWTH Aachen

(Research Institute for International Technology and Economic Cooperation of Aachen University)

Henrichistrasse 50
D-5100 Aachen
Germany
Tel: 49 241 84071 73
Geographical coverage: Germany

– 2199 – Institut de l'Environnement du Centre Commun de Recherche

T P 290
I-21020 Ispra (VA)
Italy
Tel: 39 332 78 93 04
Fax: 39 332 78 92 22
Contact name: Mr C Savatteri
Geographical coverage: Italy
Research activities: Global change and environmental chemicals

– 2200 – Institut Nacional de Engenharia e Technologia Industrial (INETI)

Azinhaga dos Lameiros à Estrada do Paço do Luminar
P-1699 Lisbon
Portugal
Tel: 351 1 716 51 41
Fax: 351 1 716 09 01
Contact name: Mrs Maria Constança Peneda
Geographical coverage: Portugal
Aims and objectives: INETI's aim is to perform research and development in the fields of environment and technology

– 2201 – International Association of Packaging Research Institutes (IAPRI)

Flamingo Straat
B-V900 Ghent
Belgium
Contact name: Dr Franz Lox, Secretary General
Geographical coverage: International
Aims and objectives: IAPRI aims to promote the exchange and collaboration of R & D and testing between packaging scientists and technologists in all parts of the world

– 2202 – Irish Science and Technology Agency (EOLAS): *see entry 2152*

Geographical coverage: Ireland

– 2203 – Kernforschungszentrum Karlsruhe GmbH (KfK)

Postfach 3640
D-7500 Karlsruhe
Germany
Tel: 49 72 47 821
Fax: 49 72 47 82 39 49
Contact name: Dr F Arendt
Geographical coverage: Germany
Research activities: Research is oriented towards environment, energy and micro systems engineering

– 2204 – Technical University of Twente

PO Box 217
NL-7500 AE Enschede
Netherlands
Tel: 31 5389 9111
Fax: 31 5335 7956
Contact name: J Weber
Geographical coverage: Netherlands
Research activities: Developing a CFC car that removes coolant and oil from discarded refrigerators

– 2205 – Territorial and Environmental Informatics Interdepartmental Research Centre

Via Ponzio 34–5
I-20133 Milan
Italy
Tel: 39 2 2399 1
Geographical coverage: Italy
Research activities: Research coordination in the fields of territorial and environmental informatics; graphics software

– 2206 – Toegepast Natuurwetenschappelijk Onderzoek (TNO)

PO Box 6013
NL-2600 JA Delft
Netherlands
Tel: 31 15 69 68 86
Fax: 31 15 61 31 86
Contact name: Prof Foppe De Walle
Geographical coverage: Netherlands
Research activities: TNO's expertise and research capacity is focused on the issue of energy and the environment and on process technology

– 2207 – Warren Spring Laboratory

Gunnels Wood Road
Stevenage
Hertfordshire SG1 2BX
United Kingdom
Tel: 0438 741122
Fax: 0438 360858
Contact name: Dr Mark Kibblewhite
Geographical coverage: United Kingdom
Aims and objectives: The Environment Technology Agency of the Department of Trade and Industry; research and development in industrial processing, and monitoring and abatement of environmental pollution
Publications: Various publications
Research activities: Monitoring environmental quality, abating pollution and contamination, optimising the use of primary materials

DATABASES

– 2208 – CORDIS

Geographical coverage: European
Coverage: Research and technological development in the EC
Database producer: CEC Telecommunications, Information and Innovation
Database host: ECHO
Coverage period: 1963–
Update frequency: variable

– 2209 – Current Technology Index

Geographical coverage: International
Coverage: Environmental technology and sciences
Database producer: Bowker Saur
Database host: Dialog
Coverage period: 1981–
Update frequency: 1,400 records monthly
Additional information: Also available in printed form

– 2210 – PASCAL

Geographical coverage: European
Coverage: Multidisciplinary bilingual database holding European information on science and technology
Database producer: Centre National de la Recherche Scientifique
Database host: Dialog; ESA–IRS; Telesystems Questel
Coverage period: 1973–
Update frequency: 40,000 records monthly

Ethical Investment

REPORTS

– 2211 – Environmental Opportunities: Building Advantage out of Uncertainty

5 Berners Road
London N1 0PW
United Kingdom
Tel: 071 354 9442
Geographical coverage: United Kingdom
Publishers: Centre for Exploitation of Science and Technology (CEST)
Coverage: Environmental investments

– 2212 – Greening of Global Investment Special Report 2108, The

Geographical coverage: United Kingdom
Publishers: Economic Intelligence Unit
Coverage: Ethical investment in the UK and world

ISBN: 0850585711
Author: Simpson, A.

RESEARCH CENTRES

– 2213 – Ethical Investment Research Information Service (EIRIS)

4.01 Bondway Business Centre
71 Bondway
London SW8 1SQ
United Kingdom
Tel: 071 735 1351
Contact name: Peter Webster, Executive Secretary
Geographical coverage: United Kingdom
Aims and objectives: EIRIS does research for shareholders who want to check their existing investments against particular ethical criteria and find suitable alternatives

12
Miscellaneous

General

GOVERNMENT ORGANISATIONS

– 2214 – Geological Survey of Ireland

Beggars Bush
Haddington Road
Dublin 4
Eire
Tel: 353 1 609511
Geographical coverage: Ireland

– 2215 – Kornyeyetvedelme Ens Teruletfejlesgtesi Minisyterium

Fo utca 44–50
H-1011 Budapest
Hungary
Tel: 36 1 201 4133
Geographical coverage: Hungary

– 2216 – Statistik Sentralbyra (SSB)
(Central Statistical Office)

PO Box 8131 Dep
N-0033 Oslo 1
Norway
Tel: 47 2 864500
Fax: 47 2 333264
Geographical coverage: Norway

NON-GOVERNMENT ORGANISATIONS

– 2217 – Environment Liaison Centre International (ELC)

PO Box 72461
Nairobi
Kenya
Kenya
Tel: 254 2 562015
Geographical coverage: International
Publications: *Ecoforum*, bi monthly; *Ecoprobe*

– 2218 – Environmental Information Service

17 St Andrew Street
Dublin 2
Eire
Tel: 353 1 6793144
Fax: 353 1 6795204
Geographical coverage: Ireland

BOOKS

– 2219 – Environment Software

Tel: 0865 512242
Geographical coverage: United Kingdom
Publishers: Elsevier Advanced Technology
Price: £695.00

– 2220 – Finland Central Statistical Office: Statistical Surveys Environmental Statistics

Tilastokeskus Annankatu 44
SF-00100 Helsinki 10
Finland
Geographical coverage: Finland
Frequency: irregular
Language: English, Finnish and Swedish

– 2221 – Information Please Environmental Almanac 1993

Geographical coverage: International
Publishers: Houghton Mifflin, 1992
Language: English
Coverage: Provides environmental profiles for every country in the world
Price: £14.99
ISBN: 0395637678

– 2222 – Statistika Centralbyraan Statistika Meddelanden Subgroup NA

(Natural Studies and the Environment)

S-701 89 Oerebro
Sweden
Geographical coverage: Sweden
Publishers: Statistika Centralbyraan Publishing Unit
Frequency: biennial
Language: Text in Swedish, summaries in English
ISSN: 0282 3500

– 2223 – Swedish Environment

Geographical coverage: Sweden
Publishers: Statistics Sweden, 1991
Editor: Solveig, Ms
Frequency: annual
Language: English
Coverage: Translated version of the publication *Miljösverige*. It presents facts and figures on the state of the environment
ISBN: 9161804371

– 2224 – What You Don't Know Will Hurt You: Environmental Information as a Basic Human Right

Geographical coverage: Norway
Publishers: Nordic Council of Ministers
Editor: Ofstad, S.
Language: Norwegian and English
Price: free
ISBN: 8272438941

DIRECTORIES/YEARBOOKS

– 2225 – Ecolinking: Everyone's Guide to Online Environmental Information

Geographical coverage: International
Publishers: Peachpit Press Publications, US, 1992
Coverage: Guide to connecting online with services and users worldwide
ISBN: 0938151355
Author: Rittner, D.

– 2226 – Environment Databases

Geographical coverage: United Kingdom
Publishers: ASLIB, 1991
Editor: Cox, J.
Coverage: Lists environment databases
ISBN: 0851422616

– 2227 – Environmental Information: A Guide to Sources

Geographical coverage: United Kingdom
Publishers: The British Library, 1992
Coverage: A survey of important and useful sources of information on environmental problems. Sources covered by the guide, include online and CD–ROM databases, printed publications and key organisations
Price: £25.00
ISBN: 0712307834
Author: Lees, N. and Woolston, H.

– 2228 – European Environmental Yearbook

Geographical coverage: European
Publishers: DocTer Istituto di Studi e Documentazione per il Territorio Milano, 1991
Language: English
Coverage: Main aspects of the environment in Europe

– 2229 – Guide to Libraries and Information Units in Government Departments and other Organisations

Geographical coverage: United Kingdom
Publishers: The British Library
Editor: Dale, P.
Price: £30.00
ISBN: 0712307729

– 2230 – International Environmental Information Sources

Geographical coverage: International
Publishers: Pira, 1990

Coverage: Entries include organisations, research centres, legislative/regulatory bodies, directories, statistics, online databases and periodicals
Price: £130.00
ISBN: 0902799584

– 2231 – Miljoarboka
(Environmental Yearbook)

Geographical coverage: Norway
Publishers: Det Norske Samlaget
Frequency: annual

– 2232 – Miljovejviseren 1991
(Danish Environmental Directory)

Geographical coverage: Denmark
Publishers: Kobenhavn: Forlaget Kommuneinformation, 1990
Editor: Horsten, V. and Stoustrup, F.
Language: Danish with English headings
Coverage: Environmental sources in Denmark
ISBN: 8773166553

– 2233 – Naturressurser Og Miljo 1991
(Natural Resources and the Environment)

Geographical coverage: Norway
Publishers: Statistik Sentralbyra Oslo–Kongsvinger 1992
Frequency: annual
Coverage: Most aspects of the environment
ISSN: 0332 8422
ISBN: 8253736517

– 2234 – State of the Environment, The

Geographical coverage: International
Publishers: OECD, 1991
ISBN: 9264134425

– 2235 – Ulrichs International Periodicals Directory

Geographical coverage: International
Publishers: Bowker-Saur
Frequency: annual
Coverage: Lists major periodicals for many countries worldwide
ISSN: 0000 0175

– 2236 – Umweltforschungskatalog
(Environment Research Directory)

Viktoriastrasse 44a
D-4800 Bielefeld 1
Germany
Tel: 49 521 58308
Geographical coverage: Germany
Publishers: Erich Schmidt Verlag GmbH & Co

Coverage: Lists environment related research establishments in Germany

– 2237 – World Environment Research Directory

Geographical coverage: International
Publishers: Pira, 1992
Coverage: Reference source of research into environmental matters worldwide
Price: £145.00
ISBN: 0902799959

– 2238 – World Peace Directory and Diary 1993

Geographical coverage: International
Publishers: Housmans, 1992
Editor: Housmans Diary Group
Frequency: annual
Coverage: Peace Diary and Directory of worldwide peace, human rights and environmental groups
Price: £5.50
ISSN: 0957 0136
ISBN: 0852832257

PERIODICALS

– 2239 – Earth Matters

Friends of the Earth
26–28 Underwood Street
London N1 7JQ
United Kingdom
Tel: 071 490 1555
Geographical coverage: United Kingdom
Publishers: Friends of the Earth (FoE)
Frequency: quarterly
Coverage: Updates on the progress of campaigns and in-depth features on certain issues
Price: £1.25
ISSN: 0956 6651

– 2240 – Environment International

Headington Hill Hall
Oxford OX3 0BW
United Kingdom
Tel: 0865 794141
Fax: 0865 743592
Geographical coverage: International
Publishers: Pergamon Press plc
Editor: Moghissi, A.
Frequency: 6 issues a year
Language: English
Annual subscription: DM700
ISSN: 0160 4120

– 2241 – Environment Research

34–38 Chapel Street
Little Germany
Bradford BD1 5DN
United Kingdom
Tel: 0274 729315
Fax: 0274 306981
Geographical coverage: European
Publishers: European Research Press Ltd
Editor: Wilson, J.
Frequency: 6 issues a year
Language: English
Annual subscription: £30.00
ISSN: 0958 9082

– 2242 – Environmental Research Newsletter

Environment Institute Joint Research Centre
I-21020 Ispra (VA)
Italy
Geographical coverage: European
Publishers: Commission of the European Communities
Editor: Rossi, Dr G.
Frequency: half yearly
Language: English
Price: free

– 2243 – Environmentalist

PO Box 81
Northwood
Middlesex HA6 3DN
United Kingdom
Tel: 09274 23586
Fax: 09274 25066
Geographical coverage: International
Publishers: Science and Technology Letters
Editor: Potter, J. F.
Frequency: quarterly
Annual subscription: $180
ISSN: 0251 1088

– 2244 – Eta Verde

Casella Postale 443
I-00100 Rome
Italy
Tel: 39 6 461 849
Geographical coverage: Italy
Publishers: Associazione l'Eta Verde
Frequency: bi monthly
Coverage: Environmental studies
Annual subscription: Lit25000

– 2245 – European Environment

34–38 Chapel Street
Little Germany
Bradford BD1 5DN
United Kingdom
Tel: 0274 729315
Fax: 0274 306981
Geographical coverage: European
Publishers: European Research Press Ltd
Editor: Gouldson, A.
Frequency: 6 issues a year
Language: English
Coverage: All aspects of the environment in Europe
Annual subscription: £25.00
ISSN: 0961 0405

– 2246 – Fresenius Environmental Bulletin

PO Box 133
CH-4010 Basel
Switzerland
Tel: 41 61 737740
Fax: 41 61 737950
Geographical coverage: Switzerland
Publishers: Birkhäuser Verlag
Editor: Korte, F.
Frequency: monthly
Language: English
Annual subscription: Sfr244
ISSN: 1018 4619

– 2247 – Governo Locale ed Economia dell' Ambiente (GEA)

Via Crimea 1
Casella Postale 290
I-47037 Rimini
Italy
Tel: 39 541 626777
Fax: 39 541 622020
Geographical coverage: Italy
Publishers: Maggioli Editore
Editor: Ruperi, A.
Frequency: bi monthly
Annual subscription: Lit105000

– 2248 – Guardian

119 Farringdon Road
London EC1R 3ER
United Kingdom
Tel: 071 278 2332
Fax: 071 837 2114
Geographical coverage: United Kingdom
Publishers: Guardian Newspapers Ltd
Editor: Preston, P.
Frequency: daily

Coverage: Every Friday covers aspects of the environment worldwide
Price: 45 pence

– 2249 – Information Bulletin

Information Bulletin
REC
Miklós tér 1
H-1035 Budapest
Hungary
Tel: 36 1 168 6284
Fax: 36 1 168 7851
Geographical coverage: European
Publishers: Regional Environmental Center
Editor: Dailey, J.
Frequency: quarterly
Language: English
Coverage: Environmental issues in central and eastern Europe
Price: free

– 2250 – Lettre de l'Environnement

11 rue de Marché St Honoré
F-75001 Paris
France
Tel: 33 1 429 667 22
Fax: 33 1 404 007 75
Geographical coverage: France
Publishers: F A Jour
Frequency: 18 issues a year

– 2251 – Miljoeprojekt

Strandgade 29
DK-1401 Copenhagen K
Denmark
Geographical coverage: Denmark
Publishers: Miljöstyrelsen
Frequency: irregular
Language: Environmental studies

– 2252 – Miljoeundersoegelser ved Marmorilik

Tagensvej 135
DK-2200 Copenhagen N
Denmark
Tel: 45 35 82 14 20
Geographical coverage: Denmark
Publishers: Groenlands Miljoundersoegelser
Language: Text in Danish, summaries in English and Greenlandic
Coverage: Environmental studies

– 2253 – Natura (Revista)

Margués de Villamagna 4
E-28001 Madrid
Spain
Tel: 34 1 435 81 00
Fax: 34 1 576 78 81
Geographical coverage: Spain
Publishers: G & J Espana SA
Editor: Huerta, A.
Coverage: Presents articles on environmental studies

– 2254 – Planeta Verde

Virgen de Nuria 23
E-28027 Madrid
Spain
Fax: 34 1 268 33 71
Geographical coverage: Spain
Editor: Cabal, E.

– 2255 – Retema–Medio Ambiente–Revista

Jacinto Verdaguer 25
2 B Esc izda
E-28019 Madrid
Spain
Tel: 34 1 471 38 98
Fax: 34 1 471 38 98
Geographical coverage: Spain
Publishers: C & M Publicaciones S L

– 2256 – Sécrétariat d'Etat Chargé de l'Environnement et de la Qualité de la Vie: Bulletin de Documentation de l'Environnement

29–31 quai Voltaire
F-75340 Paris Cedex 7
France
Tel: 33 1 401 570 00
Geographical coverage: France
Publishers: Sécrétariat d'Etat Chargé de l'Environnement
Frequency: 5 issues a year
Annual subscription: Ffr237

– 2257 – Umwelt Aktuell

Amerlingstrasse 19
A-1060 Vienna
Austria
Geographical coverage: Austria
Publishers: Oswald Moebius Verlag
Editor: Aglas, E. H.
Frequency: bi monthly
Coverage: Covers environmental issues

– 2258 – Umweltmagazin

Max-Planck-Strasse 7–9
D-8700 Würzburg 1
Germany
Tel: 49 931 418 0
Fax: 49 931 418 26 40
Geographical coverage: Germany
Publishers: Vogel Verlag und Druck K G
Editor: Jobst, J.
Frequency: monthly
Coverage: Forum for environment specialists in industry and municipalities
Price: DM12
Annual subscription: DM142
ISSN: 0341 1206

REPORTS

– 2259 – Swedish Environment, The: Annexe to Sweden; National Report to UNCED 1992

Geographical coverage: Sweden
Publishers: Ministry of the Environment, Sweden 1992
Language: English
Coverage: Overview of the Swedish environment

RESEARCH CENTRES

– 2260 – Geological Survey of Denmark

Thoravej 8
DK-2400 Copenhagen NV
Denmark
Tel: 45 31 10 66 00
Fax: 45 31 19 68 68
Geographical coverage: Denmark
Research activities: Covers the environment, energy, raw materials, ground water and other environmental concerns

– 2261 – International Institute for Environment and Society

Reichpietschufer 50
D-1000 Berlin 30
Germany
Contact name: Prof Udo Ernst Simonis, Director
Geographical coverage: Germany

DATABASES

– 2262 – EcoBase

Geographical coverage: United Kingdom
Coverage: Categorised listings of UK environmental organisations
Database producer: Keystroke Knowledge
Database host: Manchester Host; GeoNet; Poptel
Coverage period: Current
Update frequency: monthly
Additional information: Data available on diskette

– 2263 – ENREP-NEW

Geographical coverage: European
Coverage: Referral system covering two databases, ENREP and ENDOC; research in progress has details on 25,000 environmental projects
Database producer: CEC Environment, Civil Protection and Nuclear Safety
Database host: ECHO
Coverage period: 1979–
Update frequency: annually

– 2264 – Environment Digest

Geographical coverage: International
Coverage: Monthly summary of environmental news and comments
Database producer: Environment Digest
Database host: Manchester Host; GeoNet; Poptel
Coverage period: current
Update frequency: monthly

– 2265 – EUROCRON

Geographical coverage: European
Coverage: EC statistics including environment
Database producer: EUROSTAT
Database host: Eurobases
Coverage period: 1989–
Update frequency: monthly

– 2266 – GreenNet

Geographical coverage: International
Coverage: Email and conferencing; for UK and international environmental, peace and human rights issues; data is input by subscribers, managed by the Association for Progressive Communications
Database host: GreenNet
Coverage period: current
Update frequency: constant

– 2267 – HMSO Online/British Official Publications

Geographical coverage: United Kingdom
Coverage: British official publications, including EC and international organisations
Database producer: Her Majesty's Stationery Office (HMSO)
Database host: Blaise; Dialog
Coverage period: 1976–
Update frequency: 1,000 records monthly

– 2268 – Multinational Environmental Outlook

Geographical coverage: International
Coverage: Environmental issues worldwide
Database producer: Business Publishers Inc
Database host: via PTS Newsletter Database on DataStar and Dialog
Coverage period: 1981–
Update frequency: twice weekly

– 2269 – System for Information on Grey Literature in Europe (SIGLE)

Geographical coverage: European
Coverage: Covers reports and conference proceedings, including semi-published material
Database producer: European Association for Grey Literature Exploitation
Database host: Blaise; STN
Coverage period: 1980–

– 2270 – Umweltliteraturdatenbank (ULIDAT)

Umweltbundesamt
Fachgebiet Z 2.4
Bismarckplatz 1
D-1000 Berlin 33
Germany
Tel: 49 30 8903 2423
Fax: 49 30 8903 2285
Geographical coverage: Germany
Coverage: Environment literature database
Database producer: Umweltbundesamt UMPLIS
Database host: Data Star, FIZ Technik and STN International
Coverage period: 1976–

LIBRARIES

– 2271 – Umweltbibliotek

Griebenowstrasse 16
D-1058 Berlin
Germany
Geographical coverage: Germany
Aims and objectives: National environment library

Chemicals

NON-GOVERNMENT ORGANISATIONS

– 2272 – Association of Chemical Industries (VCI)

Karl Strasse 21
D-6000 Frankfurt 1
Germany
Tel: 49 69 255 64 71
Geographical coverage: Germany

– 2273 – Chemical Industries Association

King's Building
Smiths Square
London SW1 3JJ
United Kingdom
Tel: 071 834 3399
Geographical coverage: United Kingdom

– 2274 – Conseil Européen des Fédérations de l'Industrie Chimique (CEFIC)

(European Chemical Industry Council)

Avenue E Van Nieuwenhuyse 4
Box 1
B-1160 Brussels
Belgium
Tel: 32 2 676 72 11
Fax: 32 2 676 73 01
Contact name: Louis Jourdan, Director of Technical Affairs
Geographical coverage: European
Aims and objectives: CEFIC represents the National Chemical Federations and chemical companies of Europe. It covers issues such as the environment, health and safety, transport and distribution of chemicals. It is also concerned with proposed and pending EC legislation
Publications: Annual reports; position statements and monographs on selected topics

– 2275 – European Federation of Chemical and General Workers' Unions (FESCID)

Avenue Emile de Béco 109
B-1050 Brussels
Belgium
Tel: 32 2 648 24 97
Fax: 32 2 646 06 85
Contact name: Franco Bisegna, Secretary General
Geographical coverage: European
Aims and objectives: FESCID represents the views of EC Member State chemical and general workers' unions to the EC institutions

– 2276 – European Federation of Pharmaceutical Industries (EFPIA)

Avenue Louise 250
Box 91
B-1050 Brussels
Belgium
Tel: 32 2 640 68 15
Fax: 32 2 647 60 49
Contact name: Nelly Baudrihaye, Director General
Geographical coverage: European
Aims and objectives: EFPIA represents the views of pharmaceutical associations in sixteen countries

– 2277 – International Association of the Soap and Detergent Industry (AIS)

Square Marie Louise 49
B-1040 Brussels
Belgium
Tel: 32 2 230 83 71
Fax: 32 2 230 82 88
Contact name: P V Costa, Secretary General
Geographical coverage: International
Aims and objectives: AIS is a representative and educational body for the soap and detergent industry. It produces environmental information for consumers

– 2278 – Pharmaceutical Group of the EC (GPCE)

Square Ambiorix 13
B-1040 Brussels
Belgium
Tel: 32 2 736 72 81
Fax: 32 2 736 02 06
Contact name: Paul Baetens, Secretary General
Geographical coverage: European
Aims and objectives: GPCE represents the national associations of pharmacists to the EC institutions

BOOKS

– 2279 – Environmental Chemistry

Geographical coverage: United Kingdom
Publishers: Chapman and Hall, 1993
Price: £13.95
ISBN: 0412484900

PERIODICALS

– 2280 – Chemical Speciation and Bioavailability

PO Box 81
Northwood
Middlesex HA6 3AA
United Kingdom
Tel: 0923 823586
Fax: 0923 825066
Geographical coverage: United Kingdom
Publishers: Science and Technology Letters
Frequency: quarterly
Coverage: Presents papers in an interdisciplinary forum that explore the chemical physical biological and ecological effects of chemical species in the environment
Annual subscription: $140
ISSN: 0954 2299

– 2281 – Chemical Week

888 Seventh Avenue
New York NY 10106
United States of America
Tel: 1 212 621 4900
Fax: 1 212 621 4949
Geographical coverage: International
Publishers: Chemical Week Associates
Editor: Hunter, D
Frequency: weekly
Coverage: News and analysis of the chemical process industries
Price: $99
ISSN: 0009 272X

– 2282 – Chemistry in Britain

Thomas Graham Science Park
Milton Road
Cambridge CB4 4WF
United Kingdom
Tel: 0462 672555
Fax: 0462 480947
Geographical coverage: United Kingdom
Publishers: Royal Society of Chemistry
Editor: Farago, Dr P. J.
Frequency: monthly
Annual subscription: £143.00
ISSN: 0009 3106

Chemicals Miscellaneous

– 2283 – Environmental Protection Bulletin

Davis Building
165–171 Railway Terrace
Rugby
Warwickshire CV21 3HQ
United Kingdom
Tel: 0788 578214
Fax: 0788 560833
Geographical coverage: United Kingdom
Publishers: Institution of Chemical Engineers
Editor: Gardner, D.
Frequency: bi monthly
Coverage: An exchange of information and experience on environmental matters related to the chemical and process industries
Annual subscription: £105.00
ISSN: 0957 9052

– 2284 – European Chemical News

Quadrant House
The Quadrant
Sutton
Surrey SM2 5AS
United Kingdom
Tel: 081 652 3500
Geographical coverage: European
Publishers: Reed Business Publishing Ltd
Editor: Baker, J.
Frequency: weekly
Coverage: Chemistry manufacturing
Annual subscription: £210.00
ISSN: 0014 2875

– 2285 – Pigment and Resin Technology

127 Stanstead Road
London SE23 1JE
United Kingdom
Tel: 081 699 6792
Fax: 081 699 1753
Geographical coverage: United Kingdom
Publishers: Sawell Publications Ltd
Editor: Bean, J. E.
Frequency: monthly
Annual subscription: £40.25
ISSN: 0369 9420

– 2286 – Surface Coatings International

Priory House
967 Harrow Road
Wembley HA0 2SF
United Kingdom
Tel: 081 908 1086
Fax: 081 908 1219
Geographical coverage: International
Publishers: JOCCA
Editor: Gale, L.
Frequency: monthly
Coverage: News and reviews of scientific and commercial interest
Annual subscription: £88.00

RESEARCH CENTRES

– 2287 – Institut für Ökologische Chemie
(Institute for Ecological Chemistry)

Ingolstädter Landstrasse 1
Neuherberg
D-8042 Oberschleisheim
Germany
Tel: 49 89 318 72690
Fax: 49 89 318 73322
Contact name: Prof Dr A Kettrup, Director
Geographical coverage: Germany
Research activities: Compiling profiles of chemicals in the environment

– 2288 – Instituto de Química Orgánica General
(General Organic Chemistry Institute)

Juan de la Cierva 3
E-28006 Madrid
Spain
Tel: 34 1 262 29 00
Geographical coverage: Spain
Research activities: Organic chemistry (synthesis, separation techniques, environmental pollution, natural products)

DATABASES

– 2289 – Analytical Abstracts

Royal Society of Chemistry
Burlington House
London W1V 0BN
United Kingdom
Tel: 071 437 8656
Geographical coverage: International
Coverage: Analytical chemistry
Database producer: Royal Society of Chemistry
Database host: DataStar; Dialog; Orbit
Coverage period: 1980–
Update frequency: monthly

– 2290 – Beilstein Online

Geographical coverage: International
Coverage: Organic chemistry
Database producer: Beilstein Institute
Database host: Dialog; Orbit
Coverage period: 1830–1980

– 2291 – Chemical Abstracts (CA Search)

Geographical coverage: International
Coverage: Chemical literature classified in subject groups, e.g. biochemistry, organic chemical engineering
Database producer: Chemical Abstracts
Database host: BRS; DataStar; DIMDI; ESA–IRS; Orbit; Telesystems Questel
Coverage period: 1967–
Update frequency: 19,000 records fortnightly

Population

NON-GOVERNMENT ORGANISATIONS

– 2292 – European Association for Population Studies (EAPS)

Lange Houtstraat 19
PO Box 11676
NL-2502 AR The Hague
Netherlands
Contact name: Dr Charlotte Höhn, President
Geographical coverage: European
Aims and objectives: EAPS aims at promoting the study of population in Europe, fosters and stimulates interest in population matters among governments, national and international organisations and the general public
Publications: *Revue Européenne de Démographie*, quarterly

BOOKS

– 2293 – Commons Without Tragedy

Geographical coverage: United Kingdom
Publishers: Shepheard-Walwyn, 1991
Coverage: Examines the impact of population growth on the economy and the environment
Price: £10.95
ISBN: 0856831263
Author: Andelson, R. V.

– 2294 – Population Explosion, The

Geographical coverage: International
Publishers: Franklin Watts, 1990
Coverage: Looks at the growth in world population, the provision of food, pollution and efforts to plan for the future
Price: £6.95
ISBN: 0749601213
Author: Becklake, J.

– 2295 – Taking Population Seriously

Geographical coverage: United Kingdom
Publishers: Earthscan Publications, 1989
Author: Lappe, F. M. and Schurman, R.

– 2296 – Third Revolution, The: Environment, Population and a Sustainable World

Geographical coverage: United Kingdom
Publishers: I B Tauris and Co, London 1992
Author: Harrison, P.

CONFERENCE PAPERS

– 2297 – International Population Conference Proceedings

34 rue des Augustins
B-4000 Liège
Belgium
Fax: 32 41 223847
Geographical coverage: International
Publishers: International Union for the Scientific Study of Population
Frequency: every four years
Language: English
Price: varies
ISSN: 0074 9338

DIRECTORIES/YEARBOOKS

– 2298 – Vaestoentutkimuksen Vuosikirja

(Yearbook of Population Research in Finland)

Kajevankatu 16
SF-00100 Helsinki 10
Finland
Tel: 358 1 640235
Fax: 358 1 6121211
Geographical coverage: Finland
Publishers: Vaestontutkimuslaitos
Editor: Lindgren, J.
Frequency: annual
Language: Text in English

Coverage: Articles on current interest in demography in Finland. Also population data compiled at the Institute and a bibliography of Finnish population research every other year
Price: Fmk100
ISSN: 0506 3590

PERIODICALS

– 2299 – International Migration

17 route des Morillons
Case Postale 71
CH-1211 Geneva
Switzerland
Tel: 41 22 717 9111
Geographical coverage: International
Publishers: International Organisation for Migration
Editor: Appleyard, R. T.
Frequency: quarterly
Language: Text mainly in English, occasionally French or Spanish
Coverage: Covers current migration issues
Price: $20
ISSN: 0020 7985

– 2300 – Interuniversitair Demografisch Institut Publications

PO Box 17
NL-3300 AA Dordrecht
Netherlands
Tel: 31 78 334911
Fax: 31 78 334254
Geographical coverage: Netherlands
Publishers: Kluwer Academic Publishers
Frequency: irregular
Language: Text in English
Price: varies

– 2301 – Population et Sociétés

27 rue du Commandeur
F-75675 Paris Cedex 14
France
Geographical coverage: France
Publishers: Institut National d'Etudes Démographiques
Editor: Levy, M.
Frequency: monthly
Annual subscription: Ffr75
ISSN: 0184 7783

– 2302 – Population Studies

London School of Economics
Houghton Street
Aldwych
London WC2A 2AE
United Kingdom
Tel: 071 955 7666
Fax: 071 242 0392
Geographical coverage: United Kingdom
Publishers: Population Investigation Committee
Frequency: 3 issues a year
Coverage: Covers the field of demography
Annual subscription: £42.00
ISSN: 0032 4728

– 2303 – Review of Population Research

27 rue du Commandeur
F-75675 Paris Cedex 14
France
Geographical coverage: International
Publishers: Committee for International Cooperation in National Research in Demography
Editor: Bourgeois-Pichat, J
Frequency: quarterly
Language: English
Price: free
ISSN: 0377 8967

REPORTS

– 2304 – Population: An Ecofeminist Perspective

Aberdeen Studios
22 Highbury Grove
London N5 2EA
United Kingdom
Tel: 071 354 8823
Fax: 071 354 0464
Geographical coverage: United Kingdom
Publishers: Women's Environmental Network, June 1992

Transport

GOVERNMENT ORGANISATIONS

– 2305 – Bundesministerium für Öffentliche Wirtschaft und Verkehr
(Federal Ministry for Nationalised Industries and Transport)

Radetzkystrasse 2
A-1031 Vienna
Austria
Geographical coverage: Austria

– 2306 – Department of Tourism Transport and Communication

Kildare Street
Dublin 2
Eire
Tel: 353 1 789522
Geographical coverage: Ireland

– 2307 – Department of Transport (DTp)

2 Marsham Street
London SW1P 3EB
United Kingdom
Tel: 071 276 5089
Contact name: Public Enquiries Unit
Geographical coverage: United Kingdom
Aims and objectives: DTp is responsible for the motorway and trunk road network in England. It has oversight of all transport sectors and is responsible for regulating their safety. Environmental considerations are an important aspect of the DTp's activities

– 2308 – Federal Department of Transport Communications and Energy

Bundeshaus Nord
CH-3003 Berne
Switzerland
Tel: 41 31 61 41 11
Fax: 41 31 22 95 76
Geographical coverage: Switzerland

– 2309 – Ministerio de Obras Públicas y Transportes
(Ministry of Public Works Town Planning and Transport)

Nuevos Ministerios
Paseo de la Castellana 67
E-28071 Madrid
Spain
Tel: 34 1 553 16 00
Fax: 34 1 543 65 77
Contact name: José Borrell Fontelles, Minister
Geographical coverage: Spain

– 2310 – Ministry of Communications and Transport

Eteläesplanadi 16
SF-00130 Helsinki 13
Finland
Geographical coverage: Finland

– 2311 – Ministry of Equipment Housing and Transport

Grand Arche-La Défense
F-92055 Paris La Défense Cedex 04
France
Tel: 33 1 408 121 22
Geographical coverage: France

– 2312 – Ministry of Transport

Piazza della Croce Rossa
I-00161 Rome
Italy
Tel: 39 6 84901
Fax: 39 6 841 5693
Geographical coverage: Italy

– 2313 – Ministry of Transport

Postfach 20 01 00
D-5300 Bonn 2
Germany
Tel: 49 228 3000/01
Fax: 49 228 3003428/29
Geographical coverage: Germany

– 2314 – Ministry of Transport and Communications

Xenofontos 13
GR-10557 Athens
Greece
Tel: 30 1 325 1211/15
Geographical coverage: Greece

– 2315 – Ministry of Transport and Communications

S-103 33 Stockholm
Sweden
Geographical coverage: Sweden

– 2316 – **Ministry of Transport and Energy**

19–21 boulevard Royal
Luxembourg
Luxembourg
Tel: 352 47 94 1
Fax: 352 46 43 15
Geographical coverage: Luxembourg

– 2317 – **Ministry of Transport and Public Works**

Plesmanweg 1
PO Box 20901
NL-2500 EX The Hague
Netherlands
Tel: 31 70 351 6171
Fax: 31 70 351 7895
Geographical coverage: Netherlands

– 2318 – **Ministry of Transport and Shipping**

ul T Chalubinskiego 4/6
PL-00 928 Warsaw
Poland
Tel: 48 22 24 40 00
Geographical coverage: Poland

– 2319 – **Ministry of Transport and Telecommunications**

Dob u 75–81
H-1077 Budapest
Hungary
Tel: 36 1 122 0220
Geographical coverage: Hungary

– 2320 – **Ministry of Transport Communications and Public Works**

Mileticova 19
820 06 Bratislava
Slovakia
Tel: 42 7 672 36/672 44
Fax: 42 7 211 221
Geographical coverage: Slovakia

– 2321 – **Royal Ministry of Transport and Communications**

Mollergt 1-3
PO Box 8010 Dep
N-0030 Oslo
Norway
Tel: 47 2 34 90 90
Geographical coverage: Norway

– 2322 – **Trafikministeriet**
(Ministry of Transport and Public Works)

Frederiksholms Kanal 27
DK-1220 Copenhagen K
Denmark
Tel: 45 33 92 33 55
Geographical coverage: Denmark

– 2323 – **Transportforskningsberedningen (TFB)**
(Swedish Transport Research Board)

Birger Jarlstorg 5
S-111 28 Stockholm
Sweden
Tel: 46 8 796 64 00
Fax: 46 8 24 56 97
Contact name: Arne Kihblom, Information Officer
Geographical coverage: Sweden
Aims and objectives: TFB supervise and coordinate functions in Swedish transport research

NON-GOVERNMENT ORGANISATIONS

– 2324 – **Coventry City Council**

Earl Street
Coventry CV1 5RR
United Kingdom
Tel: 0293 833333
Geographical coverage: United Kingdom
Aims and objectives: The Council is helping to promote the development and use of electric cars

– 2325 – **Electric Vehicle Association (EVA)**

Leicester House
8 Leicester Street
London WC2H 7BN
United Kingdom
Tel: 071 437 0678
Geographical coverage: United Kingdom
Aims and objectives: EVA aims to represent, promote, encourage, foster and develop the interests of the electric vehicle industry

– 2326 – **Environmental Transport Association (ETA)**

The Old Post House
91 Heath Road
Weybridge KT13 8RS
United Kingdom
Tel: 0932 828882
Fax: 0932 829015
Contact name: Administrator
Geographical coverage: United Kingdom

Aims and objectives: ETA is a non-profit service and campaigning association representing transport users who are concerned about the environment
Publications: *Going Green*, four times a year

– 2327 – European Association of Automobile Manufacturers (ACEA)

20011 rue du Noyer
B-1040 Brussels
Belgium
Tel: 32 2 732 55 50
Geographical coverage: European
Aims and objectives: ACEA works in promoting community wide car recycling scheme

– 2328 – European Conference of Ministers of Transport (ECMT)

19 rue de Franqueville
F-75775 Paris Cedex 16
France
Tel: 33 1 452 482 00
Fax: 33 1 452 497 42
Contact name: G Aurbach, Secretary General
Geographical coverage: European
Aims and objectives: ECMT aim to coordinate and promote the activities of international organisations concerned with European inland transport

– 2329 – European Federation for Transport and Environment (T & E)

17 George Street
Croydon CR0 1LA
United Kingdom
Tel: 081 681 71 85
Fax: 081 666 04 22
Contact name: Chris Bowers, Coordinator
Geographical coverage: European
Aims and objectives: The T & E is an umbrella organisation for the various groups across Europe which campaign for a more environmental approach to transport. Its goal is to promote a transportation policy which has the lowest impact on the environment

– 2330 – European Transport Law

Maria Henriettalei 1
B-2018 Antwerp
Belgium
Tel: 32 3 2313655
Fax: 32 3 23442380
Geographical coverage: European
Aims and objectives: Concerned with transport of dangerous goods and transfer of dangerous waste

PRESSURE GROUPS

– 2331 – Transport 2000

Walkden House
10 Melton Street
London NW1 2EJ
United Kingdom
Tel: 071 388 8386
Fax: 071 388 2481
Contact name: Mr Stephen Joseph, Executive Director
Geographical coverage: United Kingdom
Aims and objectives: The aim of Transport 2000 is to promote environmentally sound and socially responsible transport policies that put the interests of pedestrians, public transport users and vulnerable road users to the fore

BOOKS

– 2332 – Freight Transport and the Environment

Geographical coverage: Germany
Publishers: Elsevier Science, 1991
Language: English
ISBN: 0444887709
Author: Smit, R. and Van Ham, J.

– 2333 – Freight Transport and the Environment: ECMT Seminar 1991

Geographical coverage: European
Publishers: European Conference of Ministers of Transport (ECMT), 1991 Paris
Language: English and French
Price: Ffr120
ISBN: 9282111563

– 2334 – Greening of Urban Transport, The: Planning for Walking and Cycling in Western Cities

Geographical coverage: European
Publishers: Belhaven Press, 1990
Coverage: Outlines the principles and illustrates the practice of motor traffic restraint and exclusion in urban transport planning
Price: £27.50
ISBN: 1852930926
Author: Tolley, R. D.

– 2335 – Hazardous Materials: Sources of Information on their Transportation

Geographical coverage: United Kingdom
Publishers: The British Library, 1990

Coverage: Covering the years 1979–1990, this literature guide looks at the technical, social and legal aspects of transporting hazardous materials within Europe and the UK
Price: £25.00
ISBN: 0712307737
Author: Lees, N.

– 2336 – Less Traffic Better Towns

Geographical coverage: United Kingdom
Publishers: Friends of the Earth (FoE), 1992
Coverage: Identifies key issues, options and solutions
Price: £11.95
ISBN: 18575021209

– 2337 – Multimedia Transport and Fate of Pollutants

Geographical coverage: International
Publishers: Prentice Hall, 1993
Coverage: Coverage includes the effects of transport of meteorological conditions, and modes of transport as determined by the physiochemical properties of the chemical in question
Price: £68.50
ISBN: 0136057349
Author: Cohan, Y.

– 2338 – Transport and the Environment

Geographical coverage: United Kingdom
Publishers: HMSO, 1991
Author: Department of Transport

– 2339 – Transport Policy and the Environment

Geographical coverage: European
Publishers: European Conference of Ministers of Transport (ECMT), 1990, Paris
Language: English and French
Price: Ffr225
ISBN: 9282111474
Author: Prepared in cooperation with OECD

– 2340 – Transport Policy and the Environment

Geographical coverage: International
Publishers: E & F N Spon, 1992
Coverage: Presents the current thinking from leading authorities worldwide on transport and the environment
Price: £35.00
ISBN: 0419178708
Author: Banister, D. and Button, K.

– 2341 – Transport, the Environment and Economic Policy

Geographical coverage: United Kingdom

Publishers: Edward Elgar, 1993
Coverage: Sheds new light on the environmental costs of transport. Discusses greenhouse gases, depletion of non renewable sources, urban sprawl, acid rain and oil spills
Price: £35.00
ISBN: 1852784431
Author: Button, K. J.

PERIODICALS

– 2342 – Asintra

Plaza de la Castellana 120
4 izda
E-28046 Madrid
Spain
Geographical coverage: Spain
Publishers: Association Nacional Independiente de Transportes
Coverage: Aspects of transport relating to the environment

– 2343 – Bulletin des Transports

155 rue Légendre
F-75017 Paris
France
Geographical coverage: France
ISSN: 0007 4519

– 2344 – Electric Vehicle Developments

Research Applications
City University
Northampton Square
London EC1 0HB
United Kingdom
Geographical coverage: United Kingdom
Frequency: quarterly
ISSN: 0141 9811

– 2345 – Freight

Hermes House
St Johns Road
Tunbridge Wells TN4 9UZ
United Kingdom
Tel: 0892 26171
Geographical coverage: United Kingdom
Publishers: Freight Transport Association
Frequency: monthly
Coverage: Transport policy, developments and products relating to freight
ISSN: 1016 0849

– 2346 – International Journal of Radioactive Materials Transport

PO Box 7
Ashford
Kent TN23 1YW
United Kingdom
Tel: 0233 641683
Fax: 0233 610021
Geographical coverage: International
Publishers: Nuclear Technology Publishing
Editor: Goldfinch, E. P.
Frequency: quarterly
Coverage: All aspects of transport of radioactive materials
Annual subscription: £85.00
ISSN: 0957 476X

– 2347 – Journal of Transport Economics and Policy

University of Bath
Claverton Down
Bath BA2 7AY
United Kingdom
Tel: 0225 826302
Fax: 0225 826767
Geographical coverage: United Kingdom
Editor: Glaister, S.
Frequency: 3 issues a year
Annual subscription: £41.00
ISSN: 0022 5258

– 2348 – Public Transport International

Avenue de l'Uruguay 19
B-1050 Brussels
Belgium
Tel: 32 2 673 61 00
Fax: 32 2 660 10 72
Geographical coverage: International
Publishers: International Union of Public Transport
Editor: Laconte, P.
Frequency: quarterly
Language: English, French and German
Annual subscription: Bfr2000

– 2349 – TEC Transport Environment Circulation

38 avenue Emile Zola
F-75015 Paris
France
Geographical coverage: European
Publishers: ATEC
Editor: Imbert, A. and Blesson, S.
Frequency: bi monthly
Coverage: Transportation in relation to the environment
Annual subscription: Ffr450
ISSN: 0397 6513

REPORTS

– 2350 – Acid Deposition and Vehicle Emissions: European Environmental Pressures on Britain

Geographical coverage: United Kingdom
Publishers: Royal Institute of International Affairs, Policy Studies Unit
Editor: Brackley, P.
ISBN: 0566051257

– 2351 – Air Pollution from Vehicles

Geographical coverage: United Kingdom
Publishers: HMSO
ISBN: 0115510001
Author: Transport and Road Research Laboratory

– 2352 – Characteristics of Road Transport in Hungary

Tel: 36 1 173 3454
Geographical coverage: Hungary
Publishers: Talento Foundation Budapest, 1992
Coverage: Traffic transport and mobility in Hungary
Author: Kiss, K.

– 2353 – Environment and Transport Infrastructures

Geographical coverage: European
Publishers: European Conference of Ministers of Transport (ECMT), 1988, Paris
Language: English and French
Coverage: Report of the 79th Round Table Paris
Price: Ffr110
ISBN: 9282111415

– 2354 – Environment Policy: Standards and Opportunities for Development

Geographical coverage: United Kingdom
Publishers: National Materials Handling Centre, 1991
Coverage: Environmental performance of the road freight industry

– 2355 – Green Paper on the Environment – A Community Strategy for Sustainable Mobility

Geographical coverage: European
Publishers: Commission of the EC, 1992
Coverage: The Green Paper proposes introducing a framework for a Community strategy for sustainable mobility which would set down broad guidelines to contain the impact of transport on the environment

− 2356 − International Maritime Dangerous Goods (IMDG)

IMO
4 Albert Embankment
London SE1 7SR
United Kingdom
Geographical coverage: International
Aims and objectives: IMDG Code includes regulations for the prevention of pollution by harmful substances carried by sea in packaged form

− 2357 − Transport and Climate Change: Cutting Carbon Dioxide Emissions from Cars

Geographical coverage: United Kingdom
Publishers: Friends of the Earth (FoE), 1991
Coverage: Latest fuel efficiency technologies, traffic restraint, fiscal incentives and regulations are examined
Price: £7.00
ISBN: 0905966996

− 2358 − UK Road Transport's Contribution to Greenhouse Gases

Geographical coverage: United Kingdom
Publishers: Transport and Road Research Laboratory (TRRL), 1990
Coverage: The paper is a response to a request to carry out a literature review of the contribution that UK road transport makes to global warming through the greenhouse effect
ISSN: 0266 7045
Author: Waters, M. H. L.

− 2359 − Volatile Organic Compound Emissions in Western Europe: Report 6/87

Geographical coverage: European
Publishers: CONCAWE, Brussels
Coverage: Control options and their cost effectiveness for gasoline vehicles, distribution and refining

RESEARCH CENTRES

− 2360 − Danish Packaging and Transportation Research Institute

Gregersensvej
PO Box 141
DK-2630 Taastrup
Denmark
Tel: 45 43 99 66 11
Fax: 45 43 71 37 98
Contact name: Kirsten Nielsen, Manager
Geographical coverage: Denmark
Research activities: To initiate and carry out research and development in the fields of packaging and transportation

− 2361 − Department of Political and Social Sciences

European Univ Inst
Via dei Roccettini 9
I-50016 San Domenico di Fiesole
Italy
Geographical coverage: Italy
Research activities: Transport policy

− 2362 − Institut de Recherche des Transports Centre de Documentation d'Evaluation et de Recherches Nuisances

109 avenue Salvador Allendre
F-69500 Bron
France
Tel: 33 78269093
Geographical coverage: France
Research activities: Transport; air pollution; noise pollution

− 2363 − Institut für Energie- und Transportforschung Meissen

(Energy and Transport Research Institute)

Kynastweg 57
D-8250 Meissen
Germany
Geographical coverage: Germany
Research activities: Energy conservation and the introduction of alternative energy sources to agriculture

− 2364 − Technical University of Twente

PO Box 217
NL-7500 AE Enschede
Netherlands
Tel: 31 5389 9111
Fax: 31 5335 7956
Contact name: J Weber
Geographical coverage: Netherlands
Research activities: Developing a CFC car that removes coolant and oil from discarded refrigerators

− 2365 − Transport and Road Research Laboratory (TRRL)

Old Wokingham Road
Crowthorne
Berkshire RG11 6AU
United Kingdom
Tel: 0344 773131
Geographical coverage: United Kingdom
Research activities: Aspects of transport related to the environment

DATABASES

– 2366 – International Road Research Documentation (IRRD)

Geographical coverage: International
Coverage: All aspects of road transport and land use
Database producer: OECD Road Transport Research Programme
Database host: ESA–IRS
Coverage period: 1972–
Update frequency: 1,200 records monthly

– 2367 – Toxic Materials Transport

Geographical coverage: International
Coverage: Transport of hazardous materials
Database producer: Business Publishers Inc
Database host: DataStar; Dialog
Coverage period: 1985–

– 2368 – Transdoc

Geographical coverage: European
Coverage: Transport and land use
Database producer: European Conference of Ministers of Transport
Database host: ESA–IRS
Coverage period: 1970–
Update frequency: 250 records monthly

– 2369 – Transport Research Information Services (TRIS)

Geographical coverage: International
Coverage: Transport, land use, energy, economics
Database producer: US Transportation Research Board
Database host: Dialog
Coverage period: 1968–
Update frequency: 750 records monthly

APPENDICES

Euro Info Centres 1993

Established by the Commission of the European Communities in 1987, the Euro Info Centre Network have at their disposal constantly updated Community documentation. In addition, they have access to the Commission's principal databases. EIC's are ideal sources of information

Bureau Economique de la Province de Namur (B.E.P.N.)
Euro Info Centre
Avenue Sergent Vrithoff, 2
B-5000 Namur
Belgium

Bernard Ruyssen
Tel: 32-81-73 52 09
Fax: 32-81-74 29 45

Kamer van Koophandel en Nijverheid van Antwerpen
Euro Info Centrum
Markgravestraat, 12
B-2000 Antwerp
Belgium

Mr Guy Busseniers
Tel: 32-3-232 22 19
Fax: 32-3-233 64 42

C.D.P. Idelux
Euro Info Centre
Avenue Nothomb, 8
B-6700 Arlon
Belgium

Mr Pierre Martin
Tel: 32-63-22 72 46
Fax: 32-63-22 65 84

Chambre de Commerce et d'Industrie de Bruxelles
Kamer voor Handel en Nijverheid van Brussel et/en
Fabrimetal
Euro Info Centre Brussels
Avenue Louise, 500
Louizalaan
B-1050 Brussels
Belgium

Mr Georges Mols
Tel: 32-2-648 58 73
Fax: 32-2-640 93 28

Ministerie van de Vlaamse Gemeenschap Admn. voor Economie en Werkgelegenheid
Euro Info Centrum
Markiesstraat 1
6e verdieping
B-1000 Brussels
Belgium

Mr Andre Van Haver
Tel: 32-2-507 37 30
Fax: 32-2-502 47 02

Euroguichet Hainaut-Est
Euro Info Centre
Avenue Général Michel, 1B
B-6000 Charleroi
Belgium

Mr Guy Hubaru
Tel: 32-71-33 14 60
Fax: 32 71 30 54 48

Gewestelijke Ontwikkelingsmaatschappij voor Oost-Vlaanderen
Euro Info Centrum
Floraliapaleis, bus 6
B-9000 Gent
Belgium

Mr Johan Declerck
Tel: 32-91-21 55 11
Fax: 32-91-21 55 00

Kamer voor Handel en Nijverheid van Limburg vzw.
Euro Info Centrum Limburg
Kunstlaan 20
B-3500 Hasselt
Belgium

Mrs Ingrid Fleurquin
Tel: 32-11-22 18 00
Fax: 21-11-24 16 20

National Christelijk Middenstandverbond Kortrijk
Euro Info Centrum
Lange Steenstraat, 10
B-8500 Kortrijk
Belgium

Mr Walter Borms
Tel: 32-56-22 41 23
Fax: 32-56 22 96 94

Institut Provincial des Classes Moyennes
Euro Info Centre
Bd d'Avroy 28-30
B-4000 Liege
Belgium

Mme Monique Rover
Tel: 32-41-23 38 40
Fax: 32-41-22 19 76

Bureau d'Etudes Economiques et Sociales de la Province du Hainaut
Euro Info Centre
Rue de Nimy, 50
B-7000 Mons
Belgium

Mr Christian Provost
Tel: 32-65-31 93 10/11/12
Fac: 32-65-34 80 96

Kamer voor Handel en Nijverheid van het Arrondissement Halle-Vilvoorde en arr. Leuven.
Euro Info Centrum Vlaams Brabant
Brucargo Gebouw 706
1ste verdiep lokaal 7127
B-1931 Zaventem
Belgium

Mr Patrick De Vos
Tel: 32-2-751 90 56
Fax: 32-2-751 78 11

Schulungs und Forschungszentrum SPK
Gospertstrasse, 17
B-4700 Eupen
Belgium

Mr Ernst Heeren
Mmme Virginie Smits
Tel: 32-87-74 22 12
Fax: 32-87-55 24 15

Chambre de Commerce et d'Industrie de Tournaisis
Rue Beyaert, 73-75
B-7500 Tornai
Belgium

Mlle Isabelle Walschap
Tel: 32-69-22 11 21
Fax: 32-69-21 27 84

CECOP Comité Européen de Coopératives de production et de Travail Associé
Euro Info Centre Ariés
Association Réseau d'Information de l'Economie Sociale
Av.Guillaume Tell. 59
B-1060 Brussels
Belgium

Mr Rainer Schluter
Tel: 32-2-537 57 40
Fax: 32-2-537 09 17

EF-Rådgivningskontoret for Århus Amt Regionskontoret
Århus Amts-Kommune
Euro Info Centre
Haslegaardsvænget 18-20
DK-8210 århus V
Denmark

Mr Erik Andersen
Tel: 45-86-15 03 18
Fax: 45-86-15 43 22

EF-Rådgivningskontoret for Fyn
Euro Info Centre
Norregade 51
P.O. Box 1272
DK-5000 Odense C.
Denmark

Ms Helle Knudsen
Tel: 45-66-14 60 30
Fax: 45-66-14 60 34

Sonderjyllands Erhvervsråd
Euro Info Centre
Kirkeplads 4
DK-6200 Aabenraa
Denmark

Mr Tyge Korsgaard
Tel: 45-74-62 23 84
Fax: 45-74-62 67 60

Herning Erhvervsråd
Euro Info Centre
Lykkesvej 18
DK-7400 Herning
Denmark

Mr K.O. Jensen
Tel: 45-97-12 92 00
Fax: 45-97-12 92 44

Dansk Teknisk Oplysningstjeneste
Euro Info Centre
Rygaards Alle 131 A
Postbox 1992
DK-2820 Kobenhavn (Gentofte)
Denmark

Mr Jan Stiiskjaer
Tel: 45-31-20 90 92
Fax: 45-31-18 58 04

Det Danske Handelskammer
Euro Info Centre
Borsen
DK-1217 Kobenhavn K
Denmark

Mr Bo Green
Tel: 45-33-95-05 00
Fax: 45-33-32 52 16

Håndværksrådet
Viborg A/S
Euro Info Centre
Ll. Sct. Hans Gade 20
DK-8800 Viborg
Denmark

Mr Jesper Risom Andersen
Tel: 45-86-62 77 11
Fax: 45-86-61 49 21

Storstrms Erhvervscenter
Euro Info Centre
Marienbergvej 80
DK-4760 Vordinborg
Denmark

Mr Georg Heidtmann
Tel: 45-55-34 05 55
Fax: 45-55-34 03 55

Zenit (Zentrum in Nordrhein-Westfalen für Innovation und Technik GMBH)
EG-Beratungsstelle für Unternehmen
Dohne, 54
D-W-4330 Mulheim/Ruhr 1
Germany

Mrs Hannelore Kraft
Tel: 49-208-300 04 31
Fax: 49-208-300 04 29

R.K.W. (Rationalisierungs-Kuratorium der Deutschen Wirtschaft)
EG-Beratungsstelle fur Unternehmen
Heilwigstraße, 33
D-W-2000 Hamburg 20
Germany

Mr Gerhard Menz
Tel: 49-40-460 20 87
Fax: 49-40-48 20 32

D.I.H.T. (Deutscher Industrie-und-Handelstag)-
EG-Beratungsstelle für Unternehmen
Adenauer Allee, 148
Postfach 1446
D-W-5300 Bonn 1
Germany

Ms Ellen Oesterreich
Tel: 49-228-10 46 21/2
Fax: 49-228-10 41 58

Industrie-und Handelskammer regensburg
EG-Beratungsstelle für
Unternehmen
D.Martin Luter Straße, 12
Postfach 110 355
D-W-8400 Regensburg
Germany

Mr Franz Knott
Tel: 49-941-569 42 35
Fax: 49-941-569 42 79

German National Chamber of Commerce of Craft Industries
(D.H.K.T.)
EG-Beratungsstelle für Unternehmer Haus des Deutschen Handwerks
Johanniterstraße, 1
D-W- 5300 Bonn 1
Germany

Mr Klauspeter Zanzig
Tel: 49-228-54 52 11/76
Fax: 49-228-545 205

Handwerkskammer Stuttgart
EG-Beratungsstelle für Unternehmer
Heilbronner Straße, 43
Postfach 102155
D-W-7000 Stuttgart 10
Germany

Mr Jürgen Schafer
Tel: 49-711-1657 280/252
Fax: 49-711-1657 222

Industrie und Handelskammer zu Aachen
Euro Info Centre
Theaterstraße 6-8,
Postfach 650
D-W-5100 Aachen
Germany

Mr Frank Malis
Tel: 49-241-438 223
Fax: 49-241-438 259

BAO Berlin-Marketing Service GmbH
ERIC-Berlin/offizielle EG-Beratungsstelle fur Unternehmen
Hardenbergstraße 16-18
D-W-1000 Berlin 12
Germany

Ms Monika Schulz-Strelow
Tel: 49-30-31 51 02 40/1
Fax: 49-30-31 51 03 16

Deutsches Informationszentrum fur Technische Regeln (DITR) im DIN e.v.
Euro Info Centre
Burggrafenstraße 6
Postfach 1107
D-W-1000 Berlin 30
Germany

Mr Horst-Werner Marschall
Tel: 49-30-26 01 26 05
Fax: 49-30-262 81 25

Stadt Bielefeld
Euro Info Centre
Niederwall 23
Postfach 181
D-W-4800 Bielefeld 1
Germany

Mr Berend-Jan Waterbohr
Tel: 49-521-51 67 02
Fax: 49-521-51 22 26/33 85

Bundesverband der Deutschen Volksbanken und Raiffeisenbanken
Genossenschaftliche EG Beratungs und Informationsgesellschaft GEBI mbH
Rheinweg, 67
D-W-5300 Bonn 1
Germany

Mr Ivo-Michael Zscherlich
Tel: 49-228-23 75 44
Fax: 49-228-23 75 48

EG-Beratungsstelle für Unternehmen beim Deutschen Sparkassen- und Giroverband
Simrockstraße 4
D-W-5300 Bonn 1
Germany

Mr Heinz Breier
Tel; 49-228-20 43 19/23 Fax: 49-228-20 47 25

VDI/VDE Technologiezentrum Informations-Technik GmbH
EG-Beratungsstelle für Unternehmen
Geschäftsstelle Bremen
Hanseatenhof 8
D-W-2800 Bremen 1
Germany

Dr. Klaus Lenz
Tel: 49-421-17 55 55
Fax: 49-421-17 16 86

Gesellschaft für Wirtschaftsforderung Nordrhein-Westfalen mbH
Euro Info Centre
Kavalleriestraße 8
10
Postfach 200309
D-W-4000 Dusseldorf 1
Germany

Mrs Elisabeth Altekoster
Tel: 49-211-13 00 00
Fax: 49-211-13 00 064

Industrie-und Handelskammer für München und Oberbayern
EG-Beratungsstelle fur Unternehmen
Max-Joseph-straße 2
Postfach
D-W-8000 Munich 34
Germany

Mrs Susanne Weiss
Tel: 49-89-511 62 09
Fax: 49-89-511 64 65

NATI GmbH/Niedersächsische Agentur für Technologietransfer und Innovation
Euro Info Centre
Vahrenwalderstraße 7
D-W-3000 Hannover 1
Germany

Mrs Christiane Frochtling
Tel: 49-511-935 71 20/21
Fax: 49-511-93 57 439

Investitionsbank Schleswig-Holstein
RKW, Schleswig-Holstein
EG-Beratungsstelle, EIC
Fleethörn 29-31
Postfach 1128
D-W-2300 Kiel 1
Germany

Mr Kurt Puls
Tel: 49-431-900 32 70
Fax: 49-431-900 32 07

Bundesstelle für Außenhandels-Information (BFAI)
Euro Info Centre
Agrippastraße 87-93
Postfach 108007
D-W-5000 Cologne 1
Germany

Mr Wolfgang Muller
Tel: 49-221-20 57-270
Fax: 49-221-20 57-212

EBZ- Europäisches Beratungs-Zentrum der Deutschen Wirtschaft
Gustav Heinemann ufer 84-88
D-W-5000 Cologne 51
Germany

Mrs Jutta Zemke-Heyl
Tel: 49-221-370 86 21
Fax: 49-221-370 88 40

Industrie-und Handelskammer Südlicher Oberrhein
Euro Info Centre
Lotzbeckstraße 31
D-W-7630 Lahr -Germany

Mrs Petra Steck
Tel: 49-7821-1992
Fax: 49-78-21 27 03 22

Deutsche Gesellschaft Für Mittelstandsberatung mbH
Euro Info Centre
Arabellastraße 11
D-W-8000 Munich
Germany

Mr Peter von Windau
Tel: 49-89-92 69 680
Fax: 49-89-92 69 68 39
Emanuel Leutzestraße 4
D-W-4000 Dusseldorf 11
Germany

Tel: 49-211-53 640
Fax: 49-211-53 64 150

Landesgewerbeanstalt Bayern
OTTI/WETT/LGA
Euro Info Centre
Karolinenstraße 45
D-W-8500 Nuremburg 1
Germany

Dr Monika Bias
Tel: 49-911-23 20 517
Fax: 49-911-23 20 511

Zentrale für Produktivitat und Technologie Saar e.V.
Euro Info Centre
Franz-Josef-Röder Straße 9
Postfach 163
D-W-6600 Saarbrucken 1
Germany

Mr Gerd Martin
Tel: 49-681-9520 450/1/2/5
Fax: 49-681-584 61 25

R.K.W. Rationalisierungskuratorium der Deutschen Wirtschaft/Landesgruppe Baden Württemberg
Euro Info Centre
Königstraße 49
D-W-7000 Stuttgart 1
BE Germany

Mr Dr Albrecht Fridrich
Tel: 49-711-229-98-0
Fax: 49-711-229 98 10

Industrie-und Handelskammer Trier/Handwerkskammer Trier
Euro Info Centre Trier
c/o Technologie-Zentrum Trier
Gottbillstraße 34A
D-W-5500 Trier
Germany

Mr Lothar Philippi
Tel: 49-651-1992- Fax: 49 651-810 09 19

Hessische Landesentwicklungs-und Treuhandgesellschaft mbH
Euro Info Centre
Abraham Lincoln Straße 38-42
Posfach 3107
D-W-6200 Wiesbaden
Germany

Mr Hans Joachim Werner
Tel: 49-611-774 287
Fax: 49-611-774 265

Landkreis Osnabrück Amt für Wirtschaftsforderung
Am Schölerberg 1
D-W-4500 Osnabruck
Germany

Mr Günter Dinkelmann
Tel: 49-541-501 31 04
Fax: 49-541-501 31 30

Kreis Steinfurt
Amt für Wirtschaft und Verkehr
Tecklenburger Straße 10
D-W-4430 Steinfurt
Germany

Mr Josef Wessles
Tel: 49-25-51 69 20 18
Fax: 49-25-51 69 24 00

Omnibera WirtschaftsBeratungsgesellschaft
Coburger Straße, 1C
D-W-5300 Bonn
Germany

Mr Michael Fey
Tel: 40-228-23 80 78
Fax: 49-228-23 39 22

HELABA
Landesbank Hessen-Thüringen Girozentrale
Euro Info Centre Erfurt
Am Anger 12
Postfach 167
D-O-5010 Erfurt
Germany

Mrs Sabine Janichen
Tel: 49-361-24 798
Fax: 49-361-24 565

Deutsche Gesellschaft für Mittelstandsberatung GmbH
Euro Info Centre
Zwinglistraße 36
D-O- 8020 Dresden
Germany

Mr Wolfgang Hermann
Tel: 49-351-239 11 69/81
Fax: 49-351-230 10 03

Industrie- und Handelskammer zu Leipzig (IHK)
Euro Info Centre
EG-Beratungsstelle für Unternehmen
Friedrich-Engels-Platz 5
Postfach 7000
D-O-Leipzig
Germany

Mrs Christa Friedrich
Tel: 49-341-715 31 41
Fax: 49-341-715 34 21

Industrie-und Handelskammer Rostock
EG-Beratungsstelle
Ernst Barlachstraße 7
D-O-2500 Rostock
Germany

Mr Dieter Pfliegensdorfer
Tel: 49-381-37 501
Fax: 49-381-459 11 56

Handwerkskammer Magdeburg
Euro Info Centre
Bahnhofstraße 49a
Posfach 1568
D-O-3010 Magdeburg
Germany

Mr Heinz-Dieter Domland
Tel: 49-391-561 91 61
Fax: 49-391-561 91 62

Industrie und Handelskammer Frankfurt/Oder
Euro Info Centre/EG-Beratungsstelle für Unternehmen
im Land
Bradenburg
Humboldstraße 3
Postfach 343
D-O-1200 Frankfurt/Oder
Germany

Mrs Britta Bayer
Tel: 49-335-238 63
Fax: 49-335-325 492

Wirtschaftsforderung Brandenburg GmbH
Euro Info Centre/EG-Beratungsstelle für Unternehmen
Am
Lehnitzsee
D-O 1501 Neufahrland Polux
Germany

Mr Burghard Wilcke
Tel: 49-331 23 581
Fax: 49-331-23 582

Chambre de Commerce et d'Industrie d'Athènes
Euro Info Centre
7, Akadimias Street
GR 10671 Athens
Greece

Mrs Marianthi Papadatou
Tel: 30-1-362 73 37
Fax: 30-1-1360 78 97

Hellenic Organization of Small Medium Sixe Industries
and Handicrafts (Eommex)
Euro Info Centre
Xenias Street, 16
GR-11528 Athens
Greece

Mr Photis Santrouzanos
Tel: 30-1-779 42 29
Fax: 30-1-777 86 94

Association of Industries of Northern Greece
Euro Info Centre
Morihovou Square 1
GR-54625 Thessaloniki
Greece

Mr Yanis Stavrou
Tel: 30-31 53 98 17/53 96 82
Fax: 30-31 54 14 91

Eommex Alexandroupolis
Euro Info Centre
Miaouli 15
GR-68100 Alexandroupolis
Greece

Mr Ilias Kokinis
Tel: 30-551-33 565
Fax: 30-551-33 566

Panhellenic Exporters' Association
Euro Info Centre
Kratinou, 11
GR-10552 Athens
Greece

Mr Antonis Syngellakis
Tel: 30-1-522 89 25/17 26/15 15
Fax: 30-1-524 25 68

Chamber of Iraklion
Euro Info Centre
9 Koronaeou Street
GR-71202 Iraklion
Crete
Greece

Mr Vangelis Georgantopoulos
Tel: 30-81-22 90 13
Fax: 30-81-22 29 14

Chamber of Kavala
Euro Info Centre
50 Omonias
GR-65302 Kavala
Greece

Ms Soultana Mavrommati
Tel: 30-51-83 39 64
Fax: 30-51-83 59 46

Eommex
Larissa
Euro Info Centre
Marinou Antipa & Kouma Street
GR-41222 Larissa
Greece

Mr Sotiris Blanas
Tel: 30-41-22 60 77
Fax: 30-41-25 30 19

Eommex
Mytilini
Euro Info Centre
Iktinou 2
Pl. Kyprion Agoniston
GR-81100 Mytilini
Greece

Mr Michalis Dalaklis
Tel: 30-251-24 906
Fax: 30-251-41 501

Eommex
Patras
Euro Info Centre
21, Aratou Street
GR-26221 Patras
Greece

Ms Georgia Giovrie-Skodra
Tel: 30-61-22 02 48
Fax: 30-61-22 34 96

Chambre de Commerce et d'Industrie du Pirée
Euro Info Centre
1 Rue Loudovicou
Place Roosevelt
GR-18531 Piree
Greece

Ms Catherina Kouroukli
Tel: 30-1-417 05 29
Fax: 30-1-417 46 01

Association of Industries in Thessaly and in Central Greece
Euro Info Centre
4, El. Venizelou Road
GR-38221 Volos
Greece

Ms Stella Vaina
Tel: 30-421-28 111/33622
Fax: 30-421-26 394

Chamber of Ioannina
Euro Info Centre
X. Trikoupi & O. Poutetsi St. 14
GR-45332 Ioannina
Greece

Mr John Daskalopoulos
Tel: 30-651-76 589
Fax: 30-651-25 179

CIDEM
Centro Europeo de Informacion
Empressarial-Eurofinestreta
Av. Diagonal, 403 1r
E-08008 Barcelona
Spain

Ms Amparo Lopez-Barrena
Tel: 34-3-416 08 30
Fax: 34-4-218 67 47

Camara Oficial de Comercio Industria y Navegacion de Bilbao
Centro Europeo de Informacion Empresarial
Alameda de Recalde, 50
E-48008 Bilbao (Vizcaya)
Spain

Mme Virginia Uriguen
Tel: 34-4-410 46 64
Fax: 34-4-443 41 45

Confedercion de Empresarios de Andalucia
Centro de Servicios Empresariales de Andalucia
Euro Info Centre
isla de la Cartuja, S/N
E-41010 Seville
Spain

Ms Mercedes Leon Lozano
Tel: 34-5-446 00 01/01 08
Fax: 34-5-446 16 44

Confederacion Española de Organizaciones Empresariales
Euro Info Centre
Diego de Leon, 50
E-28006 Madrid
Spain

Mr Carmen Garcia Cossio
Tel: 34-1-563 94 15
Fax: 34-1-564 01 35

IMPI
ICEX
Centro Europeo de Informacion
Empresarial-Euroventanilla
Paseo de la Castellana
141-2a Planta
E-28046 Madrid
Spain

Ms Ana Areces Estrada
Tel: 34-1-571 46 40
Fax: 34-1-571 59 12

Confederacion Regional de Empresarios de Castilla la Mancha
Centro Europeo de Informacion Empresarial
Calle del Rosario 29
4 plants
E-02001 Albacette
Spain

Ms Maria Victoria Lopez Valcarcel
Tel: 34-67-21 21 49 / 73 00
Fax: 34-67 24 02 02

Grupo Banco Popular Español
Euroventanilla
Rambla de Méndez Nuñez, 12
E-03002 Alicante
Spain

Mr Carlos Fernandez Rodriguez
Tel: 34-6-521 62 91
Fax: 34-6-520 19 54

Banco Exterior de España (BEX)
Centro Europeo de Informacion
c/Manila 56
58
E-08034 Barcelona
Spain

Mr Jorge Garcia Galceran
Tel: 34-3-204 13 66
Fax: 34-3-205 73 35

Camara Oficial de comercio, Industria y Navegacion de Barcelona
Euro Info Centre
Avenida Diagonal, 452-454
E-08006 Barcelona
Spain

Mr Rafael Garcia Santos
Tel: 34-3-415 16 00
Fax: 34-3-416 07 35

Sociedad para el Desarrollo Industrial de Extremadura (Sodiex)
Euro Info Centre
C/Dr. Maranon, 2
E-10002 Caceres
Spain

Ms Rebeca Dominguez-Cidoncha
Tel: 34-27-22 48 78
Fax: 34-27 24 33 04

Asociacion de la Industria Navarra
Euro Info Centre
P.O. Box 439
E-31191 Cordovilla
Pamplona
Spain

Ms Mae Herias Oscariz
Tel: 34-48 10 11 01/41
Fax: 34-48-10 11 00

Gobierno de Canarias Consejeria de Economie y Hacienda
Euro Info Centre
c/Nicolás Estévanez, 33
Edificio Eurocan
1 planta-despacho 11
E-35007 Las Palmas DE G. Canaria
Spain

Mr Miguel Angel Lopez Y Raygon
Tel: 34-28-27 11 42
Fax: 34-28-27 51 44

Federacion de Empresarios de la Rioja
Euro Info Centre
Calle Hermanos Moroy, 8 4
E-26001 Logrono
Spain

Mr Jesus Alloza Moya
Tel: 34-41-25 70 22
Fax: 34-41 20 25 37

Camara de Comercio e Industria de Madrid
Euro Info Centre
Plaza de la Independencia, 1
E-28001 Madrid
Spain

Carmen Verdera
Tel: 34-1-538 36 10
Fax: 34-1-538 36 43

Instituto Madrileno de Desarrollo (IMADE)
Euroventanilla
Mariano Ron 1
E-28902 Madrid (Getafe)
Spain

Mr Teresa Sotillo-Ramo
Tel: 34-1-696 11 11
Fax: 34-1-695 61 74

Banesto
Centro Europeo de Informacion
Plaza de la Constitucion 9
E-29008 Malaga
Spain

Ms Nuria Toucet Alvarez
Tel: 34-52-22 09 59
Fax: 34-52-22 09 36

Instituto de Fomento de la Region de Murcia
Euro Info Centre
Plaza San Agustin 5
E-30005 Murcia
Spain

Mr Fernando Garcia-Nieto Cano
Tel: 34-68-36 28 00/18
Fax: 34-68-29 32 45

Instituto de Fomento Regional del Principado de Asturias
Euro Info Centre
Parque Tecnologico de Asturias
E-33420 Llanera
Spain

Ms Margarita Collado Fernandez
Tel: 34-85-26 00 68
Fax: 34-85-26 44 55

Consorcio Centro de Documentacion Europea Islas Baleares
Euro Info Centre
Calle Patronato Obrero, 30
E-07006 Palma De Mallorca
Spain (Baleares)

Mr Luis Pericas O'Callaghan
Tel: 34-71-46 10 02
Fax: 34-71-46 30 70

Euroventanilla del Pais Vasco
C/Tomás Gros, No 3 Bajo
E-20001 Donostia-San Sebastian
Spain

Mr José Luis Artamendi Fernandez
Tel: 34-43-27 22 88
Fax: 34-43 27 16 57

Confederacion de Empresarios de Galicia
Euro Info Centre
C/Romero Donallo 7A
Entresuelo
E-15706 Santiago DE Compostela
Spain

Mrs Maria Ester Pereiras Neira
Tel: 34-81-59 76 54
Fax: 34-81-56 57 88

Camara de Comercio e Industria de Toledo
Euro Info Centre
Plaza San Vincente, 3
E-45001 Toledo
Spain

Mr José Maria Calvo Cirujano
Tel: 34-25-21 44 50
Fax: 34-25-21 39 00

Camara Oficial de Comercio, Industria y Navegacion de Valencia
Euro Info Centre
C/ Poeta Querol, 15
E-46002 Valencia
Spain

Mr Vicente Mompo
Tel: 34-6-351 13 01
Fax: 34-6-351 63 49/35 58

Sodical
Euro Info Centre
C/Claudio Moyano, 4-1
E-47001 Valladolid
Spain

Ms Carolina Calvo Revilla
Tel: 34-83-35 40 33
Fax: 34-83-35 47 38

Confederacion Regional de Empresarios de Aragon
Euro Info Centre
Plaza Roma, F-1, 1a Planta
E-50010 Zaragoza
Spain

Mr Juan Antonio Falcon Blasco
Tel: 34-76-32 00 00
Fax: 34-76-32 29 56

Federacion Asturiana de Empresarios
Calle Doctor Alfredo Martinez 6-2 pl
E-33005 Oviedo (Asturias)
Spain

Mme Maria José Martin
Tel: 34-85-23 21 05
Fax: 34-85-24 41 76

Chambre de Commerce er d'Industrie de Lyon
Euro Info Centre/Lyon Rhône
Alpes
16, Rue de la République
F-69289 Lyon Cedex 02
France

Mme Catherine Jamon-Servel
Mme Françoise Poulet-Froesel
Tel: 3-72-40 57 46
Fax: 33-78-37 94 00

Comité d'Expansion Aquitaine
Euro Info Centre
2, Place de la Bourse
F-33706 Bordeaux Cedex
France

Mme Martine Dronval
Tel: 33-56-01 50 10/09
Fax: 33-56-01 50 05

Région de Lorraine
Euro Info Centre
Place Gabriel Hocquard
BP 1004
F-57036 Metz Cedex 1
France

Mme Evelyne Decaux
Tel: 33-87-33 60 00
Fax: 33-87-32 89 33

Chambre de Commerce et d'Industrie de Nantes
Euroguichet
Entreprises
Centre des Salorges
16, Quai Ernest Renaud
BP 718
F-44027 Nantes Cedex 04
France

Mr Daniel Jouvenet
Tel: 33-40-44 60 60
Fax: 33-40-44 60 90

Chambre de Commerce et d'Industrie de Strasbourg et du Bas-Rhin
Euro Info Centre
10, Place Gutenberg
BP 444 R8
F-67008 Strasbourg Cedex
France

Mme Véronique Oberle
Tel: 33-88-75 25 25
Fax: 33-88-22 31 20

Chambre Régionale de Commerce et d'Industrie de Picardie
Euro Info Centre
36, Rue des Otages
F-80037 Amiens Cedex
France

Mme Marie-Françoise Duee
Tel: 33-22-82 80 93
Fax: 33-22-91 29 04

Counseil Régional de Guadeloupe
Euro Info Centre
5, Rue Victor Hugues
F-97100 Basse-Terre
La Guadeloupe
France

Mr Jean-Claude Baptistide
Tel: 590-81 16 56
Fax: 590-81 85 09

Chambre Régionale de Commerce et d'Industrie de
Franche-Comté
Euro Info Centre
Valparc
ZAC de Valentin
F-25043 Besancon
France

Mr Stephane Brugal
Tel: 33-81-80 41 11
Fax: 33-81 80 70 94

Chambre Régionale de Commerce et d'Industrie
Midi-Pyrénées
Euro Info Centre
5 Rue Dieudonné Costes, BP32
F-31701 Blagnac Cedex
France

Mr Jacques Albas
Tel: 33-62-74 20 00
Fax: 33-62-74 20 20

Chambre Régionale de Commerce et d'Industrie de
Basse Normandie
Euro Info Centre
21 Place de la République
F-14052 Caen Cedex
France

Mme Isabelle Herault
Tel: 33-31-38 31 38/67
Fax: 33-31-85 76 41

Chambre de Commerce et d'Industrie de la Guyane
Euro Info Centre
Hotel Consulaire
Place de l'Esplanade
BP 49
F-97321 Cayenne Cedex
Guyane
France

Mr Marie-Joseph Pinville
Tel: 594-30 30 00
Fax: 594-30 23 09

CCI de Chalons-sur-Marne Point Europe Champagne
Ardenne
Euro Info Centre
2, Rue de Chastillon BP 533
F-51010 Chalons-Sur-Marne
France

Mme Béatrice Decroux
Tel: 33-26-21 11 33
Fax: 33-26-64 16 84

Chambre de Commerce et d'Industrie de
Clermont-Ferrand/Issoire
Euro Info Centre Auvergne
148, Bd. Lavoisier
F-63037 Clermont-Ferrance
France

Mr Jean-Michel Chauvin
Tel: 33-73-43 32
Fax: 33-73-43 43 25

Chambre Régionale de Commerce et d'Industrie de
Bourgogne
Euro Info Centre
68, Rue Chevreul
BP 209
F-21006 Dijon
France

Mr Denis Pleux
Tel: 33-80-63 52 63
Fax: 33-80-63 52 53

Chambre de Commerce et d'Industrie de La Martinique
Euro Info Centre
50 Rue Ernest Deproge
BP 478
F-97241 Fort-de-France Cedex
La Martinique
France

Mme Cemiane Moutoucoumaro
Tel: 596-55 28 25/72
Fax: 596-60 66 68

Counseil Régional ARD/Maison des Professions/CCI
Euroguichet Nord Pas de Calais
Centre de Documentation
185 Boulevard de la Liberté
BP 2027
F-59013 Lille Cedex
France

Mme Dominique Chaussec de Lecour
Tel: 33-20-40 02 77
Fax: 33-20-40 04 33 (CR-ARD)
Tel: 33-20-99 45 00
Fax: 33-20-99 46 00 (MP)
Tel: 33-20-63 78 27
Fax: 33-20 74 82 58 (CCI)

Chambre Régionale de Commerce et d'Industrie
Limousin/Poitou/Charentes
Euro Info Centre Limousin
Boulevard des Arcades
F-87038 Limoges
France

M Martin Forst
Tel: 33-55-04 40 25/00
Fax: 33-55-04 40 40

Somecin
Euro Info Centre
2 Rue Henri-Barbusse
Mezzanine
F-13241 Marseille Cedex 1
France

Mlle Martine Liogier
Tel: 33-91-39 33 77
Fax: 33-91-39 33 60

Comité Liaison Chambres Econom. Languedoc
Roussillon
Région
Euro Info Centre
254 Rue Michel Teule
ZAC d'Alco
BP 6076
F-34030 Montpellier Cedex 1
France

Mme Dominique Guy-Chevanne
Tel: 33-67-61 81 51
Fax: 33-6761 81 59

Chambre Régionale de Commerce et d'Industrie du
Centre/C.R.C.E.
Euro Info Centre
35 Avenue de Paris
F-45000 Orleans
France

Mr Bernard Cottin
Tel: 33-38-54 58 58
Fax: 33-38-54 09 09

Chambre Régionale de Commerce et d'Industrie
Ille-de-France
Euro Info Centre Point Europe
21, Avenue de Paris
F-78021 Versailles
France

Mme Isabelle Drevet
Tel: 33-1-39 50 33 56
Fax: 33-1-39 53 78 34

Centre Français du Commerce Extérieur
Euro Info Centre
10 Avenue d'Iéna
F-75783 Paris 16
France

Mr Jean-Michel Balling
Tel: 33-1-40 73 30 00
Fax: 33-1-40 73 30 48

Ministère de l'Industrie et de l'Aménagement du Territoire
Euro Info Centre
84, Rue de Grenelle
F-75353 Paris Cedex 07
France

Mr Denis Lagniez
Tel: 33-1-43 19 28 16/19
Fax: 33-1-43 19 27 06

Chambre de Commerce et d'Industrie de Paris
Euro Info Centre
Point Europe
2, Rue de Viarmes
F-75001 Paris
France

Mme Martine Frager-Berlet
Tel: 33-1-45 08 35 90
Fax: 33-1-45 08 36 80

Association Poitou-Charentes-Europe
Euro Info Centre
47, Rue du Marché
BP 229
F-86006 Poitiers
France

Mr François Descheemaekere
Tel: 33-49-60 98 00
Fax: 33-49-41 65 72

Chambre Régionale de Commerce et d'Industrie de
Bretagne
Euro Info Centre
1, Rue du Général Guillaudot
F-35044 Rennes
France

Mr Thierry Acquitter
Tel: 33-99-25 41 57
Fax: 33-99-25 41 10

Chambre Régionale de Commerce et d'Industrie de haute
Normandie
Euro Info Centre
9, Rue Robert-Schuman
BP 124
F-76002 Rouen Cedex
France

Mr François Tahon
Tel: 33-35-88 44 42
Fax: 33-35-88 06 52

Chambre de Commerce et d'Industrie de La Réunion
Euro Info Centre
5 bis, Rue de Paris
BP 120
F-97463 Saint Denis Cedex - Ille de la Reunion
France

Mme Nathalie Joron
Tel: 262-21 53 66
Fax: 262-41 80 34

Chambre de Commerce et d'Industrie d'Annecy et de la
Haute-Savoie
2, Rue du Lac
BP 2072
F-74011 Annecy Cedex
France

Mme Claire Grange
Tel: 33-50-33 72 00
Fax: 33-1-45 56 27 06

Chambre de Commerce et d'Industrie d'Avignon et de
Vaucluse
46, Cours Jean Jaurès
BP 158
F-84008 Avignon
France

Mme Virginie Monot-Giusti
Tel: 33-90-82 40 00
Fax: 33-90-85 56 78

Chambre de Commerce et d'Industrie de Grenoble
GREX
1, Place A. Malraux
BP 297
F-38016 Grenoble
France

Mme Elisabeth Coviaux
Tel: 33-76-47 20 36
Fax: 33-76-87 73 11

Chambre de Commerce et d'Industrie
Nice Côte d'Azur/Conseil Général des Alpes Maritimes
20, Boulevard Carabacel
BP 259
F-06005 Nice Cedex
France

Mme Anne Gravoulet
Tel: 33-93 13 73 05
Fax: 33-93-13 73 99

Chambre de Commerce et d'Industrie de Pointe-à-Pitre
BP64
F-97152 Pointe-a-Pitre Cedex - la Guadeloupe
France

Mme Anne-Elisabeth Bault
Tel: 590-90 08 08
Fax: 590-90 21 87

The Irish Trade Board/An Bord Trachtala
European Business Information Centre
Merrion Hal
P.O. Box 2198
Strand Road
Sandymount
IRL-Dublin 4
Ireland

Mrs Margaret Hogan
Tel: 353-1-269 50 11
Fax: 353-1-269 58 20

Shannon Free Airport Development Company
European Businesss Information Centre
The Granary
Michael Street
IRL-Limerick
Ireland

Ms Nuala O'Carroll
Tel: 353-61-410 777
Fax: 353-61-31 56 34

Cork Chamber of Commerce
European Business Information Centre
67 South Mall
IRL-Cork
Ireland

Mr Michael Geary
Tel: 353-21-50 90 44
Fax; 353-21-27 13 47

Galway Chamber of Commerce and Industry
Euro Info Centre
Hardiman House
5 Eyre Square
IRL-Galway
Ireland

Mr Jarlath Feeney
Tel: 353-91-62 624
Fax: 353-91-61 963

Sligo Chamber of Commerce
European Business Information Centre
16 Quay Street
IRL-Sligo
Ireland

Mr Kevin McGoldryck
Tel: 353-71-612 74
Fax: 353-71-609 12

Waterford Chamber of Commerce
European Business Information Centre
ABT office
Industrial Estate
Cork Road
IRL-Waterford
Ireland

Mrs Annia Wall
Tel: 353-51-72 639
Fax: 353-51-79 220

Camera di Commercio Industria Artigianato e
Agricoltura di Milano
Eurosportello
Via Delle Orsole 4/B
L-20123 Milan
Italy

Mr Attilio Martinetti
Tel: 39-2-85 15 56 81/92/93
Fax: 39-2-85 15 56 87

Camera di Commercio Industria Artigianato e
Agricoltura di Napoli
Eurosportello
Corso Meridionale, 58
I-80143 Naples
Italy

Mr Riccardo De Falco
Tel: 3981-553 61 06/28 42 17
Fax: 39-81-285 465

Rete Artigianato Comitato Coordinamento
Confederazioni Artigiane
Eurosportello per l'impresea
Via Bruni, 17
I-25121 Brescia
Italy

Mr Antonio Carletti
Tel: 39-30-377 47 70/5
Fax: 39-30-377 48 12

Confindustria
Euro Info Centre
Direzione Comunicazione ed Immagine
Viale dell'Astronomia 30
I-00144 Rome
Italy

Mr Pier Luigi d'Agata
Tel: 39-6-590 31
Fax: 39-6-591 96 15

Associazione Degli Industriali Della Provincia di
Bologna
Eurosportello Assindustria Bologna
Via San Domenico, 4
I-40124 Bologna
Italy

Mr Marzio De Lucca
Tel: 39-51-52 96 11
Fax: 39-51-52 96 13

Camera di Commercio I.A.A. di ascoli Piceno
Eurosportello Azienda Speziale
via Luigi Mercantini, 23/25
I-63100 Ascoli Piceno
Italy

Mr Silvio Pascali
Tel: 39-736-27 92 55/03/33
Fax: 39-736-27 92 37

Istituto Finanziario Regionale Pugliese Finpuglia
Euro Info Centre
Via Lenin 2
I-70125 Bari
Italy

Ms Alessandra De Luca
Tel: 39-80-41 67 35
Fax: 39-80-41 68 09

C.C.I.A.A. Cagliari
Eurosportello
Viale Diaz 221
C/O Centro Servizi Promozionali per le Impresse
I-09126 Cagliari
Italy

Mr Giulia De Lecca
Tel: 39-70-30 68 77
Fax: 39-70-34 03 28

Camera di Commercio Industria Artigianato e
Agricoltura di
Catania
Eurosportello
Via Cappuccini 2
I-95123 Catania
Italy

Dott. Giuseppe Lanteri
Tel: 39-95-71 50 176/7
Fax: 39-95-71 50 265

Eurosportello della Camera di Commercio Industria
Agricoltura e
Artigianato
Piazza G.B. Vico, 3
I-66100 Chieti
Italy

M. Achille Renzetti
Tel: 39-871-33 12 72
Fax: 39-871-33 12 18

Promofirenze
Eurosportello
Via Faenza, 111
I-50123 Firenze
Italy

Mr Giulio Fabbri
Tel: 39-55-28 01 32
Fax: 39-55-28 33 04

Consorzio Eurosportello Confesercenti
Eurosportello
Piazza Pier Vettori 8/10
I-50143 Firenze
Italy

Dr. Lucio Scognamiglio
Tel: 39-55-27 05 247/18
Fax: 39-55-22 40 96

Camera di Commercio di Genova
Eurosportello
Torre WTC-San Benigno
Via de Marini 1, III p.
I-16149 Genoa (Sampierdarena)
Italy

Ms Maria Teresa Madama
Tel: 39-10-209 42 52/55/96
Fax: 39-10-209 42 97

C.C.I.A.A. Isernia
Euro Info Centre
Corso Risorgimento, 302
I-86170 Isernia
Italy

Mr Franco Finori
Tel: 39-865-41 29 23
Fax: 39-865-23 50 24

Associazione Industriale Lombarda
Euro Info Centre Assolombarda
Via Pantano 9
I-20122 Milan
Italy

Mr Luigi Boldrin
Tel: 39-2-58 37 03 82/04 11/04 59
Fax: 39-2-58 30 45 07

ME.SVIL. S.p.A. (Meridionale Sviluppo S.p.A.)
Euro Info Centre
Via Mariano Stabile, 160
I-90139 Palermo
Italy

Mr Stefano Vivacqua
Tel: 39-91-58 93 88
Fax: 39-91-61 11 12 1

Centro regionale commercio estero
c/o Camera di Commercio di Perugia
Euro Info Centre
Via Cacciatori delle Alpi, 40
I-06100 Perugia
Italy

Dr. Fulvio Occhiucci
Tel: 39-75-574 82 06
Fax: 39-75-280 88

Associazione "Compagnia delle Opere"
Eurosportello
Eurocdo
Via V. Rossi 2
I-61100 Pesaro
Italy

Mr Danilo Maiocchi
Tel: 39-721-41 00 88
Fax: 39-721-41 41 74

C.C.I.A.A. Ravenna
Eurosportello di Ravenna
Viale L.C. Farini, 14
I-48100 Ravenna
Italy

Mr Antonio Nannini
Tel: 39-544-303 87
Fax: 39-544-396 72

Confederazione Generale Dell'Agricoltura Italiana
Euro Info Centre
Corso Vittorio Emanuele, 101
I-00186 Rome
Italy

Mr Filippo Trifiletti
Tel: 39-6-685 22 58/23 78
Fax: 39-6-687 96 86/865 23 92

Confederazione Generale Italiana del Commercio e del Turismo
Euro Info Centre
Piazza g.G. Belli 2
I-00153 Rome
Italy

Mme Daniela Floridia
Tel: 39-6-589 89 73/76 13
Fax: 39-6-581 49 84

Istituto per l'Assistenza allo Sviluppo del Mezzogiorno (I.A.S.M.)
Eurosportello Mezzogiorno
Viale M. Pilsudski, 124
I-00197 Rome
Italy

Mr Valentino Bolic
Tel: 39-6-809 721/809 722 10/13/16/19
Fax: 39-6-809 722 12

Istituto Nazionale per il Commercio Estero
Eurosportello Lega
Corso Magenta 59
I-20123 Milan
Italy

Ms Ida Ossi
Tel: 39-2-48 13 47
Fax: 39-2-48 00 55 23

Unioncamere/Mondimpresa/Cerved
Eurosportello
Piazza Sallustio, 21
I-00187 Rome
Italy

Mr Flavio Burlizzi
Tel: 39-6-470 41
Fax: 39-6-470 42 40

Camera di Commercio Industria Artigianato e Agricoltura di Torino
Euro Info Centre
Via San Francesco da Paola, 24
I-10123 Turin
Italy

Mr Riccardo Ricotta
Tel: 39-11-571 63 70
Fax: 39-11-571 65 17

Federpiemonte
Euro Info Centre
Corso Stati Uniti, 38
I-10128 Turin
Italy

Dr. Ing. Ermano Maritano
Tel: 39-11-54 92 46
Fax: 39-11-55 75 204

Camera di Commercio I.A.A.
Euro Info Centre
Via morpurgo, 4
I-33100 Udine
Italy

Mr Fabiano Zuiani
Tel: 39-432-273 222
Fax: 39-432-509 569

Centro Estero delle Camere di Commercio del Veneto
Eurosportello Veneto
via Mestrina, 94
I-30172 Venice Mestre
Italy

Mr Gian Angelo Bellati
Tel: 39-41-98 82 00/81
Fax: 39-41-98 95 48

Associazione Industriali della Provincia di Vicenza
Euro Info Centre
Piazza Castello, 3
I-36100 Vicenzia
Italy

Mr Andrea Bertello
Tel: 39-444-54 22 11
Fax: 39-444-54 73 18

Chambre de Commerce du Grand-Duché de Luxembourg
Euroguichet
7, Rue Alcide de Gasperi
BP 1503
L-2981 Luxembourg
Luxembourg

Mr Camille Giacomelli
Tel: 352-43-58 53
Fax: 352-43-83 26

Chambre des Métieres du Grand-Duché de Luxembourg
Euroguichet-Luxembourg
2, circuit de la Foire Internationale
L-1347 Luxembourg
Luxembourg

Mme Marie Andrée Haas
Tel; 352-42 67 67-1
Fax: 352-42 67 87

Adresse postale:
BP 1604
L-1016 Luxembourg

Instituut voor het Midden-en-Kleinbedrijf
EG Adviescentrum voor Ondernemingen
9, Dalsteindreef
Postbus 12
NL-1110 AC Diemen-Zuid
Netherlands

Mr Rob Van Raaij
Tel: 31-20-690 10 71
Fax: 31-20-695 32 31

N.V. Induma / B.O.M.
EG-Adviescentrum Zuid-Nederland
Pettelaarpark 10
5216 PD Den Bosch
Netherlands

Mr Hans Aben
Tel: 31-73-87 44 20
Fax: 31-73-12 32 10

Postal address;
Postbus 70060
NL-5201 DZ Den Bosch

EVD, Economic Information and Export promotion
Euro Info Centre
Bezuidenhoutseweg, 151
NL-2594 Den Haag
Netherlands

Mr Hans Vlot
Tel: 31-70-379 88 11
Fax: 31-70-379 78 78

Euro Info Centre Zuid-Holland
Koningskade 30
NL-2596 AA Den Haag
Netherlands

Mr Maarten Rooderkerk
Tel: 31-70-379 57 10
Fax: 31-70-324 06 84

Postal address:
Postbus 29178
NL-2502 LS Den Haag

Overijselse Ontwikkelings Maatschappij N.V.
Euro Info Centre Oost Nederland
Ondernemingshuis Twente
Hengelosestraat 585
NL-7521 AG Eenschede
Netherlands

Mr Leo Van Den Bergh
Tel: 31-53-84 98 90
Fax: 31-53-84 97 11

Noordelijke Ontwillelingsmaatschappij
Euro Info Centrum Noord-Nederland
Damsport 1
Postbus 424
NL-9700 AK Groningen
Netherlands

Mr Theun J Wijbenga
Tel: 31-50-26 78 56
Fax: 31-50-26 14 75

EG-Informatie Centrum Gelderland
Postbus 38326
NL-6503 AH Nijmegen
Netherlands

Mr Harrie van Diessen
Tel: 31-80 78 00 75
Fax: 31-80 77 70 29

Kmer van Koophandel en Fabrieken voor Utrecht en Omstreken
Euro Info Centrum Midden-Nederland
Waterstraat 47
Postbus 48
NL-3500 AA Utrecht
Netherlands

Ms Karin Kerckhaert
Tel: 31-30-36 32 81/2
Fax: 31-30-36 85 41

Associaço Industrial Portuense
Eurogabinete
Exponor
Feira Internacional do Porto
Leça de Palmeria
P-4450 Matosinhos
Portugal

Mr Pedro Capucho
Tel: 351-2-996 15 29
Fax: 351-2-995 70 17

Banco de Fomento e Exterior
Eurogabinete
Av. Casal Ribeiro, 59
P-1000 Lisbon
Portugal

Mr Vasco Da Fonseca
Tel: 351-1-356 01 44
Fax: 351-1-54 85 71

Associaço Industrial do Distrito de Aveiro
Euro Info Centre
Av. Dr. Lourenço Peixinho, 146-5 A
P-3800 Aveiro
Portugal

Mr José De Matos Rodrigues
Tel: 351-34-20 095
Fax: 351-34-24 093

Eurogabinete para a Regio Centro
Rua Bernardim Ribeiro, 80
P-3000 Coimbra
Portugal

Ms Luiza Camplogaro
Tel: 351-39-40 56 88/40 01 21 ext 21
Fax: 351-39-72 37 57

Istituto de Apoia as Pequenas e Medias Empresas e ao Investimento
Eurogabinete PME
Rua do Valasco, 19 C
P-7000 Evora
Portugal

Mr Antonio Francisco Balsa Cebola
Tel: 351-66-218 75/6
Fax: 351-66-297 81

Commisso de Coordenaço da Regio do algavre
Euro Info Centre
Praça da Liberdade, 2
P-8000 Faro
Portugal

Dr. Francisco Mendonca Pinto
Tel: 351-89-80 27 09
Fax: 351-89-80 66 87

Associaço Comercial e Industrial do Funchal/C.C.I. da Madeira
Eurogabinete da Madeira
Avenida Arriaga, 41
P-9000 Funchal-Madeira
Portugal

Dr. Rui Jervis
Tel: 351-91-23 01 37/8/9
Fax: 351-91-22 20 05

Associaço Industrial Portuguesa
Eurogabinete
Praça das Industrias
P-1399 Lisbon
Portugal

Mrs Maria Manuela Costa
Tel: 351-1-362 01 00
Fax: 351-1-364 67 86

Caixa Geral de Depositos
Eurogabinete
Av. da Republica, 31
P-1000 Lisbon
Portugal

Mrs Maria José Constancio
Tel: 351-1-352 01 02
Fax: 351-1-352 02 97

Norma-Açores e Câmara de Comércio e Indústria dos Açores
Eurogabinete dos Açores
Rua Antonio Joaquim Nunes Da Silva, 55
P-9500 Ponta Delgada
Acores
Portugal

Mr José Manuel Monteiro Da Silva
Tel: 351-96-23 189
Fax: 351-96-26 808

Associaçao Comercial e Industrial de Coimbra
Avenida Sa da Bandeira, 90-92
P-3000 Coimbra
Portugal

Mrs Paula Cristina Matos Nunes
Tel: 351-39-22 843
Fax: 351-39-27 023

Camara do Comercio e Industria dos Açores
Rua da Palha 32/34
P-9700 Angra Do Heroismo
Portugal

Mrs Roselene Dores
Tel: 351-95-234 70
Fax: 351-95-271 31

Scottish Enterprise
Euro Info Centre Limited
Atrium Court
50, Waterloo Street
UK-Glasgow G2 6HQ
UK

Mr Ian Traill
Tel: 44-41-221 09 99
Fax: 44-41-221 65 39

Birmingham Chamber of Industry and Commerce
European Business Centre
75, Harborne Road
P.O. Box 360
UK-Birmingham B15 3DH
UK

Ms Sharon Clift
Tel: 44-21-454 61 71
Fax: 44-21-455 86 70

Northern Development Company
North of England Euro Info Centre
Great North House
Sandyford Road
UK-Newcastle Upon Tyne NE1 8ND
UK

Ms Lesley Boughton
Tel: 44-91-261 00 26 (NDC) 44-91-261 51 31 (EDC)
Fax: 44-91-222 17 79/261 69 11

The Greater London Business Centre Ltd
Centre for European Business Information
11, Belgrave Road
London SW1V 1RB
UK

Mrs Elizabeth Holmes
Tel: 44-71-828 62 01
Fax: 44-71-834 84 16

Local Enterprise Development Unit
Euro Info Centre
Ledu House
Upper Galwally
UK-Belfast BT8 4TB
UK

Ms Eleanor Butterwick
Tel: 44-232-49 10 31
Fax: 44-232-69 14 32

Federation of Sussex Industries and Chamber of Commerce
Euro Info Centre
169, Church Road
UK-Hove BN3 2AB
East Sussex
UK

Tel: 44-273-32 62 82
Fax: 44-273-20 79 65

Bristol Chamber of Commerce and Industry
Euro Info Centre
16, Clifton Park
UK-Bristol BS8 3BY
UK

Mrs Sarah Harris
Tel: 44-272-73 73 73
Fax: 44-272-74 53 65

Exter Enterprises Limited
University of Exeter
Euro Info Centre
Hailey Wing, Reed Hall
UK-Exeter EX4 4QR
UK

Ms Diana Letcher
Tel: 44-392-21 40 85
Fax: 44-392-26 43 75

Highland Opportunity Ltd
Development Department
Highland Regional Council
European Business Services
Business Information Source
20 Bridge Street
UK-Inverness IV1 1QR
Scotland
UK

Mr Hugh Black
Tel: 44-463-70 25 60
Fax: 44-463-71 08 48

Euro Info Centre Yorkshire & Humberside
Westgate House
100 Wellington Street
UK-Leeds LS1 4LT
UK

Mrs Christine Kenyon
Tel: 44-532-43 92 22
Fax: 44-532-43 10 88

The Leicester EIC Partnership
Euro Info Centre
The Business Centre
10 York Road
UK-Leicester LE1 5TS
UK

Mr Jeff Miller
Tel: 44-533-55 99 44
Fax: 44-533-55 34 70

North West Euro Service Ltd
Liverpool City Libraries
William Brown Street
UK-Liverpool L3 8EW
UK

Tel; 44-51-298 19 28
FaxL 44-51-207 13 42

Kent County Council Economic Development
Department
Kent European Information Centre
County Hall
UK-Maidstone ME14 1XQ
Kent
UK

Mr David Oxlade
Tel: 44-622-69 41 09
Fax; 44-622-69 41 21

Manchester Chamber of Commerce & Industry
Euro Info Centre
Oxford Street, 56
UK-Manchester M60 7HJ
UK

Mr Peter Hills
Tel: 44-61-236 32 10
Fax: 44-61 236 99 45

Norwich and Norfolk Chamber of Commerce and Industry
European Information Centre East Anglia
112 Barrack Street
UK-Norwich NR3 1UB
UK

Ms Sarah J. Abercrombie
Tel: 44-603-62 59 77
Fax: 44-603-63 30 32

Nottingham Chamber of Commerce and Industry
Euro Info Centre
First Floor
South Block
309 Haydn Road
UK-Nottingham NG5 1DG
UK

Ms Anne Pearce
Tel: 44-602-62 46 24
Fax: 44-602-85 66 12

UWCC
Wales Euro Info Centre
Guest Building
P.O. Box 430
UK-Cardiff CF1 3XT
South Glamorgan
UK

Mr Brian Wilcox
Tel: 44-222-22 95 25
Fax: 44-222-22 97 40

Thames Chiltern Chamber of Commerce & Industry
Euro Info Centre
Commerce House
2-6 Bath Road
UK-Slough SL1 3SB
Berks
UK

Ms Julia Rees
Tel: 44 753-577 877
Fax: 44-753-524 644

Shropshire Chamber of Commerce and Industry
Shropshire and Staffordshire Euro Info Centre
Industry House
Halesfield 20
UK-Telford TF7 4TA
Shropshire
UK

Mr Robert Truslove
Tel: 44-952-58 87 66
Fax: 44-952-58 25 03

The Southern Area
European Information Centre
Civic Centre
UK-Southampton SO9 4XP
Hampshire
UK

Mr David Dance
Tel: 44-703-83 28 66
Fax: 44-703-23 17 14

West Yorkshire
European Business Information Centre
Economic Development Centre
Britannia House
Broadway
UK-Bradford BD1 1JF
UK

Ms Jenny Lawson
Tel: 44-274-75 42 62
fax: 44-274-39 32 26

Staffordshire Development Association
Staffordshire European Business Centre
Martin Street, 3
UK-Stafford ST16 2LH
UK

Tel: 44-785-59 528
Fax: 44-785-21 52 86

London Chamber of Commerce and Industry
Euro Info Centre
69, Cannon Street
UK-London EC4N 5AB
UK

Ms Beth Rayney Cucala
Tel: 44-71-489 19 92
Fax: 44-71-489 03 91

Correspondence Centres

Euro Info 92
NIS
National Information Center
Havelkova 22
CS- 13000 Prague 3
Czech Republic

Mr Francis Hardy
Tel: 42-2-236 57 83
Fax: 42-2-235 97 88

The Cooperation Fund
Euro Info Correspondence Centre
Zurawia 6 / 12
PL-00-503 Warsaw
Poland

Mrs Jagoda Szonert
Tel: 48-2-625 13 19/14 26/24 66
Fax: 48-2-625 12 90

Bundeskammer der Gewerblichen Wirtschaft
Wiedner Haupstrasse 63
Postfach 150
A-1045 Wien
Austria

Herrn Dr. Walter Resl
Tel: 43-1-501 05 4191/4206/4356/4342
Fax: 43-1-502 06 255

The Finnish Foreign Trade Association
Euro Info Center
Arkadiankatu 2
P.O. Box 908
SF-00101 Helsinki
Finland

Mr Taisto Sulonen
Tel: 358-0-1992
Fax: 358-0-694 00 28

EC Information Office for Business and Industry
Drammensveien 40
N-0243 Oslo
Norway

Mr Hans J. Groll
Tel: 47-2-92-65 70
Fax: 47-2-43 16 40

OESC
Office Suisse d'Expansion Commerciale
Schweizerische Zentralentrale fur Handelforderung
Stampfenbachstrasse, 85
CH-8035 Zurich
Switzerland

Mr Marius Schneider
Tel: 41-1-365 54 43
Fax; 41-1-365 54 11

NUTEK Swedish National Transport Board for
Industrial and
Technical Development
Liljeholmsvägen 32
S-11786 Stockholm
Sweden

Mrs Monica Strom
Tel: 46-8-775 40 00
Fax: 46-8-744 40 45

Export Council of Iceland
Lagmula
108 Reykjavik
Iceland

Hungarian Chamber of Commerce
Kossuth l. tér 6-8
H-1055 Budapest
Hungary

Online Hosts

Online host services provide access through the telephone network to databases held on a central computer.

Agra Europe
25 Frant Road
Tunbridge Wells
Kent
TN2 5JT
UK

Tel: 0892 33813

American Library Association
50 East Hiran Street
Chicago IL 60611
USA

Tel: 1 312 944 6780

BELINDIS, Belgian Ministry of Economic Affairs
rue J A Mot 30
1040 Brussels
Belgium

Tel: 32 2 233 67 67

BRS Information Technologies
Achilles House
Western Avenue
London
W3 0UA
UK

Tel: 081 992 3456
Fax: 081 993 7335

Context Ltd
Tranley House
Tranley Mews
London
NW3 2QW
UK

Tel: 071 267 7055

Data Star
Plaza Suite
114 Jermyn Street
London
SW1Y 6HJ
UK

Tel: 071 930 5503
Fax: 071 930 2581

Datacentralen
Lanlystvej 40
DK-2650 Hvidövre
Copenhagen
Denmark

Tel: 45 31 758122

DIALOG Information Services
P O Box 188
Oxford
OX1 5AX
UK

Tel: 0865 730275
Fax: 0865 736354

DIMDI, Deutsches Institut fur Medizinische
Dokumentation und
Information
Weisshausstrasse 27
Postfach 420580
5000 Cologne 41
Germany

Tel: 49 221 47241
Fax: 49 221 411429

DRT Europe Services
27 Avenue des arts
B-1040 Brussels
Belgium

Tel: 32 2 230 59 80

ECHO Services, European Commission Host Organisation
BP 373
L-1023
Luxembourg

Tel: 352 488041
Fax: 352 488040

ESA-IRS European Space Agency
Via Galileo Galilei
00044 Frascati
Italy

Tel: 39 396 941801
Fax: 39 396 941803

IRS Dialtech, STIS
25 Southampton Buildings
London
WC2A 1AW
UK

Tel: 071 323 7951
Fax: 071 323 7954

EUR-OP
2 rue Mercier
L-2985
Luxembourg

Eurobases
Commission of the EC
rue de la Loi 200
B-1049 Brussels
Belgium

Tel: 32 2 235 00 01

European Parliament
Documentary Databases Division
L-2929 Luxembourg

Tel: 352 430023

Eurostat
Jean Monnet Building
L-2920
Luxembourg

Tel: 352 43011

FT Information Online Ltd
P O Box 12
Sunbury-on-Thames
Middlesex
TW16 7UD
UK

Tel: 0932 761444

GreenNet
23 Bevenden Street
London
N1 6BH
UK

Tel: 071 608 3040
Fax: 071 254 0801

London Research Centre
Parliament House
81 Black Prince Road
London
SE1 7SZ
UK

Tel: 071 735 4520
Fax: 071 627 9606

Manchester Host, Soft Solution Ltd
30 Naples Street
Manchester
M4 4DB
UK

Tel: 061 839 4212

Mead Data Central Inc.
International House
1 St Katherine's Way
London
E1 9UN
UK

Tel: 071 488 9187

National Library of Medicine, Toxicology Information Programme
8600 Rockville Pike
Bethseda
MD 20894
USA

Tel: 1 301 496 6193
Fax: 1 800 638 8480

OCLC Online Computer Library Centre Inc.
6565 Frantz Road
Dublin
OH 43017
USA

Tel: 1 614 764 6000
Fax: 1 614 764 6096

ORBIT Search Services
Achilles House
Western Avenue
London
W3 0UA
UK

Tel: 081 993 7334

PFDS Online, Pergamon Financial Data Services
Paulton House
8 Shepherdess Walk
London
N1 7LB
UK

Tel: 071 490 0049
Fax: 071 490 2979

Questel, Telesystems Questel
83-85 Boulevard Vincent Auriol
75013 Paris
France

Tel: 33 1 442 364 64
Fax: 33 1 442 364 65

Fraser Williams (q.v.)
Reuters Limited
85 Fleet Street
London
EC4P 4AJ
UK

Tel: 071 324 8024

STN, Royal Society of Chemistry
Thomas Graham House
Milton Road
Cambridge
CB4 4WF
UK

Tel: 0223 420237

CD–ROM Producers and Distributors

Bowker Electronic Publishing
245 West 17th Street
New York
NY 10011
USA

Tel: 1 212 337 6989
Fax; 1 212 645 0475

Bowker-Saur Ltd
59-60 Grosvenor Street
London
W1X 9DA
UK

Tel: 071 493 5841

Chadwyck-Healey Ltd
Cambridge Place
Cambridge
CB2 1NR
UK

Tel: 0223 311479

Context Ltd
Tranley House
Tranley Mews
London
NW3 2QW
UK

Tel: 071 267 7055
Fax: 071 267 2745

Cambridge Scientific Abstracts
7200 Wisconsin Avenue
Bethesda
MD 20814
USA

Tel: 1 301 961 6750
Fax: 1 301 961 6720

Dialog Information Services
PO Box 188
Oxford
OX1 5AX
UK

Tel: 0865 730275
Fax: 0865 736354

European Community, Joint Research Centre
Ispra Establishment
ISPRA
Italy

Tel: 39 332 789663/789111

NERIS, Maryland College
Leighton Street
Woburn
Milton Keynes
MK17 9JD
UK

Tel: 0525 290364
Fax: 0525 290288

OCLC Europe
7th Floor Tricorn House
51–53 Hagley Road
Edgbaston
Birmingham
B16 3TP
UK

Tel: 021 456 4656
Fax: 021 456 4680

Silver Platter Information Ltd
10 Barley Mow Passage
Chiswick
London
W4 4PH
UK

Tel: 081 995 8242
Fax: 081 995 5159

UMI/IPI University Microfilms International
White Swan House
Godstone
Surrey
RH9 8LW
UK

Tel: 0883 7441223

INDEXES

Geographical Index

International

Organisations

Advisory Committee for the Coordination of Information Systems (ACCIS), 2003
Agence Internationale de l'Energie Atomique (IAEA): Organismo Internacional de Energia Atómica, 1146
Agricultural University of Norway, Norwegian Centre for International Agricultural Development, 174
Asbestos International Association (AIA), 1296
Association Internationale Contre le Bruit (AICB), 1596
Batelle Institute Ltd, 1338
Bureau International de la Récupération (BIR), 589
Bureau of the Convention on Wetlands of International Importance Especially as Waterfowl Habitat: Ramsar Convention Bureau, 785
Business and Industry Advisory Committee (BIAC), 1675
Centre for Environmental Management and Planning (CEMP), 2005
Centre for International Environmental Law (CIEL), 1932
Centre for Our Common Future (COCF), 2055
Commonwealth Agricultural Bureaux International (CABI), 136
Convention on International Trade in Endangered Species of Wild Fauna and Flora (CITES), 786
Cotton Council International (CCI), 626, 1677
Earthwatch, 949, 2022
Ecological and Toxicological Association of the Dyestuff Manufacturing Industry (ETAD), 627
Environment and Development Resource Centre (EDRC), 951
Environment Liaison Centre International (ELC), 2217
Food and Agriculture Organisation of the United Nations (FAO), 157, 752, 1344
Friends of the Earth International (FoEI), 1875
Global Environmental Monitoring System, 2024
Global Environmental Technology Network (GET), 2173

Greenpeace International, 1886
Greenpeace International Atmosphere and Energy Campaign, 114
Greenpeace International EC Unit, 1887
Hazardous Export/Import Prevention Project Greenpeace International, 479
ICSU Steering Committeee for Biotechnology (COBIOTECH): Centre of Bioengineering, 1307
IEA Coal Research, 1138
Industry and Environment Programme Activity Centre (IE/PAC), 1680
Institution of Water and Environmental Management (IWEM), 349
International Agency for Research on Cancer (IARC), 1291
International Association Against Noise see Association Internationale Contre le Bruit (AICB), 1596
International Association for Ecology (INTECOL), 2124
International Association of Packaging Research Institutes (IAPRI), 523, 2201
International Association of the Soap and Detergent Industry (AIS), 2277
International Association on Water Pollution Research and Control (IAWPRC), 410, 1622
International Atomic Energy Agency see Agence Internationale de l'Energie Atomique (IAEA), 1146
International Centre for the Application of Pesticides (ICAP), 217
International Chamber of Commerce (ICC), 1727
International Cleaner Production Information Clearing House (ICPIC), 629
International Commission for Protection Against Environmental Mutagens and Carcinogens, 1389
International Commission on Radiological Protection (ICRP), 1362
International Conservation Action Analysis (ICAN), 703
International Council for Bird Preservation (ICBP), 794
International Council for the Exploration of the Seas (ICES): Conseil International pour l'Exploration de la Mer, 281
International Council of Environmental Law (ICEC): Environmental Law Center, 1939

International Council of Scientific Unions (ICSU), 1308
International Dolphin Watch (IDW), 795
International Energy Agency (IEA), 978
International Environment Bureau, 2174
International Environment Bureau (IEB), 1728
International Federation of Agricultural Producers (IFAP), 159
International Federation of Associations for the Protection of Europe's Cultural and National Heritage Europa Nostra, 866
International Federation of Recovery and Recycling see Bureau International de la Récupération (BIR), 589
International Fertiliser Industry Association (IFA), 199
International Group of National Associations of Manufacturers of Agro-Chemical Products (GIFAP), 200
International Industrial Association for Energy from Nuclear Fuel: Uranium Institute (UI), 1175
International Institute for Applied Systems Analysis (IIASA), 8, 1483, 2151
International Institute for Environment and Development (IIED), 1763
International Institute for Water: Centre International de l'Eau, 350
International Laboratory of Marine Radioactivity, 309, 493, 1590
International Lead and Zinc Study Group, 1682
International Maritime Organisation (IMO), 282
International Nuclear Power and Waste (INLA), 478, 1176
International Oil Pollution Compensation Fund (IOPC), 1116, 1573
International Organisation for Human Ecology, 2125
International Organisation for Standardisation (ISO), 2165
International Organisation of Consumer Unions (IOCU), 647, 1855
International Petroleum Industry Environmental Conservation Association (IPIECA), 720, 1117
International Physicians for the Prevention of Nuclear War (IPPNW), 1190, 1899
International Professional Association for Environment Affairs (IPRE), 1856

314 Geographical Index *International*

International Programme on Chemical Safety (IPCS), 1390
International Radiation Protection Association (IRPA), 1363
International Society for Environmental Toxicology and Cancer, 1391
International Solar Energy Society (ISES), 1037
International Solid Wastes and Public Cleansing Association (ISWA), 648
International Tanker Owners Pollution Federation Ltd (ITOPF), 1118, 1574
International Tin Research Institute, 1703
International Tropical Timber Organisation (ITTO), 754
International Union for Conservation of Nature and Natural Resources (IUCN): World Conservation Union, 904
International Water and Sanitation Centre (IRC), 411, 1623
International Water Supply Association, 351
International Water Tribunal Foundation (IWT), 352
International Waterfowl and Wetlands Research Bureau (IWRB), 860
International Whaling Commision (IWC), 796
International Wool Secretariat (IWS), 630, 1683
International Youth Federation for Environmental Studies and Conservation (IYF), 1779
OECD Environment Committee, 2051
OECD Nuclear Energy Agency (NEA), 1181
Organisation Météorologique Mondiale, 96
Packaging Industry Research Association (Pira), 560
Programme on Man and the Biosphere (MAB), 2126
Ramsar Convention Bureau, 909
Scientific Committee on Problems of the Environment (SCOPE), 2127
SOS Sea Turtles, 816
Stockholm International Peace and Research Institute (SIPRI), 1236
Textile Institute, The, 633
UNEP'S International Register of Potentially Toxic Chemicals (IRPTC), 1393
UNEP/IEO Cleaner Production Working Group on Halogenated Solvents: Department of Environmental Technology, 654
United Nations Environment Programme (UNEP), 1861
World Conservation Monitoring Centre (WCMC), 861, 2037
World Data Centre for Greenhouse Gases, 113
World Fuel Cell Council, 1044
World Health Organisation Division of Environmental Health (WHO/EHE), 1330
World Meteorological Organisation (WMO) *see* Organisation Météorologique Mondiale, 88, 96
World Ozone and Ultra-Violet Data Centre: Atmospheric Environment Service, 99
World Packaging Organisation (WPO), 509
World Radiation Data Centre, 1367
World Resources Institute (WRI), 2065
World Wide Fund for Nature (WWF), 928

Publications

1,000 Terms in Solid Waste Management, 657
Acta Toxicologica et Therapeutica, 1405
Agricultural and Veterinary Sciences International Who's Who, 4th ed, 164, 842
Agricultural Research Centres: A World Directory of Organisations and Programmes, 10th ed, 165
Air Pollution's Toll on Forests and Crops, 35, 756, 1511
Aircraft Pollution: Environmental Impacts and Future Solutions, 60, 1535
Annotated Bibliography on Environmental Auditing, 1720
Arboriculture Journal, 767
Atlas of Endangered Species, The, 834
Automotive Engineering, 472
Aviation and Space Technology, 2186
Biocycle, 666
Bowker A & I Acid Rain Abstracts Annual, 322
CD ROM Directory, 5th ed, 2010
Chemical Week, 2281
Clean Air Around the World, 37, 1513
Climate Action Network International NGO Directory 1992, 126
Climate Change, 100
Climate Change: Designing a Tradeable Permit System, 104, 1969
Climatic Change, 103
Coal Gasification for IGCC Power Generation, 1132
Coal Prospects and Policies in IEA Countries, 1126
Coal Research Projects, 1127
Death in Small Doses – The Effects of Organochlorines on Aquatic Ecosystems, 1419
Directions for Internationally Compatible Environmental Data, 2168
Eco Technics: International Pollution Control Directory, 1455
Ecolinking: Everyone's Guide to Online Environmental Information, 2225
Economic Policy Towards the Environment, 2070
Ecosystems of the World, 843, 2136
Energy Economist, 1003
Energy from Waste State-of-the-Art Report, 678, 1067
Environment Bulletin, 1747
Environment International, 2240
Environment Labelling in OECD Countries, 1668
Environment Risk, 1749
Environmental and Health Controls on Lead, 1398, 1944
Environmental Conservation, 705
Environmental Dispute Handbook, 1945
Environmental Engineering in the Process Plant, 1793
Environmental Guidelines to World Industry, 1734
Environmental Issues in Wool Processing: Are Textiles Finishing the Environment?, 634
Environmental Law in UNEP, 1947
Environmental Monitoring: Meeting the Technical Challenge, 2026
Environmental Policies for Cities in the 1990s, 2074
Environmental Pollution, 1464
Environmentalist, 2243
Essential Ecology, 2129
Eutrophication of Freshwaters, 413, 1399, 1625
FAO Yearbook 1990, 166
Global Biodiversity, 837
Global Climate Changes and Freshwater Ecosystems, 3, 329
Global Forests: Issues for Six Billion People, 759
Green Globe Yearbook, 2013
Greenhouse Earth, 130
Haznews, 670, 1413
Health Cities, 1279
Information Please Environmental Almanac 1993, 2221
INFOTERRA World Directory of Environmental Expertise, 1766
Integrated Pollution Control in Europe and North America, 1452
International Association on Water Pollution Research and Control Yearbook, 1629
International Coal Letter, 1130
International Environmental Information Sources, 2230
International Environmental Law, 1951
International Handbook of National Parks and Nature Reserves, 930
International Journal of Ambient Energy, 1055
International Journal of Environment and Pollution, 1468
International Journal of Environmental Studies, 1787
International Journal of Global Energy Issues, 1094
International Journal of Radioactive Materials Transport, 1211, 2346
International Journal of Solar Energy, 1056
International Maritime Dangerous Goods (IMDG), 299, 1585, 2356
International Migration, 2299
International Politics of Nuclear Waste, The, 481
International Population Conference Proceedings, 2297
International Protection of the Environment, 727
Journal of Hazardous Materials, 672, 1416

Journal of the Institution of Water and Environmental Management, 370
Last Chance to See..., 838
Manual of Environmental Policy: The EC and Britain, 2093
Manual of Methods in Aquatic Environment Research, 1577
Manufacturing Engineering, 1964
Mercury in Soil – Distribution Speciation and Biological Effects, 1561
Modern Plastics International, 1698
Multimedia Transport and Fate of Pollutants, 442, 2337
Nature, 2142
NEA Newsletter, 1213
New Scientist, 2143
Noise and Vibration Worldwide, 1604
North Atlantic Treaty Organisation Expert Panel on Air Pollution Modelling Proceedings, 42, 1519
NOx Control Installations on Coal Fired Plants, 1134
Nuclear Engineering International, 1215
Nuclear Law Bulletin, 1217, 1967
Occupational Exposure Limits for Airborne Toxic Substances, Third Edition, 39, 1402, 1515
OECD World Energy Statistics, 986
Plastics Technology, 577
Population Explosion, The, 2294
Product is the Poison, – The Case for a Chlorine Phase-Out, 1421
Public Transport International, 2348
Pulp and Paper Week, 553
Radioactive Waste Management, 485
Rational Use of Water and its Treatment in the Chemical Industry, 362
Recycling Lead and Zinc: The Challenge of the 1990's, 475
Recycling: Energy from Community Waste, 598
Recycling: New Materials for Community Waste, 599
Renewable Energy Sources for Fuels and Electricity, 1047
Resource Recycling, 610
Review of Population Research, 2303
Revue Générale Nucléaire: International Edition, 1219
River Conservation and Management, 330
Safety in the Use of Asbestos Report IV (2), 1298
Science of the Total Environment, 2144
Solar Energy Materials and Solar Cells, 1059
Solvents and the Environment: Industry and Environment Vol 14 No 4 1991, 444
Species, 851
State of the Environment, The, 2234
State of the World 1992: A Worldwatch Institute Report on Progress Towards a Sustainable Society, 959
Surface Coatings International, 2286
Sustainability and Environmental Policy, 956
Sustainable Development: An Imperative for Environment Protection, 960
Toxicology, 1417
Transnational Environmental Liability and Insurance, 1954
Transport Policy and the Environment, 2090, 2340
Ulrichs International Periodicals Directory, 2235
UNEP 1991 United Nations Environment Programme Environmental Data Report, 2130
Waste Management and Research, 676
Waste Management International: Vol 1 Directory of Manufacturers and Services, 664
Water Air and Soil Pollution, 1477
Water and Environment International, 375
Water Pollution by Fertilisers and Pesticides, 1404, 1628
Water Quality International, 422, 1635
Water Supply and Wastewater Disposal International Almanac, 366, 690
Who is Who in Recycling Worldwide 1991, 606
Windpower Monthly, 1066
World Directory of Environmental Organisations, 1923
World Directory of Pesticide Control Organisations, 210
World Energy and Nuclear Directory, 1194
World Environment Research Directory, 2237
World Environmental Business Handbook, 1744
World Guide to Environmental Issues and Organisations, 1924
World Nuclear Directory, 1195
World Nuclear Industry Handbook, 1196
World Peace Directory and Diary 1993, 2238
World Water, 380, 694
Worldwide Trends and Opportunities in Automotive Passenger Vehicles 1995–2005, 584

Databases

Acid Rain, 327
Acompline/Urbaline, 2019
Agricola, 181
Agris, 182, 782
Analytical Abstracts, 2289
Aqualine, 391
Aquatic Sciences and Fisheries Abstracts, 235, 862, 1647
Beilstein Online, 2290
Biobusiness, 1321
BIOSIS Previews, 2157
Biotechnology Abstracts, 1323
BRIX/FLAIR, 1101
Chemical Abstracts (CA Search), 2291
Chemical Exposure, 1432
Chemical Safety Newsbase, 1433
Coal Database, 1141
Coal Research Projects, 1142
Commonwealth and Agriculture Bureau Abstracts, 183
Compendex Plus, 1805
Current Technology Index, 2209
DOE Energy, 1021
EDF-DOC, 1022
Energy Technology Data Exchange (ETDE), 133
Energyline, 1103
Enviroline (ENVN), 2158
Environment Digest, 2264
Environmental Bibliography, 621, 1342, 1487
Food Safety Briefing (FSB), 1352
Food Science and Technology Abstracts, 1353
GeoArchive, 1566
Geobase, 110
GeoRef, 682
Global Environmental Change Report, 9
Global Resource Information Database (GRID), 2038
Greenhouse Effect Report, 134, 2118
GreenNet, 2266
Hazardous Substances Databank (HSDB), 1435
HSELine, 1294
IBSEDEX, 78, 1554
ICONDA, 1806
International Nuclear Information System (INIS), 1238
International Road Research Documentation (IRRD), 890, 2366
International Solar Energy Intelligence Report, 1080
LEXIS, 1975
Life Sciences Collection, 2159
Major Hazard Incident Data Service (MHIDAS), 683, 1436
Medline, 1437
Molars, 10
Multinational Environmental Outlook, 2268
New Scientist, 2160
Oceanic Abstracts, 314, 1592
Oil Spill Intelligence Report, 1143, 1593
Paperchem, 564
Pira Abstracts, 565
Pollution Abstracts, 1488
PTS PROMT, 1756
RAPRA Technology, 586
Registry of Toxic Effects of Chemicals on Substances (RTECS), 1438
Scarabee, 622
SciSearch, 2161
Toxic Materials Transport, 1439, 2367
Toxline, 219, 1440
Transport Research Information Services (TRIS), 892, 2369
Tulsa, 1144, 1594
UN Conference on Environment and Development (UNCED), 2119
Waste, 684
Wasteinfo, 623
Water Resources Abstracts, 433, 701, 1648

European

Organisations

Association of European Dry Battery Manufacturers (Europile), 462
Association of Plastic Manufacturers in Europe (APME), 567
Baltic Marine Environment Protection Committee Helsinki Commission (HELCOM), 1569
Bonn Commission, 277, 1570
Centre for European Policy Studies (CEPS), 2054
Climate Network Europe (CNE), 111
Combined Heat and Power Association (CHPA), 1085
Comité Européen de l'Association de l'Ozone, 112
Comité Européen de Normalisation (CEN), 496, 1707
Commission of the European Communities, 2056
Commission of the European Communities (CEC) – Joint Research Centre (JRC), 1229, 1374
Commission of the European Communities Nuclear Safety Research Directorate, 1250
Committee of Agricultural Organisations (COPA)/General Committee of Agricultural Cooperation in EC, 156
Concentration Unit for Biotechnology in Europe (CUBE), 1300, 2041
Confédération Européenne de l'Industrie des Pâtes Papiers et Cartons (CEPAC), 533
Confederation of European Agriculture (CEA), 137, 745
Confederation of European Paper Industries (CEPI), 534
Conseil Européen des Fédérations de l'Industrie Chimique (CEFIC), 2274
Conseil Européen du Droit de l'Environnement (CEDE), 1933
Cooperative Programme for Monitoring and Evaluation of Long Range Transboundary Air Pollution (EMEP), 20, 1497
Council of Europe, 899
Council of Ministers, 1817
Earthwatch Europe, 950
Ecological Studies Institute (ESI), 1935
Economic and Social Committee (ESC): Section for Protection of the Environment, 2006
Environmental Detergent Manufacturers Association (EDMA), 438, 1385, 1662
EUREAU, 407
Eurométaux, 466
Europäische Ökologische Aktion (ECOROPA), 1852
European Aluminium Association, 467
European Arctic Stratospheric Ozone Experiment (EASOE), 84
European Association for Conservation of Energy (Euro ACE), 1086
European Association for Population Studies (EAPS), 2292
European Association for the Science of Air Pollution (EURASAP), 22, 1499
European Association of Automobile Manufacturers (ACEA), 468, 2327
European Centre for Medium Range Weather Forecasts (ECMWF), 92
European Chemical Industry Council *see* Conseil Européen des Fédérations de l'Industrie Chimique (CEFIC), 2274
European Chemical Industry Ecology and Toxicology Centre (ECETOC), 1386
European Chlorinated Solvent Association (ECSA), 1387
European Committee of the European Ozone Association *see* Comité Européen de l'Association de l'Ozone, 112
European Conference of Ministers of Transport (ECMT), 2328
European Council on Environmental Law *see* Conseil Européen du Droit de l'Environnement (CEDE), 1933
European Crop Protection Association (ECPA), 192
European Environment Bureau (EEB), 1853
European Environmental Agency (EEA), 719
European Environmental Alliance (EEA), 2057
European Federation for Transport and Environment (T & E), 2058, 2329
European Federation of Animal Health (FEDESA), 790
European Federation of Chemical and General Workers' Unions (FESCID), 1678, 2275
European Federation of Pharmaceutical Industries (EFPIA), 1679, 2276
European Federation of Waste Management (FEAD), 643
European Fertiliser Manufacturers' Association (EFMA), 193
European Geographical Information Systems Foundation (EGIS), 2023
European Group for Ecological Action *see* Europäische Ökologische Aktion (ECOROPA), 1852
European Group of Automotive Recycling Association (EGARA), 469
European Institute for Transuranium Elements, 492, 1230
European Institute for Water, 408
European Institute of Ecology and Cancer (EIEC), 1290
European Investment Bank (EIB), 1725
European Network of Environmental Research Organisations (ENERO), 2172
European Nuclear Society (ENS), 1168
European Packaging Federation, 497
European Petroleum Industries Association (EUROPIA), 1113
European Recovery and Recycling Association (ERRA), 593
European Transport Law, 1388, 1937, 2330
European Union for Packaging and the Environment, 498
European Water Pollution Control Association (EWPCA), 1621
European Wind Energy Association, 1035
European Youth Forest Action (EYFA), 751
Fédération Européenne du Verre d'Emballage (FEVE), 450
Forum Atomique Européen (FORATOM), 1171
Foundation for Environmental Education in Europe (FEEE): European Office Friluftsradet, 1773
Friends of the Earth International (FoEI): European Coordination – CEAT, 1876
Global Legislators Organisation for a Balanced Environment (GLOBE): GLOBE International/EC, 1938
Green Alternative European Link (GRAEL), 1877
Informat Green Alert, 1726
Institute for European Environmental Policy (IEEP), 2059
International Council of Environmental Law (ICEL): Environmental Law Center, 1939
International Eco Meat Control *see* Internationale Scharrelvees Controle (ISC), 160, 1346, 1940
Internationale Scharrelvees Controle (ISC), 160, 1346, 1940
Network for Environmental Technology Transfer (NETT), 2176
New Energy Technologies (THERMIE), 1038
Nordic Council, 315
Nordic Council of Ministers, 2049
Oil Companies European Organisation for Environmental Health and Protection (CONCAWE), 1120, 1328, 1448
Open Air Council *see* Foundation for Environmental Education in Europe (FEEE), 1773
Organic Reclamation and Composting Association (ORCA), 254
Oslo Commission (OSCOM), 286, 1575
Pharmaceutical Group of the EC (GPCE), 2278
Regional Environment Centre for Central and Eastern Europe, 955
Secretariat for the Protection of the Mediterranean Sea, 288, 1576

Senior Advisory Group Biotechnology (SAGB): CEFIC, 1312
Society for Environmental Toxicology and Chemistry (SETAC), 440, 1392, 1710
Society for the Promotion of Life Cycle Development (SPOLD), 1711
Specific Action for Vigorous Energy Efficiency (SAVE), 1100
The European Glass Container Federation see Fédération Européenne du Verre d'Emballage (FEVE), 450
Union of Industrial and Employers' Confederation of Europe (UNICE), 1729
World Wide Fund for Nature – Central and East Europe, 915

Publications

Air Pollution Control in the European Community, 34, 1510
Blue Flag Campaign, The, 270
Business Strategy and the Environment, 1745
Chlorine Production and Use and their Environmental Risks, 1418
Climatic Change and the Mediterranean, 102, 292
Common Future of the Wadden Sea, The, 298
Corporate Responsibility Europe, 1732
Council of Europe: Farming and Wildlife, 162, 1558
Directory of European Environmental Organisations, 1916
EC Energy Monthly, 991
EC Environmental Guide, 1917
EC Waste Management Policy: Transforming from Waste Treatment to Waste Prevention, 677
Ecogenetics: Genetic Predisposition to the Toxic Effects of Chemicals, 1397
Economic Restructuring in Eastern Europe and Acid Rain Abatement Strategies 1991, 1026, 1095
EEC Environmental Policy and Britain, 2071
Effluent Treatment and Waste Disposal, 441
Energy and Environmental Conflicts in East/Central Europe: The Case of Power Generation, 1013
Energy Conscious Design: A Primer for European Architects, 1045
Energy in Europe, 1004
Energy in the European Community, 1015
Energy Taxation and Environmental Policy in EFTA Countries 1991, 1016, 2106
Environment and Transport Infrastructures, 2353
Environment Research, 2241
Environmental Management in the Soviet Union, 2009
Environmental Policy in the European Community, 2075
Environmental Research Newsletter, 2242
Environmental Technology Transfer to Eastern Europe: A Selected Bibliography 1991, 2180
Europe and the Environment, 2077
European Bulletin of Environment and Health, 1335
European Chemical News, 2284
European Community Environmental Legislation 1967–1987, 1949
European Community, The, 2078
European Directory of Renewable Energy Suppliers and Service, 1088
European Energy Report, 1017
European Environment, 2097, 2245
European Environmental Business News, 1751
European Environmental Law Review, 1961
European Environmental Yearbook, 2228
European Index of Key Plastics Recycling Schemes, 575
European Inland Fisheries Advisory Commission: Water Quality Criteria for European Freshwater Fish, 221, 424
European Integration and Environmental Policy, 2079
European Journal of Pharmacology: Environmental Toxicology and Pharmacology Section, 1412
European Water and Wastewater Industry, 365, 689
European Water Pollution Control, 1633
Examination of Water for Pollution Control: A Reference Handbook, 1626
Forest Damage in Europe – Environmental Fact Sheet No 1, 768
Forests Foregone: The EC's Trade in Tropical Timbers and the Destruction of the Rainforests, 770
Freight Transport and the Environment: ECMT Seminar 1991, 2333
Future Forest Resources of the Former European USSR, 757
Future Forest Resources of Western and Eastern Europe, 758
Glass Gazette, 452
Global Warming: The Debate 1991, 118
Green Europe, 171
Green Paper on the Environment – A Community Strategy for Sustainable Mobility, 958, 2109, 2355
Greening of Urban Transport, The: Planning for Walking and Cycling in Western Cities, 2080, 2334
IIASA Research Plan 1993, 2146
Impact of Sea Level Changes on European Coastal Lowlands, The, 267
Implications for the EC's Environmental Policy of the Treaty on European Union, The, 2110
Industry and the Environment, 1697
Information Bulletin, 2249
Initiatives for the Environment: A Publication from General Motors Europe, 1686
Management of Waste Plastic Packaging Films, 579
Monthly Environmental Policy, 2111
Natural Sources of Ionising Radiation in Europe: Radiation Atlas, 1369
Nature Conservation in Europe – Agenda 2000, 931
Naturopa, 941
Net Losses Gross Destruction – Fisheries in the European Community, 222, 853
NGO Directory for Central and Eastern Europe 1992, 1919
Nuclear Europe Worldscan, 1216
Options, 2194
Organic Produce in Europe Special Report no 2128, 262
Paradise Deferred: Environmental Policymaking in Central and Eastern Europe, 2112
Plastics Recovery in Perspective, 580
Plastics Recycling 1991, 574
Politique Européenne de l' Environnement, La, 2087
Pollution Control Policy of the European Communities, 1453
Pollution of the Mediterranean Sea, 293, 1581
POLMARK – The European Pollution Control and Waste Management Industry Directory, 663, 1458
Proposal for a Council Regulation (EC), 1723
Pulp and Paper International, 552
Radioactive Waste, 483
Realized and Projected Recycling Processes for Used Batteries, 618
Recreational Water Quality Management: Vol 1 Coastal Waters, 268, 1579
Recycling in Member Countries, 619
Research and Technological Development for the Supply and Use of Freshwater Resources, 425
Safety Assessment of Radioactive Waste Repositories: Systematic Approaches to Scenario Development, 487, 1254
Salvo Demolition Directory 93/94, 604
Silvicultural Systems, 760
Single European Dump, The: Free Trade in Hazardous and Nuclear Wastes in the New Europe, 488, 1422
Solar Energy Research and Development in the European Community: Series A–H, 1060
Solvents Digest, 445
Taxation for Environmental Protection, 728
TEC Transport Environment Circulation, 2349
Technical Innovation in the Plastics Industry and its Influence on the Environmental Problem of Plas, 583
Transport Policy and the Environment, 2089, 2339
UK Waste Law, 660, 1956
Urban Solid Waste Management, 661
Vachers European Companion and Consultants Register, 1767
Volatile Organic Compound Emissions in Western Europe: Report 6/87, 64, 1135, 1539, 2359
Who's Who in European Water 1992, 367

Wind Energy in Europe Report, 1068
WWF Baltic Bulletin, 296

Databases

ABEL, 1971
Agra Europe, 179, 2116
AGREP, 180
Biorep, 1322
Celex, 1972
CORDIS, 2208
ENREP-NEW, 2263
Environmental Chemicals Data and Information Network (ECDIN), 1434
EUREKA, 1025
EUROCRON, 2265
European Parliament Online Query Service (EPOQUE), 1973
European Update, 2117
Farm Structure Survey Retrieval System (FSSRS), 239, 889
INFOMAT, 1755
PASCAL, 2210
SESAME, 1081
SPEARHEAD, 1976
System for Information on Grey Literature in Europe (SIGLE), 2269
Transdoc, 891, 2368

Publications

Landfill Gas, 455

Austria

Organisations

Association des Régions des Alpes Centrales (ARGE ALP), 2120
Association of the Central Alps *see* Association des Régions des Alpes Centrales (ARGE ALP), 2120
Austrian Environmental Protection Association, 717
Austrian Noise Abatement Society *see* Österreichischer Arbeitsring für Lärmbekämpfung (OAL), 1600
Austrian Nuclear Society *see* Österreichische Kerntechnische Gesellschaft, 1163
Austrian Society for Nature and Environment Protection *see* Österreichisches Gesellschaft für Natur und Umweltschutz (ÖGNU), 97, 908
Bundesministerium für Land und Forstwirtschaft, 135, 744
Bundesministerium für Öffentliche Wirtschaft und Verkehr, 1670, 2305
Bundesministerium für Umwelt Jugend and Familie, 1977
Environmental Information Science Institute *see* Institut für Umwelt Informatik, 2148
Fachverband des Chemischen Industrien Österreichs, 194
Federal Chancellory Division for Nuclear Energy Coordination and Nonproliferation, 1152
Federal Forest Research Institute *see* Forstliche Bundesversuchsanstalt, 776
Federal Ministry for Agriculture and Forestry *see* Bundesministerium für Land und Forstwirtschaft, 135, 744
Federal Ministry for Nationalised Industries and Transport *see* Bundesministerium für Öffentliche Wirtschaft und Verkehr, 1670, 2305
Forstliche Bundesversuchsanstalt, 776
Freunde der Erde, 1872
Greenpeace Austria, 1878
Grüne Alternative, Die, 1830
Institut für Umwelt Informatik, 2148
Institut für Umweltwissenschaften und Naturschutz, 947
Institute of Environmental Sciences and Conservation *see* Institut für Umweltwissenschaften und Naturschutz, 947
Ministry for Health Sports and Consumer Protection, 1656
Ministry of the Environment Youth and Family *see* Bundesministerium für Umwelt Jugend und Familie, 1977
Österreichische Kerntechnische Gesellschaft, 1163
Österreichischer Arbeitsring für Lärmbekämpfung (OAL), 1600
Österreichisches Atomforum, 1183
Österreichisches Forschungszentrum Seibersdorg GmbH, 1235
Österreichisches Gesellschaft für Natur und Umweltschutz (ÖGNU), 97, 908
Österreichisches Vereinigung für das Gas und Wasserfach, 355, 1121
Österreichisches Verpackungszentrum, 507
Physikalisch Technische Versuchsanstalt für Wärme und Schalltechnik, 1612
Studiengesellschaft für die Entsorgung von Altahrrzeugen GmbH (EVA), 471
Testing Institute for Heat and Sound Technology *see* Physikalisch Technische Versuchsanstalt für Wärme und Schalltechnik, 1612
Verband Organisch–Biologisch Wirtschaftender Bauern Österreichs, 257
Vereinigung Österreichischer Papierindustrierhertsteller, 548
World Wide Fund for Nature – Austria, 913

Publications

Ludwig Boltzmann Institut für Umweltwissenschaften und Naturschutz Mitteilungen, 935
Österreichisches Tierschutzzeitung, 943
Returnable and Non-Returnable Packaging, 521
Steine Sprechen, 873
Umwelt Aktuell, 2257
Wasserwirtschaftliche Mitteilungen, 374

Belgium

Organisations

Agricultural Research Centre Ghent *see* Centrum voor Landbouwkundig Onderzoek – Gent, 178
Agronomic Centre of Applied Research Hainaut *see* Centre Agronomique de Recherches Appliquées du Hainaut, 212
Amis de la Terre, Les, 1866
Anders Gaan Leven (AGALEV), 1807
Association des Transformateurs de Matières Plastiques (Fechiplast): Fédération des Industries Chimiques, 566
Association for the Promotion of Renewable Energies *see* Association Promotion Energies Renouvelables (APERE), 1030
Association Nationale des Services d'Eau asbl (ANSEAU), 341
Association Promotion Energies Renouvelables (APERE), 1030
Association Vincotte ASBL, 641, 1444
Belgian Biotechnology Coordinating Group (BBCG), 1303, 2052
Belgian Packaging Institute (BPI), 495
Belgian Royal Meteorological Institute *see* Institut Royal Météorologique de Belgique, 94
Bestuur van de Volksgezondheid – Ministerie van Volksgezondheid, 11, 1354, 1489
Bon Beter Leefmilieu (BBL), 1847
Center for Information on the Environment, 2039
Center for Nature Protection and Education, 893
Centre Agronomique de Recherches Appliquées du Hainaut, 212
Centre de Recherche et de Contrôle Lainier et Chimique Celac, 1638
Centrum voor de Studie van Water Bodem en Lucht (Becewa Vzw), 681
Centrum voor de Studie Van Water Bodem en Lucht Becewa Vzw, 1480
Centrum voor Landbouwkundig Onderzoek – Gent, 178
Confédération Belge de la Récupération, 591
Council of Ministers, 1811
Ecole de Santé Publique – Université Libre de Bruxelles, 1339
Ecolo, 1824
Ecology and Forestry Research Council *see* Studiecentrum voor Ecologie en Bosbouw, 1485
Energik, 976
ERL Brussels, 1758
FABRIMETAL, 346, 644
Forum Nucléaire Belge (ASBL), 1173
Greenpeace Belgium, 1879
Inspection Générale de l'Environnement et des Forêts, 747
Institut Royal Météorologique de Belgique, 94
Inter Environnement Bond Beter Leefmilieu Centre d'Inform et de Documentation sur l'Environnement, 1447
Koninklijke Maatschappij voor Dierkunde van Antwerpen, 799
Minister of Small and Medium Sized Enterprises and Agriculture, 140
Ministère de la Santé Publique, Chef du Service des Nuisances, 1655
Ministère des Affaires Economiques Administration de l'Energie Service des Applications Nucléaires, 1159
Ministerie van de Vlaamse Gemeenschap Administrie voor Ruimtelijke Dienst Groen Waters en Bossen, 395
Ministry of the Environment, 1992
Nature et Progrès, 252
PHYTOPHAR, 202
Secretary of State for Public Health, 1276
Studiecentrum voor Ecologie en Bosbouw, 1485
Studiecentrum voor Kernenergie – Centre d'Etude de l'Energie Nucléaire (CEN-SCK), 1237
Technological Institute – Royal Flemish Society of Engineers, 1320
Work Group for Water, 404
World Wide Fund for Nature – Belgium, 914

Publications

Bates and Wacker Packaging Report 93, 514
Belgicatom, 1205
Geluid en Omgeving, 1602
Guide de l'Industrie Belge du Recyclage des Matières Plastiques 1991, 576
Guide to the Belgian Plastic Recycling Industry *see* Guide de l'Industrie Belge du Recyclage des Matières Plastiques 1991, 576
Milieu en Bedrijf, 731

Bosnia Hercegovina

Organisations

Skakavac, 1859

Bulgaria

Organisations

Committee on the Use of Atomic Energy for Peaceful Purposes, 1149, 2040
Ecoglasnost, 1870
Environmental Politics in Bulgaria: Ecological Studies Institute, 1851
Green Party of Bulgaria, 1827
Ministry of the Environment, 1993

Publications

Iaderna Energiia, 1210
Nuclear Energy *see* Iaderna Energiia, 1210

Byelorussia

Organisations

State Committee for Chernobyl Accident Problems, 1246

Youth Ecological Movement of Byelorussia, 1862

Croatia

Organisations

Croatian Air Pollution Prevention Association (CAPPA) *see* Hravtsko Drutvo za Zastitu Zraka, 24, 1501
Hravtsko Drutvo za Zastitu Zraka, 24, 1501

Publications

Archives of Industrial Hygiene and Toxicology *see* Arhiv za Higijenu Rada i Toksikologiju, 1407
Arhiv za Higijenu Rada i Toksikologiju, 1407

Solar Energy *see* Sunceva Energija, 1063
Sunceva Energija, 1063

Cyprus

Organisations

Cyprus Association for the Protection of the Environment, 718, 1848

Friends of the Earth (FoE), 1874

Czech Republic

Organisations

Brontosaurus Movement/EcoWatt, 1032
Ceská Asociace IUAPPA (CA IUAPPA), 18, 1495
CSFR Atomic Energy Commission, 1150
Czech Association of IUAPPA *see* Ceská Asociace IUAPPA (CA IUAPPA), 18, 1495

Energy Efficiency Centre *see* Stredisko pro Efektivni Vyuzivani Energie (SEVE), 980
Greenpeace Czechoslovakia, 1880
Hydroprojekt, 348
Institute of Handling Transport Packaging and Storage Systems (IMADOS), 504
Promysi Papiru a Celulozy Generalni Reditelstvi, 541

Stredisko pro Efektivni Vyuzivani Energie (SEVE), 980

Publications

Environment in Czechoslovakia, The, 62, 1537

Denmark

Organisations

Civil Forsvaret, 590
Council of Ministers, 1812

Danish Acoustical Institute *see* Lydteknisk Institut (LI), 1610
Danish Agrochemical Association (DAA), 191

Danish Energy Agency: Ministry of Energy, 965

Danish Institute for Fisheries and Marine Research *see* Danmarks Fiskeri–og Havundersogelser, 227, 302
Danish Laboratory for Soil and Crop Research *see* Statens Planteavis Laboratorium, 1565
Danish Meteorological Institute, 91
Danish Nuclear Society (DNS), 1166
Danish Organisation for Renewable Energy, 1034
Danish Packaging and Transportation Research Institute, 522, 2360
Danish Society for the Conservation of Nature *see* Danmarks Naturfredningsforening, 901
Danish Toxicology Centre (DTC), 1384, 1934
Danmarks Fiskeri–og Havundersogelser, 227, 302
Danmarks Naturfredningsforening, 901
Dansk Naturkost, 247
Energiministeriet, 969
Federation of Biotechnology Industries of Denmark (FBID), 1306
Geological Survey of Denmark, 2260
Greenpeace Denmark, 1881
Gronne, De, 1829
Landbrugsministeriet, 139
Lydteknisk Institut (LI), 1610
Miljoministeriet Miljostyrelsen, 1981
Miljoministeriet Miljostyrelsen Kemikaliekontrollen, 185, 1382, 1928

Ministry of Agriculture *see* Landbrugsministeriet, 139
Ministry of Energy *see* Energiministeriet, 969
Ministry of the Environment *see* Miljominsteriet Miljostyrelsen, 1981, 1994
Ministry of the Environment National Agency of Environmental Protection: Water Resources Division, 399
Ministry of Transport and Public Works *see* Trafikministeriet, 2322
National Agency for Environmental Protection, 713
National Forest and Nature Agency, 755
National Institute of Occupational Health, 1273
NOAH International, 1903
Nordic Committee for Nuclear Safety Research (NKS), 1245
Socialistik Folkeparti, 1841
Statens Planteavis Laboratorium, 1565
The State Chemical Supervision Service *see* Miljoministeriet Miljostyrelsen Kemikaliekontrollen, 185, 1382, 1928
Trafikministeriet, 2322
Vandkvalitetsinstituttet (VKI), 432, 699
Water Quality Institute *see* Vandkvalitetsinstituttet (VKI), 432, 699
World Wide Fund for Nature – Denmark, 916

Publications

Danish Environmental Directory *see* Miljovejviseren 1991, 2232
Dansk Energi Tidsskrift, 989
Eco-Labelling of Paper Products, 555, 1667
Energi och Miljö i Norden 1991, 1012, 2105
Energiministeriet Energiforskningsprogram, 1001
Energy in Denmark, 1014
Fravar ved Anmeldte Arbejdsulykker, 1283
Miljoe og Teknologi, 2193
Miljoeprojekt, 2251
Miljoeundersoegelser ved Marmorilik, 2252
Miljopolitik i Det Europaeiske Faellesskab, 2082
Miljovejviseren 1991, 2232
MST Luft, 53, 1528
Natur og Miljoe, 937
Nordic Journal of Environmental Economics Protection, 2100
Techniques in Marine Environmental Sciences, 295
Vandteknik, 372

Estonia

Organisations

Estonian Green Movement, 1871

Estonian Green Party, 1825

Finland

Organisations

Advisory Committee on Nuclear Energy, Ministry of Trade and Industry, 1145
Association of Waste Management and Entrepreneurs in Finland, 640
Baltic Marine Environment Protection Committee Helsinki Commission (HELCOM), 276, 1569
Chemical Industry Federation of Finland *see* Kemian Keskusliitto, 1309
Espoo Research Centre, 214, 1702

Federation of the Water Protection Association in Finland, 409
Finnish Air Pollution Prevention Society (FAPPS) *see* Ilmansuojeluyhdistys ry ISY, 25, 1502
Finnish Association for Nature Conservation *see* Suomen Luonnonsuojeluitto ry, 912
Finnish Centre for Radiation and Nuclear Safety *see* Sateilyturvakeskus (STUK), 1252, 1366
Finnish Forest Research Institute *see* Metsäntutkimuslaitos, 779

Finnish Institute of Marine Research *see* Merentutkimuslaitos, 311
Finnish Meteorological Institute *see* Ilmatieteen Laitos, 93
Finnish Packaging Association, 499
Finnish Pulp and Paper Research Institute *see* Oy Keskuslaboratorio – Centrallaboratorium Ab, 559
Finnish Standards Association, 1663
Green League *see* Vihreä Litto, 1845
Greenpeace Finland, 1882
Ilmansuojeluyhdistys ry ISY, 25, 1502
Ilmatieteen Laitos, 93

322 Geographical Index *France*

Kansanterveyslaitos (NPHI), 1327
Kemian Keskusliitto, 1309
Merentutkimuslaitos, 311
Metsäntutkimuslaitos, 779
Ministry for Social Affairs and Health, 1262
Ministry of Agriculture, 142
Ministry of Communications and Transport, 2310
Ministry of the Environment, 1995
National Public Health Institute *see* Kansanterveyslaitos (NPHI), 1327
Oy Keskuslaboratorio – Centrallaboratorium Ab, 559
Sateilyturvakeskus (STUK), 1252, 1366
SLVY-liitto, 256
Suomen Luonnonsuojeluitto ry, 912
Tampere University of Technology, 563

Vesien ja ympäristöntutkimuslaitos, 389, 700, 1645
Vihreä Litto, 1845
Water and Environment Research Institute *see* Vesien ja ympäristöntutkimuslaitos, 389, 700, 1645
World Wide Fund for Nature – Finland, 917

Publications

Aqua Fennica, 369
Baltic Environment Proceedings no 39: Baltic Marine Environment Protection Commission, 297, 1584

Finland Central Statistical Office: Statistical Surveys Environmental Statistics, 2220
Finland Kauppa–Jateollisuusministerioe Energiatilastot, 983
Finnish Meteorological Institute Publications on Air Quality *see* Ilmatieteen Laitos Ilmansvojelun Julkaisvja, 49
Ilmatieteen Laitos Ilmansvojelun Julkaisvja, 49
Vaestoentutkimuksen Vuosikirja, 2298
Yearbook of Population Research in Finland *see* Vaestoentutkimuksen Vuosikirja, 2298

France

Organisations

AFME, 1082
Agence de l'Environnement et de la Maîtrise de l'Energie (ADEME), 962
Agence Nationale de la Récupération et l'Elimination des Déchets (ANRED), 587
Agence Nationale pour la Gestion des Déchets Radioactifs (ANDRA), 477
Air Pollution Dept of the National Institute of Applied Chemical Research *see* Département de la Pollution Atmosphérique de l'Institut National de Récherche Chimique Appliquée, 14, 1491
Amis de la Terre, Les, 1867
Association for the Control and Saving of Energy *see* AFME, 1082
Association for the Development of Agricultural Fuel (ADECA), 155
Association for the Prevention of Atmospheric Pollution *see* Association pour la Prevention de la Pollution Atmosphérique (APPA), 16, 1493
Association for the Promotion of Recycled Paper, 525
Association pour la Prevention de la Pollution Atmosphérique (APPA), 16, 1493
Association Technique de l'Industrie du Gaz en France, 1110
Atmospheric Pollution Technical Research Centre *see* Centre Interprofessionnel Technique d'Etudes de la Pollution Atmosphérique (CITEPA), 66
Biofranc (FNAB), 243
Blue Bubble *see* Bulle Bleue, 80
Bulle Bleue, 80

Cellule Environnement Institut National de la Recherche Agronomique (INRA), 176
Centre d'Enseignement et de Recherche pour la Gestion des Resources Naturelles et l' Environnement, 2018
Centre d'Enseignement et de Recherche pour la Gestion des Ressources Naturelles et l' Environnement, 382
Centre d'Information du Plomb, 1325, 1445
Centre Interprofessionnel Technique d'Etudes de la Pollution Atmosphérique (CITEPA), 66, 1541
Centre National pour l'Exploitation des Océans, 278, 1571
Commissariat à l'Energie Atomique (CEA), 1148
Compagnie Générale des Eaux, 343
Copacel, 536
Council of Ministers, 1813
Département de la Pollution Atmosphérique de l'Institut National de Recherche Chimique Appliquée, 14, 1491
Département Pollution des Eaux de l'Institut National de Recherche Chimique Appliquée, 1639
Economy and Performance Service (ECO-PER), 437
French Association for the Protection of Water, 394
French Environment and Energy Management Agency *see* Agence de l'Environnement et de la Maîtrise de l'Energie (ADEME), 962
French Federation of the Paper Board and Cellulose Industry *see* Copacel, 536
French Research Institute for Ocean Utilisation *see* Institut Français de Recherche pour l'Exploitation de la Mer (IFREMER), 305
Greenpeace France, 1883
Industrielle de l'Environnement et du Recyclage (FEDEREC), 594
Institut Atlantique, 280, 1572
Institut de Recherche des Transports Centre de Documentation d'Evaluation et de Recherches Nuisances, 70, 1546, 2362
Institut Français de l'Energie, 977
Institut Français de Recherche pour l'Exploitation de la Mer (IFREMER), 305
Institut National de l'Environnement Industriel et des Risques (INERIS), 645
Institut National de la Santé et de la Recherche Medicale, 72, 1427, 1548
Institut Pasteur de Lille, 1428
Interprofessional Technical Centre for Studies in Atmospheric Pollution *see* Centre Interprofessionnel Technique d'Etudes de la Pollution Atmosphérique (CITEPA), 1541
Management of Natural Resources and Environment Study Centre *see* Centre d'Enseignement et de Recherche pour la Gestion des Resources Naturelles et l' Environnement, 382, 2018
Ministère de l'Environnement, 337
Direction de la Prévention des Pollutions de l'Eau, 1619, 1654
Ministère de l'Environnement et du Cadre de Vie, 1984
Ministry of Agriculture and Forestry, 148, 749
Ministry of Equipment Housing and Transport, 2311
Ministry of Health and Humanitarian Policies, 1267, 2047
National Federation for the Defence of the Environment, 702

National Institute for Industrial Environment and Hazards see Institut National de l'Environnement Industriel et des Risques (INERIS), 645
Office National d'Etudes Recherches Aerospatiales, 1618
Office National de la Chasse (ONC), 808
Organbio, 1311
Panos Institute, 953
Secrétaire d'Etat auprès du Premier Ministre de la Prevention des Risques Technologiques, 402
Service Central de Protection Contre les Rayonnements Ionisants, 476
SOLEREC Département Solaire, 1041
Syndicat National des Engrais (SNIE), 204
Union of Industries of the Protection of Plants (UIPP), 206
Verts, 1844
Water Pollution Dept of the National Institute of Applied Chemical Research see Département Pollution des Eaux de l'Institut National de Recherche Chimique Appliquée, 1639
World Wide Fund for Nature – France, 918

Publications

Amenagement et Nature, 932
Annuaire de l'Administration des DRIR, 987
Bulletin des Transports, 2343
Code Permanent Environnement et Nuisances, 1459
Crisis in the French Nuclear Power Industry, 1224
Directory of the French Nuclear Industry, 1193
Droit Nucléaire, 1206
Du Sol à la Table: formerly Agriculture et Vie, 168
Enerpresse, 1005
Environmental Protection and Antipollution Directory see Guide de l'Environnement et des Techniques Antipollution, 1457
Environnement: Le Guide des Sources d'Information, 1918
Etudes de Pollution Atmosphérique à Paris et dans les Départements Periphériques, 47, 1524
France Commissariat à l'Energie Atomique Notes d'Information, 1209
Guide de l'Environnement et des Techniques Antipollution, 1457

Horticulture Française, 845
Industrie et Environnement, 1696
Info Déchets Environnement et Technique, 671
Journal de Toxicologie Clinique et Experimentale, 1414
Lettre de l'Environnement, 2250
Nature et Mieux-Vivre, 940
Packaging in Europe: France Special Report no 2046, 516
Politique de l'Environnement dans la Communauté Européenne, La, 2086
Pollustop, 1471
Pollution Atmosphérique, 54, 1529
Population et Sociétés, 2301
Radiprotection, 1372
Recyclage des Matières Plastiques, Le, 582
Revue d'Ecologie: La Terre et la Vie, 850
Revue Générale Nucléaire, 1218
Revue Juridique de l'Environnement, 1968
Sécrétariat d'Etat Chargé de l'Environnement et de la Qualité de la Vie: Bulletin de Documentation de l'Environnement, 2256
Spirit of Versailles: The Business of Environmental Management, 1743
Technologie Appropriée de Déminéralisation de l'Eau Potable, La, 415

Germany

Organisations

Arbeitsgemeinschaft Information Meeresforschung Meerestechnik, 275, 1568
ARC, 1808
Association of Chemical Industries (VCI), 1673, 2272
Association of Petroleum Industry see Mineralöl Wirtschafts Verband eV (MWV), 1119
Bund, 1868
Bundesamt für Wirtschaft, 1105
Bundesantstalt für Arbeitsschutz und Unfallforschung, 12, 1490
Bundesarbeitgeberverband Chemie, 434
Bundesdeutscher Arbeitskreis für Umweltbewusstes Management (BAUM), 2004
Bundesforschungs Anstalt für Fischerei, 226, 741, 1586
Bundesforschungsanstalt für Naturschutz und Landschaftsökologie, 946
Bundesministerium für Forschung und Technologie (BMFT), 1147
Bundesministerium für Umwelt Naturschutz und Reaktorsicherheit, 711, 1239, 1649

Bundesverband Glasindustrie, 449
Commission on Air Pollution Prevention in VDI and DIN see Kommission Reinhaltung der Luft (KRdL) in VDI und DIN, 28, 1505
Council of Ministers, 1814
Deutsche Naturschutzring (DNR), 1849
Deutsche Verkehrswissenschaftliche Gesellschaft eV, 406, 1613
Deutscher Wetterdienst (DWD), 1072
Deutsche Wissenschaftliche Gesellschaft für Erdöl, 1137
Deutscher Wetterdienst (DWD), 106
Deutscherverein des Gas und Wasserfaches eV (DVGW), 1112
Deutscherverein des Gas- und Wasserfaches eV (DVGW), 344
Deutsches Atomforum eV (DAIF), 1167
Deutsches Hydrographisches Institut, 272, 490, 1587
Duales System Deutschland GmbH (DSD), 592
Energy and Transport Research Institute see Institut für Energie- und Transportforschung Meissen, 1075, 2363
ERL Umwelt Consult: Rhein-Main-Neckar GmbH, 1762

Faunistisch-Ökologische Arbeitsgemeinschaft, 791
Federal Ministry for the Environment Federal Republic of Germany, 1980
Federal Office for Trade and Industry Branch for Mineral Oil and Gas see Bundesamt für Wirtschaft, 1105
Federal Research Centre for Fisheries see Bundesforschungs Anstalt für Fischerei, 226, 741, 1586
Federal Research Centre for Nature Conservation and Landscape Ecology see Bundesforschungsanstalt für Naturschutz und Landschaftsökologie, 946
Food Technology and Packaging see Fraunhofer-Institut für Lebensmitteltechnologie und Verpackung, 501, 1708
Forschungsinstitut für Internationale Technische und Wirtschaftliche Zusammenarbeit der RWTH Aachen, 2198
Forschungsring für Biologisch-Dynamische Landwirtschaft, 249
Forschungszentrum Jülich GmbH (KFA), 1231

324 Geographical Index *Germany*

Fraunhofer Institute for Atmospheric Environmental Research *see* Fraunhofer-Institut für Atmosphärische Umweltforschung, 85
Fraunhofer Institute for Environmental Chemistry and Toxicology *see* Fraunhofer-Institut für Umweltchemie und Ökotoxikologie (IUCT), 215, 1426
Fraunhofer Institute for Toxicology and Aerosol Research *see* Fraunhofer-Institut für Toxicologie und Aerosolforschung, 1425
Fraunhofer-Institut für Atmosphärische Umweltforschung, 85
Fraunhofer-Institut für Lebensmitteltechnologie und Verpackung, 501, 1708
Fraunhofer-Institut für Toxicologie und Aerosolforschung, 1425
Fraunhofer-Institut für Umweltchemie und Ökotoxikologie (IUCT), 215, 1426
German Environmental Agency *see* Umweltbundesamt, 723, 2179
German Hydrographic Institute *see* Deutsches Hydrographisches Institut, 272, 490, 1587
German Meteorological Service *see* Deutscher Wetterdienst (DWD), 106, 1072
German Nappy Service Association *see* Verband Deutscher Windelservice Lücke GmbH, 547
German Scientific Society for Oil Gas and Coal *see* Deutsche Wissenschaftliche Gesellschaft für Erdöl, 1137
Gesellschaft für Strahlen und Umweltforschung GmbH (GSF), 1376
Greenpeace Germany, 1884
Grüne Liga, 1831
Grüne Partei – Ost, 1832
Hahn-Meitner Institut Berlin GmbH (HMI), 1074
Hessische Gesellschaft für Ornithologie und Naturschutz eV, 793
Industrieverband Agrar (IVA), 197
Institut für Betriebswirtschaft Bundesforschungsanstalt für Landwirtschaft, 238, 428
Institut für Energie- und Transportforschung Meissen, 1075, 2363
Institut für Immission- Arbeits- und Strahlenschutz, 71, 1377, 1547
Institut für Ökologische Chemie, 2287
Institut für Strahlenhygiene (ISH), 1378
Institut für Wasser- Boden- und Lufthygiene, 429, 697, 1643
Institute for Ecological Chemistry *see* Institut für Ökologische Chemie, 2287
Institute for Emission Labour and Radiation Protection *see* Institut für Immission- Arbeits- und Strahlenschutz, 71, 1377, 1547
Institute for Farm Economics Federal Agricultural Research Centre *see* Institut für Betriebswirtschaft Bundesforschungsanstalt für Landwirtschaft, 238, 428

Institute for Radiation Hygiene *see* Institut für Strahlenhygiene (ISH), 1378
Institute for Water Soil and Air Hygiene *see* Institut für Wasser- Boden- und Lufthygiene, 429, 697, 1643
International Institute for Environment and Safety *see* Wissenschaftszentrum Berlin (WZB), 2064
International Institute for Environment and Society, 2261
Kernforschungszentrum Karlsruhe GmbH (KfK), 2203
Kerntechnische Ausschuss (KTA), 1251
Kommission Reinhaltung der Luft (KRdL) in VDI und DIN, 28, 1505
Landesanstalt für Immissionsschutz Nordrhein-Westfalen, 74, 1550
Medical Institute for Environmental Hygiene *see* Medizinisches Institut für Umwelthygiene, 75, 1341, 1551
Medizinisches Institut für Umwelthygiene, 75, 1341, 1551
Mineralöl Wirtschafts Verband eV (MWV), 1119
Ministerium für Soziale Gesundheit und Energie des Landes Schleswig-Holstein, 970, 2045
Ministry for the Environment Nature Conservation and Nuclear Reactors Safety, 896, 1244
Ministry of Food Agriculture and Forestry, 154, 750
Ministry of Health, 1263
Ministry of Transport, 2313
North Rhine Westphalia State Centre of Air Quality, Noise and Vibration Control *see* Landesanstalt für Immissionsschutz Nordrhein-Westfalen, 74, 1550
Nuclear Safety Standards Commission *see* Kerntechnische Ausschuss (KTA), 1251
Ornithologische Arbeitsgemeinschaft für Schleswig-Holstein und Hamburg, 809
Paper Technology Foundation for Paper Production and Conversion *see* Papiertechnische Stiftung für Papierezeugung und Papierverarbeitung, 561
Papiertechnische Stiftung für Papierezeugung und Papierverarbeitung, 561
Radiation and Environmental Research Centre *see* Gesellschaft für Strahlen und Umweltforschung GmbH (GSF), 1376
Research Institute for International Technology and Economic Cooperation of Aachen University *see* Forschungsinstitut für Internationale Technische und Wirtschaftliche Zusammenarbeit der RWTH Aachen, 2198
Schutzgemeinschaft Deutscher Nordseeküste, 287
Standing Committee on Urban and Building Climatology, 98
Umweltbibliotek, 2271
Umweltbundesamt, 723, 2179
Verband Deutscher Papierfabriken (VDP), 546

Verband Deutscher Windelservice Lücke GmbH, 547
Waste Watchers Deutschland eV, 656
Wissenschaftszentrum Berlin (WZB), 2064
World Wide Fund for Nature – Germany, 919
Zoologische Gesellschaft Frankfurt von 1858 eV, 833

Publications

Abwassertechnik (AWT), 691, 1631
Achema Handbook Pollution Control, 1454
Addressbook of Environmental Experts *see* Adressbuch Umwelt-Experten, 1765
Adressbuch Umwelt-Experten, 1765
Atom-Informationen, 1200
Atomwirtschaft-Atomtechnik, 1203
ATW News, 1204
Brennstoff-Wärme-Kraft (BWK), 988
Controlling Pollution in the Round: Change and Choice in Environmental Regulation in UK and Germany, 2103
Deutsche Dependance, 607
Dictionary of Ecology *see* Wörterbuch Ökologie, 2135
Energieanwendung, 1000, 2188
Engineering and Automation, 1006
Entsorgung 92, 662
Entsorgungspraxis, 2189
Environment Research Directory *see* Umweltforschungskatalog, 2236
Environmental Liability and Environmental Criminal Law *see* Umwelthaftung und Umweltstrafrecht, 1957
Environmental Protection in Germany *see* Umweltschutz in Deutschland, 729, 737
Forum Städte-Hygiene, 1282
Freight Transport and the Environment, 2332
Fusion, 1008
German Chamber of Industry and Commerce: German Packaging Laws, 515, 1700
Gesunde Pflanzen, 170
GSF Mensch und Umwelt, 1337
Journal for Water and Wastewater Research *see* Zeitschrift für Wasser und Abwasserforschung, 381, 695
Journal of Biochemical Toxicology, 1415
Kernenergie, 1212
Korrespondez Abwasser, 371
Literaturberichte über Wasser, Abwasser, Luft und Feste Abfallstoffe, 1469
Lufthygienischer Monatsbericht, 52, 1527
Man and the Environment: Product Life Cycle Analysis – What does it mean?, 1713
Müllmagazin, 673
Ökolgiepolitik, 1926
Packaging in Europe: West Germany Special Report no 2047, 519

Packaging Ordinance A Practical
    Handbook *see*
    Verpackungsverbordnung, 511
Präparator, Der, 709
Protecting the Earth: A Status Report with
    Recommendations for a New Energy
    Policy, 131, 1538, 2113
Realism in Green Politics, 1913
Recycling, 609
Sonnenenergie und Wärmepumpe, 1062
Tableaux Numeriques des Analyses
    Physico Chimiques des Eaux du Rhin
    see Zahlentafeln der Physikalisch
    Chemischen Untersuchungen des
    Rheinwassers, 331
Übersetzungen-Kerntechnische Regeln, 1253
Umwelt, 735, 1699
Umwelt Technologie Aktuell, 2195
Umweltforschungskatalog, 2236
Umwelthaftung und Umweltstrafrecht, 1957
Umweltmagazin, 2258
Umweltpolitik in der Europäischen
    Gemeinschaft, Die, 2091
Umweltschutz in Deutschland, 729
Verpackungsverordnung, 511
Wasser Abwasser GWF, 1634
Wasser Luft und Boden, 674, 1476
Wasser und Abwasser in Forschung und
    Praxis, 420
Wasserwirtschaft, 373, 692
Waste Management '92 *see* Entsorgung
    92, 662
Waste, Wastewater, Air Laws and
    Technology, 40, 687, 1627, 1958
Wind-Energie Jahrbuch, 1049
Windkraft Journal, 1065
Wörterbuch Ökologie, 2135
Zahlentafeln der Physikalisch Chemischen
    Untersuchungen des Rheinwassers, 331
Zeitschrift für Wasser und
    Abwasserforschung, 381, 695

Databases

ELFIS, 184, 783
Energie, 1024
Umweltforschungsdatenbank
    (UFORDAT), 2162
Umweltliteraturdatenbank (ULIDAT), 2270

Greece

Organisations

Centre for Occupational Hygiene and
    Safety *see* Kentron Yginis Keasfalias tis
    Ergassias (KYAE), 1292
Commission for Municipal Solid Waste:
    Department of Environmental Studies,
    642
Council of Ministers, 1815
Environmental Law Association, 1936
Environmental Pollution Control Project –
    Athens, 1481
Federation of Ecologist Alternative
    Organisations, 1854
Greek Agrochemical Association (GAA),
    196
Greek Atomic Energy Commission, 1153,
    1357
Greek National Centre for Marine
    Research, 231, 303, 336, 1642
Greenpeace Greece, 1885
Hellenic Association on Environmental
    Pollution, 1446
Hellenic Nuclear Society, 1174
Hellenic Ornithological Society, 792
Institute of Meteorology and Physics of the
    Atmospheric Environment, 7, 1076
Kentron Yginis Keasfalias tis Ergassias
    (KYAE), 1292
Ministry of Health Welfare and Social
    Security, 1270
Ministry of Industry, Energy and
    Technology and Commerce, 973
Ministry of the Environment Housing and
    Public Works, 2000
Ministry of the Environment Town
    Planning and Public Works, 400
Ministry of Transport and
    Communications, 2314
Panellinio Kentro Oikologikon Erevnon
    (PAKOE), 722
Panhellenic Centre of Environmental
    Studies *see* Panellinio Kentro
    Oikologikon Erevnon (PAKOE), 722
Sea Turtle Protection Society, 814
Water Resources Division National
    Technical University of Athens, 412
Water Supply and Sewage Committee of
    Major Athens (EYDAP), 359, 686
World Wide Fund for Nature – Greece, 920

Hungary

Organisations

Association of Hungarian Chemical Works
    (AHCW), 189
Biokultura Association, 244
Biokultura Egyelulet, 245
East European Environmental Research
    (ISTER), 961, 1769
Erdészeti Tudományos Intézet, 775
Forest Research Institute *see* Erdészeti
    Tudományos Intézet, 775
Független Ökological Központ (FÖK), 2123
Göncöl Alliance, 1036
Hungarian Institute for Materials Handling
    and Packaging, 502, 1774
Hungarian Society of Nature
    Conservationists, 903
Independent Ecological Centre *see* Független
    Ökological Központ (FÖK), 2123
Kornyeyetvedelme Ens Teruletfejlesgtesi
    Minisyterium, 2215
Ministry for Environment and Water
    Management, 339, 1989
Ministry of Transport and
    Telecommunications, 2319
National Atomic Energy Commission *see*
    Orszagos Atomenergia Bizottsag, 1182
National Water Authority Department for
    International Relations, 340
Orszagos Atomenergia Bizottsag, 1182
Papir–es Nyomdaipari Muszaki Egyesulet,
    539
Technical Association of the Paper and
    Printing Industry *see* Papir–es
    Nyomdaipari Muszaki Egyesulet, 539

Publications

Characteristics of Road Transport in
    Hungary, 2352
Energia es Atomtechnika, 1207
Energiagazdalkodas, 998

Iceland

Organisations

Hollustuvernd Rikisins, 1340
Iceland Forestry Research Station *see* Rannsóknarstöd Skógraektar Rikisins, 781
Icelandic Fisheries Laboratories *see* Rannsóknastofnun Fiskidnadarins, 233
Icelandic Meteorological Office *see* Vedurstofa Islands, 109
Ministry of Agriculture, 145
Ministry of Health and Social Security, 1268
Ministry of the Environment, 1996
National Centre for Hygiene Food Control and Environmental Protection *see* Hollustuvernd Rikisins, 1340
National Energy Authority *see* Orkustonfnun, 1019
Nature Conservation Council, 898
Orkustonfnun, 1019
Rannsóknarstöd Skógraektar Rikisins, 781
Rannsóknastofnun Fiskidnadarins, 233
Vedurstofa Islands, 109

Ireland

Organisations

Agricultural and Food Development Authority, 173
Bioresearch Ireland (EOLAS), 1305
Comhaontas Glas, 1809
Council of Ministers, 1816
Department of Agriculture and Food, 138
Department of Energy, 966
Department of Health, 1260
Department of the Environment, 1978
Department of Tourism Transport and Communication, 2306
Earthwatch, 1869
Environment Policy Committee: Confederation of Irish Industry, 2043
Environment Policy Section, 1652
Environmental Information Service, 2218
Environmental Management and Auditing Services Ltd, 1718
Environmental Research Unit, 68, 1544
Environmental Research Unit Water Resources, 427
Federation of Irish Chemical Industries (FICI), 195
Geological Survey of Ireland, 2214
Greenpeace Ireland, 1888
Institution of Engineers in Ireland, 1791
Irish Coastal Environment Group, 265
Irish Environmental Conservation: Organisation for Youth (ECO), 704
Irish Forestry Board, 777
Irish Industry Confederation – Environmental Policy Committee, 1684
Irish Organic Farmers and Growers Association, 250
Irish Science and Technology Agency (EOLAS), 2152
Irish Sea Fisheries Board, 232, 310
Irish Wildbird Conservancy (IWC), 797
Irish Wildlife Federation (IWF), 798
Irish Women's Environmental Network, 1900
Meteorological Service, 95
Nuclear Energy Board, 1180
Office for the Protection of the Environment, 715
Office of Public Works, 2002
Principal Water Authority Corporation of Dublin, 401
Radiological Protection, Institute of Ireland, 1365
Solar Energy Society of Ireland, 1040
The Green Party *see* Comhaontas Glas, 1809

Publications

Chemico–Biological Interactions, 1408
Irish Journal of Environmental Science, 2140

Italy

Organisations

Air Pollution Study Committee *see* Comitato di Studio perl'Inquinamento Atmosferico (CSIA), 19, 1496
Amici della Terre, 1863
Assobiotec, 1302
Association Industry Difesa Produzioni Agricole (AGROFARMA), 188
Associazione Italiana di Protezione Contro le Radiazioni (AIRP), 1361
Associazione Italiana fra gli Industriali della Carta Cartoni e Paste per Carta, 531
Associazione Nazionale di Ingegneria Nucleare (ANDIN), 1164
Associazione Nazionale Fra Fabbricanti di Imballaggi Metallici (ANFIMA), 463
Associazione per l'Agricultura Biodinamica, 242
Centre for Informatic Application in Agriculture *see* Centro Studi per l'Applicazione dell'Informatica in Agricoltura, 177
Centre for the Study of Man and the Environment *see* Centro Studi l'Uomo e l'Ambiente, 1317
Centre of Ecology and Climatology/Geophysical Experimental Observatory *see* Centro di Ecologia e Climatologia/Osservatorio Geofisico Sperimentale Macerata, 5, 1070
Centro di Ecologia e Climatologia/Osservatorio Geofisico Sperimentale Macerata, 5, 1070
Centro Studi l'Uomo e l'Ambiente, 1317
Centro Studi per l'Applicazione dell'Informatica in Agricoltura, 177
Comitato di Studio perl'Inquinamento Atmosferico (CSIA), 19, 1496

Comitato Nazionale per la Ricerca e per lo Sviluppo dell'Energia Nucleare e dell' Energie Alternativ, 1071, 1228
Council of Ministers, 1817
Department of Political and Social Sciences, 2114, 2361
Ecology Research Groups *see* Gruppi Ricerca Ecologica (GRE), 2147
Energy Conservation in Buildings Institute *see* Istituto per l'Edilizia ed il Risparmio Energetico, 1099
ERL Italia, 1760
Federazione Italiana Impresa Pubbliche Gas Acqua e Varie (FEDERGASACQUA), 347, 1114
Federazione Italiana Pro Natura, 902
Forum Italiano dell'Energia Nucleare (FIEN), 1172
Greenpeace Italy, 1889
Gruppi Ricerca Ecologica (GRE), 2147
Institut de l'Environnement du Centre Commun de Recherche, 6, 2199
Institute of Atmospheric Pollution *see* Istituto Sull'Inquinamento Atmosferico, 73, 1549
Istituto del Terziario per l'Ambiente, 649
Istituto di Ricerca sulle Acque (IRSA), 387, 1644
Istituto di Ricerche e Sperimentazione Laniera 'O. Rivetti, 638
Istituto per l'Edilizia ed il Risparmio Energetico, 1099
Istituto per le Piante da Legno e l'Ambiente SpA, 778
Istituto Sperimentale per lo Studio e la Difesa del Suolo, 1563
Istituto Sull'Inquinamento Atmosferico, 73, 1549
Italian Animal Rights Association *see* Lega Italiana dei Diritti dell'Animale (LIDA), 800
Italian Bird Protection Society *see* Lega Italiana Protezione Uccelli, 801
Italian Commission for Nuclear and Alternative Energy Sources *see* Comitato Nazionale per la Ricerca e per lo Sviluppo dell'Energia Nucleare e delle Energie Alternativ, 1071
Italian Federation for the Protection of Nature *see* Federazione Italiana Pro Natura, 902
Italian Radiation Protection Association *see* Associazione Italiana di Protezione Contro le Radiazioni (AIRP), 1361
Joint Research Centre Ispra Establishment, 494, 1259
League for the Environment *see* Lega per l'Ambiente, 439, 721
Lega Italiana dei Diritti dell'Animale (LIDA), 800
Lega Italiana Protezione Uccelli, 801
Lega per l'Ambiente, 439, 721
Ministero dei Lavori Pubblici, 396
Ministero per i Beni Culturali e Ambientali, 1987
Ministry of Agriculture, 144
Ministry of Health, 1264
Ministry of the Environment, 1997
Ministry of Transport, 2312
O. Rivetti Wool Industry Research and Experimental Institute *see* Istituto di Ricerche e Sperimentazione Laniera 'O. Rivetti, 638
Ozono Elettronica Internazionale, 631
Research Institute for the Study and Conservation of the Soil *see* Istituto Sperimentale per lo Studio e la Difesa del Suolo, 1563
Territorial and Environmental Informatics Interdepartmental Research Centre, 2205
Timber and Environment Institute *see* Istituto per le Piante da Legno e l'Ambiente SpA, 778
Verdi, 1842
Verdi Arcobaleno, 1843
Water Research Institute *see* Istituto di Ricerca sulle Acque (IRSA), 387, 1644
World Wide Fund for Nature – Italy, 921

Publications

A E S Ambiente e Sicurezza: Rivista Dell'Antiquinamento, 332, 1630
Acqua Aria, 325
Atomo Petrolio Elettricita, 1202
Difesa Ambientale, 46, 1523
Dimensione Energia, 990
Economia Delle Fonti di Energia, 1090
ENEA Notiziario Energia e Innovazione, 993
Energia, 996
Energie Alternative: Habitat Territorio Energia, 1052
Eta Verde, 2244
Governo Locale ed Economia dell' Ambiente (GEA), 2247
Informazione Innovativa, 1011, 1286, 2192
Inquinamento, 1467
Natura e Societa, 939, 1965
Packaging in Europe: Italy Special Report no 2049, 517
Politica Ambientale nella Comunità Europea, La, 2083
Protec, 733
Smog, 56, 1531
Studi Parlamentari e di Politica, 2102
Vanishing Tuscan Landscapes, 878

Databases

ENEL, 1023

Latvia

Organisations

Latvian Environment Protection Club (VAK), 725

Lithuania

Organisations

Lithuania Ministry of Energy, 1155

Lithuanian Green Party, 1836

Luxembourg

Organisations

Council of Ministers, 1818
Déi Gréne Alternativ, 1823
Green Alternative Party *see* Déi Gréne Alternativ, 1823
Greenpeace Luxembourg, 1890
Ligue Luxembourgeoise pour la Protection de la Nature et des Oiseaux, 802
Luxembourg Ecological Movement *see* Mouvement Ecologique Luxembourg (MECO), 1902
Ministère de l'Environnement et des Eaux et Forêts, 338, 748, 1983
Ministry of Agriculture and Viticulture, 149
Ministry of Environment and Territorial Administration, 397
Ministry of Public Health, 1271
Ministry of Transport and Energy, 974, 2316
Mouvement Ecologique Luxembourg (MECO), 1902

Netherlands

Organisations

Association of Netherlands Paper and Board Manufacturers (VNP), 530
Center for Energy Saving and Clean Technology, 1083
Centre for Agrobiological Research *see* Centruum voor Agrobiologisch Onderzoek (CABO), 213
Centre for Energy Conservation and Environmental Technologies *see* Centrum voor Energiebesparing en Schone Technologie, 1096
Centrum voor Energiebesparing en Schone Technologie, 1096
Centruum voor Agrobiologisch Onderzoek (CABO), 213
Consumentenbond, 1326
Consumers Association *see* Consumentenbond, 1326
Council of Ministers, 1819
Department of Air Pollution, 67, 1542
Department of Toxicology, 1424
Dutch Society for the Wadden Sea Conservation *see* Landelijke Vereniging tot Behoud van de Waddenzee (LVBW), 283
Environmental Management and Auditing Services Ltd, 1719
ERL Nederland, 1761
Federation Seas at Risk, 279
Foundation for Environmental Education, 1771
Foundation for Packaging and the Environment (SNV), 500
Friends of the Earth (FoE), 290
Geological Survey of the Netherlands, 1115
Green Party *see* Groenen, De, 1828
Greenpeace Netherlands, 1891
Groenen, De, 1828
Institute for Ecological Research *see* Instituut voor Oecologisch Onderzoek, 2150
Instituut voor Natuurbeschermingseducatie (IVN), 1775
Instituut voor Oecologisch Onderzoek, 2150
Instituut voor Plantenziektenkundig Onderzoek (IPO), 216
Koninklijk Nederlands Meteorologisch Instituut, 107
Landelijke Vereniging tot Behoud van de Waddenzee (LVBW), 283
Ministry of Agriculture, Nature Conservation and Fisheries, 153, 897
Ministry of Economic Affairs: Directorate General for Energy, 971, 1657
Ministry of Housing Physical Planning and Environment, 398, 1659: Central Dep for Information and Interior Relations, 1991
Ministry of Transport and Public Works, 2317
Ministry of Welfare Health and Cultural Affairs, 1272
National Environmental Policy Plan Plus (NEPP): Ministry of Housing Planning and the Environment, 2048
National Institute of Public Health and Environmental Protection *see* Rijksinstituut voor Volksgezondheid en Milieuhygiene (RIVM), 1274
Natuurmonumenten, 906
Nederlands Atoomforum, 1177
Nederlands Verpakkingscentrum, 505
Nederlandse Industriële en Agarische Biotechnologie Associatie (NIABA), 1310
Nederlandse Stichting Geluidshinder, 1599
Nederlandse Stichting voor Fytofarmacie (NEFYTO), 201
Netherlands Energy Research Foundation ECN, 1140, 1233
Netherlands Institute for Fishery Investigations *see* Rijksinstituut voor Visserijonderzoek (RIVO), 234
Netherlands Packaging Industry *see* Nederlands Verpakkingscentrum, 505
Netherlands Society for Nature and the Environment *see* Stichting Natuur en Milieu (SNM), 911 *see* Stichting Natuur en Milieu (SNM), 1942
Noise Abatement Society *see* Nederlandse Stichting Geluidshinder, 1599
North Sea Foundation *see* Werkgroep Noordzee, 291
NOVEM, 979
NV Vereenigde Glasfabrieken, 451
Plant Protection Research Institute *see* Instituut voor Plantenziektenkundig Onderzoek (IPO), 216
Rijksinstituut voor Visserijonderzoek (RIVO), 234
Rijksinstituut voor Volksgezondheid en Milieuhygiene (RIVM), 1274
Royal Netherlands Meteorological Institute *see* Koninklijke Nederlands Meteorologisch Instituut, 107
Society for Clean Air in the Netherlands (CLAN) *see* Vereniging der Lucht, 1, 1509
Stichting Natuur en Milieu (SNM), 911, 1942
Studies en Consultancy voor Milieu en Omgeving – TNO, 1486
Technical University of Twente, 2204, 2364
The Society for the Preservation of Nature in the Netherlands *see* Natuurmonumenten, 906
TNO Study and Information Centre for Environment *see* Studies en Consultancy voor Milieu en Omgeving – TNO, 1486
Toegepast Natuurwetenschappelijk Onderzoek (TNO), 1020, 2206
Vereniging der Lucht, 1, 1509
Vereniging Milieudefensie, 1908
Vereniging Voor Biologische-Dynamische Landbouw, 258
Wereld Natuur Fonds, 289
Werkgroep Noordzee, 291

World Wide Fund for Nature – Netherlands, 922

Publications

Adresboek voor Milieuhygiene Veiligheids-Techniek en Recuperatie, 1280
Agricultural Research Department *see* SCOOP Dienst Landbouwkundig Onderzoek Staring Centrum, 172, 418
Allicht, 1197
Ambient Air Pollutants from Industrial Sources, 36, 1512
Aquatic Toxicology, 1406, 1632
Directory of Environmental Health and Safety *see* Adresboek voor Milieuhygiene Veiligheids-Techniek en Recuperatie, 1280
Ecological Engineering, 237, 1460, 1796
Ecological Modelling, 1797
Energie & Milieutechnologie, 999, 2187
Energiespectrum, 1208
Energy and Environmental Technology *see* Energie & Milieutechnologie, 999, 2187
Environmental News from the Netherlands, 2095
Excerpta Medica Section 46: Environmental Health and Pollution Control, 1336, 1465
Fens and Bogs in the Netherlands: Vegetation, History, Nutrient Dynamics and Conservation, 875
Forum, 1007
Fundamental Aspects of Pollution Control and Environmental Science, 1466
Interuniversitair Demografisch Instituut Publications, 2300
Journal of Aquatic Ecosystem Health, 333
Journal of Atmospheric Chemistry, 51, 1526
Landscape Ecology, 881
Mapping Critical Loads for Europe, 323
Milieubeleid in de Europese Gemeenschap, Het, 2081
Milieudefensie, 936
Natuur en Milieu, 942
Netherlands Centraal Bureau voor de Statistiek: Nederlandse Energiehuishouding, 985
Prevalence of Fish Diseases with Reference to Dutch Coastal Waters, 224, 300
Reinwater, 1475
Resources Conservation and Recycling, 611
SCOOP Dienst Landbouwkundig Onderzoek Staring Centrum: Instituut voor Onderzoek van het Landelijk Gebied, 172, 418
Studies in Environmental Science, 2145
Water Resources Management, 377
Waterschapsbelangen, 423, 736
Wetlands Ecology and Management, 379

Norway

Organisations

Central Office of Historic Monuments, 864
Central Statistical Office *see* Statistik Sentralbyra (SSB), 2216
DEBIO, 248
Department of Consumer Affairs: Ministry of Family and Consumer Affairs, 1650
Directorate for Nature Management, The, 894
Directorate for Wildlife and Freshwater Fish *see* Direktoratet for Vilt og Ferskvannsfisk, 858
Direktoratet for Vilt og Ferskvannsfisk, 858
Division for Air Monitoring and Industrial Compliance, 13, 1492, 2021
Fisheries Directorate Nutrition Institute *see* Fiskeridirektoratets Ernaeringsinstitutt, 229
Fiskeridirektoratets Ernaeringsinstitutt, 229
Fiskeridirektoratets Havforskningsinstitutt, 230
Greenpeace Norway, 1892
IKU Continental Shelf and Petroleum Technology Research Institute, 1139
Institut for Energiiteknikk (IFE), 1018, 1232
Marine Research Institute *see* Fiskeridirektoratets Havforskningsinstitutt, 230
Miljopartiet de Gronne, 1839
Miljoverndepartementet, 1982
Ministry of Environment *see* Miljoverndepartementet, 1982
Ministry of Industry, 1107
Natur og Ungdom, 1781
Nature and Youth *see* Natur og Ungdom, 1781
NIPH Department of Toxicology, 430, 1430
Norges Jeger og Fiskerforbund, 804
Norges Naturvernforbund (NNV), 907, 1904
Norsk Atomforum, 1178
Norsk Institut for Luftforskning (NILU), 76, 1552
Norsk Institut for Skogforskning (NISK), 780
Norsk Institut for Vannforskning (NIVA), 431
Norsk Ornitologisk Forening, 805
Norsk Polar Institut, 887
Norske Emballasjeforening, Den, 506
Norske Videnskaps-Akademi, Det, 2154
Norwegain Forest Research Institute *see* Norsk Institut for Skogforskning (NISK), 780
Norwegian Association of Anglers and Hunters *see* Norges Jeger og Fiskerforbund, 804
Norwegian Clean Air Association *see* Ren Luft Foreningen (RLF), 33 *see* Ren Luft Foreningen (RLF), 1508
Norwegian Clean Air Campaign, 31, 1556
Norwegian Foundation for Environmental Labelling *see* Stiftelsen Miljomerking i Norge, 1664
Norwegian Institute for Air Research *see* Norsk Institut for Luftforskning (NILU), 76 *see* Norsk Institut for Luftforskning (NILU), 1552
Norwegian Institute for Water Research *see* Norsk Institut for Vannforskning (NIVA), 431
Norwegian Institute of Technology, 353, 1624
Norwegian Mapping Authority, 874
Norwegian Meteorological Institute: Meteorological Synthesising Centre–West of EME, 108
Norwegian Ornithological Society *see* Norsk Ornitologisk Forening, 805
Norwegian Packaging Association *see* Norske Emballasjeforening, Den, 506
Norwegian Pulp and Paper Association *see* Papirindustriens Sentralforbund, 540
Norwegian Pulp and Paper Research Institute *see* Papirindustriens Forskningsinstitutt, 562
Oslo Yann Og Avlop, 354, 651
Papirindustriens Forskningsinstitutt, 562
Papirindustriens Sentralforbund, 540
Ren Luft Foreningen (RLF), 33, 1508
Royal Ministry of Health and Social Affairs, 1275
Royal Ministry of Petroleum and Energy, 1108
Royal Ministry of Transport and Communications, 2321
State Pollution Control Authority (SFT): Oil Pollution Control Department, 1109, 1567
Statens Forurensningstilsyn (SFT), 1443
Statistik Sentralbyra (SSB), 2216
Stiftelsen Miljomerking i Norge, 1664

The Norwegian Science Academy *see* Norske Videnskaps-Akademi, Det, 2154
The Norwegian Society for the Conservation of Nature *see* Norges Naturvernforbund (NNV), 907
The Norwegian State Pollution Control Authority *see* Statens Forurensningstilsyn (SFT), 1443
World Wide Fund for Nature – Norway, 923

Publications

Agriculture and Fertilisers: Fertilisers in Perspective – Their Role in Feeding the World, 207
Environmental Yearbook *see* Miljoarboka, 2231
Forest Management in a time of emissions *see* Skogsskjötsel i en Utslippstid, 761 *see* Skogsskjötscl i en Utslippstid, 1516
Forurensning i Norge 1991, 1456
Future is Now, The, 1450

Miljoarboka, 2231
Natur og Miljoe, 938
Natural Resources and the Environment *see* Naturressurser Og Miljo 1991, 2233
Naturressurser Og Miljo 1991, 2233
Pollution in Norway *see* Forurensning i Norge 1991, 1456
Skogsskjötsel i en Utslippstid, 761, 1516
What You Don't Know Will Hurt You: Environmental Information as a Basic Human Right, 2224

Poland

Organisations

Environmental Engineering Institute *see* Instytut Podstaw Inzynierii Srodowiska PAN, 885, 1482, 1804
Environmental Protection Institute, Laboratory of Environmental Monitoring *see* Instytut Ksztaltowania Srodowiska, 742 2034
Fundacja na Rzecz Efelatywnego Wykorzystanos Energii, 1098
Information Centre for Air Protection, Polish Ecological Club, 26, 1503
Institute for Sustainable Development, 952
Instytut Celulozowo-Papierniczy, 558
Instytut Ksztaltowania Srodowiska, 742, 2034
Instytut Meteorologii i Gospodarki Wodnej, 308, 386
Instytut Podstaw Inzynierii Srodowiska PAN, 885, 1482, 1804
International Environmental Service Centre (IESC), 2175

Liga Ochrony Przyrody, 29, 905
Meteorology and Water Management Institute *see* Instytut Meteorologii i Gospodarki Wodnej, 308, 386
Ministerstwo Ochrony Srodowiska i Zasobow Naturalriych, 1988
Ministry of Agriculture and Food Industry, 147
Ministry of Health and Social Welfare, 1269
Ministry of Transport and Shipping, 2318
National Atomic Energy Agency, 1160
Polish Ecological Club/Information Centre for Air Protection, 32, 1507
Polish Foundation for Energy Efficiency *see* Fundacja na Rzecz Efelatywnego Wykorzystanos Energii, 1098
Polish Party of Greens *see* Polska Partia Zielonych (PPZ), 1840
Polska Partia Zielonych (PPZ), 1840
Polski Klub Ekologiczny, 1905
Polskie Zresenie Inazynierowi Technikow Sanitarnych (PZITS), 356

Pulp and Paper Research Institute *see* Instytut Celulozowo-Papierniczy, 558
Stowarzyszenie Inzynierow i Technikow Rezemyslu Papierniczego w Polsce, 543
Technical Association of the Polish Paper Industry *see* Stowarzyszenie Inzynierow i Technikow Rezemyslu Papierniczego w Polsce, 543
The Nature Protection League *see* Liga Ochrony Przyrody, 29, 905

Publications

Energetyka, 995
Environment Protection Engineering, 1798
Ochrona Przyrody, 708
Polska Akademia Nauk Instytut Podstaw Inzynierii Srodowiska Prace i Studia, 1474

Portugal

Organisations

AGROBIO, 240
Amigos da Terra, 1864
Aquário Vasco da Gama, 225, 301
Associação Portuguesa das Empresas Indústrias de Produtos Químicos, 187
Associação Portuguesa de Fabricantes de Papel e Cartão, 529
Associação Portuguesa dos Recursos Hidricos (APRH), 405
Associação Portuguesa para Estudos de Saneamento Basico (APESB), 1277

Camara Municipal de Lisboa Direcção Municipal de Infraestructuras e Saneamento, 392
Centro de Investigacões Florestais, 773
Centro Ecologico, 2121
Council of Ministers, 1820
Direção Geral de Energia, 968, 1927
Direção Geral de Energia Departamento de Energia Nuclear, 1151
Ecological Centre *see* Centro Ecologico, 2121
Empresa Publica das Aguas Livres (EPAL), 345

Forestry Research Centre *see* Centro de Investigacões Florestais, 773
Gabinete de Protecção e Segurança Nuclear, 1356
General Directorate for Energy *see* Direção Geral de Energia, 968, 1927
Institut Nacional de Engenharia e Technologia Industrial (INETI), 2200
Laboratório Nacional de Engenharia e Tecnologia Industrial (LNETI), 1379
MDP/CDE–Green Party, 1837
Ministerio de Ambiente e Recursos Naturais, 1985
Ministry of Agriculture and Fisheries, 146

Ministry of Environment and Natural Resources, 1658
National Laboratory for Engineering and Industrial Technology *see* Laboratório Nacional de Engenharia e Tecnologia Industrial (LNETI), 1379
Nucleo Portugues de Estudio e Protecção da Vida Selvagem, 807

Portuguese Core Group for Wildlife Studies and Protection *see* Nucleo Portugues de Estudio e Protecção da Vida Selvagem, 807
Secretaria Estado Ambiente Hidrologicos, 357, 653
Vasco da Gama Aquarium *see* Aquário Vasco da Gama, 225, 301

Publications

Ministerio da Qualidade de Vida Comissão Nacional do Ambiente Boletim, 1470, 2016
Política de Ambiente na Comunidade Europeia, A, 2084

Romania

Organisations

Institutl Roman de Cercetari Marine, 307, 1589
Institutul de Cercetäri pentru Pedologie si Agrochimie, 385, 698
Institutul de Cercetäri si Projectäri pentru Gospodärirea Apelor, 384
Marine Research Institute of Romania *see* Institutl Roman de Cercetari Marine, 307, 1589
Ministry of the Environment, 1998

National Commission for the Control of Nuclear Activities, 1161
National Council for Environmental Protection, 714
Research and Design Institute for Water Resources Engineering *see* Institutl de Cercetäri si Projectäri pentru Gospodärirea Apelor, 384
Romanian Society of Hygiene and Public Health *see* Societatea Româna de Igiena si Sanatate Publica, 1278
Societatea Româna de Igiena si Sanatate Publica, 1278

Soil Science and Agrochemistry Research Institute *see* Institutul de Cercetäri pentru Pedologie si Agrochimie, 385, 698

Publications

Energetica, 994, 2137
Institutul de Studii si Proiectari Energetice Bulletinul, 1093

Russia

Organisations

All-Union Research Institute for Reactor Operators (VNIIAES), 1225
Greenpeace Russia, 1893
Minelsbumprom, 526
Ministry of Atomic Energy and Industry, 1156
Ministry of Atomic Energy of the Russian Federation (MINATOM), 1157

Ministry of Timber, Pulp and Paper and Wood Processing *see* Minelsbumprom, 526
State Committee for Environmental Protection, 716
State Committee for the Supervision of Nuclear and Radiation Safety under the President of Russia, 1248, 1360

Publications

Atomnaya Energiya, 1201
Ecos, 934
Ekspress–Informatsiya Pryamoe Preobrazovanie Teplovoi i Khimicheskoi Energii V Elekricheskuyu, 992
Problems of Large Metropolitan Areas *see* Problemy Bolshikh Gorodov, 732
Problemy Bolshikh Gorodov, 732

Slovakia

Organisations

Ministry of Health, 1265
Ministry of the Environment of the Slovak Republic, 2001

Ministry of Transport Communications and Pubiic Works, 2320
Nuclear Power and Plant Research Institute (VUJE), 1234

Slovak Union of Nature and Landscape Protectors (SZOPK), 929, 1907
Slovakian Society for Conservation of Nature (SZOPK), 910

Slovenia

Organisations

Greens of Slovenia *see* Zeleni Slovenije, 1846
Union of the Societies for the Protection of the Environment in Slovenia, 724
Zeleni Slovenije, 1846

Publications

Varstvo Narave, 945

Spain

Organisations

Amigos de la Tierra, 1865
Asociación de Bioindustria, 1301
Asociación de Investigación Técnica de la Industria Papelera Española, 557
Asociación de Recuperadores de Papel y Cartón (REPACAR), 528
Asociación Ecologista de Defensa de la Naturaleza (AEDENAT), 1029
Asociación Española de Fabricantes de Agroquímicos para la Protección de las Plantas, 186
Asociación Nacional de Recuperadores de Vidrio, 447
Asociación Vida Sana, 241
Centre de Recerca Ecológica i Aplicacions Forestals, 772
Centre of Ecological Research and Forestry Application *see* Centre de Recerca Ecológica i Aplicacions Forestals, 772
Centro de Estudios de la Energía Solar (CENSOLAR), 1069
Centro de Investigación Energética Medioambientale y Tecnológica (CIEMAT), 964, 2197
Centro Ecológico Nacional SA, 2122
Consejo de Seguridad Nuclear, 1240, 1355
Coordinadora de Organizaciones de Defensa Ambiental (CODA), 21, 1498
Council of Ministers, 1821
ERL Espana, 1759
Forum Atómico Español, 1170
Fundación Para la Ecología y la Protección del Medio Ambiente (FEPMA), 23, 1500
General Organic Chemistry Institute *see* Instituto de Química Orgánica General, 2288
Greenpeace Spain, 1894
Grupo Ecologista de Agrónomos (GEDEA), 158
Instituto de Plásticos y Caucho, 585
Instituto de Química Orgánica General, 2288
Izquierda de los Pueblos, 1835
Liga Para la Defensa del Patrimonio Natural (DEPANA), 867
Ministerio de Agricultura Pesca y Alimentación, 141
Ministerio de Industria Comercio y Turismo, 1106, 1929
Ministerio de Obras Públicas y Transportes, 2309
Ministerio de Obras Públicas y Urbanismo – Dirección General de Medio Ambiente (DGMA), 1986
Ministry of Agriculture, Fisheries and Food *see* Ministerio de Agricultura Pesca y Alimentación, 141
Ministry of Health and Consumer Affairs, 1266
Ministry of Public Works Town Planning and Transport *see* Ministerio de Obras Públicas y Transportes, 2309
Plastics and Rubber Institute *see* Instituto de Plásticos y Caucho, 585
Secretaria de Estado para las Politicas del Agua y del Medio Ambiente, 15, 403
Sociedad Española de Ornitologia (SEO), 815
Sociedad Española de Protección Radiológica (SEPR), 1358
Sociedad Nuclear Española (SNE), 1185
Spanish Paper Industry Research Association *see* Asociación de Investigación Técnica de la Industria Papelera Española, 557
Spanish Plastics Foundation for the Protection of the Environment (SPFPE), 572
Spanish Society for Radiological Protection *see* Sociedad Española de Protección Radiológica (SEPR), 1358
World Wide Fund for Nature – Spain, 924

Publications

Anuario Profesional del Medio Ambiente 1993, 1915
Asintra, 2342
Economia Industrial, 1691
Energia, 997
Era Solar, 1053
Instituto Nacional de Investigaciónes Agrarias Comunicaciónes Serie: Recursos Naturales, 706
Instituto Nacional Para la Conservación de la Naturaleza Monografías, 707
Natura (Revista), 2253
Nova – Revista de Salud, 1287
Packaging in Europe: Spain and Portugal Special Report no R162, 518
Planeta Verde, 2254
Política de medio ambiente en la Communidad Europea, 2085
Quercus, 734
Retema–Medio Ambiente–Revista, 2255
Vida Silvestre, 710

Databases

Base Relacional de la Industria y Servicios Ambientales (BRISA), 1704
Centro de Documentación del Medio Ambiente: Escuela de Organización Industrial, 2020

Sweden

Organisations

Arbetsmiljöinstitutet, 1289
Department of Environmental Forestry, 774
Department of Plant Ecology, 856
Department of Wildlife Ecology, 857
Foreningen Karnteknik, 1169
Geological Survey of Sweden, Department of Forest Soils, 1562
Green Party *see* Miljopartiet de Gröna (MP), 1838
Greenpeace Sweden, 1895
Industrin for Vaxt–Och Traskyddsnedel (IVT), 198
Jordens Vanner, 1901
Kontrollforeningen for Alternativ Odling (KRAV), 251
Miljopartiet de Gröna (MP), 1838
Ministry of Agriculture, 143
Ministry of the Environment, 1660
Ministry of the Environment and Natural Resources, 1999
Ministry of Transport and Communications, 2315
National Institute of Occupational Health *see* Arbetsmiljöinstitutet, 1289
Svensk Polska Miljoforeningen (SPM): Szwedsko Polskie Towarzystwo Ochrony Srodowiska, 1860
Svenska Solenergiföreningen, 1042
Swedish Atomic Forum (SAFO), 1186
Swedish Environmental Research Institute (IVL), 2156
Swedish NGO Secretariat on Acid Rain, 317
Swedish Nuclear Society *see* Foreningen Karnteknik, 1169
Swedish Packaging Research Institute – PACKFORSK, 524
Swedish Pulp and Paper Association (SCPF), 544
Swedish Solar Energy Society *see* Svenska Solenergiföreningen, 1042
Swedish Standards Institution Environmental Labelling Programme, 1665
Swedish Transport Research Board *see* Transportforskningsberedningen (TFB), 2323
The Swedish Polish Association for Environment Protection *see* Svensk Polska Miljoforeningen (SPM), 1860
Transportforskningsberedningen (TFB), 2323
World Wide Fund for Nature – Sweden, 925

Publications

Acid News, 324
Ambio: A Journal of the Human Environment Research and Management, 1332, 2014
Effects of Acidification on Bird and Mammal Populations, 835
Enviro, 1462
Human Environment in Sweden, 1285
Information Sources for Construction and the Environment *see* Informationskaller Bygg and Miljo, 1689
Informationskaller Bygg and Miljo, 1689
Natural Studies and the Environment *see* Statistika Centralbyraan Statistika Meddelanden Subgroup NA, 2222
Procedures for Enhancing the Use of Environmentally Friendly Vehicles, 1617
Risk Assessment of Ecological Effects and Economic Impacts of Acidification on Forestry in Sweden, A, 771
Skogsutsikter, 762, 1517
Statistika Centralbyraan Statistika Meddelanden Subgroup NA, 2222
Sveriges Djurskyddsfoereningars Riksfoerbund, 852
Sveriges Natur, 944
Swedish Environment, 2223
Swedish Environment, The: Annexe to Sweden; National Report to UNCED 1992, 2259
Swedish Environmental Protection Agency Report, 738
Swedish Nuclear News, 1222
Vatten, 419

Databases

Baltic, 313, 1591
Swedish Environment Research Index (SERIX), 743
VANYTT, 79, 1104, 1555

Switzerland

Organisations

Biotecno SA, 625
Bundesamt für Energiewirtschaft, 963
Bundesamt für Umweltschutz, 680, 740, 1478
Eidgenössische Anstalt für Wasserversorgung Abwasserreinigung und Gewässerschutz, 383, 1641
Eidgenössische Kommission für die Sicherheit von Kernanlagen, 491, 1257
Eidgenössische Kommission zur Überwachung der Radioaktivität, 1375, 2033
Fachverband Schweizerischer Becherhersteller (FSBH), 538
Federal Commission for the Safety of Nuclear Installations *see* Eidgenössische Kommission für die Sicherheit von Kernanlagen, 491, 1257
Federal Department of Transport Communications and Energy, 2308
Federal Office of Energy *see* Bundesamt für Energiewirtschaft, 963
Federal Office of Environment, Forests and Landscape, 1653
Federal Office of Environmental Protection *see* Bundesamt für Umweltschutz, 680, 740, 1478
Green Party of Switzerland *see* Grüne Partei der Schweiz/Parti Ecologiste Suisse (GPS), 1833
Greenpeace Switzerland, 1896
Grüne Partei der Schweiz/Parti Ecologiste Suisse (GPS), 1833
Schweizer Arbeitsgemeinschaft für Umweltforschung (SAGUF), 2155
Schweizerische Gesellschaft für Biologischen Landbau, 255
Schweizerische Meteorologische Anstalt, 1079
Schweizerische Vereinigung für Atomenergie (SVA), 1184
Société Suisse de l'Industrie de Gaz et des Eaux (SSIGE), 358, 1122
Société Suisse Industries Chimiques (SSIC), 203
Station Fédérale de Recherches Agronomiques de Zurich Reckenholz, 218
Swiss Association for Atomic Energy *see* Schweizerische Vereinigung für Atomenergie (SVA), 1184

Swiss Association for Environmental Research *see* Schweizer Arbeitsgemeinschaft für Umweltforschung (SAGUF), 2155
Swiss Association for Health Techniques, 1329
Swiss Centre for Appropriate Technology (SKAT), 2178
Swiss Federal Commission of Radioactivity Surveillance *see* Eidgenössische Kommission zur Überwachung der Radioaktivität, 1375, 2033
Swiss Federal Institute for Water Resources and Water Pollution Control *see* Eidgenössische Anstalt für Wasserversorgung Abwasserreinigung und Gewässerschutz, 383, 1641
Swiss Federal Research Station for Agronomy Zurich *see* Station Fédérale de Recherches Agronomiques de Zurich Reckenholz, 218
Swiss League for the Protection of National Heritage, 870
Swiss Meteorological Institute *see* Schweizerische Meteorologische Anstalt, 1079
Swiss Nuclear Society, 1187
Swiss Society of Chemical Industries *see* Société Suisse Industries Chimiques (SSIC), 203
Verband der Schweizerischen Sellstoff-, Papier- und Kartonindustrie, 545
Women's Environmental Network (WEN), 1909
World Wide Fund for Nature – Switzerland, 926
Zentralstelle für Gesamtverteidigung, 1431

Publications

Feld Wald Wasser, 417
Fresenius Environmental Bulletin, 2246
Indoor Environment, 50
Schweizerische Vereinigung für Atomenergie Bulletin, 1221
Solaria, 1061
Umwelttechnik, 1288

Turkey

Organisations

Cekmece Nuclear Research and Training Centre *see* Cekmece Nükleer Arastirma ve Egitim Merkezi, 1227, 1479
Cekmece Nükleer Arastirma ve Egitim Merkezi, 1227, 1479
Green Party of Izmir *see* Izmir Yesiller Partisi, 1834
Hava Kirlenmesi Arastirmalari ve Denetimi Türk Milli Komitesi, 69, 1545
Izmir Yesiller Partisi, 1834
Ministry of Energy and Natural Resources, 972, 1990
TISIT, 205
Turkish Atomic Energy Authority *see* Türkiye Atom Enerjisi Kurumu, 1188
Turkish Ministry of Energy and Natural Resources, 975
Turkish National Committee for Air Pollution Research and Control (TUNCAP) *see* Hava Kirlenmesi Arastirmalari ve Denetimi Türk Milli Komitesi, 69, 1545
Turkish Nuclear Energy Forum *see* Türkiye Nukleer Enerji Kurumu, 1189
Türkiye Atom Enerjisi Kurumu, 1188
Türkiye Nukleer Enerji Kurumu, 1189

Publications

Turkish Journal of Nuclear Sciences, 1223

United Kingdom

Organisations

Agriculture and Food Research Council (AFRC), 175
Aluminium Can Recycling Association (ACRA), 459
Aluminium Federation Ltd, 460
Aluminium Foil Recycling Campaign, 461
Architectural Heritage Fund, 863
Asbestos Information Centre (AIC), 1295
ASH Partnership, 1706
Association of Reclaimed Textile Processors (ARTP), 624
Atmospheric Research and Information Centre (ARIC), 17, 316, 1494
Atomic Energy Technology, 1226
Bioindustry Association (BIA), 1304
Biotechnology Unit, 1299
British Agrochemicals Association Ltd (BAA), 190
British Battery Manufacturers Association (BBMA), 464
British Effluent and Water Association (BEWA), 342, 685
British Glass Recycling Department, 448
British Industrial Biological Research Association, 1423
British Non-Ferrous Metals Association, 1674
British Nuclear Forum (BNF), 1165
British Paper and Board Industry Federation (BPBIF), 532
British Plastics Federation (BPF), 568
British Plastics Reclamation Centre, 569
British Scrap Federation, 588, 1931
British Standards Institution (BSI), 2163
British Textile Technology Group, 637, 2030
British Trust for Ornithology (BTO), 854
British Wind Energy Association (BWEA), 1031
Building Energy Efficiency Division, 1084
Building Research Establishment (BRE), 1803
Business and Environment Research Unit, 1669
Can Makers, The, 465
Centre for Alternative Technology (CAT), 246, 1033, 2170
Centre for Environmental Technology (ICCET), 2196
Centre for Exploitation of Science and Technology (CEST), 2171
Chemical and Oil Recycling Association, 435, 1111
Chemical Industries Association, 1676, 2273
Chemicals Recovery Association (CRA), 436

United Kingdom **Geographical Index** 335

Climate Action Network UK (CAN UK), 90
Coal Research Establishment (CRE) Technical Services, 1136
Communities Against Toxics, 1394
Conservation Papers Ltd, 535
Council for Occupational Standards and Qualifications in Environmental Conservation (COSQUEC), 2164
Council of Ministers, 1822
Countryside Commission, 900
Coventry City Council, 2324
Cranfield Biotechnology Centre, 1318, 1789
Department of Energy, 967
Department of Environmental Science, 2032
Department of Health, 1380
Department of the Environment (DoE), 1651, 1979
Department of the Environment, Endangered Species Branch, 784
Department of the Environment, Waste Management Division, 639
Department of the Environment, Water Directorate, 393
Department of Trade and Industry Environment Unit, 1671
Department of Trade and Industry Environmental Enquiry Point, 1672
Department of Transport (DTp), 2307
Directorate of Fisheries Research, 228, 1640
ECO Environmental Education Trust, 1790
Electric Vehicle Association (EVA), 2325
Energy Technology Support Unit (ETSU), 1097
English Nature, 895
Environment and Technology Association for Paper Sacks (ETAPS), 537
Environment Council, 1850
Environmental Information Service: The British Library, 1770
Environmental Management and Auditing Services Ltd, 1717
Environmental Resources Ltd, 1757
Environmental Transport Association (ETA), 2326
Ethical Investment Research Information Service (EIRIS), 2213
Evergreen Recycled Fashions, 628
Food Commission: Formerly Food Additives Campaign Team (FACT), 1345
Forestry Commission, 746
Forests Forever Campaign (FFC), 753
Friends of the Earth (FoE), 1873
Green Party, 1826
Greenpeace UK, 1897
Group of Experts on the Scientific Aspects of Marine Pollution (GESAMP), 304, 1588
Health and Safety Executive, 1261
Henry Doubleday Research Association (HDRA), 263
Her Majesty's Inspectorate of Pollution (HMIP), 1441
HM Nuclear Installations Inspectorate Health and Safety Executive, 1154
Industry Council for Electronic Recycling (ICER), 595
Industry Council for Packaging and the Environment (INCPEN), 503
Institute for Marine Environmental Research, 306
Institute of Materials Management (IMM), 1753
Institute of Sound and Vibration Research, 1608
Institute of Terrestrial Ecology (ITE), 859, 884, 2149
Institute of Wastes Management (IWM), 646
Institution of Mechanical Engineers, 1792
Institution of Water and Environmental Management (IWEM), 2007
Intergovernmental Panel for Climatic Change (IPCC), 86
International Council on Monuments and Sites (ICOMOS UK), 865
International Union of Air Pollution Prevention Associations (IUAPPA), 27, 1504
Inveresk Research International Ltd, 1429
Linking Environment and Farming (LEAF), 161, 1347
Marine Conservation Society (MCS), 284: Let Coral Reefs Live, 803
Media Natura (MN), 1857
Medical Research Council, Toxicology Unit, 1381
Meteorological Office, The (The 'Met Office'), 87
Ministry of Agriculture Fisheries and Food (MAFF), 150
  Conservation Policy and Environmental Protection Division, 151, 712, 2046
  Fisheries Radiological Laboratory, 152, 1383
  Marine Environmental Protection Division, 273, 1343
Monitoring and Assessment Research Centre (MARC), 2035
National Association for Environmental Education (NAEE), 1780
National Association of Waste Disposal Contractors (NAWDC), 650
National Radiological Protection Board (NRPB), 1364
National Rivers Authority (NRA), 328, 1620
National Society for Clean Air and Environmental Protection (NSCA), 30, 1506
National Trust (NT), 868
National Westminster Bank, 1941
Natural Environment Research Council (NERC), 2153
Natural Resources Research (NRR), 1709
North Atlantic Salmon Conservation Organisation (NASCO), 285, 806
Nuclear Electric plc, 1179
Nuclear Free Local Authorities National Steering Committee, 1162, 2050
Organic Advisory Service, 264
Organic Food and Farming Centre, 253
Otter Trust, 810
Packaging and Industrial Film Association (PIFA), 508
Paint Research Association, 1484
Panos Institute, 954
Photographic and Waste Management Association, 652
Pulp and Paper Information Centre, 542
RAPRA Technology, 2177
RECOUP, 571
Renewable Energy Advisory Group (REAG), 1028
Renewable Energy Enquiries Bureau, 1039
Royal Commission on Environmental Pollution (RCEP), 1442
Royal Society for the Protection of Birds (RSPB), 811
RSNC Wildlife Trusts Partnership: Formerly Royal Society for Nature Conservation, 812
Save-a-Can, 470
Scientists for Global Responsibility (SGR), 1858
Scottish Campaign to Resist the Atomic Menace (SCRAM), 1906
Scottish Marine Biological Association, 312
Sea Mammal Research Unit (SMRU), 813
Society for Prevention of Asbestosis and Industrial Diseases (SPAID), 1297
Society of Dyers and Colourists, 632
Soil Association, The, 1348
Soil Survey and Land Research Centre (SSLRC), 388, 888, 1564
Tidy Britain Group, The (TBG), 527
Transport 2000, 2067, 2331
Transport and Road Research Laboratory (TRRL), 2365
Warren Spring Laboratory, 2207
Waste Management Information Bureau (WMIB), 655
Waste Processing Association, 1043
Waste Watch, 596
Wastes Technical Policy Unit, 453
Water Directorate Marine Branch Department of the Environment, 274
Water Research Centre (WRC), 390
WATT Committee on Energy, 981
WATT Committee's Working Group on Methane, 454
Women's Environmental Network (WEN), 1910
World Action for Recycling Materials and Energy from Waste: WARMER, 597
World Wide Fund for Nature – UK, 927

Publications

1992/93 ENDS Directory and Market Analysis, 1764
Acid Deposition and Vehicle Emissions: European Environmental Pressures on Britain, 58, 1533, 2350

336 Geographical Index *United Kingdom*

Acid Deposition: Volume 1 Sources Effects and Controls, 318, 1557
Acid Deposition: Volume 2 Origins Impacts and Abatement Strategies, 319
Acid Politics, 320, 2068
Acid Rain and the Environment, Volume 3: 1988–91, 321
Agricultural and Food Research Council Handbook, 125, 163
Air Pollution and Noise Bulletin, 43, 1520
Air Pollution from Vehicles, 59, 1614, 2351
Alternative Times, 1050, 1198
Annals of Occupational Hygiene, 44, 1521, 2028
Applied Acoustics, 1601
Applied Energy, 1089
Assessing Biological Risks of Biotechnology, 1313
Atmospheric Environment Part A, 82
Atmospheric Environment Part B, 83
Atom, 1199
Automotive Consortium on Recycling and Disposal Proposals, 614
Bikes Not Fumes, 61, 1536
Biodegradable Polymers and Plastics, 573
Biomass and Bioenergy, 1051
Biosphere, The, 81
Biotechnology in Industry Healthcare and the Environment: Special Report no 2178, 1316
Biotechnology: The Science and the Business, 1314
Britain's Nuclear Waste: Safety/Sting, 480
British Coal Research Association (BCRA) Quarterly, 1125
BS 7750: What the New Environmental Management Standards Mean for Your Business, 2167
C is for Chemicals, 1395
CFCs and Halons: Alternatives and the Scope for Recovery for Recycling and Destruction, 615
Changing Atmosphere, The, 2
Chemical Speciation and Bioavailability, 2280
Chemistry in Britain, 2282
Clean Air, 45, 1522
Climate Change and Human Impact on the Landscape, 101
Colliery Guardian, 1128
Commons Without Tragedy, 2293
Concise Oxford Dictionary of Earth Sciences, The, 2131
Confederation of British Industry News, 1746
Construction Materials and the Environment, 2169
Contaminated Land: Counting the Cost of Our Past, 1560
Continuous Emission Monitoring, 38, 1514
Corporate Environmental Policy Statements, 2104
Corporate Environmental Responsibility: Law and Practice, 1943
Country Studies and Technical Options, Volume 2, 115

Croners Environmental Management, 2008
Croners Waste Management, 667
Crop Protection Chemicals, 208
Crop Protection Directory 1988–89, 209
Design for Recyclability, 473
Desulphurisation, 41, 1518
Dictionary of Ecology and Environment, 2132
Dictionary of Environment and Development, 1782
Dictionary of Environmental Science and Technology 1991, 2133
Dictionary of the Environment, 1783
Digest of Environmental Protection and Water Statistics, 360, 726
Digest of UK Energy Statistics, 982
Dioxins in the Environment Pollution Paper 27, 1396
Director's Guide to Environmental Issues, The, 1733
Directory of Corporate Environmental Policy, 2092
Directory of Emissions Monitoring Equipment Suppliers 1992, 2027
Directory of Environmental Journals and Media Contacts, 1784
Directory of Environmental Technology, 2183
Directory of the Environment, 2011
Disposal of Vehicles: Issues and Actions, 474
E is for Additives, 1349
Earth Matters, 2239
ECO Directory of Environmental Databases in the UK 1992, 2012
Eco Labelling, 1666
Econews, 1925
Ecos, 933
Ecotoxicology, 1409
Electric Vehicle Developments, 2344
Elephants, Ivory and Economics, 836
Emissions for Heavy Duty Vehicles, 1616
Encyclopedia of Environmental Science and Engineering, 2134
ENDS, 668, 1461, 2094
Energy and the Environment, 124
Energy Digest, 1002
Energy Policies and the Greenhouse Effect Volume 1: Policy Appraisal, 116, 2072
Energy Without End, 1046
ENTEC Directory of Environmental Technology, 2184
Environment and Industry Digest, 2190
Environment Business, 1748
Environment Contacts: A Guide for Business (Who does what in Government Departments), 1741
Environment Databases, 2226
Environment Industry Yearbook, 1687
Environment Policy: Standards and Opportunities for Development, 2107, 2354
Environment Software, 2219
Environmental Action in Eastern Europe, 2073
Environmental Assessment Report, 1716
Environmental Assessment: A Guide to the Procedures, 1712

Environmental Auditing, 1721
Environmental Auditing Handbook: A Guide to Corporate and Environmental Risk Management, 1722
Environmental Chemistry, 2279
Environmental Education – Journal of the NAEE, 1786
Environmental Effects of Conventional and Organic Biological Farming Systems, The, 259
Environmental Engineering, 1799
Environmental Health, 1333
Environmental Health News, 1334
Environmental Impact of Paper Recycling, 556
Environmental Impact of Recycling, The, 616
Environmental Information: A Guide to Sources, 2227
Environmental Issues in the Pulp and Paper Industries, 549
Environmental Law, 1946
Environmental Law Brief, 1960
Environmental Monitoring and Analytical Specimen Banking, 2025
Environmental Opportunities: Building Advantage out of Uncertainty, 2211
Environmental Policy and Practice, 669, 1463, 2096
Environmental Practice in Local Government, 2076
Environmental Protection Bulletin, 730, 1695, 2283
Environmental Protection through Sound Water Management in the Pulp and Paper Industry, 361
Environmental Respiratory Diseases, 1331, 1449
Environmental Software, 2138
Environmental Technology, 2191
Environmental Toxicology and Chemistry, 1410
Environmental Toxicology and Water Quality, 416, 1411
Environmetrics, 2139
Farmers Weekly, 169
Farming and the Countryside, 236
Fifth Fuel, 1092
Finishing, 48, 1525
Freight, 2098, 2345
Furniture Recycling Network, 602
Future for UK Environmental Policy Special Report no 2182, The, 2108
Garners Environmental Law, 1950
Global Environmental Change – UK Research Framework, 4
Global Status of Peatlands and their Role in Carbon Recycling, The, 129, 1133
Global Warming, 117
Global Warming Forum, A, 119
Good Beach Guide to over 170 of Britain's Best Beaches 1990, 266
Great Britain Department of Energy Publications, 1009
Green Belt, Green Fields and the Urban Fringe: The Pressures of Land in the 1980's, 879

United Kingdom Geographical Index 337

Green Book, The (Authority on Tractors Agricultural and Forestry), 167, 766
Green Energy Matters, 1054
Green Engineering, 1800
Green Engineering – A Current Awareness Bulletin, 1801
Green Magazine, 1010
Green PC, The, 1451
Greenbits, 844
Greener Marketing, 1735
Greenhouse Gases Bulletin, 127
Greenhouse Issues, 128, 1129
Greening of Business, The, 1736
Greening of Global Investment Special Report 2108, The, 2212
Guardian, 2248
Guide to Libraries and Information Units in Government Departments and other Organisations, 2229
Guide to Resources in Environmental Education, 1785
Hazardous Materials: Sources of Information on their Transportation, 658, 2335
Health and Safety at Work, 1284
Health and Safety Directory, 5th revised ed, 1281
Horticulture Week, 846
Hothouse Earth, 120
In the Company of Green: Corporate Communications for the New Environment, 1737
Incineration and the Environment – A Source Book, 659
Industrial Environmental Services Directory, 1688
Information Sources in Biotechnology, 1315
Institute of Materials Management Members' Reference Book and Buyers Guide 1993, The, 1742
Interpretation Heritage, 872
Introduction to Aquaculture, 220, 414
Introduction to Energy, 984
Introduction to Environmental Engineering, 1794
Journal of Environmental Law, 1962
Journal of Environmental Management, 2015
Journal of Planning and Environmental Law, 1963
Journal of Radiological Protection, 1371
Journal of Transport Economics and Policy, 2099, 2347
Journal of Tropical Ecology, 769
Kind Food Guide, The, 260
Kirkgate Market: A Feasibility Study, 617
Land Use Change: The Causes and Consequences, 876
Land Use Policy, 880
Landscape Changes in Britain, 877
Legal 500, The, 5th ed, 1952
Less Traffic Better Towns, 2336
Living in the Greenhouse, 121
Living with Radiation: National Radiological Protection Board, 1368
Management for a Small Planet, 1738
Managing the Environment, 2017
Managing the Environment: The Greening of European Business, 1739
Marine Environmental Research, 294
Marine Pollution, 1578
Marine Pollution Bulletin, 1582
Marine Pollution Research Titles, 1583
Marketing, 1752
Materials Reclamation Weekly, 608
Metals and their Compounds in the Environment, 1400
NAMAS Concise Directory of Accredited Laboratories 1991, 2185
National Directory of Recycling Information, 603
Natural Environment Research Council News, 2141
Neurotoxicity, 1401, 1953
New Farmer and Grower, 261
Noise and Vibration Bulletin, 1603
North Sea Oil and Gas Directory, 1124
Nuclear Energy, 1214
Nuclear Power and the Greenhouse Effect, 122, 1191
Oryx, 848
Packaging, 513
Packaging in the Environment, 510
Packaging Industry Directory 1993, 512, 1690
Packaging Waste and the Polluter Pays Principle: A Taxation Solution, 520
Paper and Packaging Analyst, 551
Petroleum Review, 1131
Pigment and Resin Technology, 2285
Poisonous Fish, 223, 335, 1420
Pollution, 1472
Pollution Prevention, 1473
Population Studies, 2302
Population: An Ecofeminist Perspective, 2304
Practical Alternatives, 1057
Progress in the Recycling of Plastics in the UK, 581
Radioactive Aerosols, 482
Radioactive Waste Management, 484
Radon in Dwellings, 1370
Railway Noise Standards: Let's Get Them Right, 1605
Recovery and Rehabilitation of Contaminated Land, 1559
Recycled Papers: The Essential Guide, 550
Recycling: Public Attitudes and Market Reality, 620
Red Data Birds in Britain, 839
Refrigeration and Air Conditioning, 55, 1530
Regulated Rivers Research and Management, 334
Reinforced Plastics, 578
Reprocessing of Tyres and Rubber Wastes, 600
Risks from Radon in Homes: Report of a Working Group of the Institute of Radiation Protection, 1373
Role of Environmental Assessment in the Planning Process, The, 1714
Safe Energy, 1058, 1220
Sanitary Landfill, 456
SAVE Action Guide, The, 871
Seaside Award, 271
Sitefile Digest: A Digest of Authorised Waste Treatment and Disposal Sites in Great Britain, 457
Society of Dyers and Colourists, 635
Solar Power via Satellite, 1048
Solvent Recovery Handbook, 443
States and Anti-Nuclear Movements, 1192, 1914
Stratospheric Ozone 1991, 123
Streetwise, 1788
Submerging Coasts, 269
Survey of Gassing Landfill Sites in England and Wales, A, 458
Taking Population Seriously, 2295
Talking Conservation, 554
Textiles and the Environment Special Report no 2150, 636
Third Revolution, The: Environment, Population and a Sustainable World, 2296
This Common Inheritance: Britain's Environmental Strategy, 2088
Tourism Industry and the Environment, The, 1701
Trace Element Occurrence in British Groundwaters, 426, 1636
Transport and Climate Change: Cutting Carbon Dioxide Emissions from Cars, 105, 2357
Transport and the Environment, 2338
Transport, the Environment and Economic Policy, 2341
Tropical Rain Forest, 763
Tropical Rain Forest Ecology, 764
UK Centre for Economic and Environmental Development Bulletin, 1768
UK Marine Oil Pollution Legislation, 1123, 1580, 1955
UK Recycling Directory, The, 605
UK Road Transport's Contribution to Greenhouse Gases, 132, 2358
UK Waste Management Industry, 679
Vital Signs 1992/93: The Trends that are Shaping our Future, 1715
Warmer Bulletin, 612
Warren Spring Laboratory UK Smoke and Sulphur Dioxide Monitoring Networks, 57, 1532, 2029
Waste Disposal and Recycling Bulletin, 613
Waste Management, 486, 675
Waste Management Yearbook, 665
Waste Not Want Not: The Production and Dumping of Toxic Waste, 601, 1403
Water and Drainage Law, 363, 1959
Water and the Environment, 364, 688
Water and Waste Treatment, 376, 693
Water Bulletin, 421
Water Pollution Incidents in England and Wales 1990, 211, 1637
Water Supply, 378
WATT Committee on Energy: Air Pollution Acid Rain and the Environment, 65, 326, 1540

Who's Hand on the Chainsaw? UK Government Policy and the Tropical Rain Forests, 765
Who's Who in the Environment – England, 1920
Who's Who in the Environment – Scotland, 1921
Who's Who in the Environment – Wales, 1922
Who's Who in the Water Industry, 368
Wind Engineering, 1064
Working for Public Safety and Environmental Protection, 489, 739, 1255

Databases

Air/Water Pollution Report, 77, 1553
Current Biotechnology Abstracts, 1324
EcoBase, 2262
EMBASE, 1293
Energy Conservation News, 1102
HMSO Online/British Official Publications, 2267

Ukraine

Organisations

Greenpeace Ukraine, 1898

Ministry of Chernobyl, 1158
State Committee for Nuclear and Radiation Safety, 1247, 1359

Zelenyi Svit, 1911

Organisations – Alphabetical Index

Advisory Committee for the Coordination of Information Systems (ACCIS), 2003
Advisory Committee on Nuclear Energy, Ministry of Trade and Industry, 1145
AFME, 1082
Agence de l'Environnement et de la Maîtrise de l'Energie (ADEME), 962
Agence Internationale de l'Energie Atomique (IAEA): Organismo Internacional de Energia Atómica, 1146
Agence Nationale de la Récupération et l'Elimination des Déchets (ANRED), 587
Agence Nationale pour la Gestion des Déchets Radioactifs (ANDRA), 477
Agricultural and Food Development Authority, 173
Agricultural Research Centre Ghent see Centrum voor Landbouwkundig Onderzoek – Gent, 178
Agricultural University of Norway, Norwegian Centre for International Agricultural Development, 174
Agriculture and Food Research Council (AFRC), 175
AGROBIO, 240
Agronomic Centre of Applied Research Hainaut see Centre Agronomique de Recherches Appliquées du Hainaut, 212
Air Pollution Dept of the National Institute of Applied Chemical Research see Département de la Pollution Atmosphérique de l'Institut National de Recherche Chimique Appliquée, 14, 1491
Air Pollution Study Committee see Comitato di Studio perl'Inquinamento Atmsferico (CSIA), 19, 1496
All-Union Research Institute for Reactor Operators (VNIIAES), 1225
Aluminium Can Recycling Association (ACRA), 459
Aluminium Federation Ltd, 460
Aluminium Foil Recycling Campaign, 461
Amici della Terre, 1863
Amigos da Terra, 1864
Amigos de la Tierra, 1865
Amis de la Terre, Les, 1866, 1867
Anders Gaan Leven (AGALEV), 1807
Aquário Vasco da Gama, 225, 301
Arbeitsgemeinschaft Information Meeresforschung Meerestechnik, 275, 1568
Arbetsmiljöinstitutet, 1289
ARC, 1808
Architectural Heritage Fund, 863
Asbestos Information Centre (AIC), 1295

Asbestos International Association (AIA), 1296
ASH Partnership, 1706
Asociación de Bioindustria, 1301
Asociación de Investigación Técnica de la Industria Papelera Española, 557
Asociación de Recuperadores de Papel y Cartón (REPACAR), 528
Asociación Ecologista de Defensa de la Naturaleza (AEDENAT), 1029
Asociación Española de Fabricantes de Agroquímicos para la Protección de las Plantas, 186
Asociación Nacional de Recuperadores de Vidrio, 447
Asociación Vida Sana, 241
Assobiotec, 1302
Associação Portuguesa das Empresas Indústrias de Produtos Químicos, 187
Associação Portuguesa de Fabricantes de Papel e Cartão, 529
Associação Portuguesa dos Recursos Hidricos (APRH), 405
Associação Portuguesa para Estudos de Saneamento Basico (APESB), 1277
Association des Régions des Alpes Centrales (ARGE ALP), 2120
Association des Transformateurs de Matières Plastiques (Fechiplast): Fédération des Industries Chimiques, 566
Association for the Control and Saving of Energy see AFME, 1082
Association for the Development of Agricultural Fuel (ADECA), 155
Association for the Prevention of Atmospheric Pollution see Association pour la Prevention de la Pollution Atmosphérique (APPA), 16, 1493
Association for the Promotion of Recycled Paper, 525
Association for the Promotion of Renewable Energies see Association Promotion Energies Renouvelables (APERE), 1030
Association Industry Difesa Produzioni Agricole (AGROFARMA), 188
Association Internationale Contre le Bruit (AICB), 1596
Association Nationale des Services d'Eau asbl (ANSEAU), 341
Association of Chemical Industries (VCI), 1673, 2272
Association of European Dry Battery Manufacturers (Europile), 462
Association of Hungarian Chemical Works (AHCW), 189

Association of Netherlands Paper and Board Manufacturers (VNP), 530
Association of Petroleum Industry see Mineralöl Wirtschafts Verband eV (MWV), 1119
Association of Plastic Manufacturers in Europe (APME), 567
Association of Reclaimed Textile Processors (ARTP), 624
Association of the Central Alps see Association des Régions des Alpes Centrales (ARGE ALP), 2120
Association of Waste Management and Entrepreneurs in Finland, 640
Association pour la Prevention de la Pollution Atmosphérique (APPA), 16, 1493
Association Promotion Energies Renouvelables (APERE), 1030
Association Technique de l'Industrie du Gaz en France, 1110
Association Vincotte ASBL, 641, 1444
Associazione Italiana di Protezione Contro le Radiazioni (AIRP), 1361
Associazione Italiana fra gli Industriali della Carta Cartoni e Paste per Carta, 531
Associazione Nazionale di Ingegneria Nucleare (ANDIN), 1164
Associazione Nazionale Fra Fabbricanti di Imballaggi Metallici (ANFIMA), 463
Associazione per l'Agricultura Biodinamica, 242
Atmospheric Pollution Technical Research Centre see Centre Interprofessionnel Technique d'Etudes de la Pollution Atmosphérique (CITEPA), 66
Atmospheric Research and Information Centre (ARIC), 17, 316, 1494
Atomic Energy Technology, 1226
Austrian Environmental Protection Association, 717
Austrian Noise Abatement Society see Österreichischer Arbeitsring für Lärmbekämpfung (OAL), 1600
Austrian Nuclear Society see Österreichische Kerntechnische Gesellschaft, 1163
Austrian Society for Nature and Environment Protection see Österreichisches Gesellschaft für Natur und Umweltschutz (ÖGNU), 97, 908
Baltic Marine Environment Protection Committee Helsinki Commission (HELCOM), 276, 1569
Batelle Institute Ltd, 1338
Belgian Biotechnology Coordinating Group (BBCG), 1303, 2052

340 Organisations – Alphabetical Index

Belgian Packaging Institute (BPI), 495
Belgian Royal Meteorological Institute *see* Institut Royal Météorologique de Belgique, 94
Bestuur van de Volksgezondheid – Ministerie van Volksgezondheid, 11, 1354, 1489
Biofranc (FNAB), 243
Bioindustry Association (BIA), 1304
Biokultura Association, 244
Biokultura Egyelulet, 245
Bioresearch Ireland (EOLAS), 1305
Biotechnology Unit, 1299
Biotecno SA, 625
Blue Bubble *see* Bulle Bleue, 80
Bon Beter Leefmilieu (BBL), 1847
Bonn Commission, 277, 1570
British Agrochemicals Association Ltd (BAA), 190
British Battery Manufacturers Association (BBMA), 464
British Effluent and Water Association (BEWA), 342, 685
British Glass Recycling Department, 448
British Industrial Biological Research Association, 1423
British Non-Ferrous Metals Association, 1674
British Nuclear Forum (BNF), 1165
British Paper and Board Industry Federation (BPBIF), 532
British Plastics Federation (BPF), 568
British Plastics Reclamation Centre, 569
British Scrap Federation, 588, 1931
British Standards Institution (BSI), 2163
British Textile Technology Group, 637, 2030
British Trust for Ornithology (BTO), 854
British Wind Energy Association (BWEA), 1031
Brontosaurus Movement/EcoWatt, 1032
Building Energy Efficiency Division, 1084
Building Research Establishment (BRE), 1803
Bulle Bleue, 80
Bund, 1868
Bundesamt für Energiewirtschaft, 963
Bundesamt für Umweltschutz, 680, 740, 1478
Bundesamt für Wirtschaft, 1105
Bundesanstalt für Arbeitsschutz und Unfallforschung, 12, 1490
Bundesarbeitgeberverband Chemie, 434
Bundesdeutscher Arbeitskreis für Umweltbewusstes Management (BAUM), 2004
Bundesforschungs Anstalt für Fischerei, 226, 741, 1586
Bundesforschungsanstalt für Naturschutz und Landschaftsökologie, 946
Bundesministerium für Forschung und Technologie (BMFT), 1147
Bundesministerium für Land und Forstwirtschaft, 135, 744
Bundesministerium für Öffentliche Wirtschaft und Verkehr, 1670, 2305
Bundesministerium für Umwelt Jugend and Familie, 1977
Bundesministerium für Umwelt Naturschutz und Reaktorsicherheit, 711, 1239, 1649
Bundesverband Glasindustrie, 449
Bureau International de la Récupération (BIR), 589
Bureau of the Convention on Wetlands of International Importance Especially as Waterfowl Habitat: Ramsar Convention Bureau, 785
Business and Environment Research Unit, 1669
Business and Industry Advisory Committee (BIAC), 1675
Camara Municipal de Lisboa Direcção Municipal de Infraestructuras e Saneamento, 392
Can Makers, The, 465
Cekmece Nuclear Research and Training Centre *see* Cekmece Nükleer Arastirma ve Egitim Merkezi, 1227, 1479
Cekmece Nükleer Arastirma ve Egitim Merkezi, 1227, 1479
Cellule Environnement Institut National de la Recherche Agronomique (INRA), 176
Center for Energy Saving and Clean Technology, 1083
Center for Information on the Environment, 2039
Center for Nature Protection and Education, 893
Central Office of Historic Monuments, 864
Central Statistical Office *see* Statistik Sentralbyra (SSB), 2216
Centre Agronomique de Recherches Appliquées du Hainaut, 212
Centre d'Enseignement et de Recherche pour la Gestion des Ressources Naturelles et l'Environnement, 382, 2018
Centre d'Information du Plomb, 1325, 1445
Centre de Recerca Ecológica i Aplicacions Forestals, 772
Centre de Recherche et de Contrôle Lainier et Chimique Celac, 1638
Centre for Agrobiological Research *see* Centrum voor Agrobiologisch Onderzoek (CABO), 213
Centre for Alternative Technology (CAT), 246, 1033, 2170
Centre for Energy Conservation and Environmental Technologies *see* Centrum voor Energiebesparing en Schone Technologie, 1096
Centre for Environmental Management and Planning (CEMP), 2005
Centre for Environmental Technology (ICCET), 2196
Centre for European Policy Studies (CEPS), 2054
Centre for Exploitation of Science and Technology (CEST), 2171
Centre for Informatic Application in Agriculture *see* Centro Studi per l'Applicazione dell'Informatica in Agricoltura, 177
Centre for International Environmental Law (CIEL), 1932
Centre for Occupational Hygiene and Safety *see* Kentron Yginis Keasfalias tis Ergassias (KYAE), 1292
Centre for Our Common Future (COCF), 2055
Centre for the Study of Man and the Environment *see* Centro Studi l'Uomo e l'Ambiente, 1317
Centre Interprofessionnel Technique d'Etudes de la Pollution Atmosphérique (CITEPA), 66, 1541
Centre National pour l'Exploitation des Océans, 278, 1571
Centre of Ecological Research and Forestry Application *see* Centre de Recerca Ecológica i Aplicacions Forestals, 772
Centre of Ecology and Climatology/Geophysical Experimental Observatory *see* Centro di Ecologia e Climatologia/Osservatorio Geofisico Sperimentale Macerata, 5, 1070
Centro de Estudios de la Energía Solar (CENSOLAR), 1069
Centro de Investigações Florestais, 773
Centro de Investigación Energética Medioambientale y Tecnológica (CIEMAT), 964, 2197
Centro di Ecologia e Climatologia/Osservatorio Geofisico Sperimentale Macerata, 5, 1070
Centro Ecologico, 2121
Centro Ecológico Nacional SA, 2122
Centro Studi l'Uomo e l'Ambiente, 1317
Centro Studi per l'Applicazione dell'Informatica in Agricoltura, 177
Centrum voor de Studie van Water Bodem en Lucht (Becewa Vzw), 681, 1480
Centrum voor Energiebesparing en Schone Technologie, 1096
Centrum voor Landbouwkundig Onderzoek – Gent, 178
Centruum voor Agrobiologisch Onderzoek (CABO), 213
Ceská Asociace IUAPPA (CA IUAPPA), 18, 1495
Chemical and Oil Recycling Association, 435, 1111
Chemical Industries Association, 1676, 2273
Chemical Industry Federation of Finland *see* Kemian Keskusliitto, 1309
Chemicals Recovery Association (CRA), 436
Civil Forsvaret, 590
Climate Action Network UK (CAN UK), 90
Climate Network Europe (CNE), 111
Coal Research Establishment (CRE) Technical Services, 1136
Combined Heat and Power Association (CHPA), 1085
Comhaontas Glas, 1809
Comitato di Studio perl'Inquinamento Atmosferico (CSIA), 19, 1496

Comitato Nazionale per la Ricerca e per lo Sviluppo dell'Energia Nucleare e dell' Energie Alternativ, 1071, 1228
Comité Européen de l'Association de l'Ozone, 112
Comité Européen de Normalisation (CEN), 496, 1707
Commissariat à l'Energie Atomique (CEA), 1148
Commission for Municipal Solid Waste: Department of Environmental Studies, 642
Commission of the European Communities, 2056
Commission of the European Communities (CEC) – Joint Research Centre (JRC), 1229, 1374
Commission of the European Communities Nuclear Safety Research Directorate, 1250
Commission on Air Pollution Prevention in VDI and DIN see Kommission Reinhaltung der Luft (KRdL) in VDI und DIN, 28, 1505
Committee of Agricultural Organisations (COPA)/General Committee of Agricultural Cooperation in EC, 156
Committee on the Use of Atomic Energy for Peaceful Purposes, 1149, 2040
Commonwealth Agricultural Bureaux International (CABI), 136
Communities Against Toxics, 1394
Compagnie Générale des Eaux, 343
Concentration Unit for Biotechnology in Europe (CUBE), 1300, 2041
Confédération Belge de la Récupération, 591
Confédération Européenne de l'Industrie des Pâtes Papiers et Cartons (CEPAC), 533
Confederation of European Agriculture (CEA), 137, 745
Confederation of European Paper Industries (CEPI), 534
Conseil Européen des Fédérations de l'Industrie Chimique (CEFIC), 2274
Conseil Européen du Droit de l'Environnement (CEDE), 1933
Consejo de Seguridad Nuclear, 1240, 1355
Conservation Papers Ltd, 535
Consumentenbond, 1326
Consumers Association see Consumentenbond, 1326
Convention on International Trade in Endangered Species of Wild Fauna and Flora (CITES), 786
Cooperative Programme for Monitoring and Evaluation of Long Range Transboundary Air Pollution (EMEP), 20, 1497
Coordinadora de Organizaciones de Defensa Ambiental (CODA), 21, 1498
Copacel, 536
Cotton Council International (CCI), 626, 1677
Council for Occupational Standards and Qualifications in Environmental Conservation (COSQUEC), 2164

Council of Europe, 899
Council of Ministers, 1810, 1811, 1812, 1813, 1814, 1815, 1816, 1817, 1818, 1819, 1820, 1821, 1822
Countryside Commission, 900
Coventry City Council, 2324
Cranfield Biotechnology Centre, 1318, 1789
Croatian Air Pollution Prevention Association (CAPPA) see Hravtsko Drutvo za Zastitu Zraka, 24, 1501
CSFR Atomic Energy Commission, 1150
Cyprus Association for the Protection of the Environment, 718, 1848
Czech Association of IUAPPA see Ceská Asociace IUAPPA (CA IUAPPA), 18, 1495
Danish Acoustical Institute see Lydteknisk Institut (LI), 1610
Danish Agrochemical Association (DAA), 191
Danish Energy Agency: Ministry of Energy, 965
Danish Institute for Fisheries and Marine Research see Danmarks Fiskeri–og Havundersogelser, 227, 302
Danish Laboratory for Soil and Crop Research see Statens Planteavis Laboratorium, 1565
Danish Meteorological Institute, 91
Danish Nuclear Society (DNS), 1166
Danish Organisation for Renewable Energy, 1034
Danish Packaging and Transportation Research Institute, 522, 2360
Danish Society for the Conservation of Nature see Danmarks Naturfredningsforening, 901
Danish Toxicology Centre (DTC), 1384, 1934
Danmarks Fiskeri–og Havundersogelser, 227, 302
Danmarks Naturfredningsforening, 901
Dansk Naturkost, 247
DEBIO, 248
Déi Gréne Alternativ, 1823
Département de la Pollution Atmosphérique de l'Institut National de Recherche Chimique Appliquée, 14, 1491
Département Pollution des Eaux de l'Institut National de Recherche Chimique Appliquée, 1639
Department of Agriculture and Food, 138
Department of Air Pollution, 67, 1542
Department of Consumer Affairs: Ministry of Family and Consumer Affairs, 1650
Department of Energy, 966, 967
Department of Environmental Forestry, 774
Department of Environmental Science, 2032
Department of Health, 1260, 1380
Department of Plant Ecology, 856
Department of Political and Social Sciences, 2114, 2361
Department of the Environment, 1978
Department of the Environment (DoE), 1651, 1979

Department of the Environment, Endangered Species Branch, 784
Department of the Environment, Waste Management Division, 639
Department of the Environment, Water Directorate, 393
Department of Tourism Transport and Communication, 2306
Department of Toxicology, 1424
Department of Trade and Industry Environment Unit, 1671
Department of Trade and Industry Environmental Enquiry Point, 1672
Department of Transport (DTp), 2307
Department of Wildlife Ecology, 857
Deutsche Naturschutzring (DNR), 1849
Deutsche Verkehrswissen-schaftliche Gesellschaft eV, 406, 1613
Deutsche Wissenschaftliche Gesellschaft für Erdöl, 1137
Deutscher Wetterdienst (DWD), 106, 1072
Deutscherverein des Gas- und Wasserfaches eV (DVGW), 344, 1112
Deutsches Atomforum eV (DAIF), 1167
Deutsches Hydrographisches Institut, 272, 490, 1587
Direção Geral de Energia, 968, 1927
Direção Geral de Energia Departamento de Energia Nuclear, 1151
Directorate for Nature Management, The, 894
Directorate for Wildlife and Freshwater Fish see Direktoratet for Vilt og Ferskvannsfisk, 858
Directorate of Fisheries Research, 228, 1640
Direktoratet for Vilt og Ferskvannsfisk, 858
Division for Air Monitoring and Industrial Compliance, 13, 1492, 2021
Duales System Deutschland GmbH (DSD), 592
Dutch Society for the Wadden Sea Conservation see Landelijke Ver tot Behoud van de Waddenzee (LVBW), 283
Earthwatch, 949, 1869, 2022
Earthwatch Europe, 950
East European Environmental Research (ISTER), 961, 1769
ECO Environmental Education Trust, 1790
Ecoglasnost, 1870
Ecole de Santé Publique – Université Libre de Bruxelles, 1339
Ecolo, 1824
Ecological and Toxicological Association of the Dyestuff Manufacturing Industry (ETAD), 627
Ecological Centre see Centro Ecologico, 2121
Ecological Studies Institute (ESI), 1935
Ecology and Forestry Research Council see Studiecentrum voor Ecologie en Bosbouw, 1485
Ecology Research Groups see Gruppi Ricerca Ecologica (GRE), 2147

Organisations – Alphabetical Index

Economic and Social Committee (ESC): Section for Protection of the Environment, 2006
Economy and Performance Service (ECO-PER), 437
Eidgenössische Anstalt für Wasserversorgung Abwasserreinigung und Gewässerschutz, 383, 1641
Eidgenössische Kommission für die Sicherheit von Kernanlagen, 491, 1257
Eidgenössische Kommission zur Überwachung der Radioaktivität, 1375, 2033
Electric Vehicle Association (EVA), 2325
Empresa Publica das Aguas Livres (EPAL), 345
Energik, 976
Energiministeriet, 969
Energy and Transport Research Institute see Institut für Energie- und Transportforschung Meissen, 1075, 2363
Energy Conservation in Buildings Institute see Istituto per l'Edilizia ed il Risparmio Energetico, 1099
Energy Efficiency Centre see Stredisko pro Efektivni Vyuzivani Energie (SEVE), 980
Energy Technology Support Unit (ETSU), 1097
English Nature, 895
Environment and Development Resource Centre (EDRC), 951
Environment and Technology Association for Paper Sacks (ETAPS), 537
Environment Council, 1850
Environment Liaison Centre International (ELC), 2217
Environment Policy Committee: Confederation of Irish Industry, 2043
Environment Policy Section, 1652
Environmental Detergent Manufacturers Association (EDMA), 438, 1385, 1662
Environmental Engineering Institute see Instytut Podstaw Inzynierii Srodowiska PAN, 885, 1482, 1804
Environmental Information Science Institute see Institut für Umwelt Informatik, 2148
Environmental Information Service: The British Library, 1770, 2218
Environmental Law Association, 1936
Environmental Management and Auditing Services Ltd, 1717, 1718, 1719
Environmental Politics in Bulgaria: Ecological Studies Institute, 1851
Environmental Pollution Control Project – Athens, 1481
Environmental Protection Institute, Laboratory of Environmental Monitoring see Instytut Ksztaltowania Srodowiska, 742, 2034
Environmental Research Unit, 68, 1544
Environmental Research Unit Water Resources, 427
Environmental Resources Ltd, 1757
Environmental Transport Association (ETA), 2326

Erdészeti Tudományos Intézet, 775
ERL Brussels, 1758
ERL Espana, 1759
ERL Italia, 1760
ERL Nederland, 1761
ERL Umwelt Consult: Rhein–Main–Neckar GmbH, 1762
Espoo Research Centre, 214, 1702
Estonian Green Movement, 1871
Estonian Green Party, 1825
Ethical Investment Research Information Service (EIRIS), 2213
EUREAU, 407
Eurométaux, 466
Europäische Ökologische Aktion (ECOROPA), 1852
European Aluminium Association, 467
European Arctic Stratospheric Ozone Experiment (EASOE), 84
European Association for Conservation of Energy (Euro ACE), 1086
European Association for Population Studies (EAPS), 2292
European Association for the Science of Air Pollution (EURASAP), 22, 1499
European Association of Automobile Manufacturers (ACEA), 468, 2327
European Centre for Medium Range Weather Forecasts (ECMWF), 92
European Chemical Industry Council see Conseil Européen des Fédérations de l'Industrie Chimique (CEFIC), 2274
European Chemical Industry Ecology and Toxicology Centre (ECETOC), 1386
European Chlorinated Solvent Association (ECSA), 1387
European Committee of the European Ozone Association see Comité Européen de l'Association de l'Ozone, 112
European Conference of Ministers of Transport (ECMT), 2328
European Council on Environmental Law see Conseil Européen du Droit de l'Environnement (CEDE), 1933
European Crop Protection Association (ECPA), 192
European Environment Bureau (EEB), 1853
European Environmental Agency (EEA), 719
European Environmental Alliance (EEA), 2057
European Federation for Transport and Environment (T & E), 2058, 2329
European Federation of Animal Health (FEDESA), 790
European Federation of Chemical and General Workers' Unions (FESCID), 1678, 2275
European Federation of Pharmaceutical Industries (EFPIA), 1679, 2276
European Federation of Waste Management (FEAD), 643
European Fertiliser Manufacturers' Association (EFMA), 193
European Geographical Information Systems Foundation (EGIS), 2023

European Group for Ecological Action see Europäische Ökologische Aktion (ECOROPA), 1852
European Group of Automotive Recycling Association (EGARA), 469
European Institute for Transuranium Elements, 492, 1230
European Institute for Water, 408
European Institute of Ecology and Cancer (EIEC), 1290
European Investment Bank (EIB), 1725
European Network of Environmental Research Organisations (ENERO), 2172
European Nuclear Society (ENS), 1168
European Packaging Federation, 497
European Petroleum Industries Association (EUROPIA), 1113
European Recovery and Recycling Association (ERRA), 593
European Transport Law, 1388, 1937, 2330
European Union for Packaging and the Environment, 498
European Water Pollution Control Association (EWPCA), 1621
European Wind Energy Association, 1035
European Youth Forest Action (EYFA), 751
Evergreen Recycled Fashions, 628
FABRIMETAL, 346, 644
Fachverband des Chemischen Industrien Österreichs, 194
Fachverband Schweizerischer Becherhersteller (FSBH), 538
Faunistisch–Ökologische Arbeitsgemeinschaft, 791
Federal Chancellory Division for Nuclear Energy Coordination and Nonproliferation, 1152
Federal Commission for the Safety of Nuclear Installations see Eidgenössische Kommission für die Sicherheit von Kernanlagen, 491, 1257
Federal Department of Transport Communications and Energy, 2308
Federal Forest Research Institute see Forstliche Bundesversuchsanstalt, 776
Federal Ministry for Agriculture and Forestry see Bundesministerium für Land und Forstwirtschaft, 135, 744
Federal Ministry for Nationalised Industries and Transport see Bundesministerium für Öffentliche Wirtschaft und Verkehr, 1670, 2305
Federal Ministry for the Environment Federal Republic of Germany, 1980
Federal Office for Trade and Industry Branch for Mineral Oil and Gas see Bundesamt für Wirtschaft, 1105
Federal Office of Energy see Bundesamt für Energiewirtschaft, 963
Federal Office of Environment, Forests and Landscape, 1653
Federal Office of Environmental Protection see Bundesamt für Umweltschutz, 680, 740, 1478
Federal Research Centre for Fisheries see Bundesforschungs Anstalt für

Organisations – Alphabetical Index

Fischerei, 226 *see* Bundesforschungs Anstalt für Fischerei, 74, 1586
Federal Research Centre for Nature Conservation and Landscape Ecology *see* Bundesforschungsanstalt für Naturschutz und Landschaftsökologie, 946
Fédération Européenne du Verre d'Emballage (FEVE), 450
Federation of Biotechnology Industries of Denmark (FBID), 1306
Federation of Ecologist Alternative Organisations, 1854
Federation of Irish Chemical Industries (FICI), 195
Federation of the Water Protection Association in Finland, 409
Federation Seas at Risk, 279
Federazione Italiana Impresa Pubbliche Gas Acqua e Varie (FEDERGASACQUA), 347, 1114
Federazione Italiana Pro Natura, 902
Finnish Air Pollution Prevention Society (FAPPS) *see* Ilmansuojeluyhdistys ry ISY, 25, 1502
Finnish Association for Nature Conservation *see* Suomen Luonnonsuojeluitto ry, 912
Finnish Centre for Radiation and Nuclear Safety *see* Sateilyturvakeskus (STUK), 1252, 1366
Finnish Forest Research Institute *see* Metsäntutkimuslaitos, 779
Finnish Institute of Marine Research *see* Merentutkimuslaitos, 311
Finnish Meteorological Institute *see* Ilmatieteen Laitos, 93
Finnish Packaging Association, 499
Finnish Pulp and Paper Research Institute *see* Oy Keskuslaboratorio – Centrallaboratorium Ab, 559
Finnish Standards Association, 1663
Fisheries Directorate Nutrition Institute *see* Fiskeridirektoratets Ernaeringsinstitutt, 229
Fiskeridirektoratets Ernaeringsinstitutt, 229
Fiskeridirektoratets Havforskningsinstitutt, 230
Food and Agriculture Organisation of the United Nations (FAO), 157, 752, 1344
Food Commission: Formerly Food Additives Campaign Team (FACT), 1345
Food Technology and Packaging *see* Fraunhofer-Institut für Lebensmitteltechnologie und Verpackung, 501, 1708
Foreningen Karnteknik, 1169
Forest Research Institute *see* Erdészeti Tudományos Intézet, 775
Forestry Commission, 746
Forestry Research Centre *see* Centro de Investigacões Florestais, 773
Forests Forever Campaign (FFC), 753
Forschungsinstitut für Internationale Technische und Wirtschaftliche Zusammenarbeit der RWTH Aachen, 2198

Forschungsring für Biologisch–Dynamische Landwirtschaft, 249
Forschungszentrum Jülich GmbH (KFA), 1231
Forstliche Bundesversuchsanstalt, 776
Forum Atómico Español, 1170
Forum Atomique Européen (FORATOM), 1171
Forum Italiano dell'Energia Nucleare (FIEN), 1172
Forum Nucléaire Belge (ASBL), 1173
Foundation for Environmental Education, 1771
Foundation for Environmental Education in Europe (FEEE): European Office Friluftsradet, 1773
Foundation for Packaging and the Environment (SNV), 500
Fraunhofer Institute for Atmospheric Environmental Research *see* Fraunhofer-Institut für Atmosphärische Umweltforschung, 85
Fraunhofer Institute for Environmental Chemistry and Toxicology *see* Fraunhofer-Institut für Umweltchemie und Ökotoxikologie (IUCT), 215, 1426
Fraunhofer Institute for Toxicology and Aerosol Research *see* Fraunhofer-Institut für Toxicologie und Aerosolforschung, 1425
Fraunhofer-Institut für Atmosphärische Umweltforschung, 85
Fraunhofer-Institut für Lebensmitteltechnologie und Verpackung, 501, 1708
Fraunhofer-Institut für Toxicologie und Aerosolforschung, 1425
Fraunhofer-Institut für Umweltchemie und Ökotoxikologie (IUCT), 215, 1426
French Association for the Protection of Water, 394
French Environment and Energy Management Agency *see* Agence de l'Environnement et de la Maîtrise de l'Energie (ADEME), 962
French Federation of the Paper Board and Cellulose Industry *see* Copacel, 536
French Research Institute for Ocean Utilisation *see* Institut Français de Recherche pour l'Exploitation de la Mer (IFREMER), 305
Freunde der Erde, 1872
Friends of the Earth (FoE), 290, 1873, 1874
Friends of the Earth International (FoEI), 1875: European Coordination – CEAT, 1876
Független Ökológical Központ (FÖK), 2123
Fundación Para la Ecología y la Protección del Medio Ambiente (FEPMA), 23, 1500
Fundacja na Rzecz Efelatywnego Wykorzystanos Energii, 1098
Gabinete de Protecção e Segurança Nuclear, 1356

General Directorate for Energy *see* Direção Geral de Energia, 968, 1927
General Organic Chemistry Institute *see* Instituto de Química Orgánica General, 2288
Geological Survey of Denmark, 2260
Geological Survey of Ireland, 2214
Geological Survey of Sweden, Department of Forest Soils, 1562
Geological Survey of the Netherlands, 1115
German Environmental Agency *see* Umweltbundesamt, 723, 2179
German Hydrographic Institute *see* Deutsches Hydrographisches Institut, 272, 490, 1587
German Meteorological Service *see* Deutscher Wetterdienst (DWD), 106, 1072
German Nappy Service Association *see* Verband Deutscher Windelservice Lücke GmbH, 547
German Scientific Society for Oil Gas and Coal *see* Deutsche Wissenschaftliche Gesellschaft für Erdöl, 1137
Gesellschaft für Strahlen und Umweltforschung GmbH (GSF), 1376
Global Environmental Monitoring System, 2024
Global Environmental Technology Network (GET), 2173
Global Legislators Organisation for a Balanced Environment (GLOBE): GLOBE International/EC, 1938
Göncöl Alliance, 1036
Greek Agrochemical Association (GAA), 196
Greek Atomic Energy Commission, 1153, 1357
Greek National Centre for Marine Research, 231, 303, 336, 1642
Green Alternative European Link (GRAEL), 1877
Green Alternative Party *see* Déi Gréne Alternativ, 1823
Green League *see* Vihreä Litto, 1845
Green Party, 1826 *see* Groenen, De, 1828 *see* Miljopartiet de Gröna (MP), 1838
Green Party of Bulgaria, 1827
Green Party of Izmir *see* Izmir Yesiller Partisi, 1834
Green Party of Switzerland *see* Grüne Partei der Schweiz/Parti Ecologiste Suisse (GPS), 1833
Greenpeace Austria, 1878
Greenpeace Belgium, 1879
Greenpeace Czechoslovakia, 1880
Greenpeace Denmark, 1881
Greenpeace Finland, 1882
Greenpeace France, 1883
Greenpeace Germany, 1884
Greenpeace Greece, 1885
Greenpeace International, 1886
Greenpeace International Atmosphere and Energy Campaign, 114
Greenpeace International EC Unit, 1887
Greenpeace Ireland, 1888
Greenpeace Italy, 1889

Greenpeace Luxembourg, 1890
Greenpeace Netherlands, 1891
Greenpeace Norway, 1892
Greenpeace Russia, 1893
Greenpeace Spain, 1894
Greenpeace Sweden, 1895
Greenpeace Switzerland, 1896
Greenpeace UK, 1897
Greenpeace Ukraine, 1898
Greens of Slovenia *see* Zeleni Slovenije, 1846
Groenen, De, 1828
Gronne, De, 1829
Group of Experts on the Scientific Aspects of Marine Pollution (GESAMP), 304, 1588
Grüne Alternative, Die, 1830
Grüne Liga, 1831
Grüne Partei – Ost, 1832
Grüne Partei der Schweiz/Parti Ecologiste Suisse (GPS), 1833
Grupo Ecologista de Agrónomos (GEDEA), 158
Gruppi Ricerca Ecologica (GRE), 2147
Hahn-Meitner Institut Berlin GmbH (HMI), 1074
Hava Kirlenmesi Arastirmalari ve Denetimi Türk Milli Komitesi, 69, 1545
Hazardous Export/Import Prevention Project Greenpeace International, 479
Health and Safety Executive, 1261
Hellenic Association on Environmental Pollution, 1446
Hellenic Nuclear Society, 1174
Hellenic Ornithological Society, 792
Henry Doubleday Research Association (HDRA), 263
Her Majesty's Inspectorate of Pollution (HMIP), 1441
Hessische Gesellschaft für Ornithologie und Naturschutz eV, 793
HM Nuclear Installations Inspectorate Health and Safety Executive, 1154
Hollustuvernd Rikisins, 1340
Hravtsko Drutvo za Zastitu Zraka, 24, 1501
Hungarian Institute for Materials Handling and Packaging, 502, 1774
Hungarian Society of Nature Conservationists, 903
Hydroprojekt, 348
Iceland Forestry Research Station *see* Rannsóknarstöd Skógraektar Rikisins, 781
Icelandic Fisheries Laboratories *see* Rannsóknastofnun Fiskidnadarins, 233
Icelandic Meteorological Office *see* Vedurstofa Islands, 109
ICSU Steering Committeee for Biotechnology (COBIOTECH): Centre of Bioengineering, 1307
IEA Coal Research, 1138
IKU Continental Shelf and Petroleum Technology Research Institute, 1139
Ilmansuojeluyhdistys ry ISY, 25, 1502
Ilmatieteen Laitos, 93

Independent Ecological Centre *see* Független Ökológical Központ (FÖK), 2123
Industrielle de l'Environnement et du Recyclage (FEDEREC), 594
Industrieverband Agrar (IVA), 197
Industrin for Vaxt–Och Traskyddsnedel (IVT), 198
Industry and Environment Programme Activity Centre (IE/PAC), 1680
Industry Council for Electronic Recycling (ICER), 595
Industry Council for Packaging and the Environment (INCPEN), 503
Informat Green Alert, 1726
Information Centre for Air Protection, Polish Ecological Club, 26, 1503
Inspection Générale de l'Environnement et des Forêts, 747
Institut Atlantique, 280, 1572
Institut de l'Environnement du Centre Commun de Recherche, 6, 2199
Institut de Recherche des Transports Centre de Documentation d'Evaluation et de Recherches Nuisances, 70, 1546, 2362
Institut for Energiiteknikk (IFE), 1018, 1232
Institut Français de l'Energie, 977
Institut Français de Recherche pour l'Exploitation de la Mer (IFREMER), 305
Institut für Betriebswirtschaft Bundesforschungsanstalt für Landwirtschaft, 238, 428
Institut für Energie- und Transportforschung Meissen, 1075, 2363
Institut für Immission- Arbeits- und Strahlenschutz, 71, 1377, 1547
Institut für Ökologische Chemie, 2287
Institut für Strahlenhygiene (ISH), 1378
Institut für Umwelt Informatik, 2148
Institut für Umweltwissenschaften und Naturschutz, 947
Institut für Wasser- Boden- und Lufthygiene, 429, 697, 1643
Institut Nacional de Engenharia e Technologia Industrial (INETI), 2200
Institut National de l'Environnement Industriel et des Risques (INERIS), 645
Institut National de la Santé et de la Recherche Medicale, 72, 1427, 1548
Institut Pasteur de Lille, 1428
Institut Royal Météorologique de Belgique, 94
Institute for Ecological Chemistry *see* Institut für Ökologische Chemie, 2287
Institute for Ecological Research *see* Instituut voor Oecologisch Onderzoek, 2150
Institute for Emission Labour and Radiation Protection *see* Institut für Immission- Arbeits- und Strahlenschutz, 71, 1377, 1547
Institute for European Environmental Policy (IEEP), 2059
Institute for Farm Economics Federal Agricultural Research Centre *see*

Institut für Betriebswirtschaft Bundesforschungsanstalt für Landwirtschaft, 238, 428
Institute for Marine Environmental Research, 306
Institute for Radiation Hygiene *see* Institut für Strahlenhygiene (ISH), 1378
Institute for Sustainable Development, 952
Institute for Water Soil and Air Hygiene *see* Institut für Wasser- Boden- und Lufthygiene, 429, 697, 1643
Institute of Atmospheric Pollution *see* Istituto Sull'Inquinamento Atmosferico, 73, 1549
Institute of Environmental Sciences and Conservation *see* Institut für Umweltwissenschaften und Naturschutz, 947
Institute of Handling Transport Packaging and Storage Systems (IMADOS), 504
Institute of Materials Management (IMM), 1753
Institute of Meteorology and Physics of the Atmospheric Environment, 7, 1076
Institute of Sound and Vibration Research, 1608
Institute of Terrestrial Ecology (ITE), 859, 884, 2149
Institute of Wastes Management (IWM), 646
Institution of Engineers in Ireland, 1791
Institution of Mechanical Engineers, 1792
Institution of Water and Environmental Management (IWEM), 349, 2007
Institutl Roman de Cercetari Marine, 307, 1589
Instituto de Plásticos y Caucho, 585
Instituto de Química Orgánica General, 2288
Institutul de Cercetäri pentru Pedologie si Agrochimie, 385, 698
Institutul de Cercetäri si Projectäri pentru Gospodärirea Apelor, 384
Instituut voor Natuurbeschermingseducatie (IVN), 1775
Instituut voor Oecologisch Onderzoek, 2150
Instituut voor Plantenziektenkundig Onderzoek (IPO), 216
Instytut Celulozowo-Papierniczy, 558
Instytut Ksztaltowania Srodowiska, 742, 2034
Instytut Meteorologii i Gospodarki Wodnej, 308, 386
Instytut Podstaw Inzynierii Srodowiska PAN, 885, 1482, 1804
Inter Environnement Bond Beter Leefmilieu Centre d'Inform et de Documentation sur l'Environnement, 1447
Intergovernmental Panel for Climatic Change (IPCC), 86
International Agency for Research on Cancer (IARC), 1291
International Association Against Noise *see* Association Internationale Contre le Bruit (AICB), 1596

International Association for Ecology (INTECOL), 2124
International Association of Packaging Research Institutes (IAPRI), 523, 2201
International Association of the Soap and Detergent Industry (AIS), 2277
International Association on Water Pollution Research and Control (IAWPRC), 410, 1622
International Atomic Energy Agency see Agence Internationale de l'Energie Atomique (IAEA), 1146
International Centre for the Application of Pesticides (ICAP), 217
International Chamber of Commerce (ICC), 1727
International Cleaner Production Information Clearing House (ICPIC), 629
International Commission for Protection Against Environmental Mutagens and Carcinogens, 1389
International Commission on Radiological Protection (ICRP), 1362
International Conservation Action Analysis (ICAN), 703
International Council for Bird Preservation (ICBP), 794
International Council for the Exploration of the Seas (ICES): Conseil International pour l'Exploration de la Mer, 281
International Council of Environmental Law (ICEL): Environmental Law Center, 1939
International Council of Scientific Unions (ICSU), 1308
International Council on Monuments and Sites (ICOMOS UK), 865
International Dolphin Watch (IDW), 795
International Eco Meat Control see Internationale Scharrelvees Controle (ISC), 160, 1346, 1940
International Energy Agency (IEA), 978
International Environment Bureau, 2174
International Environment Bureau (IEB), 1728
International Environmental Service Centre (IESC), 2175
International Federation of Agricultural Producers (IFAP), 159
International Federation of Associations for the Protection of Europe's Cultural and National Heritage Europa Nostra, 866
International Federation of Recovery and Recycling see Bureau International de la Récupération (BIR), 589
International Fertiliser Industry Association (IFA), 199
International Group of National Associations of Manufacturers of Agro-Chemical Products (GIFAP), 200
International Industrial Association for Energy from Nuclear Fuel: Uranium Institute (UI), 1175
International Institute for Applied Systems Analysis (IIASA), 8, 1483, 2151
International Institute for Environment and Development (IIED), 1763

International Institute for Environment and Safety see Wissenschaftszentrum Berlin (WZB), 2064
International Institute for Environment and Society, 2261
International Institute for Water: Centre International de l'Eau, 350
International Laboratory of Marine Radioactivity, 309, 493, 1590
International Lead and Zinc Study Group, 1682
International Maritime Organisation (IMO), 282
International Nuclear Power and Waste (INLA), 478, 1176
International Oil Pollution Compensation Fund (IOPC), 1116, 1573
International Organisation for Human Ecology, 2125
International Organisation for Standardisation (ISO), 2165
International Organisation of Consumer Unions (IOCU), 647, 1855
International Petroleum Industry Environmental Conservation Association (IPIECA), 720, 1117
International Physicians for the Prevention of Nuclear War (IPPNW), 1190, 1899
International Professional Association for Environment Affairs (IPRE), 1856
International Programme on Chemical Safety (IPCS), 1390
International Radiation Protection Association (IRPA), 1363
International Society for Environmental Toxicology and Cancer, 1391
International Solar Energy Society (ISES), 1037
International Solid Wastes and Public Cleansing Association (ISWA), 648
International Tanker Owners Pollution Federation Ltd (ITOPF), 1118, 1574
International Tin Research Institute, 1703
International Tropical Timber Organisation (ITTO), 754
International Union for Conservation of Nature and Natural Resources (IUCN): World Conservation Union, 904
International Union of Air Pollution Prevention Associations (IUAPPA), 27, 1504
International Water and Sanitation Centre (IRC), 411, 1623
International Water Supply Association, 351
International Water Tribunal Foundation (IWT), 352
International Waterfowl and Wetlands Research Bureau (IWRB), 860
International Whaling Commision (IWC), 796
International Wool Secretariat (IWS), 630, 1683
International Youth Federation for Environmental Studies and Conservation (IYF), 1779
Internationale Scharrelvees Controle (ISC), 160, 1346, 1940

Interprofessional Technical Centre for Studies in Atmospheric Pollution see Centre Interprofessionnel Technique d'Etudes de la Pollution Atmosphérique (CITEPA), 1541
Inveresk Research International Ltd, 1429
Irish Coastal Environment Group, 265
Irish Environmental Conservation: Organisation for Youth (ECO), 704
Irish Forestry Board, 777
Irish Industry Confederation – Environmental Policy Committee, 1684
Irish Organic Farmers and Growers Association, 250
Irish Science and Technology Agency (EOLAS), 2152
Irish Sea Fisheries Board, 232, 310
Irish Wildbird Conservancy (IWC), 797
Irish Wildlife Federation (IWF), 798
Irish Women's Environmental Network, 1900
Istituto del Terziario per l'Ambiente, 649
Istituto di Ricerca sulle Acque (IRSA), 387, 1644
Istituto di Ricerche e Sperimentazione Laniera 'O. Rivetti, 638
Istituto per l'Edilizia ed il Risparmio Energetico, 1099
Istituto per le Piante da Legno e l'Ambiente SpA, 778
Istituto Sperimentale per lo Studio e la Difesa del Suolo, 1563
Istituto Sull'Inquinamento Atmosferico, 73, 1549
Italian Animal Rights Association see Lega Italiana dei Diritti dell'Animale (LIDA), 800
Italian Bird Protection Society see Lega Italiana Protezione Uccelli, 801
Italian Commission for Nuclear and Alternative Energy Sources see Comitato Nazionale per la Ricerca e per lo Sviluppo dell'Energia Nucleare e delle Energie Alternativ, 1071
Italian Federation for the Protection of Nature see Federazione Italiana Pro Natura, 902
Italian Radiation Protection Association see Associazione Italiana di Protezione Contro le Radiazioni (AIRP), 1361
Izmir Yesiller Partisi, 1834
Izquierda de los Pueblos, 1835
Joint Research Centre Ispra Establishment, 494, 1259
Jordens Vanner, 1901
Kansanterveyslaitos (NPHI), 1327
Kemian Keskusliitto, 1309
Kentron Yginis Keasfalias tis Ergassias (KYAE), 1292
Kernforschungszentrum Karlsruhe GmbH (KfK), 2203
Kerntechnische Ausschuss (KTA), 1251
Kommission Reinhaltung der Luft (KRdL) in VDI und DIN, 28, 1505
Koninklijk Nederlands Meteorologisch Instituut, 107
Koninklijke Maatschappij voor Dierkunde van Antwerpen, 799

346 Organisations – Alphabetical Index

Kontrollforeningen for Alternativ Odling (KRAV), 251
Kornyeyetvedelme Ens Teruletfejlesgtesi Minisyterium, 2215
Laboratório Nacional de Engenharia e Tecnologia Industrial (LNETI), 1379
Landbrugsministeriet, 139
Landelijke Vereniging tot Behoud van de Waddenzee (LVBW), 283
Landesanstalt für Immissionsschutz Nordrhein-Westfalen, 74, 1550
Latvian Environment Protection Club (VAK), 725
League for the Environment *see* Lega per l'Ambiente, 439 *see* Lega per l'Ambiente, 721
Lega Italiana dei Diritti dell'Animale (LIDA), 800
Lega Italiana Protezione Uccelli, 801
Lega per l'Ambiente, 439, 721
Liga Ochrony Przyrody, 29, 905
Liga Para la Defensa del Patrimonio Natural (DEPANA), 867
Ligue Luxembourgeoise pour la Protection de la Nature et des Oiseaux, 802
Linking Environment and Farming (LEAF), 161, 1347
Lithuania Ministry of Energy, 1155
Lithuanian Green Party, 1836
Luxembourg Ecological Movement *see* Mouvement Ecologique Luxembourg (MECO), 1902
Lydteknisk Institut (LI), 1610
Management of Natural Resources and Environment Study Centre *see* Centre d'Enseignement et de Recherche pour la Gestion des Resources Naturelles et l'Environnement, 382, 2018
Marine Conservation Society (MCS), 284: Let Coral Reefs Live, 803
Marine Research Institute *see* Fiskeridirektoratets Havforskningsinstitutt, 230
Marine Research Institute of Romania *see* Institutl Roman de Cercetari Marine, 307, 1589
MDP/CDE–Green Party, 1837
Media Natura (MN), 1857
Medical Institute for Environmental Hygiene *see* Medizinisches Institut für Umwelthygiene, 75, 1341, 1551
Medical Research Council, Toxicology Unit, 1381
Medizinisches Institut für Umwelthygiene, 75, 1341, 1551
Merentutkimuslaitos, 311
Meteorological Office, The (The 'Met Office'), 87
Meteorological Service, 95
Meteorology and Water Management Institute *see* Instytut Meteorologii i Gospodarki Wodnej, 308, 386
Metsäntutkimuslaitos, 779
Miljoministeriet Miljostyrelsen, 1981
Miljoministeriet Miljostyrelsen Kemikaliekontrollen, 185, 1382, 1928
Miljopartiet de Gröna (MP), 1838
Miljopartiet de Gronne, 1839

Miljoverndepartementet, 1982
Minelsbumprom, 526
Mineralöl Wirtschafts Verband eV (MWV), 1119
Minister of Small and Medium Sized Enterprises and Agriculture, 140
Ministère de l'Environnement: Direction de la Prévention des Pollutions de l'Eau, 337, 1619, 1654
Ministère de l'Environnement et des Eaux et Forêts, 338, 748, 1983
Ministère de l'Environnement et du Cadre de Vie, 1984
Ministère de la Santé Publique, Chef du Service des Nuisances, 1655
Ministère des Affaires Economiques Administration de l'Energie Service des Applications Nucléaires, 1159
Ministerie van de Vlaamse Gemeenschap Administrie voor Ruimtelijke Dienst Groen Waters en Bossen, 395
Ministerio de Agricultura Pesca y Alimentación, 141
Ministerio de Ambiente e Recursos Naturais, 1985
Ministerio de Industria Comercio y Turismo, 1106, 1929
Ministerio de Obras Públicas y Transportes, 2309
Ministerio de Obras Públicas y Urbanismo – Dirección General de Medio Ambiente (DGMA), 1986
Ministerium für Soziale Gesundheit und Energie des Landes Schleswig-Holstein, 970, 2045
Ministero dei Lavori Pubblici, 396
Ministero per i Beni Culturali e Ambientali, 1987
Ministerstwo Ochrony Srodowiska i Zasobow Naturalriych, 1988
Ministry for Environment and Water Management, 339, 1989
Ministry for Health Sports and Consumer Protection, 1656
Ministry for Social Affairs and Health, 1262
Ministry for the Environment Nature Conservation and Nuclear Reactors Safety, 896, 1244
Ministry of Agriculture (Denmark) *see* Landbrugsministeriet, 139
Ministry of Agriculture (Finland), 142
Ministry of Agriculture (Iceland), 143
Ministry of Agriculture (Italy), 144
Ministry of Agriculture (Sweden), 145
Ministry of Agriculture and Fisheries, 146
Ministry of Agriculture and Food Industry, 147
Ministry of Agriculture and Forestry, 148, 749
Ministry of Agriculture and Viticulture, 149
Ministry of Agriculture Fisheries and Food (MAFF), 150
  Conservation Policy and Environmental Protection Division, 151, 712, 2046
  Fisheries Laboratory: Fisheries Radiological Laboratory, 152, 1383

Marine Environmental Protection Division, 273, 1342
Ministry of Agriculture, Fisheries and Food *see* Ministerio de Agricultura Pesca y Alimentación, 141
Ministry of Agriculture, Nature Conservation and Fisheries, 153, 897
Ministry of Atomic Energy and Industry, 1156
Ministry of Atomic Energy of the Russian Federation (MINATOM), 1157
Ministry of Chernobyl, 1158
Ministry of Communications and Transport, 2310
Ministry of Economic Affairs: Directorate General for Energy, 971, 1657
Ministry of Energy *see* Energiministeriet, 969
Ministry of Energy and Natural Resources, 972, 1990
Ministry of Environment *see* Miljoverndepartementet, 1982
Ministry of Environment and Natural Resources, 1658
Ministry of Environment and Territorial Administration, 397
Ministry of Equipment Housing and Transport, 2311
Ministry of Food Agriculture and Forestry, 154, 750
Ministry of Health, 1263, 1264, 1265
Ministry of Health and Consumer Affairs, 1266
Ministry of Health and Humanitarian Policies, 1267, 2047
Ministry of Health and Social Security, 1268
Ministry of Health and Social Welfare, 1269
Ministry of Health Welfare and Social Security, 1270
Ministry of Housing Physical Planning and Environment, 398, 1659: Central Dep for Information and Interior Relations, 1991
Ministry of Industry, 1107
Ministry of Industry, Energy and Technology and Commerce, 973
Ministry of Public Health, 1271
Ministry of Public Works Town Planning and Transport *see* Ministerio de Obras Públicas y Transportes, 2309
Ministry of the Environment (Norway) *see* Miljoverndepartement, 1982
Ministry of the Environment (Sweden), 1659
Ministry of the Environment *see* Miljoministeriet Miljostyrelsen, 1981
Ministry of the Environment (Belgium), 1992
Ministry of the Environment (Bulgaria), 1993
Ministry of the Environment (Denmark), 1994
Ministry of the Environment (Finland), 1995
Ministry of the Environment , (Iceland), 1996

Ministry of the Environment (Italy), 1997
Ministry of the Environment (Romania), 1998
Ministry of the Environment and Natural Resources, 1999
Ministry of the Environment Housing and Public Works, 2000
Ministry of the Environment National Agency of Environmental Protection: Water Resources Division, 399
Ministry of the Environment of the Slovak Republic, 2001
Ministry of the Environment Town Planning and Public Works, 400
Ministry of the Environment Youth and Family *see* Bundesministerium für Umwelt Jugend and Familie, 1977
Ministry of Timber, Pulp and Paper and Wood Processing *see* Minelsbumprom, 526
Ministry of Transport, 2312, 2313
Ministry of Transport and Communications, 2314, 2315
Ministry of Transport and Energy, 974, 2316
Ministry of Transport and Public Works, 2317 *see* Trafikministeriet, 2322
Ministry of Transport and Shipping, 2318
Ministry of Transport and Telecommunications, 2319
Ministry of Transport Communications and Public Works, 2320
Ministry of Welfare Health and Cultural Affairs, 1272
Monitoring and Assessment Research Centre (MARC), 2035
Mouvement Ecologique Luxembourg (MECO), 1902
National Agency for Environmental Protection, 713
National Association for Environmental Education (NAEE), 1780
National Association of Waste Disposal Contractors (NAWDC), 650
National Atomic Energy Agency, 1160
National Atomic Energy Commission *see* Orszagos Atomenergia Bizottsag, 1182
National Centre for Hygiene Food Control and Environmental Protection *see* Hollustuvernd Rikisins, 1340
National Commission for the Control of Nuclear Activities, 1161
National Council for Environmental Protection, 714
National Energy Authority *see* Orkustonfnun, 1019
National Environmental Policy Plan Plus (NEPP): Ministry of Housing Planning and the Environment, 2048
National Federation for the Defence of the Environment, 702
National Forest and Nature Agency, 755
National Institute for Industrial Environment and Hazards *see* Institut National de l'Environnement Industriel et des Risques (INERIS), 645
National Institute of Occupational Health, 1273 *see* Arbetsmiljöinstitutet, 1289

National Institute of Public Health and Environmental Protection *see* Rijksinstituut voor Volksgezundheid en Milieuhygiene (RIVM), 1274
National Laboratory for Engineering and Industrial Technology *see* Laboratório Nacional de Engenharia e Tecnologia Industrial (LNETI), 1379
National Public Health Institute *see* Kansanterveyslaitos (NPHI), 1327
National Radiological Protection Board (NRPB), 1364
National Rivers Authority (NRA), 328, 1620
National Society for Clean Air and Environmental Protection (NSCA), 30, 1506
National Trust (NT), 868
National Water Authority Department for International Relations, 340
National Westminster Bank, 1941
Natur og Ungdom, 1781
Natural Environment Research Council (NERC), 2153
Natural Resources Research (NRR), 1709
Nature and Youth *see* Natur og Ungdom, 1781
Nature Conservation Council, 898
Nature et Progrès, 252
Natuurmonumenten, 906
Nederlands Atoomforum, 1177
Nederlands Verpakkingscentrum, 505
Nederlandse Industriële en Agarische Biotechnologie Associatie (NIABA), 1310
Nederlandse Stichting Geluidshinder, 1599
Nederlandse Stichting voor Fytofarmacie (NEFYTO), 201
Netherlands Energy Research Foundation ECN, 1140, 1233
Netherlands Institute for Fishery Investigations *see* Rijksinstituut voor Visserijonderzoek (RIVO), 234
Netherlands Packaging Industry *see* Nederlands Verpakkingscentrum, 505
Netherlands Society for Nature and the Environment *see* Stichting Natuur en Milieu (SNM), 911, 1942
Network for Environmental Technology Transfer (NETT), 2176
New Energy Technologies (THERMIE), 1038
NIPH Department of Toxicology, 430, 1430
NOAH International, 1903
Noise Abatement Society *see* Nederlandse Stichting Geluidshinder, 1599
Nordic Committee for Nuclear Safety Research (NKS), 1245
Nordic Council, 315
Nordic Council of Ministers, 2049
Norges Jeger og Fiskerforbund, 804
Norges Naturvernforbund (NNV), 907, 1904
Norsk Atomforum, 1178
Norsk Institut for Luftforskning (NILU), 76, 1552

Norsk Institut for Skogforskning (NISK), 780
Norsk Institut for Vannforskning (NIVA), 431
Norsk Ornitologisk Forening, 805
Norsk Polar Institut, 887
Norske Emballasjeforening, Den, 506
Norske Videnskaps-Akademi, Det, 2154
North Atlantic Salmon Conservation Organisation (NASCO), 285, 806
North Rhine Westphalia State Centre of Air Quality, Noise and Vibration Control *see* Landesanstalt für Immissionsschutz Nordrhein-Westfalen, 74, 1550
North Sea Foundation *see* Werkgroep Noordzee, 291
Norwegain Forest Research Institute *see* Norsk Institut for Skogforskning (NISK), 780
Norwegian Association of Anglers and Hunters *see* Norges Jeger og Fiskerforbund, 804
Norwegian Clean Air Association *see* Ren Luft Foreningen (RLF), 33, 1508
Norwegian Clean Air Campaign, 31, 1556
Norwegian Foundation for Environmental Labelling *see* Stiftelsen Miljomerking i Norge, 1664
Norwegian Institute for Air Research *see* Norsk Institut for Luftforskning (NILU), 76, 1552
Norwegian Institute for Water Research *see* Norsk Institut for Vannforskning (NIVA), 431
Norwegian Institute of Technology, 353, 1624
Norwegian Mapping Authority, 874
Norwegian Meteorological Institute: Meteorological Synthesising Centre–West of EME, 108
Norwegian Ornithological Society *see* Norsk Ornitologisk Forening, 805
Norwegian Packaging Association *see* Norske Emballasjeforening, Den, 506
Norwegian Pulp and Paper Association *see* Papirindustriens Sentralforbund, 540
Norwegian Pulp and Paper Research Institute *see* Papirindustriens Forskningsinstitutt, 562
NOVEM, 979
Nuclear Electric plc, 1179
Nuclear Energy Board, 1180
Nuclear Free Local Authorities National Steering Committee, 1162, 2050
Nuclear Power and Plant Research Institute (VUJE), 1234
Nuclear Safety Standards Commission *see* Kerntechnische Ausschuss (KTA), 1251
Nucleo Portugues de Estudio e Protecção da Vida Selvagem, 807
NV Vereenigde Glasfabrieken, 451
O. Rivetti Wool Industry Research and Experimental Institute *see* Istituto di Ricerche e Sperimentazione Laniera 'O. Rivetti, 638
OECD Environment Committee, 2051
OECD Nuclear Energy Agency (NEA), 1181

348 Organisations – Alphabetical Index

Office for the Protection of the Environment, 715
Office National d'Etudes Recherches Aerospatiales, 1618
Office National de la Chasse (ONC), 808
Office of Public Works, 2002
Oil Companies European Organisation for Environmental Health and Protection (CONCAWE), 1120, 1328, 1448
Open Air Council *see* Foundation for Environmental Education in Europe (FEEE), 1773
Organbio, 1311
Organic Advisory Service, 264
Organic Food and Farming Centre, 253
Organic Reclamation and Composting Association (ORCA), 254
Organisation Météorologique Mondiale, 96
Orkustonfnun, 1019
Ornithologische Arbeitsgemeinschaft für Schleswig-Holstein und Hamburg, 809
Orszagos Atomenergia Bizottsag, 1182
Oslo Commission (OSCOM), 286, 1575
Oslo Yann Og Avlop, 354, 651
Österreichische Kerntechnische Gesellschaft, 1163
Österreichischer Arbeitsring für Lärmbekämpfung (OAL), 1600
Österreichisches Atomforum, 1183
Österreichisches Forschungszentrum Seibersdorf GmbH, 1235
Österreichisches Gesellschaft für Natur und Umweltschutz (ÖGNU), 97, 908
Österreichisches Vereinigung für das Gas und Wasserfach, 355, 1121
Österreichisches Verpackungszentrum, 507
Otter Trust, 810
Oy Keskuslaboratorio – Centrallaboratorium Ab, 559
Ozono Elettronica Internazionale, 631
Packaging and Industrial Film Association (PIFA), 508
Packaging Industry Research Association (Pira), 560
Paint Research Association, 1484
Panellinio Kentro Oikologikon Erevnon (PAKOE), 722
Panhellenic Centre of Environmental Studies *see* Panellinio Kentro Oikologikon Erevnon (PAKOE), 722
Panos Institute, 953, 954
Paper Technology Foundation for Paper Production and Conversion *see* Papiertechnische Stiftung für Papierezeugung und Papierverarbeitung, 561
Papiertechnische Stiftung für Papierezeugung und Papierverarbeitung, 561
Papir–es Nyomdaipari Muszaki Egyesulet, 539
Papirindustriens Forskningsinstitutt, 562
Papirindustriens Sentralforbund, 540
Pharmaceutical Group of the EC (GPCE), 2278
Photographic and Waste Management Association, 652

Physikalisch Technische Versuchsanstalt für Wärme und Schalltechnik, 1612
PHYTOPHAR, 202
Plant Protection Research Institute *see* Instituut voor Plantenziektenkundig Onderzoek (IPO), 216
Plastics and Rubber Institute *see* Instituto de Plásticos y Caucho, 585
Polish Ecological Club/Information Centre for Air Protection, 32, 1507
Polish Foundation for Energy Efficiency *see* Fundacja na Rzecz Efelatywnego Wykorzystanos Energii, 1098
Polish Party of Greens *see* Polska Partia Zielonych (PPZ), 1840
Polska Partia Zielonych (PPZ), 1840
Polski Klub Ekologiczny, 1905
Polskie Zresenie Inazynierowi Technikow Sanitarnych (PZITS), 356
Portuguese Core Group for Wildlife Studies and Protection *see* Nucleo Portugues de Estudio e Protecção da Vida Selvagem, 807
Principal Water Authority Corporation of Dublin, 401
Programme on Man and the Biosphere (MAB), 2126
Promysi Papiru a Celulozy Generalni Reditelstvi, 541
Pulp and Paper Information Centre, 542
Pulp and Paper Research Institute *see* Instytut Celulozowo-Papierniczy, 558
Radiation and Environmental Research Centre *see* Gesellschaft für Strahlen und Umweltforschung GmbH (GSF), 1376
Radiological Protection, Institute of Ireland, 1365
Ramsar Convention Bureau, 909
Rannsóknarstöd Skógraektar Rikisins, 781
Rannsóknastofnun Fiskidnadarins, 233
RAPRA Technology, 2177
RECOUP, 571
Regional Environment Centre for Central and Eastern Europe, 955
Ren Luft Foreningen (RLF), 33, 1508
Renewable Energy Advisory Group (REAG), 1028
Renewable Energy Enquiries Bureau, 1039
Research and Design Institute for Water Resources Engineering *see* Institutl de Cercetäri si Projectäri pentru Gospodärirea Apelor, 384
Research Institute for International Technology and Economic Cooperation of Aachen University *see* Forschungsinstitut für Internationale Technische und Wirtschaftliche Zusammenarbeit der RWTH Aachen, 2198
Research Institute for the Study and Conservation of the Soil *see* Istituto Sperimentale per lo Studio e la Difesa del Suolo, 1563
Rijksinstituut voor Visserijonderzoek (RIVO), 234
Rijksinstituut voor Volksgezondheid en Milieuhygiene (RIVM), 1274

Romanian Society of Hygiene and Public Health *see* Societatea Româna de Igiena si Sanatate Publica, 1278
Royal Commission on Environmental Pollution (RCEP), 1442
Royal Ministry of Health and Social Affairs, 1275
Royal Ministry of Petroleum and Energy, 1108
Royal Ministry of Transport and Communications, 2321
Royal Netherlands Meteorological Institute *see* Koninklijke Nederlands Meteorologisch Instituut, 107
Royal Society for the Protection of Birds (RSPB), 811
RSNC Wildlife Trusts Partnership: Formerly Royal Society for Nature Conservation, 812
Sateilyturvakeskus (STUK), 1252, 1366
Save-a-Can, 470
Schutzgemeinschaft Deutsche Nordseeküste, 287
Schweizer Arbeitsgemeinschaft für Umweltforschung (SAGUF), 2155
Schweizerische Gesellschaft für Biologischen Landbau, 255
Schweizerische Meteorologische Anstalt, 1079
Schweizerische Vereinigung für Atomenergie (SVA), 1184
Scientific Committee on Problems of the Environment (SCOPE), 2127
Scientists for Global Responsibility (SGR), 1858
Scottish Campaign to Resist the Atomic Menace (SCRAM), 1906
Scottish Marine Biological Association, 312
Sea Mammal Research Unit (SMRU), 813
Sea Turtle Protection Society, 814
Secrétaire d'Etat auprès du Premier Ministre de la Prevention des Risques Technologiques, 402
Secretaria de Estado para las Politicas del Agua y del Medio Ambiente, 15, 403
Secretaria Estado Ambiente Hidrologicos, 357, 653
Secretariat for the Protection of the Mediterranean Sea, 288, 1576
Secretary of State for Public Health, 1276
Senior Advisory Group Biotechnology (SAGB): CEFIC, 1312
Service Central de Protection Contre les Rayonnements Ionisants, 476
Skakavac, 1859
Slovak Union of Nature and Landscape Protectors (SZOPK), 929, 1907
Slovakian Society for Conservation of Nature (SZOPK), 910
SLVY-liitto, 256
Socialistik Folkeparti, 1841
Sociedad Española de Ornitologia (SEO), 815
Sociedad Española de Protección Radiológica (SEPR), 1358
Sociedad Nuclear Española (SNE), 1185

Organisations – Alphabetical Index

Societatea Româna de Igiena si Sanatate Publica, 1278
Société Suisse de l'Industrie de Gaz et des Eaux (SSIGE), 358, 1122
Société Suisse Industries Chimiques (SSIC), 203
Society for Clean Air in the Netherlands (CLAN) see Vereniging Lucht, 1, 1509
Society for Environmental Toxicology and Chemistry (SETAC), 440, 1392, 1710
Society for Prevention of Asbestosis and Industrial Diseases (SPAID), 1297
Society for the Promotion of Life Cycle Development (SPOLD), 1711
Society of Dyers and Colourists, 632
Soil Association, The, 1348
Soil Science and Agrochemistry Research Institute see Institutul de Cercetäri pentru Pedologie si Agrochimie, 385, 698
Soil Survey and Land Research Centre (SSLRC), 388, 888, 1564
Solar Energy Society of Ireland, 1040
SOLEREC Département Solaire, 1041
SOS Sea Turtles, 816
Spanish Paper Industry Research Association see Asociación de Investigación Técnica de la Industria Papelera Española, 557
Spanish Plastics Foundation for the Protection of the Environment (SPFPE), 572
Spanish Society for Radiological Protection see Sociedad Española de Protección Radiológica (SEPR), 1358
Specific Action for Vigorous Energy Efficiency (SAVE), 1100
Standing Committee on Urban and Building Climatology, 98
State Committee for Chernobyl Accident Problems, 1246
State Committee for Environmental Protection, 716
State Committee for Nuclear and Radiation Safety, 1247, 1359
State Committee for the Supervision of Nuclear and Radiation Safety under the President of Russia, 1248, 1360
State Pollution Control Authority (SFT): Oil Pollution Control Department, 1109, 1567
Statens Forurensningstilsyn (SFT), 1443
Statens Planteavis Laboratorium, 1565
Station Fédérale de Recherches Agronomiques de Zurich Reckenholz, 218
Statistik Sentralbyra (SSB), 2216
Stichting Natuur en Milieu (SNM), 911, 1942
Stiftelsen Miljomerking i Norge, 1664
Stockholm International Peace and Research Institute (SIPRI), 1236
Stowarzyszenie Inzynierow i Technikow Rezemyslu Papierniczego w Polsce, 543
Stredisko pro Efektivni Vyuzivani Energie (SEVE), 980
Studiecentrum voor Ecologie en Bosbouw, 1485

Studiecentrum voor Kernenergie – Centre d'Etude de l'Energie Nucléaire (CEN-SCK), 1237
Studiengesellschaft für die Entsorgung von Altahrrzeugen GmbH (EVA), 471
Studies en Consultancy voor Milieu en Omgeving – TNO, 1486
Suomen Luonnonsuojeluitto ry, 912
Svensk Polska Miljoforeningen (SPM): Szwedsko Polskie Towarzystwo Ochrony Srodowiska, 1860
Svenska Solenergiföreningen, 1042
Swedish Atomic Forum (SAFO), 1186
Swedish Environmental Research Institute (IVL), 2156
Swedish NGO Secretariat on Acid Rain, 317
Swedish Nuclear Society see Foreningen Karnteknik, 1169
Swedish Packaging Research Institute – PACKFORSK, 524
Swedish Pulp and Paper Association (SCPF), 544
Swedish Solar Energy Society see Svenska Solenergiföreningen, 1042
Swedish Standards Institution Environmental Labelling Programme, 1665
Swedish Transport Research Board see Transportforskningsberedningen (TFB), 2323
Swiss Association for Atomic Energy see Schweizerische Vereinigung für Atomenergie (SVA), 1184
Swiss Association for Environmental Research see Schweizer Arbeitsgemeinschaft für Umweltforschung (SAGUF), 2155
Swiss Association for Health Techniques, 1329
Swiss Centre for Appropriate Technology (SKAT), 2178
Swiss Federal Commission of Radioactivity Surveillance see Eidgenössische Kommission zur Überwachung der Radioaktivität, 1375, 2033
Swiss Federal Institute for Water Resources and Water Pollution Control see Eidgnössische Anstalt für Wasserversorgung Abwasserreinigung und Gewässerschutz, 383, 1641
Swiss Federal Research Station for Agronomy Zurich see Station Fédérale de Recherches Agronomiques de Zurich Reckenholz, 218
Swiss League for the Protection of National Heritage, 870
Swiss Meteorological Institute see Schweizerische Meteorologische Anstalt, 1079
Swiss Nuclear Society, 1187
Swiss Society of Chemical Industries see Société Suisse Industries Chimiques (SSIC), 203
Syndicat National des Engrais (SNIE), 204
Tampere University of Technology, 563

Technical Association of the Paper and Printing Industry see Papir–es Nyomdaipari Muszaki Egyesulet, 539
Technical Association of the Polish Paper Industry see Stowarzyszenie Inzynierow i Technikow Rezemyslu Papierniczego w Polsce, 543
Technical University of Twente, 2204, 2364
Technological Institute – Royal Flemish Society of Engineers, 1320
Territorial and Environmental Informatics Interdepartmental Research Centre, 2205
Testing Institute for Heat and Sound Technology see Physikalisch Technische Versuchsanstalt für Wärme und Schalltechnik, 1612
Textile Institute, The, 633
The European Glass Container Federation see Fédération Européenne du Verre d'Emballage (FEVE), 450
The Green Party see Comhaontas Glas, 1809
The Nature Protection League see Liga Ochrony Przyrody, 29, 905
The Norwegian Science Academy see Norske Videnskaps-Akademi, Det, 2154
The Norwegian Society for the Conservation of Nature see Norges Naturvernforbund (NNV), 907
The Norwegian State Pollution Control Authority see Statens Forurensningstilsyn (SFT), 1443
The Society for the Preservation of Nature in the Netherlands see Natuurmonumenten, 906
The State Chemical Supervision Service see Miljoministeriet Miljostyrelsen Kemikaliekontrollen, 185, 1382, 1928
The Swedish Polish Association for Environment Protection see Svensk Polska Miljoforeningen (SPM), 1860
Tidy Britain Group, The (TBG), 527
Timber and Environment Institute see Istituto per le Piante da Legno e l'Ambiente SpA, 778
TISIT, 205
TNO Study and Information Centre for Environment see Studies en Consultancy voor Milieu en Omgeving – TNO, 1486
Toegepast Natuurwetenschappelijk Onderzoek (TNO), 1020, 2206
Trafikministeriet, 2322
Transport 2000, 2067, 2331
Transport and Road Research Laboratory (TRRL), 2365
Transportforskningsberedningen (TFB), 2323
Turkish Atomic Energy Authority see Türkiye Atom Enerjisi Kurumu, 1188
Turkish Ministry of Energy and Natural Resources, 975
Turkish National Committee for Air Pollution Research and Control (TUNCAP) see Hava Kirlenmesi

Arastirmalari ve Denetimi Türk Milli Komitesi, 69, 1545
Turkish Nuclear Energy Forum *see* Türkiye Nukleer Enerji Kurumu, 1189
Türkiye Atom Enerjisi Kurumu, 1188
Türkiye Nukleer Enerji Kurumu, 1189
Umweltbibliotek, 2271
Umweltbundesamt, 723, 2179
UNEP'S International Register of Potentially Toxic Chemicals (IRPTC), 1393
UNEP/IEO Cleaner Production Working Group on Halogenated Solvents: Department of Environmental Technology, 654
Union of Industrial and Employers' Confederation of Europe (UNICE), 1729
Union of Industries of the Protection of Plants (UIPP), 206
Union of the Societies for the Protection of the Environment in Slovenia, 724
United Nations Environment Programme (UNEP), 1861
Vandkvalitetsinstituttet (VKI), 432, 699
Vasco da Gama Aquarium *see* Aquário Vasco da Gama, 225, 301
Vedurstofa Islands, 109
Verband der Schweizerischen Sellstoff-, Papier- und Kartonindustrie, 545
Verband Deutscher Papierfabriken (VDP), 546
Verband Deutscher Windelservice Lücke GmbH, 547
Verband Organisch–Biologisch Wirtschaftender Bauern Österreichs, 257
Verdi, 1842
Verdi Arcobaleno, 1843
Vereniging der Lucht, 1, 1509
Vereniging Milieudefensie, 1908
Vereniging Voor Biologische–Dynamische Landbouw, 258
Vereinigung Österreichischer Papierindustrierhertsteller, 548
Verts, 1844
Vesien ja ympäristöntutkimuslaitos, 389, 700, 1645
Vihreä Litto, 1845
Warren Spring Laboratory, 2207
Waste Management Information Bureau (WMIB), 655
Waste Processing Association, 1043
Waste Watch, 596
Waste Watchers Deutschland eV, 656
Wastes Technical Policy Unit, 453
Water and Environment Research Institute *see* Vesien ja ympäristöntutkimuslaitos, 389, 700, 1645
Water Directorate Marine Branch Department of the Environment, 274
Water Pollution Dept of the National Institute of Applied Chemical Research *see* Département Pollution des Eaux de l'Institut National de Recherche Chimique Appliquée, 1639
Water Quality Institute *see* Vandkvalitetsinstituttet (VKI), 432, 699
Water Research Centre (WRC), 390
Water Research Institute *see* Istituto di Ricerca sulle Acque (IRSA), 387, 1644
Water Resources Division National Technical University of Athens, 412
Water Supply and Sewage Committee of Major Athens (EYDAP), 359, 686
WATT Committee on Energy, 981
WATT Committee's Working Group on Methane, 454
Wereld Natuur Fonds, 289
Werkgroep Noordzee, 291
Wissenschaftszentrum Berlin (WZB), 2064
Women's Environmental Network (WEN), 1909, 1910
Work Group for Water, 404
World Action for Recycling Materials and Energy from Waste: WARMER, 597
World Conservation Monitoring Centre (WCMC), 861, 2037
World Data Centre for Greenhouse Gases, 113
World Fuel Cell Council, 1044
World Health Organisation Division of Environmental Health (WHO/EHE), 1330
World Meteorological Organisation (WMO) *see* Organisation Météorologique Mondiale, 88, 96
World Ozone and Ultra-Violet Data Centre: Atmospheric Environment Service, 99
World Packaging Organisation (WPO), 509
World Radiation Data Centre, 1367
World Resources Institute (WRI), 2065
World Wide Fund for Nature (WWF), 928
World Wide Fund for Nature – Austria, 913
World Wide Fund for Nature – Belgium, 914
World Wide Fund for Nature – Central and East Europe, 915
World Wide Fund for Nature – Denmark, 916
World Wide Fund for Nature – Finland, 917
World Wide Fund for Nature – France, 918
World Wide Fund for Nature – Germany, 919
World Wide Fund for Nature – Greece, 920
World Wide Fund for Nature – Italy, 921
World Wide Fund for Nature – Netherlands, 922
World Wide Fund for Nature – Norway, 923
World Wide Fund for Nature – Spain, 924
World Wide Fund for Nature – Sweden, 925
World Wide Fund for Nature – Switzerland, 926
World Wide Fund for Nature – UK, 927
Youth Ecological Movement of Byelorussia, 1862
Zeleni Slovenije, 1846
Zelenyi Svit, 1911
Zentralstelle für Gesamtverteidigung, 1431
Zoologische Gesellschaft Frankfurt von 1858 eV, 833

Publications – Alphabetical Index

1,000 Terms in Solid Waste Management, 657
1992/93 ENDS Directory and Market Analysis, 1764
A E S Ambiente e Sicurezza: Rivista Dell'Antiquinamento, 332, 1630
Abwassertechnik (AWT), 691, 1631
Achema Handbook Pollution Control, 1454
Acid Deposition and Vehicle Emissions: European Environmental Pressures on Britain, 58, 1533, 2350
Acid Deposition: Volume 1 Sources Effects and Controls, 318, 1557
Acid Deposition: Volume 2 Origins Impacts and Abatement Strategies, 319
Acid News, 324
Acid Politics, 320, 2068
Acid Rain and the Environment, Volume 3: 1988–91, 321
Acqua Aria, 325
Acta Toxicologica et Therapeutica, 1405
Addressbook of Environmental Experts see Adressbuch Umwelt-Experten, 1765
Adresboek voor Milieuhygiene Veiligheids-Techniek en Recuperatie, 1280
Adressbuch Umwelt-Experten, 1765
Agricultural and Food Research Council Handbook, 125, 163
Agricultural and Veterinary Sciences International Who's Who, 4th ed, 164, 842
Agricultural Research Centres: A World Directory of Organisations and Programmes, 10th ed, 165
Agricultural Research Department see SCOOP Dienst Landbouwkundig Onderzoek Staring Centrum, 172, 418
Agriculture and Fertilisers: Fertilisers in Perspective – Their Role in Feeding the World, 207
Air Pollution and Noise Bulletin, 43, 1520
Air Pollution Control in the European Community, 34, 1510
Air Pollution from Vehicles, 59, 1614, 2351
Air Pollution's Toll on Forests and Crops, 35, 756, 1511
Aircraft Pollution: Environmental Impacts and Future Solutions, 60, 1535
Allicht, 1197
Alternative Times, 1050, 1198
Ambient Air Pollutants from Industrial Sources, 36, 1512
Ambio: A Journal of the Human Environment Research and Management, 1332, 2014
Amenagement et Nature, 932
Annals of Occupational Hygiene, 44, 1521, 2028
Annotated Bibliography on Environmental Auditing, 1720
Annuaire de l'Administration des DRIR, 987
Anuario Profesional del Medio Ambiente 1993, 1915
Applied Acoustics, 1601
Applied Energy, 1089
Aqua Fennica, 369
Aquatic Toxicology, 1406, 1632
Arboriculture Journal, 767
Archives of Industrial Hygiene and Toxicology see Arhiv za Higijenu Rada i Toksikologiju, 1407
Arhiv za Higijenu Rada i Toksikologiju, 1407
Asintra, 2342
Assessing Biological Risks of Biotechnology, 1313
Atlas of Endangered Species, The, 834
Atmospheric Environment Part A, 82
Atmospheric Environment Part B, 83
Atom, 1199
Atom-Informationen, 1200
Atomnaya Energiya, 1201
Atomo Petrolio Elettricita, 1202
Atomwirtschaft-Atomtechnik, 1203
ATW News, 1204
Automotive Consortium on Recycling and Disposal Proposals, 614
Automotive Engineering, 472
Aviation and Space Technology, 2186
Baltic Environment Proceedings no 39: Baltic Marine Environment Protection Commission, 297:, 1584
Bates and Wacker Packaging Report 93, 514
Belgicatom, 1205
Bikes Not Fumes, 61, 1536
Biocycle, 666
Biodegradable Polymers and Plastics, 573
Biomass and Bioenergy, 1051
Biosphere, The, 81
Biotechnology in Industry Healthcare and the Environment: Special Report no 2178, 1316
Biotechnology: The Science and the Business, 1314
Blue Flag Campaign, The, 270
Bowker A & I Acid Rain Abstracts Annual, 322
Brennstoff-Wärme-Kraft (BWK), 988
Britain's Nuclear Waste: Safety/Sting, 480
British Coal Research Association (BCRA) Quarterly, 1125
BS 7750: What the New Environmental Management Standards Mean for Your Business, 2167
Bulletin des Transports, 2343
Business Strategy and the Environment, 1745
C is for Chemicals, 1395
CD ROM Directory, 5th ed, 2010
CFCs and Halons: Alternatives and the Scope for Recovery for Recycling and Destruction, 615
Changing Atmosphere, The, 2
Characteristics of Road Transport in Hungary, 2352
Chemical Speciation and Bioavailability, 2280
Chemical Week, 2281
Chemico–Biological Interactions, 1408
Chemistry in Britain, 2282
Chlorine Production and Use and their Environmental Risks, 1418
Clean Air, 45, 1522
Clean Air Around the World, 37, 1513
Climate Action Network International NGO Directory 1992, 126
Climate Change, 100
Climate Change and Human Impact on the Landscape, 101
Climate Change: Designing a Tradeable Permit System, 104, 1969
Climatic Change, 103
Climatic Change and the Mediterranean, 102, 292
Coal Gasification for IGCC Power Generation, 1132
Coal Prospects and Policies in IEA Countries, 1126
Coal Research Projects, 1127
Code Permanent Environnement et Nuisances, 1459
Colliery Guardian, 1128
Common Future of the Wadden Sea, The, 298
Commons Without Tragedy, 2293
Concise Oxford Dictionary of Earth Sciences, The, 2131
Confederation of British Industry News, 1746
Construction Materials and the Environment, 2169
Contaminated Land: Counting the Cost of Our Past, 1560
Continuous Emission Monitoring, 38, 1514
Controlling Pollution in the Round: Change and Choice in Environmental Regulation in UK and Germany, 2103

Corporate Environmental Policy Statements, 2104
Corporate Environmental Responsibility: Law and Practice, 1943
Corporate Responsibility Europe, 1732
Council of Europe: Farming and Wildlife, 162, 1558
Country Studies and Technical Options, Volume 2, 115
Crisis in the French Nuclear Power Industry, 1224
Croners Environmental Management, 2008
Croners Waste Management, 667
Crop Protection Chemicals, 208
Crop Protection Directory 1988–89, 209
Danish Environmental Directory see Miljovejviseren 1991, 2232
Dansk Energi Tidsskrift, 989
Death in Small Doses – The Effects of Organochlorines on Aquatic Ecosystems, 1419
Design for Recyclability, 473
Desulphurisation, 41, 1518
Deutsche Dependance, 607
Dictionary of Ecology see Wörterbuch Ökologie, 2135
Dictionary of Ecology and Environment, 2132
Dictionary of Environment and Development, 1782
Dictionary of Environmental Science and Technology 1991, 2133
Dictionary of the Environment, 1783
Difesa Ambientale, 46, 1523
Digest of Environmental Protection and Water Statistics, 360, 726
Digest of UK Energy Statistics, 982
Dimensione Energia, 990
Dioxins in the Environment Pollution Paper 27, 1396
Directions for Internationally Compatible Environmental Data, 2168
Director's Guide to Environmental Issues, The, 1733
Directory of Corporate Environmental Policy, 2092
Directory of Emissions Monitoring Equipment Suppliers 1992, 2027
Directory of Environmental Health and Safety see Adresboek voor Milieuhygiene Veiligheids-Techniek en Recuperatie, 1280
Directory of Environmental Journals and Media Contacts, 1784
Directory of Environmental Technology, 2183
Directory of European Environmental Organisations, 1916
Directory of the Environment, 2011
Directory of the French Nuclear Industry, 1193
Disposal of Vehicles: Issues and Actions, 474
Droit Nucléaire, 1206
Du Sol à la Table: formerly Agriculture et Vie, 168
E is for Additives, 1349
Earth Matters, 2239

EC Energy Monthly, 991
EC Environmental Guide, 1917
EC Waste Management Policy: Transforming from Waste Treatment to Waste Prevention, 677
ECO Directory of Environmental Databases in the UK 1992, 2012
Eco Labelling, 1666
Eco Technics: International Pollution Control Directory, 1455
Eco-Labelling of Paper Products, 555, 1667
Ecogenetics: Genetic Predisposition to the Toxic Effects of Chemicals, 1397
Ecolinking: Everyone's Guide to Online Environmental Information, 2225
Ecological Engineering, 237, 1460, 1796
Ecological Modelling, 1797
Econews, 1925
Economia Delle Fonti di Energia, 1090
Economia Industrial, 1691
Economic Policy Towards the Environment, 2070
Economic Restructuring in Eastern Europe and Acid Rain Abatement Strategies 1991, 1026, 1095
Ecos, 933, 934
Ecosystems of the World, 843, 2136
Ecotoxicology, 1409
EEC Environmental Policy and Britain, 2071
Effects of Acidification on Bird and Mammal Populations, 835
Effluent Treatment and Waste Disposal, 441
Ekspress–Informatsiya Pryamoe Preobrazovanie Teplovoi i Khimicheskoi Energii V Elekricheskuyu, 992
Electric Vehicle Developments, 2344
Elephants, Ivory and Economics, 836
Emissions for Heavy Duty Vehicles, 1616
Encyclopedia of Environmental Science and Engineering, 2134
ENDS, 668, 1461, 2094
ENEA Notiziario Energia e Innovazione, 993
Energetica, 994, 2137
Energetyka, 995
Energi och Miljö i Norden 1991, 1012, 2105
Energia, 996, 997
Energia es Atomtechnika, 1207
Energiagazdalkodas, 998
Energie & Milieutechnologie, 999, 2187
Energie Alternative: Habitat Territorio Energia, 1052
Energieanwendung, 1000, 2188
Energiespectrum, 1208
Energiministeriet Energiforskningsprogram, 1001
Energy and Environmental Conflicts in East/Central Europe: The Case of Power Generation, 1013
Energy and Environmental Technology see Energie & Milieutechnologie, 999, 2187
Energy and the Environment, 124

Energy Conscious Design: A Primer for European Architects, 1045
Energy Digest, 1002
Energy Economist, 1003
Energy from Waste State-of-the-Art Report, 678, 1067
Energy in Denmark, 1014
Energy in Europe, 1004
Energy in the European Community, 1015
Energy Policies and the Greenhouse Effect Volume 1: Policy Appraisal, 116, 2072
Energy Taxation and Environmental Policy in EFTA Countries 1991, 1016, 2106
Energy Without End, 1046
Enerpresse, 1005
Engineering and Automation, 1006
ENTEC Directory of Environmental Technology, 2184
Entsorgung 92, 662
Entsorgungspraxis, 2189
Enviro, 1462
Environment and Industry Digest, 2190
Environment and Transport Infrastructures, 2353
Environment Bulletin, 1747
Environment Business, 1748
Environment Contacts: A Guide for Business (Who does what in Government Departments), 1741
Environment Databases, 2226
Environment in Czechoslovakia, The, 62, 1537
Environment Industry Yearbook, 1687
Environment International, 2240
Environment Labelling in OECD Countries, 1668
Environment Policy: Standards and Opportunities for Development, 2107, 2354
Environment Protection Engineering, 1798
Environment Research, 2241
Environment Research Directory see Umweltforschungskatalog, 2236
Environment Risk, 1749
Environment Software, 2219
Environmental Action in Eastern Europe, 2073
Environmental and Health Controls on Lead, 1398, 1944
Environmental Assessment Report, 1716
Environmental Assessment: A Guide to the Procedures, 1712
Environmental Auditing, 1721
Environmental Auditing Handbook: A Guide to Corporate and Environmental Risk Management, 1722
Environmental Chemistry, 2279
Environmental Conservation, 705
Environmental Dispute Handbook, 1945
Environmental Education – Journal of the NAEE, 1786
Environmental Effects of Conventional and Organic Biological Farming Systems, The, 259
Environmental Engineering, 1799
Environmental Engineering in the Process Plant, 1793

Environmental Guidelines to World Industry, 1734
Environmental Health, 1333
Environmental Health News, 1334
Environmental Impact of Paper Recycling, 556
Environmental Impact of Recycling, The, 616
Environmental Information: A Guide to Sources, 2227
Environmental Issues in the Pulp and Paper Industries, 549
Environmental Issues in Wool Processing: Are Textiles Finishing the Environment?, 634
Environmental Law, 1946
Environmental Law Brief, 1960
Environmental Law in UNEP, 1947
Environmental Liability and Environmental Criminal Law see Umwelthaftung und Umweltstrafrecht, 1957
Environmental Management in the Soviet Union, 2009
Environmental Monitoring and Analytical Specimen Banking, 2025
Environmental Monitoring: Meeting the Technical Challenge, 2026
Environmental News from the Netherlands, 2095
Environmental Opportunities: Building Advantage out of Uncertainty, 2211
Environmental Policies for Cities in the 1990s, 2074
Environmental Policy and Practice, 669, 1463, 2096
Environmental Policy in the European Community, 2075
Environmental Pollution, 1464
Environmental Practice in Local Government, 2076
Environmental Protection and Antipollution Directory see Guide de l'Environnement et des Techniques Antipollution, 1457
Environmental Protection Bulletin, 730, 1695, 2283
Environmental Protection in Germany see Umweltschutz in Deutschland, 729, 737
Environmental Protection through Sound Water Management in the Pulp and Paper Industry, 361
Environmental Research Newsletter, 2242
Environmental Respiratory Diseases, 1331, 1449
Environmental Software, 2138
Environmental Technology, 2191
Environmental Technology Transfer to Eastern Europe: A Selected Bibliography 1991, 2180
Environmental Toxicology and Chemistry, 1410
Environmental Toxicology and Water Quality, 416, 1411
Environmental Yearbook see Miljoarboka, 2231
Environmentalist, 2243
Environmetrics, 2139

Environnement: Le Guide des Sources d'Information, 1918
Era Solar, 1053
Essential Ecology, 2129
Eta Verde, 2244
Etudes de Pollution Atmosphérique à Paris et dans les Départements Periphériques, 47, 1524
Europe and the Environment, 2077
European Bulletin of Environment and Health, 1335
European Chemical News, 2284
European Community Environmental Legislation 1967–1987, 1949
European Community, The, 2078
European Directory of Renewable Energy Suppliers and Service, 1088
European Energy Report, 1017
European Environment, 2097, 2245
European Environmental Business News, 1751
European Environmental Law Review, 1961
European Environmental Yearbook, 2228
European Index of Key Plastics Recycling Schemes, 575
European Inland Fisheries Advisory Commission: Water Quality Criteria for European Freshwater Fish, 221, 424
European Integration and Environmental Policy, 2079
European Journal of Pharmacology: Environmental Toxicology and Pharmacology Section, 1412
European Water and Wastewater Industry, 365, 689
European Water Pollution Control, 1633
Eutrophication of Freshwaters, 413, 1399, 1625
Examination of Water for Pollution Control: A Reference Handbook, 1626
Excerpta Medica Section 46: Environmental Health and Pollution Control, 1336, 1465
FAO Yearbook 1990, 166
Farmers Weekly, 169
Farming and the Countryside, 236
Feld Wald Wasser, 417
Fens and Bogs in the Netherlands: Vegetation, History, Nutrient Dynamics and Conservation, 875
Fifth Fuel, 1092
Finishing, 48, 1525
Finland Central Statistical Office: Statistical Surveys Environmental Statistics, 2220
Finland Kauppa–Jateollisuusministerioe Energiatilastot, 983
Finnish Meteorological Institute Publications on Air Quality see Ilmatieteen Laitos Ilmansvojelun Julkaisvja, 49
Forest Damage in Europe – Environmental Fact Sheet No 1, 768
Forest Management in a time of emissions see Skogsskjötsel i en Utslippstid, 761, 1516

Forests Foregone: The EC's Trade in Tropical Timbers and the Destruction of the Rainforests, 770
Forum, 1007
Forum Städte-Hygiene, 1282
Forurensning i Norge 1991, 1456
France Commissariat à l'Energie Atomique Notes d'Information, 1209
Fravar ved Anmeldte Arbejdsulykker, 1283
Freight, 2098, 2345
Freight Transport and the Environment, 2332
Freight Transport and the Environment: ECMT Seminar 1991, 2333
Fresenius Environmental Bulletin, 2246
Fundamental Aspects of Pollution Control and Environmental Science, 1466
Furniture Recycling Network, 602
Fusion, 1008
Future for UK Environmental Policy Special Report no 2182, The, 2108
Future Forest Resources of the Former European USSR, 757
Future Forest Resources of Western and Eastern Europe, 758
Future is Now, The, 1450
Garners Environmental Law, 1950
Geluid en Omgeving, 1602
German Chamber of Industry and Commerce: German Packaging Laws, 515, 1700
Gesunde Pflanzen, 170
Glass Gazette, 452
Global Biodiversity, 837
Global Climate Changes and Freshwater Ecosystems, 3, 329
Global Environmental Change – UK Research Framework, 4
Global Forests: Issues for Six Billion People, 759
Global Status of Peatlands and their Role in Carbon Recycling, The, 129, 1133
Global Warming, 117
Global Warming Forum, A, 119
Global Warming: The Debate 1991, 118
Good Beach Guide to over 170 of Britain's Best Beaches 1990, 266
Governo Locale ed Economia dell' Ambiente (GEA), 2247
Great Britain Department of Energy Publications, 1009
Green Belt, Green Fields and the Urban Fringe: The Pressures of Land in the 1980's, 879
Green Book, The (Authority on Tractors Agricultural and Forestry), 167, 766
Green Energy Matters, 1054
Green Engineering, 1800
Green Engineering – A Current Awareness Bulletin, 1801
Green Europe, 171
Green Globe Yearbook, 2013
Green Magazine, 1010
Green Paper on the Environment – A Community Strategy for Sustainable Mobility, 958, 2109, 2355
Green PC, The, 1451
Greenbits, 844

354 Publications – Alphabetical Index

Greener Marketing, 1735
Greenhouse Earth, 130
Greenhouse Gases Bulletin, 127
Greenhouse Issues, 128, 1129
Greening of Business, The, 1736
Greening of Global Investment Special Report 2108, The, 2212
Greening of Urban Transport, The: Planning for Walking and Cycling in Western Cities, 2080, 2334
GSF Mensch und Umwelt, 1337
Guardian, 2248
Guide de l'Environnement et des Techniques Antipollution, 1457
Guide de l'Industrie Belge du Recyclage des Matières Plastiques 1991, 576
Guide to Libraries and Information Units in Government Departments and other Organisations, 2229
Guide to Resources in Environmental Education, 1785
Guide to the Belgian Plastic Recycling Industry see Guide de l'Industrie Belge du Recyclage des Matières Plastiques 1991, 576
Hazardous Materials: Sources of Information on their Transportation, 658, 2335
Haznews, 670, 1413
Health and Safety at Work, 1284
Health and Safety Directory, 5th revised ed, 1281
Health Cities, 1279
Horticulture Française, 845
Horticulture Week, 846
Hothouse Earth, 120
Human Environment in Sweden, 1285
Iaderna Energiia, 1210
IIASA Research Plan 1993, 2146
Ilmatieteen Laitos Ilmansvojelun Julkaisvja, 49
Impact of Sea Level Changes on European Coastal Lowlands, The, 267
Implications for the EC's Environmental Policy of the Treaty on European Union, The, 2110
In the Company of Green: Corporate Communications for the New Environment, 1737
Incineration and the Environment – A Source Book, 659
Indoor Environment, 50
Industrial Environmental Services Directory, 1688
Industrie et Environnement, 1696
Industry and the Environment, 1697
Info Déchets Environnement et Technique, 671
Information Bulletin, 2249
Information Please Environmental Almanac 1993, 2221
Information Sources for Construction and the Environment see Informationskaller Bygg and Miljo, 1689
Information Sources in Biotechnology, 1315
Informationskaller Bygg and Miljo, 1689

Informazione Innovativa, 1011, 1286, 2192
INFOTERRA World Directory of Environmental Expertise, 1766
Initiatives for the Environment: A Publication from General Motors Europe, 1686
Inquinamento, 1467
Institute of Materials Management Members' Reference Book and Buyers Guide 1993, The, 1742
Instituto Nacional de Investigaciónes Agrarias Comunicaciónes Serie: Recursos Naturales, 706
Instituto Nacional Para la Conservación de la Naturaleza Monografías, 707
Institutul de Studii si Proiectari Energetice Bulletinul, 1093
Integrated Pollution Control in Europe and North America, 1452
International Association on Water Pollution Research and Control Yearbook, 1629
International Coal Letter, 1130
International Environmental Information Sources, 2230
International Environmental Law, 1951
International Handbook of National Parks and Nature Reserves, 930
International Journal of Ambient Energy, 1055
International Journal of Environment and Pollution, 1468
International Journal of Environmental Studies, 1787
International Journal of Global Energy Issues, 1094
International Journal of Radioactive Materials Transport, 1211, 2346
International Journal of Solar Energy, 1056
International Maritime Dangerous Goods (IMDG), 299, 1585, 2356
International Migration, 2299
International Politics of Nuclear Waste, The, 481
International Population Conference Proceedings, 2297
International Protection of the Environment, 727
Interpretation Heritage, 872
Interuniversitair Demografisch Institut Publications, 2300
Introduction to Aquaculture, 220, 414
Introduction to Energy, 984
Introduction to Environmental Engineering, 1794
Irish Journal of Environmental Science, 2140
Journal de Toxicologie Clinique et Experimentale, 1414
Journal for Water and Wastewater Research see Zeitschrift für Wasser und Abwasserforschung, 381, 695
Journal of Aquatic Ecosystem Health, 333
Journal of Atmospheric Chemistry, 51, 1526
Journal of Biochemical Toxicology, 1415
Journal of Environmental Law, 1962

Journal of Environmental Management, 2015
Journal of Hazardous Materials, 672, 1416
Journal of Planning and Environmental Law, 1963
Journal of Radiological Protection, 1371
Journal of the Institution of Water and Environmental Management, 370
Journal of Transport Economics and Policy, 2099, 2347
Journal of Tropical Ecology, 769
Kernenergie, 1212
Kind Food Guide, The, 260
Kirkgate Market: A Feasibility Study, 617
Korrespondez Abwasser, 371
Land Use Change: The Causes and Consequences, 876
Land Use Policy, 880
Landfill Gas, 455
Landscape Changes in Britain, 877
Landscape Ecology, 881
Last Chance to See..., 838
Legal 500, The, 5th ed, 1952
Less Traffic Better Towns, 2336
Lettre de l'Environnement, 2250
Literaturberichte über Wasser, Abwasser, Luft und Feste Abfallstoffe, 1469
Living in the Greenhouse, 121
Living with Radiation: National Radiological Protection Board, 1368
Ludwig Boltzmann Institut für Umweltwissenschaften und Naturschutz Mitteilungen, 935
Lufthygienischer Monatsbericht, 52, 1527
Man and the Environment: Product Life Cycle Analysis – What does it mean?, 1713
Management for a Small Planet, 1738
Management of Waste Plastic Packaging Films, 579
Managing the Environment, 2017
Managing the Environment: The Greening of European Business, 1739
Manual of Environmental Policy: The EC and Britain, 2093
Manual of Methods in Aquatic Environment Research, 1577
Manufacturing Engineering, 1964
Mapping Critical Loads for Europe, 323
Marine Environmental Research, 294
Marine Pollution, 1578
Marine Pollution Bulletin, 1582
Marine Pollution Research Titles, 1583
Marketing, 1752
Materials Reclamation Weekly, 608
Mercury in Soil – Distribution Speciation and Biological Effects, 1561
Metals and their Compounds in the Environment, 1400
Milieu en Bedrijf, 731
Milieubeleid in de Europese Gemeenschap, Het, 2081
Milieudefensie, 936
Miljoarboka, 2231
Miljoe og Teknologi, 2193
Miljoeprojekt, 2251
Miljoeundersoegelser ved Marmorilik, 2252

Publications – Alphabetical Index

Miljopolitik i Det Europaeiske Faellesskab, 2082
Miljovejviseren 1991, 2232
Ministerio da Qualidade de Vida Comissão Nacional do Ambiente Boletim, 1470, 2016
Modern Plastics International, 1698
Monthly Environmental Policy, 2111
MST Luft, 53, 1528
Müllmagazin, 673
Multimedia Transport and Fate of Pollutants, 442, 2337
NAMAS Concise Directory of Accredited Laboratories 1991, 2185
National Directory of Recycling Information, 603
Natur og Miljoe, 937, 938
Natura (Revista), 2253
Natura e Societa, 939, 1965
Natural Environment Research Council News, 2141
Natural Resources and the Environment *see* Naturressurser Og Miljo 1991, 2233
Natural Sources of Ionising Radiation in Europe: Radiation Atlas, 1369
Natural Studies and the Environment *see* Statistika Centralbyraan Statistika Meddelanden Subgroup NA, 2222
Nature, 2142
Nature Conservation in Europe – Agenda 2000, 931
Nature et Mieux-Vivre, 940
Naturopa, 941
Naturressurser Og Miljo 1991, 2233
Natuur en Milieu, 942
NEA Newsletter, 1213
Net Losses Gross Destruction – Fisheries in the European Community, 222, 853
Netherlands Centraal Bureau voor de Statistiek: Nederlandse Energiehuishouding, 985
Neurotoxicity, 1401, 1953
New Farmer and Grower, 261
New Scientist, 2143
NGO Directory for Central and Eastern Europe 1992, 1919
Noise and Vibration Bulletin, 1603
Noise and Vibration Worldwide, 1604
Nordic Journal of Environmental Economics Protection, 2100
North Atlantic Treaty Organisation Expert Panel on Air Pollution Modelling Proceedings, 42, 1519
North Sea Oil and Gas Directory, 1124
Nova – Revista de Salud, 1287
NOx Control Installations on Coal Fired Plants, 1134
Nuclear Energy *see* Iaderna Energiia, 1210, 1214
Nuclear Engineering International, 1215
Nuclear Europe Worldscan, 1216
Nuclear Law Bulletin, 1217, 1967
Nuclear Power and the Greenhouse Effect, 122, 1191
Occupational Exposure Limits for Airborne Toxic Substances, Third Edition, 39, 1402, 1515
Ochrona Przyrody, 708

OECD World Energy Statistics, 986
Ökolgiepolitik, 1926
Options, 2194
Organic Produce in Europe Special Report no 2128, 262
Oryx, 848
Österreichisches Tierschutzzeitung, 943
Packaging, 513
Packaging in Europe: France Special Report no 2046, 516
Packaging in Europe: Italy Special Report no 2049, 517
Packaging in Europe: Spain and Portugal Special Report no R162, 518
Packaging in Europe: West Germany Special Report no 2047, 519
Packaging in the Environment, 510
Packaging Industry Directory 1993, 512, 1690
Packaging Ordinance A Practical Handbook *see* Verpackungsverbordnung, 511
Packaging Waste and the Polluter Pays Principle: A Taxation Solution, 520
Paper and Packaging Analyst, 551
Paradise Deferred: Environmental Policymaking in Central and Eastern Europe, 2112
Petroleum Review, 1131
Pigment and Resin Technology, 2285
Planeta Verde, 2254
Plastics Recovery in Perspective, 580
Plastics Recycling 1991, 574
Plastics Technology, 577
Poisonous Fish, 223, 335, 1420
Politica Ambientale nella Comunità Europea, La, 2083
Política de Ambiente na Comunidade Europeia, A, 2084
Política de medio ambiente en la Communidad Europea, 2085
Politique de l'Environnement dans la Communauté Européenne, La, 2086
Politique Européenne de l' Environnement, La, 2087
Pollustop, 1471
Pollution, 1472
Pollution Atmosphérique, 54, 1529
Pollution Control Policy of the European Communities, 1453
Pollution in Norway *see* Forurensning i Norge 1991, 1456
Pollution of the Mediterranean Sea, 293, 1581
Pollution Prevention, 1473
POLMARK – The European Pollution Control and Waste Management Industry Directory, 663, 1458
Polska Akademia Nauk Instytut Podstaw Inzynierii Srodowiska Prace i Studia, 1474
Population et Sociétés, 2301
Population Explosion, The, 2294
Population Studies, 2302
Population: An Ecofeminist Perspective, 2304
Practical Alternatives, 1057
Präparator, Der, 709

Prevalence of Fish Diseases with Reference to Dutch Coastal Waters, 224, 300
Problems of Large Metropolitan Areas *see* Problemy Bolshikh Gorodov, 732
Problemy Bolshikh Gorodov, 732
Procedures for Enhancing the Use of Environmentally Friendly Vehicles, 1617
Product is the Poison, – The Case for a Chlorine Phase-Out, 1421
Progress in the Recycling of Plastics in the UK, 581
Proposal for a Council Regulation (EC), 1723
Protec, 733
Protecting the Earth: A Status Report with Recommendations for a New Energy Policy, 131, 1538, 2113
Public Transport International, 2348
Pulp and Paper International, 552
Pulp and Paper Week, 553
Quercus, 734
Radioactive Aerosols, 482
Radioactive Waste, 483
Radioactive Waste Management, 484, 485
Radiprotection, 1372
Radon in Dwellings, 1370
Railway Noise Standards: Let's Get Them Right, 1605
Rational Use of Water and its Treatment in the Chemical Industry, 362
Realism in Green Politics, 1913
Realized and Projected Recycling Processes for Used Batteries, 618
Recovery and Rehabilitation of Contaminated Land, 1559
Recreational Water Quality Management: Vol 1 Coastal Waters, 268, 1579
Recyclage des Matières Plastiques, Le, 582
Recycled Papers: The Essential Guide, 550
Recycling, 609
Recycling in Member Countries, 619
Recycling Lead and Zinc: The Challenge of the 1990's, 475
Recycling: Energy from Community Waste, 598
Recycling: New Materials for Community Waste, 599
Recycling: Public Attitudes and Market Reality, 620
Red Data Birds in Britain, 839
Refrigeration and Air Conditioning, 55, 1530
Regulated Rivers Research and Management, 334
Reinforced Plastics, 578
Reinwater, 1475
Renewable Energy Sources for Fuels and Electricity, 1047
Reprocessing of Tyres and Rubber Wastes, 600
Research and Technological Development for the Supply and Use of Freshwater Resources, 425
Resource Recycling, 610
Resources Conservation and Recycling, 611
Retema–Medio Ambiente–Revista, 2255

Returnable and Non-Returnable Packaging, 521
Review of Population Research, 2303
Revue d'Ecologie: La Terre et la Vie, 850
Revue Générale Nucléaire, 1218
Revue Générale Nucléaire: International Edition, 1219
Revue Juridique de l'Environnement, 1968
Risk Assessment of Ecological Effects and Economic Impacts of Acidification on Forestry in Sweden, A, 771
Risks from Radon in Homes: Report of a Working Group of the Institute of Radiation Protection, 1373
River Conservation and Management, 330
Role of Environmental Assessment in the Planning Process, The, 1714
Safe Energy, 1058, 1220
Safety Assessment of Radioactive Waste Repositories: Systematic Approaches to Scenario Development, 487, 1254
Safety in the Use of Asbestos Report IV (2), 1298
Salvo Demolition Directory 93/94, 604
Sanitary Landfill, 456
SAVE Action Guide, The, 871
Schweizerische Vereinigung für Atomenergie Bulletin, 1221
Science of the Total Environment, 2144
SCOOP Dienst Landbouwkundig Onderzoek Staring Centrum: Instituut voor Onderzoek van het Landelijk Gebied, 172, 418
Seaside Award, 271
Sécrétariat d'Etat Chargé de l'Environnement et de la Qualité de la Vie: Bulletin de Documentation de l'Environnement, 2256
Silvicultural Systems, 760
Single European Dump, The: Free Trade in Hazardous and Nuclear Wastes in the New Europe, 488, 1422
Sitefile Digest: A Digest of Authorised Waste Treatment and Disposal Sites in Great Britain, 457
Skogsskjötsel i en Utslippstid, 761, 1516
Skogsutsikter, 762, 1517
Smog, 56, 1531
Society of Dyers and Colourists, 635
Solar Energy see Sunceva Energija, 1063
Solar Energy Materials and Solar Cells, 1059
Solar Energy Research and Development in the European Community: Series A–H, 1060
Solar Power via Satellite, 1048
Solaria, 1061
Solvent Recovery Handbook, 443
Solvents and the Environment: Industry and Environment Vol 14 No 4 1991, 444
Solvents Digest, 445
Sonnenenergie und Wärmepumpe, 1062
Species, 851
Spirit of Versailles: The Business of Environmental Management, 1743
State of the Environment, The, 2234
State of the World 1992: A Worldwatch Institute Report on Progress Towards a Sustainable Society, 959
States and Anti-Nuclear Movements, 1192, 1914
Statistika Centralbyraan Statistika Meddelanden Subgroup NA, 2222
Steine Sprechen, 873
Stratospheric Ozone 1991, 123
Streetwise, 1788
Studi Parlamentari e di Politica, 2102
Studies in Environmental Science, 2145
Submerging Coasts, 269
Sunceva Energija, 1063
Surface Coatings International, 2286
Survey of Gassing Landfill Sites in England and Wales, A, 458
Sustainability and Environmental Policy, 956
Sustainable Development: An Imperative for Environment Protection, 960
Sveriges Djurskyddsfoereningars Riksfoerbund, 852
Sveriges Natur, 944
Swedish Environment, 2223
Swedish Environment, The: Annexe to Sweden; National Report to UNCED 1992, 2259
Swedish Environmental Protection Agency Report, 738
Swedish Nuclear News, 1222
Tableaux Numeriques des Analyses Physico Chimiques des Eaux du Rhin see Zahlentafeln der Physikalisch Chemischen Untersuchungen des Rheinwassers, 331
Taking Population Seriously, 2295
Talking Conservation, 554
Taxation for Environmental Protection, 728
TEC Transport Environment Circulation, 2349
Technical Innovation in the Plastics Industry and its Influence on the Environmental Problem of Plas, 583
Techniques in Marine Environmental Sciences, 295
Technologie Appropriée de Déminéralisation de l'Eau Potable, La, 415
Textiles and the Environment Special Report no 2150, 636
Third Revolution, The: Environment, Population and a Sustainable World, 2296
This Common Inheritance: Britain's Environmental Strategy, 2088
Tourism Industry and the Environment, The, 1701
Toxicology, 1417
Trace Element Occurrence in British Groundwaters, 426, 1636
Transnational Environmental Liability and Insurance, 1954
Transport and Climate Change: Cutting Carbon Dioxide Emissions from Cars, 105, 2357
Transport and the Environment, 2338
Transport Policy and the Environment, 2089, 2090, 2339, 2340
Transport, the Environment and Economic Policy, 2341
Tropical Rain Forest, 763
Tropical Rain Forest Ecology, 764
Turkish Journal of Nuclear Sciences, 1223
Übersetzungen-Kerntechnische Regeln, 1253
UK Centre for Economic and Environmental Development Bulletin, 1768
UK Marine Oil Pollution Legislation, 1123, 1580, 1955
UK Recycling Directory, The, 605
UK Road Transport's Contribution to Greenhouse Gases, 132, 2358
UK Waste Law, 660, 1956
UK Waste Management Industry, 679
Ulrichs International Periodicals Directory, 2235
Umwelt, 735, 1699
Umwelt Aktuell, 2257
Umwelt Technologie Aktuell, 2195
Umweltforschungskatalog, 2236
Umwelthaftung und Umweltstrafrecht, 1957
Umweltmagazin, 2258
Umweltpolitik in der Europäischen Gemeinschaft, Die, 2091
Umweltschutz in Deutschland, 729
Umwelttechnik, 1288
UNEP 1991 United Nations Environment Programme Environmental Data Report, 2130
Urban Solid Waste Management, 661
Vachers European Companion and Consultants Register, 1767
Vaestoentutkimuksen Vuosikirja, 2298
Vandteknik, 372
Vanishing Tuscan Landscapes, 878
Varstvo Narave, 945
Vatten, 419
Verpackungsverordnung, 511
Vida Silvestre, 710
Vital Signs 1992/93: The Trends that are Shaping our Future, 1715
Volatile Organic Compound Emissions in Western Europe: Report 6/87, 64, 1135, 1539, 2359
Warmer Bulletin, 612
Warren Spring Laboratory UK Smoke and Sulphur Dioxide Monitoring Networks, 57, 1532, 2029
Wasser Abwasser GWF, 1634
Wasser Luft und Boden, 674, 1476
Wasser und Abwasser in Forschung und Praxis, 420
Wasserwirtschaft, 373, 692
Wasserwirtschaftliche Mitteilungen, 374
Waste Disposal and Recycling Bulletin, 613
Waste Management, 486, 675
Waste Management '92 see Entsorgung 92, 662
Waste Management and Research, 676

Publications – Alphabetical Index

Waste Management International: Vol 1 Directory of Manufacturers and Services, 664
Waste Management Yearbook, 665
Waste Not Want Not: The Production and Dumping of Toxic Waste, 601, 1403
Waste, Wastewater, Air Laws and Technology, 40, 687, 1627, 1958
Water Air and Soil Pollution, 1477
Water and Drainage Law, 363, 1959
Water and Environment International, 375
Water and the Environment, 364, 688
Water and Waste Treatment, 376, 693
Water Bulletin, 421
Water Pollution by Fertilisers and Pesticides, 1404, 1628
Water Pollution Incidents in England and Wales 1990, 211, 1637
Water Quality International, 422, 1635
Water Resources Management, 377
Water Supply, 378
Water Supply and Wastewater Disposal International Almanac, 366, 690
Waterschapsbelangen, 423, 736
WATT Committee on Energy: Air Pollution Acid Rain and the Environment, 65, 326, 1540

Wetlands Ecology and Management, 379
What You Don't Know Will Hurt You: Environmental Information as a Basic Human Right, 2224
Who is Who in Recycling Worldwide 1991, 606
Who's Hand on the Chainsaw? UK Government Policy and the Tropical Rain Forests, 765
Who's Who in European Water 1992, 367
Who's Who in the Environment – England, 1920
Who's Who in the Environment – Scotland, 1921
Who's Who in the Environment – Wales, 1922
Who's Who in the Water Industry, 368
Wind Energy in Europe Report, 1068
Wind Engineering, 1064
Wind-Energie Jahrbuch, 1049
Windkraft Journal, 1065
Windpower Monthly, 1066
Working for Public Safety and Environmental Protection, 489, 739, 1255
World Directory of Environmental Organisations, 1923

World Directory of Pesticide Control Organisations, 210
World Energy and Nuclear Directory, 1194
World Environment Research Directory, 2237
World Environmental Business Handbook, 1744
World Guide to Environmental Issues and Organisations, 1924
World Nuclear Directory, 1195
World Nuclear Industry Handbook, 1196
World Peace Directory and Diary 1993, 2238
World Water, 380, 694
Worldwide Trends and Opportunities in Automotive Passenger Vehicles 1995–2005, 584
Wörterbuch Ökologie, 2135
WWF Baltic Bulletin, 296
Yearbook of Population Research in Finland *see* Vaestoentutkimuksen Vuosikirja, 2298
Zahlentafeln der Physikalisch Chemischen Untersuchungen des Rheinwassers, 331
Zeitschrift für Wasser und Abwasserforschung, 381, 695

Databases – Alphabetical Index

ABEL, 1971
Acid Rain, 327
Acompline/Urbaline, 2019
Agra Europe, 179, 2116
AGREP, 180
Agricola, 181
Agris, 182, 782
Air/Water Pollution Report, 77, 1553
Analytical Abstracts, 2289
Aqualine, 391
Aquatic Sciences and Fisheries Abstracts, 235, 862, 1647
Baltic, 313, 1591
Base Relacional de la Industria y Servicios Ambientales (BRISA), 1704
Beilstein Online, 2290
Biobusiness, 1321
Biorep, 1322
BIOSIS Previews, 2157
Biotechnology Abstracts, 1323
BRIX/FLAIR, 1101
Celex, 1972
Centro de Documentación del Medio Ambiente: Escuela de Organización Industrial, 2020
Chemical Abstracts (CA Search), 2291
Chemical Exposure, 1432
Chemical Safety Newsbase, 1433
Coal Database, 1141
Coal Research Projects, 1142
Commonwealth and Agriculture Bureau Abstracts, 183
Compendex Plus, 1805
CORDIS, 2208
Current Biotechnology Abstracts, 1324
Current Technology Index, 2209
DOE Energy, 1021
EcoBase, 2262
EDF-DOC, 1022
ELFIS, 184, 783
EMBASE, 1293
ENEL, 1023
Energie, 1024
Energy Conservation News, 1102

Energy Technology Data Exchange (ETDE), 133
Energyline, 1103
ENREP-NEW, 2263
Enviroline (ENVN), 2158
Environment Digest, 2264
Environmental Bibliography, 621, 1342, 1487
Environmental Chemicals Data and Information Network (ECDIN), 1434
EUREKA, 1025
EUROCRON, 2265
European Parliament Online Query Service (EPOQUE), 1973
European Update, 2117
Farm Structure Survey Retrieval System (FSSRS), 239, 889
Food Safety Briefing (FSB), 1352
Food Science and Technology Abstracts, 1353
GeoArchive, 1566
Geobase, 110
GeoRef, 682
Global Environmental Change Report, 9
Global Resource Information Database (GRID), 2038
Greenhouse Effect Report, 134, 2118
GreenNet, 2266
Hazardous Substances Databank (HSDB), 1435
HMSO Online/British Official Publications, 2267
HSELine, 1294
IBSEDEX, 78, 1554
ICONDA, 1806
INFOMAT, 1755
International Nuclear Information System (INIS), 1238
International Road Research Documentation (IRRD), 890, 2366
International Solar Energy Intelligence Report, 1080
LEXIS, 1975
Life Sciences Collection, 2159

Major Hazard Incident Data Service (MHIDAS), 683, 1436
Medline, 1437
Molars, 10
Multinational Environmental Outlook, 2268
New Scientist, 2160
Oceanic Abstracts, 314, 1592
Oil Spill Intelligence Report, 1143, 1593
Paperchem, 564
PASCAL, 2210
Pira Abstracts, 565
Pollution Abstracts, 1488
PTS PROMT, 1756
RAPRA Technology, 586
Registry of Toxic Effects of Chemicals on Substances (RTECS), 1438
Scarabee, 622
SciSearch, 2161
SESAME, 1081
SPEARHEAD, 1976
Swedish Environment Research Index (SERIX), 743
System for Information on Grey Literature in Europe (SIGLE), 2269
Toxic Materials Transport, 1439, 2367
Toxline, 219, 1440
Transdoc, 891, 2368
Transport Research Information Services (TRIS), 892, 2369
Tulsa, 1144, 1594
Umweltforschungsdatenbank (UFORDAT), 2162
Umweltliteraturdatenbank (ULIDAT), 2270
UN Conference on Environment and Development (UNCED), 2119
VANYTT, 79, 1104, 1555
Waste, 684
Wasteinfo, 623
Water Resources Abstracts, 433, 701, 1648

Learning Resources Centre